Asbestos

Volume 1

Frontispiece

(a) Chrysotile

(b) Crocidolite

(Photographs by Susan Maher, Department of Geology, University of Manchester)

(c) Amosite

Hand specimens of asbestos as they occur in nature

Asbestos

Volume 1

Properties, Applications, and Hazards

Edited by
L. Michaels
*Institute of Laryngology and Otology,
University of London*
S. S. Chissick
*Department of Chemistry,
University of London King's College*

A Wiley–Interscience Publication

JOHN WILEY & SONS

Chichester · New York · Brisbane · Toronto

Library of Congress Cataloging in Publication Data:
Main entry under title:

Asbestos.

'A Wiley–Interscience publication.'
1. Asbestos. I. Michaels, Leslie. II. Chissick, Seymour S.
TA455.A6A78 614.8'3 78–16535
ISBN 0 471 99698 X (v. 1)

Printed in Northern Ireland at The Universities Press, Belfast

Contributors

S. T. Beckett — *Institute of Occupational Medicine, Roxburgh Place, Edinburgh EH8 9SU*

W. D. Buchanan — *134 The Drive, Rickmansworth, Herts. WD3 4DP.*

L. H. Capel — *London Chest Hospital, London, E2. 9JX*

M. Chesney — *Center for Occupational and Environmental Safety and Health, SRI International, Menlo Park, California 94025, U.S.A.*

S. S. Chissick — *University of London King's College, Strand, London, WC2R 2LS*

J. D. Cook — *Hazardous Materials Service, Harwell Laboratory, Harwell, Didcot, Oxon. OX11 ORA*

R. Derricott — *Technical Policy Architect's Department, The Greater London Council, County Hall, London, SE1 7PB.*

J. S. Felton — *Long Beach Naval Shipyard, Long Beach, California, U.S.A.*

M. Feuerstein — *Center for Occupational and Environmental Safety and Health, SRI International, Menlo Park, California 94025, U.S.A.*

M. D. Gidley — *Center for Occupational and Environmental Safety and Health, SRI International, Menlo Park, California 94025, U.S.A.*

P. N. Giever — *Center for Occupational and Environmental Safety and Health, SRI International, Menlo Park, California 94025, U.S.A.*

A. A. Hodgson — *Cape Asbestos Fibres Ltd., Uxbridge, Middlesex UB8 2JQ*

J. S. P. Jones — *City Hospital, Hucknall Road, Nottingham NG5 1PB.*

R. J. Levine — *Chemical Institute of Toxicology, P.O. Box 12137, Research Triangle Park, North Carolina 27709, U.S.A.*

L. Michaels — *Institute of Laryngology and Otology, Gray's Inn Road, London WC1X 8EE.*

A. P. Middleton — *Institute of Occupational Medicine, Roxburgh Place, Edinburgh EH8 9SU.*

M. Newhouse — *TUC Centenary Institute of Occupational Health, London School of Tropical Medicine and Hygiene, Keppel Street, London WC1E 7HT.*

A. M. PYE — *Fulmer Research Institute, Stoke Poges, Slough, Bucks SL2 4QD.*

E. T. SMITH — *Hazardous Materials Service, Harwell Laboratory, Harwell, Didcot, Oxon. OX11 ORA*

B. E. SUTA — *Center for Resource and Environmental Systems Studies, SRI International, Menlo Park, California 94025, U.S.A.*

P. J. WARREN — *The London Hospital Medical College, London, E1 2AD.*

J. ZUSSMAN — *Department of Geology, University of Manchester, Manchester, M13 9PL.*

Contents

Preface

'In spite of the large expansion of asbestos textiles in the U.K. over the years, it was only towards the end of the 1920*s that the special biological effects of asbestos were recognized. At this time, some machines were ventilated and the general conditions equalled those in the cotton textile mills of the period but they left a lot to be desired. Fibre was blended on the floor by hand and most machines were fed manually. Looms were only partially ventilated and a worker who remembers those days has said that asbestos fibre was literally knee-deep under them. The dust and stout fibre that dropped on to the floor under the cording machines had to be cleared by hand; this, and the stripping (or cleaning) of the cords, also by hand, was a very dusty and dirty operation and whilst it was being carried out a man could not be recognized at a distance of six or eight feet. Fibre was screened on open vibratory screens, the short fibres being allowed to fall on the floor while the "overs" were bagged by hand. The ventilation system exhausted into settling chambers which were cleaned out by a gang of men every Saturday morning. The men, wearing respirators, went into the settling chamber, shovelled the dust into wheelbarrows which were then wheeled along and their contents tipped down a shute where it was mixed with water to form a sludge for disposal. The dust clouds formed during this tipping operation can be imagined'.*

D. W. Hills
(From a paper in *Biological Effects of Asbestos*, published by the New York Academy of Science, 1965)

This is vivid description of conditions in an asbestos textile factory of about 50–60 years ago, which, in the light of the medical knowledge of the time, were considered to be acceptable. The ventilation of some machines and the wearing of respirators during the performance of particularly dirty operations were presumably to facilitate breathing in the dust-laden atmosphere—the causal relationship between airborne dust and death from lung disease among asbestos workers only being established in the late 1920s. Since that time there have been considerable and significant advances in the understanding of the health hazards associated with asbestos fibres and the working conditions within the asbestos industry. However, it is only during the 1970s that there has been sustained public attention towards and concern about the health hazards of asbestos both for asbestos workers and for the population at large.

This book had its genesis during 1976 when one of us (S.S.C.) was required to find some information about asbestos, following the spate of public concern about the material at the time. An attempt to find a comprehensive source book on all aspects of asbestos to provide the foundation for a literature survey met with failure. It was rapidly realized that although there was a very considerable body of literature on asbestos, this was scattered and not readily obtainable without very considerable effort. For example, the Chemical Society Library subject index lists only two publications under the heading 'Asbestos', both of them more than 10 years old. The Editors decided to attempt to fill this gap in the literature by providing a convenient book covering in detail those areas of the subject that they felt would have been useful to them had the information been available in this format in the mid-1970s. The coverage has been limited to areas of particular interest to the Editors but it is hoped that the topics are of wide current interest. Certainly all those concerned with the asbestos industry (employers, employees, trade unions, safety representatives, etc.), the enforcement and application of the Health and Safety at Work etc. Act 1974 (Government Health and Safety Inspectors and industrial safety officers and safety advisers), architects and other designers, public and environmental health scientists, doctors concerned with the asbestos illnesses, and the general public should find this book of value in whole or in part. Some duplication of material between the chapters has been allowed so that each chapter stands on its own in providing the reader with information essential to the understanding of the subject matter.

The contributors were selected for their expertise in the subject [e.g. Dr. W. D. Buchanan, O.B.E., was H.M. Deputy Senior Medical Inspector of Factories (H.M. Factory Inspectorate) and Senior Employment Medical Advisor (Employment Medical Advisory Service), Dr. A. A. Hodgson is the Research Director of Cape Asbestos Fibres and author of the Royal Institute of Chemistry monograph *Chemistry of the Fibrous Silicates*, and Professor J. Zussman is a world authority on mineralogy and co-author of the treatise *Rock Forming Minerals* (Longmans, London) by Deer, Howie and Zussman]. Considerable care was taken not to involve members of the asbestos industry or its declared opponents. The one exception is the chapter on the chemistry of asbestos, which is clearly a non-controversial subject, for which it was felt that a member of the asbestos industry would be an appropriate contributor. Also, no contact was made with the asbestos industry or its declared opponents until the work on the book was substantially complete—and then only to obtain permission to reproduce photographs, statements and other previously published items. In this respect we should like to express our thanks to Mr. W. Penney of the Asbestos Information Committee for his assistance in obtaining these permissions from members of the industry in Britain and America. We should also like to express our thanks to the British Post-Graduate Medical Federation and the University of London King's College for the facilities provided, to Mr. A. Hughes (Faculty of Laws, King's College) and Dr. J. L. Cordingly (Faculty of Medicine, King's College) for helpful discussion and comments, the staff of the Health and Safety Executive Library for their very willing help and co-operation, and to Miss P. D. Tschan and Miss M. E. Young for filing and typing and endless supplies of black coffee.

LESLIE MICHAELS, M.D.
Professor of Pathology in the University of London and Dean of the Medical School of the Institute of Laryngology & Otology (British Post-Graduate Medical Federation)

SEYMOUR CHISSICK, Ph.D.
Lecturer in Chemistry in the University of London, Assistant Director of the Chemical Laboratories and Safety Officer, King's College, London

April 1978

CHAPTER 1

INTRODUCTION

Leslie Michaels,

Institute of Laryngology & Otology, London

Seymour S. Chissick,

University of London King's College

(1) INTRODUCTION

Asbestos is the common name given to a number of naturally occurring, hydrated mineral (inorganic) silicates that possess a crystalline structure and that are incombustible in air and separable into filaments. There are four main types of asbestos, all of which are chemically different (and hence have different properties and applications); these are:

1. chrysotile ($3MgO.2SiO_2.2H_2O$) or white asbestos, which occurs as fine silky flexible white fibres and is mined mainly in Canada, Russia and Rhodesia;
2. amosite [$(FeMg)SiO_3$], a straight brittle fibre, light grey to pale brown in colour and found in South Africa;
3. crocidolite [$NaFe(SiO_3)_2.FeSiO_3.H_2O$] or blue asbestos, which is found as a straight blue fibre in South Africa, Western Australia, and Bolivia; and
4. anthophyllite [$(MgFe)_7.Si_8O_{22}.(OH)_2$], a brittle white fibre mined in Finland and Africa.

Other types of asbestos include tremolite [$Ca_2Mg_5Si_8O_{22}.(OH)_2$] and actinolite [$CaO.3(MgFe)O.4SiO_2$].

The type of asbestos used in a manufactured product cannot be identified reliably by colour because the natural colours tend to change through the action of heat. Moreover, the presence of asbestos may be concealed by decorative coatings, finishing cements, paint or cladding. In order to determine thc presence and type of asbestos used, it is necessary to cxamine the bulk sample of the material by microscopy or X-ray analysis.

Asbestos is widely used because it is a relatively cheap material with special chemical and physical properties which make it virtually indestructible [chemical resistance (particularly to acids), fire resistance, mechanical strength, high length to diameter (aspect) ratio, flexibility, and good friction and wear characteristics]. It readily lends itself to a variety of manufacturing processes on account of its wet strength, ease of formation of slurries with water, and good drying characteristics. Crocidolite (blue asbestos) is a fibrous material containing approximately 50% of combined silica and nearly 40% of combined iron(II) and iron(III) (as oxides). Crocidolite is the strongest of the asbestos fibres and its high mechanical strength combined with its acid resistance makes it a valuable industrial material. It does, however, tend to fuse at high temperatures. Today its use is largely in the manufacture (not in the U.K.) of asbestos–cement pressure pipes.

Amosite is a fibrous material containing approximately 50% of silica and 40% of iron(II) oxide. Its characteristic features are its resistance to corrosion, the springiness of the fibres, and the high bulk density volume that they produce when separated by processing. Amosite is widely used for heat insulation applications, e.g. fire-resistant insulation board.

Chrysotile (white asbestos) is a fibrous material containing approximately 40% each of silica and magnesium oxide and it currently accounts for over 90% of world asbestos consumption. Grades of chrysotile asbestos with long fibres are spun into yarn which is used in asbestos cloth for protective clothing, heat insulation, etc. While the fibres have very good heat resistance, they are destroyed by acids.

Anthophyllite is a rare fibrous material of variable composition containing up to 60% of silica, 17–31% of magnesium oxide, and up to 20% of iron (II) oxide. It finds use as an expensive filler and in specialized applications on account of its good heat and chemical resistance.

Asbestos has been known and used in small amounts for thousands of years, but it was not widely used prior to the latter part of the 19th Century (Wright, 1969). Asbestos was first mined in the Quebec (Canada) chrysotile fields in 1878, followed by Russia in 1885 and South Africa in 1906. Its use grew slowly during the early part of the 20th Century and by 1930 a cumulative world total of about 5×10^9 Kg of asbestos had been mined. Since the decade 1930–40 production has accelerated such that the production of asbestos for 1976 alone (approximately 5.2×10^9 Kg) exceeded the 1930 cumulative total. Some 3000 uses of asbestos have been recorded (Rosato, 1959), mainly in the construction industry (about two thirds of the total), including asbestos cement sheet and pipe, flooring and roofing products, electrical and thermal insulation materials, friction products, coatings, compounds and textiles, and also heat shields for spacecraft. In most of the asbestos products the fibres are bound in a matrix or are encapsulated and do not contribute a health risk in these states. Potential health risks arise during the drilling, sawing, etc., of asbestos products—activities which can result in the liberation of free asbestos fibres. The production of free fibres can affect people other than the workers directly concerned (Harries, 1976; Skidmore and Jones, 1975).

Within 20 years of the first factory production of asbestos, the public health hazards associated with asbestos started to come to light. Between 1890 and 1895 sixteen out of seventeen workers in a French asbestos weaving factory had died and by 1899 eleven men who had worked in an asbestos spinning factory in the U.K. had died at about the age of 30, having spent the whole of their working lives in this occupation ('cording'). The last of the deaths was reported by Dr. Montague Murray to the 1906 Departmental Committee for Compensation for Industrial Diseases, and this is the first recorded case of the disease which was to become known as asbestosis (HMSO, 1970). The first complete description of asbestosis appeared in 1927 (Cooke, 1927; McDonald, 1927). It is a disease resulting from the inhalation of very small particles of asbestos dust, which are deposited in the lungs, causing the formation of scar tissue (fibrosis) and resulting in fatigue and breathlessness after some years. Other diseases associated with exposure to asbestos dust are cancer and mesothelioma. About 50% of asbestosis sufferers develop lung cancer and the probability of this happening is greatly increased if the asbestosis sufferer is also a cigarette smoker. There is also evidence that cancers in other sites of the body may occasionally be asbestos-linked. Mesothelioma is a cancer of the membrane lining of the chest or abdomen which has been identified during the past 20 years. It is almost exclusively associated with asbestos exposure (Wright, 1969) and may have a long latent period.

The widespread commercial use of asbestos over approximately the past 100 years has led to its uncontrolled distribution throughout much of the industrialized world and its appearance in the general environment. The general concentration of asbestos fibres in the urban atmosphere is actually in the lower part of the range 10 ng M^{-3} to 100 ng M^{-3} (Holt and Young, 1973; Nicholson and Pundsack, 1973; Sebastien and Bignon, 1974; Sebastien *et al.*, 1976; Selikoff *et al.*, 1972), but this can rise considerably in certain locations. Thus, in the vicinity of factories using asbestos, concentrations of asbestos fibres in the external atmosphere of up to 5000 ng M^{-3} have been reported (Nicholson *et al.*, 1975; Rickards, 1973) and in buildings where the asbestos present is being worked with or is damaged, the concentration of asbestos in the air can be as high as 800 ng M^{-3} (Sebastien *et al.*, 1976). Low concentrations of asbestos are also found in water

(Nicholson and Pundsack, 1973; Cunningham and Pontefract, 1971; American Water Works Association, 1974; Cook *et al.*, 1974; Zielhuis, 1977), food (Merliss, 1971; Wolff and Oehme, 1974), beverages (Wehman and Plantholt, 1974) and pharmaceuticals and other medicinal products (Nicholson *et al.*, 1972; Infante and Lemen, 1976).

During about the first 60 years of the commercial exploitation of asbestos, the industry did not sufficiently appreciate the risks and the exposure of workers to relatively high levels of dust was common. Over the years standards have progressively been improved and working conditions are more strictly controlled today than they were, say, 10 years ago. The group with the highest risk of contracting asbestos related disease is insulation workers ('laggers'). The incidence of asbestosis was very high among asbestos textile workers in the 1920s but has been greatly reduced over the years following improvements in dust conditions (Figures 1.1 and 1.2). In those parts of the asbestos industry where the fibres are firmly bound in products (asbestos cement and friction materials), the incidence of asbestos-related disease has in general been lowest. It is interesting that, despite the very high airborne dust concentrations which exist in chrysotile mining and milling,

Figure 1.1 The carding of asbestos fibre into silver (1925–1935). The photograph shows a twin card with the hopper feed to the left of the operator and the finished spools of silver emerging on the right with a cross feed between the two parts of the machine. The arrangement of the dust exhaust can be seen leading up to the main duct. Reproduced by kind permission of Scandura Ltd. (A.B.B.A Group Company)

Figure 1.2 Enclosed carding (1966–1978). Each individual carding machine is housed in an enclosure. The comprehensive dust exhaust ducting can be seen inside the enclosure but, in addition, the enclosures are exhausted from front to back at the rate of 7500 ft^3/min. The enclosures can be opened for access to any part of the machine. Reproduced by kind permission of Scandura Ltd. (AB.B.A Group Company)

the incidence of severe asbestos related disease (especially mesothelioma) has been low.

Because of the pathological effects of airborne asbestos dust, attempts have been made to provide less harmful substitutes which could be used in place of asbestos. Developments to date indicate that the continued use of asbestos fibre as such will be essential in a number of industries for the foreseeable future. Even if it were possible to stop using asbestos completely now, it will be very many years before all the asbestos currently in use for thermal insulation, etc., has been removed and it will be necessary to ensure that workers engaged in the asbestos industry can carry out their various activities in such a way that their health is not impaired.

(2) POINTS OF NOTE IN THE DEVELOPMENT OF THE BRITISH ASBESTOS INDUSTRY

It is interesting to consider the main events in the development of the British asbestos industry over about the past 60 years.

Early years of the 20th Century — At this time the dustiest process (as reported by Merewether, below) was blending, which involved laying the various grades of fibres in horizontal layers on the floor and then cutting off a vertical slice of the layers by hand which was then fed down a shute into an operating machine. The next dustiest process was dry measuring and 14 lb of fibre and dust per week were removed by hand from under each loom.

1920s — The medical profession began to suspect a causal effect between the presence of asbestos fibres and fatality among asbestos workers who contracted lung disease.

1928 — Factory Department of the Home Office initiated investigation into above.

1930 — Merewether Report and concept that full development of asbestosis could be prevented if the concentration of asbestos particles in the air breathed remained below the 'datum' level found to be normally present at the 'spinning' process in a factory processing raw asbestos. Major conclusion of Report: *improved ventilation and dust suppression in places giving rise to asbestos dust is the principal safeguard against the ill-effects of the dust.*

Subsequent Home Office–Asbestos Textile Manufacturers Joint Conference and agreement that everything possible must be done to suppress dust.

1931 — Manufacturers undertook to render every possible assistance. Five-member Committee (two Engineering Inspectors of Factories and representatives of Cape Asbestos Co. Ltd., British Belting Asbestos Ltd., and Turner Bros. Asbestos Co. Ltd.) set up and agreements made which were subsequently to be given statutory effect under the Asbestos Regulations. Suggestions of the General Council of the Trades Union Congress were taken into account in drafting the Regulations. The agreements were aimed at interfering as little as possible with the then existing working methods or layout of premises, e.g. substitution of mechanical methods for handwork, separation of processes and wider spacing of plant were not specifically urged and apart from some aspects of hand cleaning the prohibition of any of the working methods was not suggested. The representatives of the asbestos industry stated that they could not commit the whole of the trade to the agreements reached; that they expected a sufficient period of time to be allowed before requirements based on the agreements were enforced; and the British factories might be subject to unfair competition if restrictive requirements were enforced under Section 79 of the Factories and Workshops Act 1901.

Shortly after the Report on the Conference (Bellhouse, 1931) Asbestos Industries Regulations published (December 1931), which required: the use of exhaust ventilation systems in such a way as to ensure that no asbestos dust was allowed to escape into the air of any room in which people worked; and the provision of breathing apparatus and special clothing for the use of workers where such dust was unavoidable (e.g. handling asbestos at certain stages during manufacture and the hand cleaning of certain equipment). Certain processes were exempt from the Regulations on the grounds that they did not give rise to a concentration of asbestos dust higher than the 'datum' level of the Merewether Report. Asbestos monitoring was not required.

1933 Ventilation systems in cording and weaving introduced.

1939 Damping techniques established in the weaving process.

1942 Dust settling chambers in the ventilation system replaced by sleeve filters (removing the need to clean out the chambers–a hazardous operation)

1951 Routine dust sampling commenced. Prior to this time scientific methods of air sampling were not sufficiently advanced to reflect accurately the actual concentrations of asbestos dust in the workers' breathing zone. Also, there was no scientific knowledge of what concentration of dust could be accepted as 'safe' and there was little data relating asbestos dust concentration with asbestosis.

1953–1957 Card extraction system improved and oiling of fibres (with oil emulsion to suppress dust) introduced at the mixing stage.

1959 Industrial Health Advisory Committee of the Ministry of Labour carries out a special survey of medicinal services at a variety of factories.

1960 'Safe' standard of 177 *particles* of asbestos dust per cubic centimetre of air accepted by the Industrial Health Advisory Committee (based on level set by the American Conference of Governmental Industrial Hygienists).

1964 Indications of an abnormally high incidence of lung cancer among asbestosis sufferers and of the development of mesothelioma in some people following short periods of exposure to certain types of asbestos. (Buchanan, 1965). Factory Inspectorate considers revision of the Asbestos Industry Regulations 1931 in consultation with the Industry, the TUC, the Asbestosis Research Council, etc.

1965 Medical Advisory Panel set up to report on the medical aspects of the asbestos problem.

1966 Existing standards of dust control brought into question as evidence accumulates that asbestos is more dangerous than had been thought. Detailed country-wide survey by Engineering and Chemical Inspectors of Factories to gather information on the effectiveness of dust control methods and the standards of protection which should be provided. It is interesting that up to this time the industry had spent a large amount of money on dust control, but there seems to have been a lack of appreciation of the standards of dust control needed in the circumstances. There are also indications that the Chemical and Engineering branches of the Factory Inspectorate, who were required to advise on the industry, were not in accord over the appropriate standards of dust control and how these might be achieved.

1968 Medical Advisory Panel Report published findings: the number of new cases of asbestosis was increasing throughout the industry, i.e. in those areas of the industry where the 1931 Regulations applied and to a greater extent in those areas of the industry where the Regulations did not apply; crocidolite (blue asbestos) was a particular risk and should be replaced with other types of asbestos as far as possible; there was not a fully satisfactory method of measuring asbestos dust concentrations; the 177 particles of asbestos per cubic centimetre of air standard to be replaced with a system of fibre counts or a gravimentric method of sampling; no medically 'safe' limit of exposure to asbestos dust could be defined, although its establishment should be a long-term goal; a provisional standard for airborne asbestos dust exposure of 1.9–7.7 fibre counts/cm^3 (depending on the process involved and based on what could currently be obtained in the best factories) was recommended; regular medical supervision of asbestos workers to be established, linked with a continuing record of their dust exposure and morbidity and mortality.

Hygiene Standard for Chrysotile Asbestos Dust published by the British Occupational Hygiene Society, based on the concept that so long as there is any airborne chrysotile dust in the work environment there may be some small health risk and that exposure up to certain limits can be tolerated for a lifetime without incurring undue risks. The limit was set at a maximum concentration of 2 fibres/cm^3 over a period of 50 years (or correspondingly higher concentrations for shorter periods), at which level the risk of being affected by the dust to the extent of showing the earliest demonstrable effects on the lung due to asbestos was believed to be less that 1%.

1969 Asbestos Regulations 1969 made under the Factories Act 1961 and were based on a study of 290 workers over periods ranging from 10 to over 30 years (*N.B.*, a group of 58 workers who have worked in the industry for at least 10 years since 1951 under conditions that conform to those of the 1969 Regulations have been carefully monitored and all were free from asbestos-related disease). The Regulations were meant to compel the employers concerned to take specified measures to prevent the entry of asbestos dust into the air of the workplace, and to provide individual protection against its inhalation where such prevention was not possible. The 1969 Regulations covered activities which had been excluded from the 1931 Regulations and defined asbestos dust as *dust consisting of or containing asbestos to such an extent as is liable to cause danger to the health of employed persons.*

1970 Employment Medical Advisory Service commences continuing study of the health of asbestos workers.

1976 Advisory Committee on Asbestos set up by the Health and Safety Commission.

1977 Selected evidence to the Advisory Committee on Asbestos published.

1978 First and second reports of the Advisory Committee on Asbestos published (*Asbestos: Work on thermal and accoustic insulation and sprayed coatings*, and *Asbestos: Measurement and monitoring of asbestos in air*).

(3) BIBLIOGRAPHY OF SOME TERMS USED IN THIS BOOK AND IN THE LITERATURE ON THE SUBJECT

Asbestos bodies: Microscopic fibres found in the lungs of people who have been exposed to asbestos particles in the atmosphere. Their presence does not necessarily denote that the person has 'asbestosis' which is the disease produced by asbestos inhalation.

Asbestosis: A non-malignant, progressive, irreversible lung disease caused by the inhalation of asbestos dust and characterized by diffuse intestinal fibrosis (see Wagner, 1965; Gilson, 1971). Asbestosis develops slowly and is detectable by a combination of clinical, radiographic and lung function tests. In the early stages there is uncertainty about diagnosis. The characteristic symptoms are progressive breathlessness and an unproductive cough. Once diagnosis of the disease is certain the fibrosis tends to increase despite removal of the patient from further dust exposure. The degree of risk of developing asbestosis appears to be related to the duration and level of asbestos dust exposure and it is not yet known if there are detectable early stages at which removal from further dust exposure markedly influences the course of the disease. Cases of asbestosis have occurred in people many years after their last known exposure to the dust. Asbestosis has been a recognized industrial disease in the U.K. since 1930.

Asbestosis Research Council: A cooperative research organization established in 1957 by the major U.K. companies engaged in the manufacture of asbestos products. The fundamental objective of the Asbestosis Research Council is to foster research into the causation and prevention of asbestosis and any other diseases possibly associated with exposure to asbestos. The objective is achieved by: the finance and sponsorship of agreed medical research projects; the promotion of improvements; dust sampling and counting methods aimed at the establishment of standard methods for general use in industry; attempts to establish within practical limits the maximum airborne concentrations that can be safely tolerated in operations and processes involving asbestos; and promotion of improvements in the manufacture, handling, and use of asbestos products so as to reduce or eliminate the dangers of inhaling asbestos.

Bronchial cancer: Bronchial cancer (carcinoma of the bronchus, squamous carcinoma of the bronchus) associated with exposure to airborne asbestos dust cannot be distinguished from that caused in other ways, e.g. by cigarette smoking. Workers exposed to airborne asbestos dust who also smoke cigarettes have a materially greater risk of developing bronchial cancer than members of the general public with the same smoking habits.

Bronchiectasis: A condition of permanent dilation of one or more bronchi.

Cancer: See Tumour.

Carcinogen: An agent or process which significantly increases the yield of malignant neoplasms in a population.

Cause specific death rate: The incidence of all identified deaths in the population (see Death rate).

Collapse of the Lungs: In this condition the terminal alveoli are completely devoid of air.

Death rate: The population's total mortality experience (i.e. the incidence of all deaths in the population).

Degree of risk: Quantification of dose–response relationship.

Differential thermal analysis (DTA): A technique of recording the difference in temperature between a substance and a reference material (such as alumina or glass beads) against either time or temperature as the two specimens are subjected to identical temperature regimes in an environment heated or cooled at a controlled rate.

Emphysema of the Lungs: In this condition there is enlargement of the air spaces of the lungs accompanied by destructive changes of the walls of some of those air spaces. Emphysema often accompanies a pneumoconiosis such as asbestosis. It may result from a variety of causes.

Incidence (of a disease): The number of new cases of the disease in the population during a specified period of time (e.g. 5 years, 1 year, 7 days).

Incidence rate (of a disease): The ratio of incidence to total number of people in the population (and who might have contracted the disease—the number at risk).

Medical effects of inhaling asbestos dust:

(a) Pleural thickening: A thickening of the surface of the lungs may occur after exposure to asbestos dust. This has been found in some workers and also in some people living near asbestos tips. In the absence of asbestosis, pleural thickening does not itself have any ill effects and is not necessarily associated with cancer of the lung or mesothelioma.

(b) Asbestosis: See Asbestosis.

(c) Cancer: (i) Cancer of the larynx, stomach, colon and rectum may be associated with asbestos; (ii) there is an increase in the incidence of cancer of the bronchial tubes or lungs in people who have asbestosis. It is not yet certain if asbestos can contribute to cancer when asbestosis is not already present. Cigarette smoking leads to a much greater incidence of lung cancer among asbestos workers than among the population as a whole.

Asbestos-related diseases can appear from several years to tens of years after first exposure to airborne asbestos dust and the recent increasing incidence of cases in many countries is the result of past dust exposure which was obviously too high. The current incidence of asbestos related disease is not regarded as a measure of the effects of present dust levels.

Mesothelioma: A malignant tumour (exceptionally rare in the general population) of the lining of the thoracic (pleural) or more rarely abdominal cavity (peritoneum). Mesothelioma is usually associated with exposure to blue asbestos, but has also occurred among workers exposed to other types of asbestos. The duration of the exposure required to cause mesothelioma may be brief and the appearance of the disease can occur some considerable time later. Cases of mesothelioma have been reported in which no exposure to asbestos can be traced. Most cases of mesothelioma appear to be related to past exposure to asbestos and present evidence indicates that insulation workers as a group are at highest risk.

Pleurisy: An inflammation of the pleural membrane covering the lung or of its parietal (outer) reflections.

Pneumoconiosis: Pulmonary lesions arising from the inhalation of dust during the course of work. The following classification is based on the pathological reaction of the lung tissue to the dust.

1. Benign: iron, tin, barium, antimomy.
2. Fibrotic: (i) silica, asbestos and other silicates. (ii) Mixed dusts e.g. coal ore, iron ore.
3. Inflammatory: vanadium, manganese, beryllium, hard metal, cadmium, aluminium.
4. Allergic: cotton dust (byssinosis), mouldy hay (farmer's lung), osmium, platinum salts, hard wood dust.

Population: See Study chart.

Prevalence (of a disease): The total number of cases (both new and already existing) in the population during a specified period of time (e.g. 5 years, 1 year, 7 days) or at a specific time (i.e. date).
Prevalence rate (of a disease): The ratio of the number of cases of a disease to the total number of people in the population (and who might have contracted the disease—the number at risk).
Progression: A term used to indicate development of a tumour by way of permanent, irreversible qualitative change in one or more of its characteristics. The essential feature of progression is the gaining of a proliferative advantage by the increasingly malignant tumour over both the normal cells of the tissue and the less malignant cells of the tumour.
Prospective study: The observation of a population through time into the future. This process allows for extensive planning to ensure that the data collected is adequate to investigate all hypotheses of interest and merit.
Pulmonary rales: See Rales.
Rales: Dry, crackling sounds in the lung.
Radiological classification of pneumoconiosis: An agreed international classification of radiological changes due to pneumoconiosis is given in Table 1.1 (International Labour Office, Geneva, 1958).

Table 1.1 International classification of persistent radiological opacities in the lung fields provoked by the inhalation of mineral dust*
(*Geneva Classification*, 1958)

<table>
<tr><td></td><td>No peneumo-coniosis</td><td>Suspect</td><td colspan="13">Pneumoconiosis</td></tr>
<tr><td>Type of opacity</td><td></td><td></td><td>Linear opacities</td><td colspan="9">Small opacities</td><td colspan="3">Large opacities</td></tr>
<tr><td>Qualitative features</td><td rowspan="2">O</td><td rowspan="2">Z</td><td rowspan="2">L</td><td colspan="3">p</td><td colspan="3">m</td><td colspan="3">n</td><td rowspan="2">A</td><td rowspan="2">B</td><td rowspan="2">C</td></tr>
<tr><td>Quantitative features</td><td>1</td><td>2</td><td>3</td><td>1</td><td>2</td><td>3</td><td>1</td><td>2</td><td>3</td></tr>
<tr><td>Additional symbols</td><td>(co)/(cp)</td><td>(cv)</td><td>(di)</td><td colspan="3">(em)</td><td colspan="3">(hi)</td><td colspan="3">(pl)</td><td colspan="2">(px)</td><td>(tb)</td></tr>
</table>

* Including coal and carbon dusts.

Definitions and comments

The object of the classification is to codify the radiological appearances of the pneumoconioses in a simple, easily reproducible way. It is intended to describe the radiographic appearances of the persistent opacities associated with pneumoconiosis, not to define pathological entities, nor to take into account the question of working capacity.

Where there is an appreciable difference in the appearance of the two lungs, the two appearances may be described separately, beginning with the right lung.

Table 1.1 (*Contd.*)

No pneu-noconiosis	O	No radiographic evidence of pneumoconiosis.
Suspect opacities	Z	Increased lung markings.
		Pneumoconiosis
Linear opacities	L	Numerous linear or reticular opacities, the lung pattern being normal, accentuated or obscured.
Small opacities†		*The following types are defined according to the greatest diameter of the predominant opacities.* — *The categorization depends on the extent and the profusion of the opacities.*
	p	Punctiform opacities. Size up to 1.5 mm. — *Category* 1: A small number of opacities in an area equivalent to at least two anterior rib spaces and at the most not greater than one-third of the two lung fields.
	m	Micro-nodular or miliary opacities. Greatest diameter between 1.5 and 3 mm. — *Category* 2: Opacities more numerous and diffuse than in category 1 and distributed over most of the lung fields.
	n	Nodular opacities. Size between 3 and 10 mm. — *Category* 3: Very numerous profuse opacities covering the whole or nearly the whole of the lung fields.
Large opacities‡	A	An opacity having a longest diameter of between 1 and 5 cm, or several opacities each greater than 1 cm, the sum of whose longest diameters does not exceed 5 cm.
	B	One or more opacities, larger or more numerous than those in category A, whose combined area does not exceed one-third of one lung field.
	C	One or more large opacities, whose combined area exceeds one-third of one lung field.
		Additional symbols
Recom-mended additional symbols§	(co)	abnormalities of the cardiac outline. To be replaced by (cp): cor pulmonale, if this condition is strongly suspected.
	(cv)	cavity.
	(di)	significant distortion of the intra-thoracic organs.
	(em)	marked emphysema.
	(hi)	marked abnormalities of the hilar shadows.
	(pl)	significant pleural abnormalities.
	(px)	pneumothorax
	(tb)	opacities suggestive of active tuberculosis.

† The choice of order of the symbols is left to the convenience of the physician. ‡ The background of small opacities should be specified as far as possible. § the use of these symbols is optional.

Respiratory function tests: *Vital Capacity* (V.C.) Volume of air which can be expelled after deep inspiration. Normally between 3 and 5 litres.
Forced Expiratory Volume ($F.E.V_1$) Most of the vital capacity can be expired in the first second. This amount is the $F.E.V_1$.
Functional Residual Capacity (F.R.C.) Volume of air in the lungs at the end of a normal expiration. Normally between 2 and 3 litres.
Peak Flow Rate (P.F.R.) A measurement of the rate of the initial peak of respiration lasting about 1/100th of a second. Normally 500 to 700 litres per minute in adults.
Respiratory tract, anatomical parts of: Air passes through the nose and nasopharynx (back of nose) and then enters the larynx (voice box). From there it passes through the trachea and enters the bronchi on each side, which subdivide into smaller bronchi finally becoming minute tubes, bronchioles, which give off the terminal alveoli, tiny sacs in which respiratory exchange between blood and air takes place.
Restrospective investigation: The observation of a population through time into the past, i.e. looking at historical evidence. The latter may be incomplete, unattainable, or unreliable.
Study cohort: The group or population identified for study, e.g. a given factory labour force.
Thermogravimetric analysis (TGA): A technique of recording the difference in weight between a substance and a reference material (such as alumina or glass beads) against either time or temperature as the two specimens are subjected to identical temperature regimes in an environment heated or cooled at a controlled rate.
Tumour: A tumour is an abnormal mass of tissue, the growth of which exceeds and is uncoordinated with that of the normal tissues, and persists in the same excessive manner after the cessation of the stimuli which evoked the change (Willis, 1960). Malignant tumours are those with the power to invade adjacent tissues and to spread to other parts of the body. Benign tumours do not have this power.
Tumours, Neoplasms: New formations of tissue usually leading to a mass of tissue composed of a single cell type. Neoplasms may be benign, e.g. lipoma composed of adult fat cells, or malignant, e.g. mesothelioma of the pleura or squamous carcinoma of the bronchus. Malignant tumours invade extensively and may metastasize, i.e. spread by lymphatic or blood vessels to distant sites.

(4) COMMERCIALLY AVAILABLE PRODUCTS AND SERVICES

The following information is included for the benefit of readers who may have need to seek information on the great range and variety of products and services available for dealing with asbestos-related problems.

The details refer to only a limited number of products and services and the list of organizations is not meant to be exhaustive. Many other products and services are available and the non-inclusion of these does not imply that they are regarded as unsatisfactory in any way. The inclusion of a particular product or service, etc., is not meant to be an endorsement of that product or service. None of the items mentioned has been tested in any formal sense, but inclusion does indicate that there is no known or suspected reason why the use of the product or service should not prove to be satisfactory.

4(i) Respiratory equipment

Care must be exercised in the selection of respiratory protection equipment. Where doubt exists as to the most suitable form of equipment to use, all of the organizations listed will provide expert advice and guidance. Cartridge-type respirators (complying with BS 2091) may only be used in atmospheres not immediately dangerous to life and containing a sufficient supply of oxygen. For work with crocidolite a positive-pressure air line system should be used (BS 4275).

ARCO Safety Clothing and Equipment
Waverley Street,
Hull HU1 2SJ

Telephone: 0482-27678

ARCO Safety Clothing and Equipment supply the Polimask respirators manufactured by Pirelli (Figure 1.3). The masks are made of polychoroprene rubber with a double frame and chin protection and hooks for easy fitting of the harness. There are two basic models—for one or two filters, i.e. Polimask 200 and Polimask 200-2, both approved by the Health and Safety Executive under the Asbestos Regulations 1969.

British American Optical Company
Radlett Road,
Watford,
Herts. WD2 4LJ

Telephone: 0923-33522 (Mrs. E. Bristow)

AO Safety Products are manufacturers of Approved Respirators for use against asbestos dust, reference the Duralair Facemask, Code 610000, and Cartridge 061110.

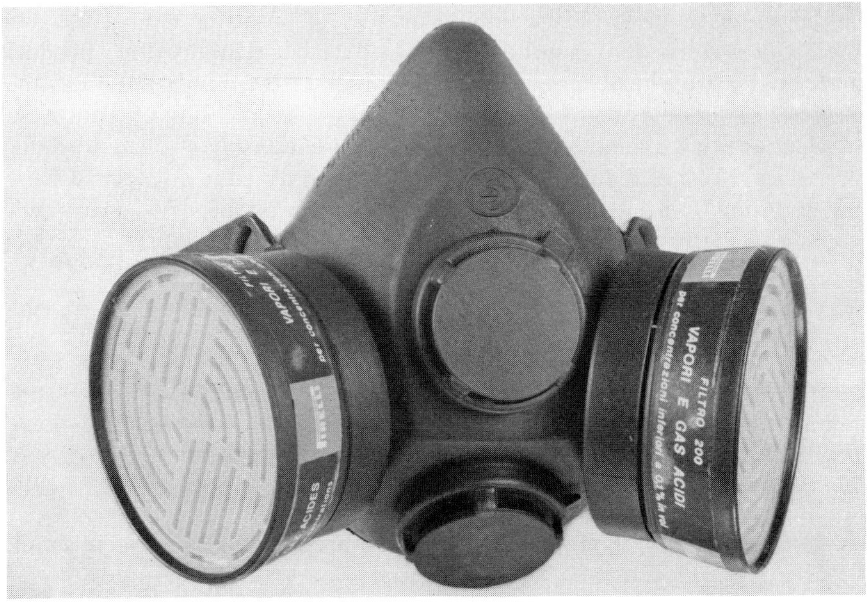

Figure 1.3 Pirelli Polimask 200-2. (Photograph by courtesy of Sekur Personal Protection Products)

James North & Sons Ltd.
P.O. Box 3,
Hyde,
Cheshire SK14 1RL

Telephone: 061-368 5811

James North & Sons Ltd. produce three respirators and two dust cartridges, all which have been approved by the Health and Safety Executive and are tested by them annually and a licence to this effect issued. The respirators also conform to BS 2091:1969 and are suitable for use in atmospheres not immediately dangerous to life and containing adequate oxygen.

The red colour-coded (to check against wrong catridge being used) cartridges screw directly into the filter holder(s) on to a foam disc and assure a positive gas-tight seal. All moulded parts are made from high-impact polystyrene to withstand rough usage: there are no metal parts. The masks are held on the face by headbands which are fitted with nylon quick-release buckles having a reverse lever action. This enables the respirator to be drawn firmly against the face with minimum tension, simply by pulling on the ends of the head harness. The mask is instantly released by pressing on the protruding ends of the nylon buckles. Contaminated hands or gloves need not touch the head.

Martindale Protection Ltd.
Neasden Lane,
London NW10 1RN

Telephone: 01–450 8561

Martindale Protection Ltd. manufacture a range of dust respirators which comply with the appropriate British Standard specifications and are approved by the Health and Safety Executive for use by operators who are handling all types of asbestos.

The negative pressure dust respirators (not to be used in an oxygen-deficient atmosphere) are approved and kite marked to BS 2091:1969 and every filter is individually tested, as specified in BS 4400, on a sodium chloride penetometer. The filter medium is electrostatically charged, resin-impregnated merino wool, encapsulated in a plastic case, and gives protection against dusts in the size range 0.5–0.8 μm in diameter.

The Martindale Mark IV positive-pressure powered respirators (approved and kite marked to BS 4558) are compact, lightweight, completely self-contained units, free of trailing wires and hoses, and allow the operator unrestricted mobility. The 6-V d.c. motor is driven by a sealed rechargeable nickel–cadmium battery which has sufficient power for at least a 7-h shift. The motor draws air through the filters and delivers it, via flexible hoses, to the facepiece (used, e.g., with blue asbestos dust), hood, or blouse.

Standard Efficiency (SF) or High Efficiency (HEF) filters can be supplied and Martindale will advise on the correct type. At an air flow-rate of 120 l/min the HEF Filters have >99.95% efficiency initially and the SE filters have >98% efficiency initially against 0.5 μm diameter dust particles.

Phoenix Accessories (Safety) Ltd.
Waterloo Mills,
Waterloo Road,
Pudsey,
Yorks. LS28 8DQ

Telephone: 0532-574475

Phoenix Accessories (Safety) Ltd. are manufacturers and distributors of a range of safety equipment, with special emphasis on respiratory equipment including protective work wear for operators concerned with the removal of asbestos. The Feenair respirator is approved by the asbestos industry and conforms to all the specifications of BS 2091:1969. The yellow colour-coded cartridge is approved by H.M. Factory Inspectorate and complies to BS 2091.

Protector Safety Products (UK) Ltd.
Protector House,
719 Banbury Avenue,
Slough SL1 4LL

Telephone: 0753-38317 (Mr. S. Elliot)

The Protector-manufactured range includes a series of ori-nasal respirators fitted with particulate filters of merino wool, which are approved by the Health and Safety Executive for use against white asbestos. Approval has also been received from Standards Authorities all over the world and also from specialized industrial authorities. The respirators are available in a choice of sizes to give a close fit to any user, and are held in position by a durable, adjustable head harness incorporating a quick release for ease of operation and comfort. The cartridge filters (which exceed BS 2091 requirements) are subjected to rigorous laboratory testing to ensure compliance with British and Australian Standards requirements.

For work with crocidolite a half-face or full-face (if eye protection is also required) air-fed mask is provided, usually coupled in the air line to a filter regulator unit to ensure a supply of respirable air.

A range of air supplied hoods made of polyesterized nylon and PVC-coated terylene, suitable for protection against all types of asbestos, is also manufactured.

4(ii) Dust Control Equipment

The Asbestos Regulations 1969 led to an increase in the safety aspects of working with asbestos. Indeed, these Regulations have had a greater effect on the industry than any other single piece of legislation within the past 25 years. Dust control equipment manufacturers, in consulation with the asbestos users and the Factory Inspectorate, have made considerable efforts to develop techniques and equipment which would satisfy the requirements of the Regulations efficiently and economically. Over the years considerable improvements have been effected in capture hood design and in filtering devices.

Bivac Consultants Ltd.
Beehive Works,
Marsland Street,
Stockport SKI 20Q

Telephone: 061-480 3468/9 (Mr. E. A. Green)

Bivac Consultants Ltd. are designers and manufacturers of central vacuum cleaning plant and are able to supply industrial vacuum cleaning equipment (both fixed installations and mobile vacuum cleaners) specially designed to comply with the regulations issued by the Asbestosis Research Council. The fixed installations convey the collected dust to a central receiving hopper with a special bagging device individually made to suit conditions with exhaust air filtering to keep fibre emission well within the prescribed limits (Figure 1.4). The systems are widely used in asbestos factories throughout the U.K.

The mobile industrial air cleaner Type ACL (formerly called Industrial Vacuum Cleaner Type CL/CB) was submitted to independent test by the University of Manchester Occupational Hygiene Service in 1970. Under the extreme conditions of the tests, the asbestos fibre content of the exhaust air from the equipment was <0.02 fibres/cm^3, i.e. less than one tenth the current industrial air standard for white asbestos.

B.V.C. Ltd.
Goblin Works,
Leatherhead,
Surrey KT22 8TZ

Telephone: Ashtead-76121 (R. L. Allen)

B.V.C. Ltd. manufacture a range of portable industrial vacuum cleaners having three-stage filtration to standards laid down by the Asbestosis Research Council. The filtration system includes throw-away paper bags, which are easily removed from the cleaner for loading into an approved type of impermeable plastic bag for safe disposal of collected asbestos dust and fibres, a fabric filter, and a high efficiency outlet filter.

B.V.C. Ltd. also design, manufacture and instal piped-in central vacuum cleaning systems incorporating special high-efficiency filters for dealing with asbestos work. A free advice service is available.

The 'Drumclene' equipment also produced enables the effective method of high-pressure air blast cleaning of brake drums to be carried out without contamination of the workshop atmosphere and without health hazard to the workers. The unit is trolley mounted and is simply pushed over the brake shoes and backplate so that its flexible rubber skirt forms an effective seal against escape of airborn dust and asbestos fibres when a compressed air line is used to remove loose dust from the brake mechanism. The negative pressure inside the unit ensures that there is no escape into the atmosphere of asbestos-laden dust.

Dust Control Equipment Ltd.
Humberstone Lane,
Thurmaston,
Leicester LE4 8HP

Telephone: 0533-696161

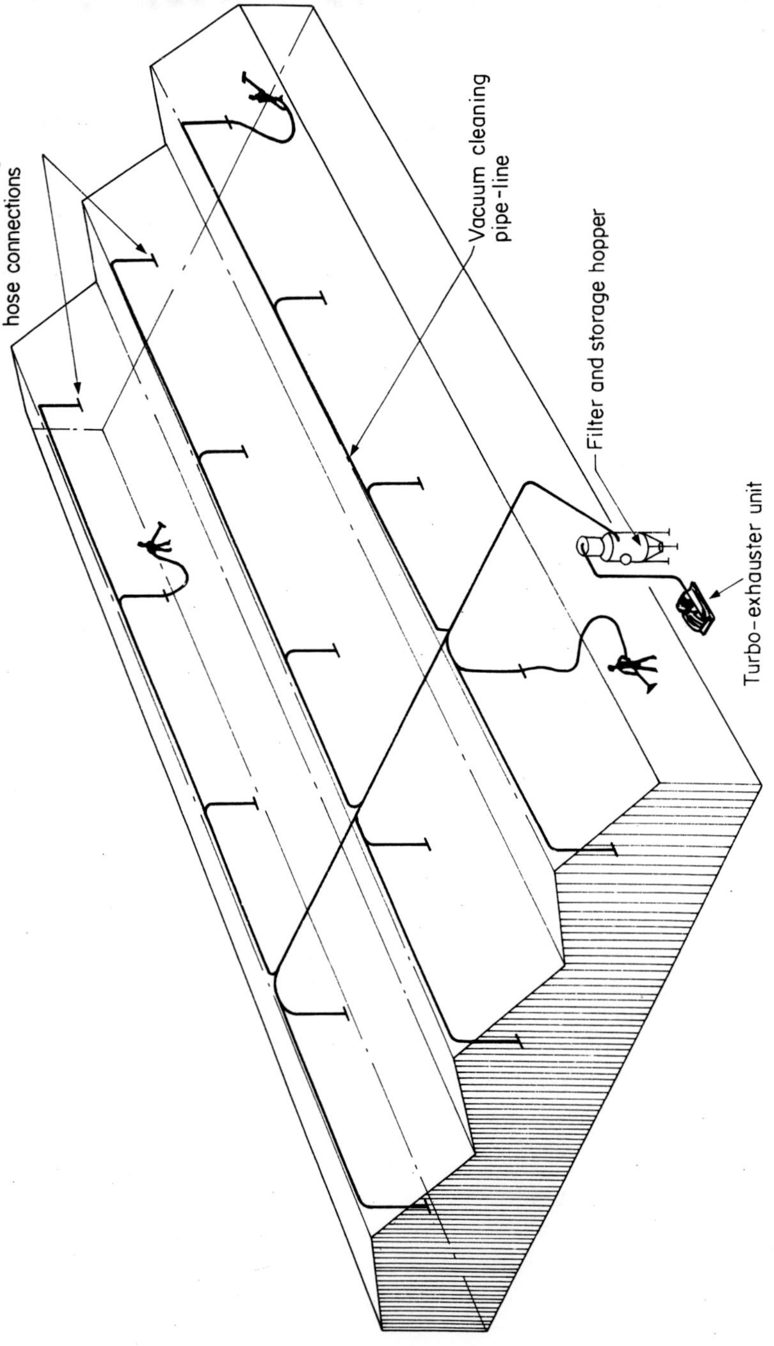

Figure 1.4 Typical arrangement of central vacuum cleaning plant. (Diagram by courtesy of Bivac Consultants Ltd.)

Dust Control Equipment Ltd. (DCE) has been closely involved in the development of equipment for asbestos dust control for many years, both with the Factory Inspectorate (now known as the Health and Safety Inspectorate) and the asbestos manufacturers. There are DCE associate companies in West Germany, France, Holland, Australia, South Africa, Scandinavia, the United States and Japan, and there is a network of agents throughout the world.

The DCE Unimaster range of dust-control units provides economical and effective solutions to a wide variety of dust problems throughout industry. To give this essential versatility in application, the Unimaster has five basic constructions, coupled with a standard range of fan sizes, filtration areas, and dust container capacities. Assembled in different combinations, these give a choice of over 500 Unimasters.

The DCE Unimaster is designed for compactness, simplicity in operation and ease of servicing. Flat, parallel, multipad design of filter element provides maximum fabric area for minimum of floor space. Choice of inlet positions enables the unit to fit more easily into the required layout. The multipad fabric filter element is made up into one piece to minimize the number of seals between fabric and frame. A flexible wire-mesh insert fits into each pad to ensure maximum effective use of the fabric area and to assist thorough cleaning. Filter cleaning is motorized and automatically activated. This reduces dependence on the operator, avoids unproductive tasks and maintains the operational efficiency of the unit. A foolproof quick-release sealer gear permits quick inspection and removal of the pull-out dust container. Dust containers are of a sensible size and designed for easy handling. Flush fitting, full-width access doors and anchored quick-release catches simplify routine maintenance of the fan and slide-out filter assembly. No tools are needed.

DCE Dalamatic reverse jet fabric filters are designed for continuous operation on applications involving heavy dust burdens and requiring high collection efficiencies. They are available as multi-module cased filters and in a range of 28 single module 'Insertable' filters for silo venting, tipping points, bagging, conveying systems, and integrating with process equipment. The Dalamatic can filter heavy dust burdens at a high filtration velocity and constant level of resistance. Collection efficiency often exceeds 99.99%. The flat envelope configuration of filter elements makes the Dalamatic extremely compact and insures maximum filtration area in a given space. The top inlet of this filter insures a downward flow and a lower pressure loss for a given filtration velocity. Full width access from the clean air side makes inspections and changing of filter bags easier and safer. Filter elements are designed so that one man can change a filter bag without help. Filter elements are cleaned in turn by a brief burst of compressed air in the reverse direction of the main air flow. This is electronically controlled, automatic,

Figure 1.5 Cutting asbestos board into strips. The dust is collected by a DCE Dalamatic filter outside. (Photograph by courtesy of DCE Vokes Group Ltd.)

Figure 1.6 DEC Dalamatic filters collecting dust from asbestos board cutting and trimming. (Photograph by courtesy of DCE Vokes Group Ltd.)

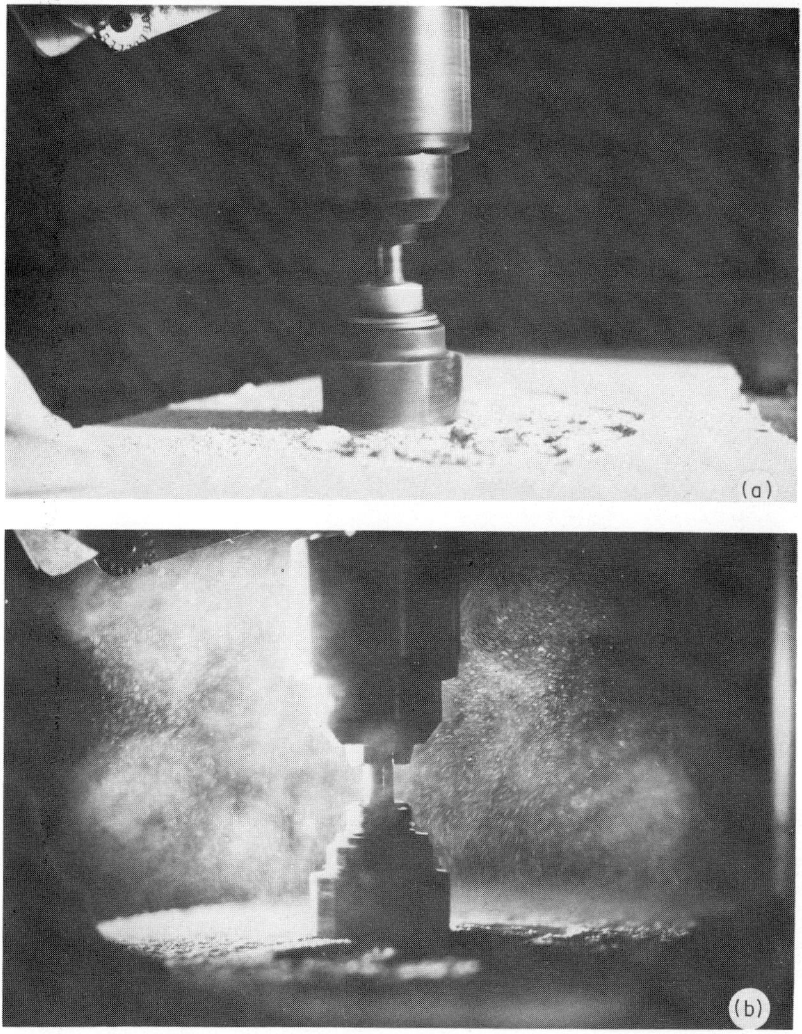

Figure 1.7 Drilling asbestos sheet (a) under normal lighting (no dust visible) and (b) under special lighting showing dust cloud. (Photograph by courtesy of DCE Vokes Group Ltd.)

and continuous. With no moving parts, filter reliability is greater than with mechanical cleaning systems. The method of sealing each filter bag by compressing an integral sealing ring between the insert header and the seal frame insures a tight seal, without screws or toggle bolts. The controller and filter cleaning assembly are located below the clean air chamber for easy access and adjustment.

The use of DCE equipment is illustrated in Figures 1.5–1.10.

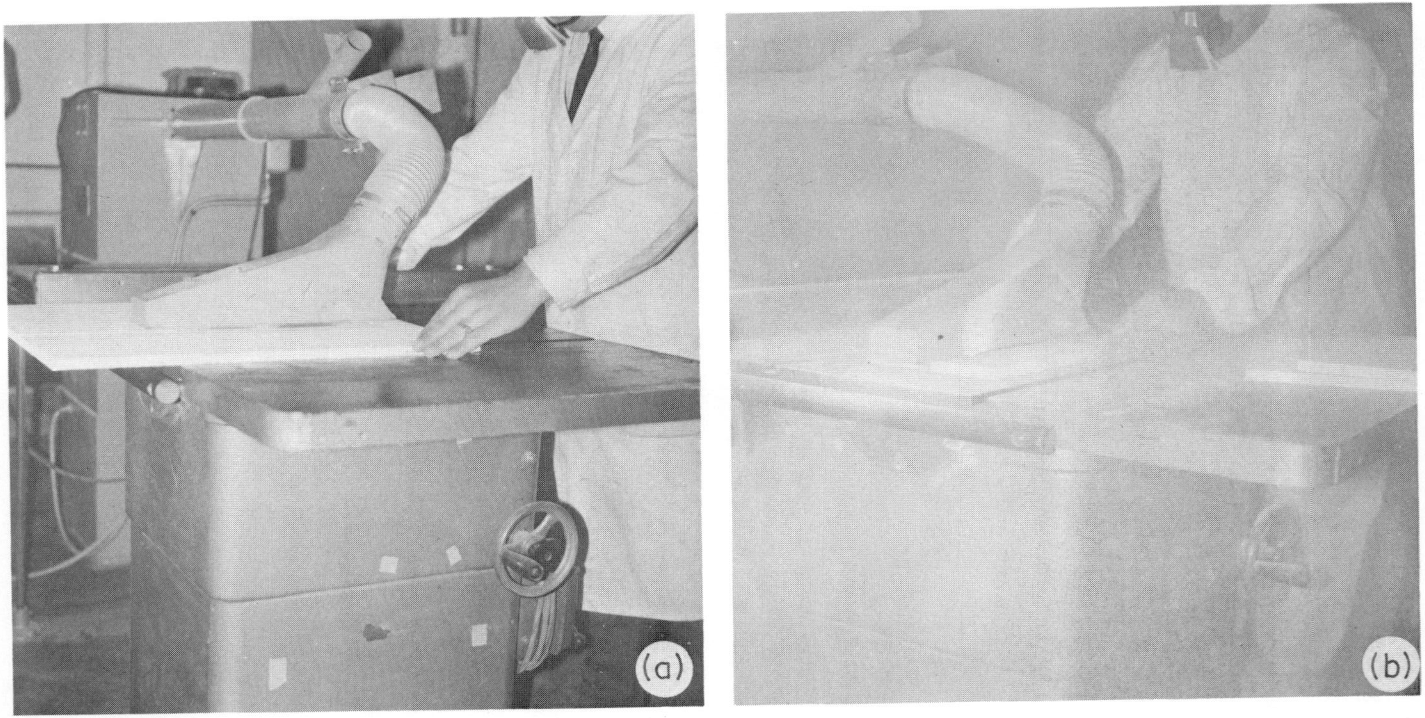

Figure 1.8 Cutting asbestos board (a) with dust control, (b) without dust control. Special lighting shows the absence of dust pollution and demonstrates the effectiveness of the capture hood and associated dust control unit in (a). Note the high level of dust pollution in the working area revealed by the special lighting in (b). (Photograph by courtesy of DCE Vokes Group Ltd.)

Figure 1.9 DCE Unimaster collecting dust from asbestos handling. (Photograph by courtesy of DCE Vokes Group Ltd.)

Figure 1.10 DCE Unimaster collecting dust from brake shoe and clutch lining machining and finishing. (Photograph by courtesy of DCE Vokes Group Ltd.)

Nilfisk Ltd.
Newmarket Road,
Bury St. Edmunds,
Suffolk IP33 3SR

Telephone: 0284-63163

Nilfisk produce a range of industrial and commercial vacuum cleaning equipment with 3- and 4-stage filter systems. The 4-stage system, fitted with a glass-fibre absolute filter, will retain 100% particles in the size range 0.6–0.9 μm (and larger) and particles as small as 0.3 μm are 99.99% retained. Special versions of the Nilfisk (GA72 and GA73) have been approved by the National Occupational Hygiene Service, Manchester, for use with asbestos dust. Both machines include an absolute filter and a shakable main filter, and are provided with a manometer which indicates when the filter needs to be shaken.

Portable Factory Equipment Ltd.
Summit Works,
Smith Street,
Hockley,
Birmingham B19 3EW

Telephone: 021-554 7241/2/3

An extremely powerful suction cleaner for handling the very fine powders found within the asbestos industry is manufactured. The Superfine ULTIMAT incorporates a 6-stage filter system and has been tested by the Chemical Defence Establishment at Porton Down and by the Industrial Hygiene Division of H.M. Factory Inspectorate Chemical Branch, both of whom have confirmed the efficiency of the system.

Sturtevant Engineering Co. Ltd.
Westergate Road,
Moulescoomb Way,
Brighton BN2 4QB

Telephone: 0273-601666 (A. J. Coombe)

This company supplies equipment for the collection by suction of materials such as asbestos, including both normal vacuum cleaning in industrial premises, and also airborne asbestos particle collection from shrouded portable tools, e.g. drills, saws, and grinders, and from asbestos-milling apparatus. The equipment supplied for these applications are of two different types: (1) mobile equipment used within a workplace into which the conveyed air is recirculated, requiring high-efficiency filtration to remove all harmful particles; and (2) static installations, comprising an air mover and filter set located outside the workplace (in the open air or in some conveniently sited plant room). Several of the portable and mobile vacuum cleaners have been specifically approved by bodies such as the Asbestosis Research Council and the Central Electricity Generating Board. Recent trials by the latter organization on two different Sturtevant vacuum cleaners fitted with 3-stage filtration gave outlet concentrations of crocidolite of 4×10^{-3} fibres/cm^3, well below the 2×10^{-1} fibres/cm^3 specified in the 1969 Asbestos Regulations.

Trend Machinery and Cutting Tools Ltd.
Unit 'N,'
Penfold Works,
Imperial Way,
Watford, WD2 4YY

Telephone: 0923-49911

Trend Machinery & Cutting Tools Ltd. provide an extensive range of power tools with dust-extraction facilities fitted to them. The tools have been tested by the Asbestosis Research Council and strobe light tests have been applied by the appropriate department. The equipment is illustrated in Figures 1.11–1.13.

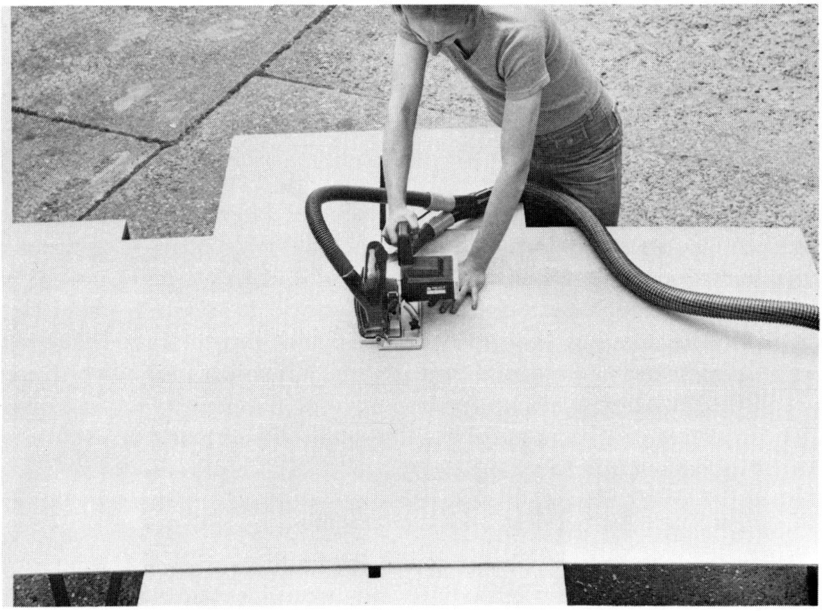

Figure 1.11 Cutting asbestos sheet with the Elu MH65D/E. (Photograph by courtesy of Trend Machinery and Cutting Tools Ltd.)

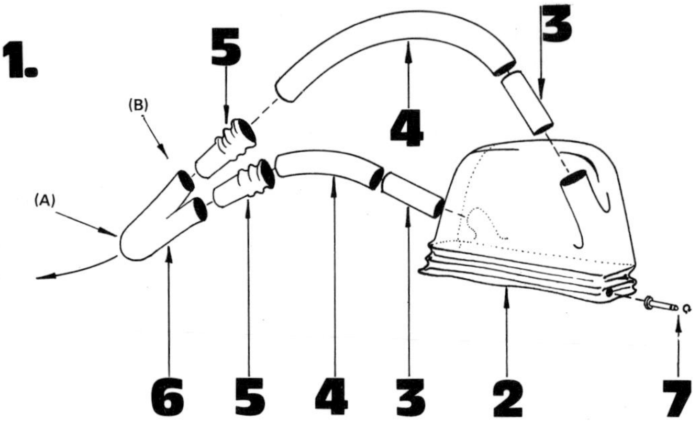

Figure 1.12 Cowel set for Trend Machinery and Cutting Tools Ltd. rip saw MH65D/E. 1, Cowl set for MH65D/E rip saw (up to and including 'Y' junction piece). Part Ref.: X/24/001. 2, Plastic cowl. Part Ref.: X/24/003. 3, Two aluminum connecting tubes ($1\frac{1}{4}$ in. i.d., 32 mm. Part Ref.: X/27 008. 4, Two lengths of hose, 19 and 7 in. Part Ref.: X/24/006. 5, Two PVC end pieces ($1\frac{1}{2}$ in. i.d.). Part Ref.: X/27/006A. 6, 'Y' junction piece. Part Ref.: X/24/004. A = 56 mm i.d. main, 61 mm o.d. outlet. B = 32 mm i.d. twin, 37 mm o.d. small outlets. 7, 6 mm diameter securing pin, with circlip to suit. Part Ref.: X/24/008.

Universal Industrial Appliances
Wellingborough Road,
Rushden,
Northamptonshire NN10 9BE

Telephone: 0933-588116

Universal Industrial Appliances, a Division of Norris Industries Rushden Ltd., produce vacuum cleaning equipment for the recovery of asbestos and other hazardous dust. Models 201A and 745A (Mark 2 and Mark 3) have been tested by the Asbestosis Research Council, meet Government Health and Safety Regulations and satisfy H.M. Factory Inspectorate. Independent tests have also been carried out by the Chemical Research Defence Establishment (copy available for inspection if required).

The cleaners can be used for routine industrial cleaning or as local dust extractors for circular saws, drills, etc.

4(iii) Miscellaneous Services and Equipment

Other services and equipment dealt with in this section are laundering of contaminated clothing, asbestos fibre collection, identification and counting equipment and services, protective clothing, insulation sealants and removal services, and pulmonary function instrumentation.

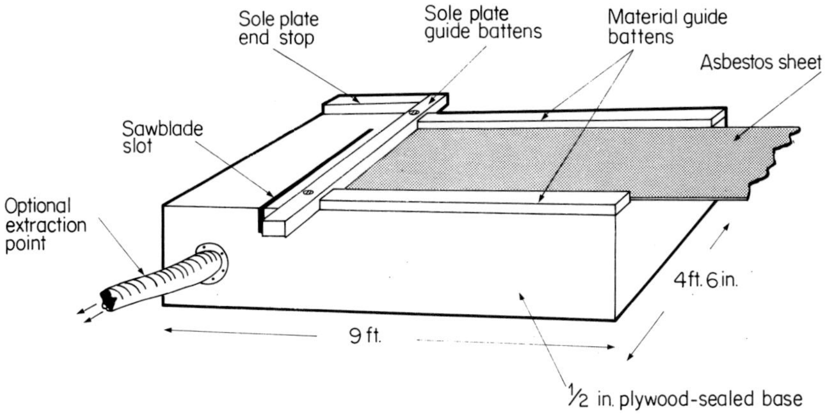

Figure 1.13 Suggested extraction box/table for cutting asbestos. (By courtesy of Trend Machinery and Cutting Tools Ltd.)

Association of British Launderers and Cleaners Ltd.
Lancaster Gate House,
319 Pinner Road,
Harrow,
Middlesex HA1 4HX

Telephone 01-863 7755

There is a list of companies operating in various parts of the country who provide a service for the cleaning/laundering of asbestos-contaminated clothing, and any person/company requiring such a service should write to the ABLC who will put them in touch with a local laundry/cleaner (N.B. Members of the ABLC represent about 95% of the laundries and about 75% of the dry cleaners in the U.K.).

The contaminated clothing must be sent to the laundry/cleaner concerned in a sealed polythene bag, clearly marked with the words 'Asbestos Contaminated Clothing', as specified in Section 3.3 of the Control and Safety Guide (No. 1) issued by the Asbestosis Research Council. Special notification must be made where the work is contaminated with blue asbestos. Customers should use polythene bags which are either soluble in hot water or stitched with alginate thread. The former can be obtained from Enak Ltd., Foundry Close, Horsham, Sussex, and the alginite-stitched bags from Simpla Plastics Ltd., Phoenix Estate, Caerphilly Road, Cardiff, S. Wales.

Blue Dragon (Hillingdon) Ltd.
Uxbridge,
Middlesex UB10 0N2

Telephone: 89-36571/2 (Mr. D. Brown)

Blue Dragon (Hillingdon) Ltd. specialise in the laundering of asbestos garments for a number of industrial organizations and local government departments over a wide area. Garments must be sent in sealed and marked polythene bags and they must be wet down before the bag is sealed. Each order is washed independently and is supervised by a member of the supervisory staff.

Cambridge Instruments Ltd.
Melbourn,
Royston,
Herts. SG8 6EJ

Telephone: 0763-60611

The Quantimet 720 Image Analyser (Figure 1.14) provides geometric and densitometric information from virtually any optical or electro-optical input. The image of the specimen is focused on to the face of a TV camera which scans the image and transforms it into an electronic signal which is then analysed. The apparatus overcomes the drawbacks—slow, tedious, prone to error (due to subjectivity and operator fatigue)—of manual counting.

C. F. Casella & Co. Ltd.
Regent House,
Britannia Walk,
London N1 7ND

Telephone: 01-253 8581

The Casella T13051/2 pump is a robust, light, and reliable pump which has

Figure 1.14 Quantimet 720 Image Analyser. (Photograph by courtesy of Cambridge Scientific Instruments Ltd.)

an adjustable flow-rate over the range 0.05–2.5 l/min and can be used with a variety of sampling heads. The pump, which weighs 580 g, measures 12.4×5 cm and can operate for up to 10 h (less at high flow-rates). Recharging of the internal nickel–cadmium battery can be done overnight or over the weekend. The pump can be obtained as an individual item, as can any of the accessories, and is incorporated in the Casella Asbestos Sampling Kit, reference number T13023. The kit comprises the T13051/2 pump, a charger for the internal battery of the pump, seven open-fronted filter holders with caps, lapel clip, box of 5.0-μm membrane filters, box of 0.8-μm membrane filters, harness, circular damper, connecting tube, static sampler stand, flow meter (5.0–100 cm^3/min), and spare battery, all fitted in a compact carrying case.

Cooperative Laundries Society Ltd.
P.O. Box 9,
Equitable House,
Wallsend,
Tyne and Wear NE28 8LR

Telephone: 0632-625281/5

A water-soluble poly(vinyl alcohol) bag is provided which eliminates any risk of the operator being contaminated by the garments prior to washing. The asbestos-contaminated overalls, etc., in the bag are placed in a washing machine in which the bag dissolves in the warm water. An extended cleaning process is used.

A. Gallenkamp & Co. Ltd.
P.O. Box 290,
Technico House,
Christopher Street,
London EC2P 2ER

Telephone: 01-247 3211

Gallenkamp offer a microscope outfit suitable for both the identification of asbestos fibres and for the counting of fibres on membrane filters. The outfit is based on the Olympus Model BH microscope, which is fitted with phase-contrast accessories for counting asbestos fibres and dispersion staining and polarization are used for identifying asbestos fibres. A closed-circuit TV system is also available.

Gelman Hawksley Ltd.
10 Harrowden Road,
Brackmills,
Northampton NN4 OEB

Telephone: 0604-65141

Gelman Hawksley are suppliers of Gelman membrane filters, Andersen aerodynamic samplers, Royco particle counters and the Hawksley air sampler.

The Gelman GN-4 0.8-μm mixed cellulose ester membranes in pure white have been developed primarily for asbestos sampling and conform to the ACGIH specifications. The filters are available in packs of 100 in three diameters: 25, 37, and 47 mm.

The Andersen Minisampler collects and sizes airborne particles into ranges ≥ 4.7, 4.7–3.3, 3.3–2.1, 2.1–0.65, and <0.65 μm.

The Royco Type 225 counter is suitable for the accurate, automatic sizing and counting of asbestos particles.

The Hawksley Air Sampler is light and compact. It is fitted with a carbon vane pump with integral exhaust filter, producing a constant pulse-free vacuum. There is a flow meter and an automatic timer fitted.

Histon Overalls Ltd.
Abbey Mill,
Neath Abbey,
Glamorgan

Telephone: 0639-4711

Histon Overalls Ltd. are manufacturers of overall clothing and protective clothing and equipment. The garment currently manufactured for use by asbestos workers is shown in Figure 1.15, and can be obtained in polyester–cotton (67.33), nylon or terylene.

Institute of Occupational Medicine,
Roxburgh Place,
Edinburgh EH8 9SU

Telephone: 031-667 5131

The Institute of Occupational Medicine is primarily a research centre, but also undertakes occupational hygiene service work. This includes monitoring, on either a short- or a long-term basis, of airborne asbestos concentrations, evaluation of membrane filter samples, and identification of asbestos in insulation materials. The service is available in most parts of the U.K. directly from the Institute in Edinburgh or through field stations at Whitburn (Tyne and Wear), Mansfield Woodhouse (Midlands) and Pontypridd (South Wales). Asbestos fibre monitoring and counting are carried out by all of the centres and a proportion of the samples are check-counted by the Institute's asbestosis research staff, who are in direct contact with governmental and industrial organizations. Insulation samples are normally supplied by the client. Identification for the asbestos type is carried out in

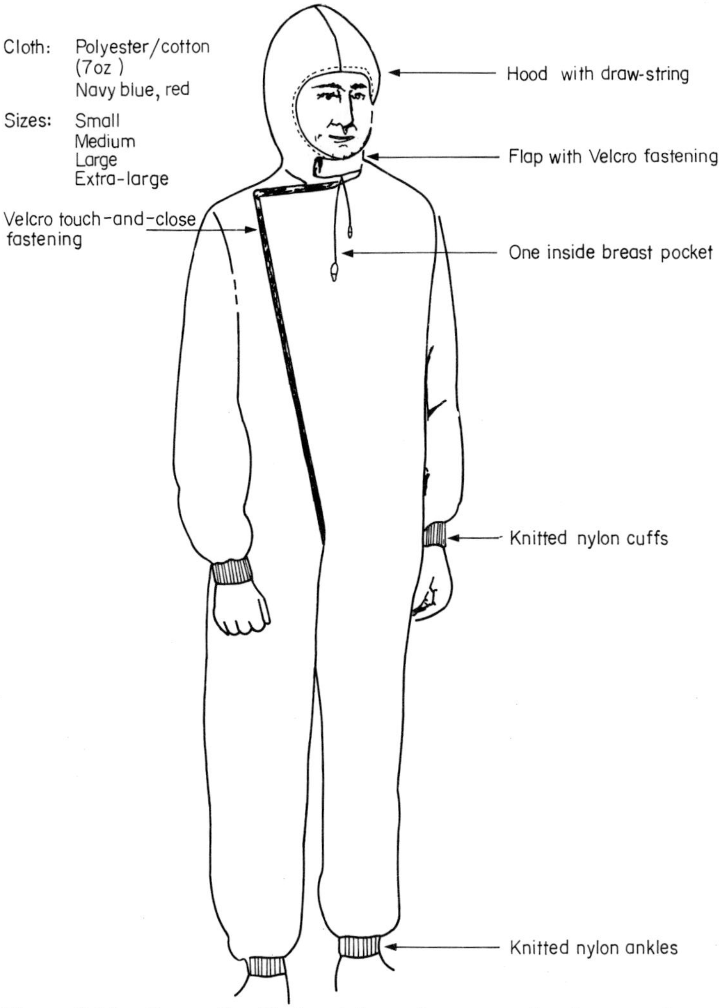

Figure 1.15 Coverall with hood for asbestos workers' protection as manufactured by Histon Overalls Ltd

Edinburgh, where advanced analytical techniques are available (electron microscopy, X-ray diffraction, infrared spectroscopy, etc.) and research specialists can be consulted.

Idenden Adhesives Ltd,
Blackwater Way Industrial Estate,
Ash Road,
Aldershot,
Hants. GU12 4DN
Telephone: 0252-311608

Idenden ET-150 is a tough, durable, fire-retardant mastic coating suitable for the repair, renovating, and protection of old asbestos insulation so as to bring it up to acceptable safety levels. The material is supplied as a thick, soft paste and it can be applied by float, brush, paint roller, or spray gun. It is water-based, non-flammable and non-toxic and is safe and simple to use. Two coats should be applied, giving a dry film thickness of 1 mm for outdoor lagging and 0.5 mm for indoor lagging. Brushes and tools can be washed with cold water instantly after use.

International Research & Development Co. Ltd.
Fossway,
Newcastle upon Tyne NE6 2YD

Telephone: 0632-650451 (Dr. R. W. Gale or Mr. D. F. Gibbs, Department of Biotechnology)

IRD has considerable experience and expertise in the monitoring and quantification of asbestos, both in bulk form and as airborne dust.

The approved techniques used by IRD for sampling and quantification of airborne asbestos are laid down in the Asbestosis Research Council's Technical Notes No. 1, *The Measurement of Airborne Asbestos Dust by the Membrane Filter Method*, and No. 2, *Dust Sampling Procedures for use with the Asbestos Regulations.*

Approved X-ray diffraction and advanced microscopy techniques are used in the identification of asbestos types present in samples. Such techniques overcome, amongst other things, the otherwise difficult task of differentiating between amosite and crocidolite.

A full on-site asbestos survey is offered whereby IRD staff thoroughly investigate potential asbestos hazards.

Clients are advised of any potential hazard indicated from the test results and further advice is available regarding ways to overcome safety problems. Findings are presented in report form following surveys. Reports can often be submitted within 48 h of a client's request.

Liquid Plastics Ltd.
P.O. Box 7,
London Road,
Preston PR1 4AJ

Telephone: 0772-59781

Liquid Plastics Ltd. produce Decadex Firecheck for the encapsulation of exposed asbestos. It is currently in use in 100 countries worldwide and is a highly efficient sealer for all types of exposed asbestos. Easily applied by brush or spray, the product totally encapsulates the asbestos with a non-deteriorating dust seal which is proof against a wide range of chemical and temperature conditions.

Of particular importance is Firecheck's ability to resist impact fracture and maintain or even enhance the original fire classification of the substrate. The thick, tightly adherent skin formed on curing is completely watertight and non-toxic with an elongation factor in excess of $\times 8$.

Decadex Firecheck is recognized and recommended by the Department of the Environment, the Asbestos Information Committee and the Factory Inspectorate. It is supplied in a wide range of attractive colours and provides a highly decorative finish.

Before encapsulating the exposed asbestos, it is necessary only to remove all loose fibres and repair and rebuild any damaged areas by use of a proprietary filler. On pipework and ducting, LPL's Flexitape is used to enbalm any areas where the insulation, while intact, is not adhering to the substrate.

The surface to be treated is first primed with a brush or spray application of LPL Sealer, Decadex Firecheck is then applied at not more than 2.5 m^2/l and allowed to cure. A second coat of similar thickness is then applied.

McCrone Research Associates Ltd.
2 McCrone Mews,
Belsize Lane,
London NW3 5BG

Telephone: 01-435 2282

McCrone Research Associates Ltd. supply a range of items for microscopial identification of asbestos, including a dispersion staining objective and an asbestos reference set. Courses are held in the U.K. on identification and atmospheric monitoring (details from the Registrar, McCrone Research Institute). Comprehensive and confidential scientific services for identification and atmospheric monitoring of asbestos are also offered—contact Jean Prentice. McCrone Research Associates are listed as analysts by the Health and Safety Executive.

MDA Scientific, Inc.
808 Busse Highway,
Park Ridge,
Ill. 60068, U.S.A.

and

MDA Scientific (UK) Ltd.
Ferndown Industrial Estate,
86 Cobham Road,
Wimborne,
Dorset BH21 7PQ

Telephone: 0202-872106

The BDX Super Sampler is a versatile pump that can be used for a variety of sampling applications, having flow-rate ranges of 25–200 cm^3/min and 1–3 l/min. For asbestos determination the BDX 44 configuration is used. The flow meter variation with a 0.8 μm filter is in the range ±5% and when sampling at 2 l/min the BDX 44 has sufficient power to pump for a full 8-h shift.

Nicholson's (Overalls) Ltd.
Georges Road,
Stockport,
Cheshire SK4 1DP

Telephone: 061-480 7318 (R. A. Firmin)

Coverall, Style 92, meets the required specification of the Asbestosis Research Council. It is a sealed suit without access vents to inclothing, designed for complete protection of wearer and product. The cuffs and anklets are made of stretch nylon; there is a separate hood and a zip front; a breast pocket and mandarin collar are optional. Materials available are: polyester-cotton (67:33) (meeting the requirements of the Asbestosis Research Council), 100% Terylene, and 100% cotton. There are several colours and sizes range from 34 to 50 in. chest in regular and tall fittings.

P. K. Morgan Ltd.
10 Manor Road,
Chatham,
Kent ME4 6AL

Telephone: 0634-44384 and 47949

P. K. Morgan manufacture pulmonary function instrumentation. Their Transfer Test apparatus for the measurement of single breath diffusions and total lung capacities have been used by hospitals and commercial companies all over the world.

The Model B Transfer Test system, which provides single-breath only, is designed for use where other lung function tests are adequately served by existing equipment and only the addition of a single-breath diffusion capacity is required. The Model C covers all normal lung function tests (single-breath: vital capacity, fast expired volume, single-breath diffusion capacity, membrane diffusion capacity, pulmonary capillary blood volume, etc.; multi-breath closed-circuit: resting ventilation, oxygen uptake, vital capacity and subdivisions, functional residual capacity, residual volume, total lung capacity, maximum breathing capacity; open-circuit multi-breath: steady-state diffusing capacity at rest and upon exercise).

An automated lung volume spirometer for single-breath spirometry, resting ventilation, oxygen uptake, vital capacity, fast expired volume, maximal voluntary ventilation, and flow volume loops is also available.

Perkin-Elmer Ltd.
Maxwell Road,
Beaconsfield,
Bucks. HP9 1QA

Telephone: 04946-6161

The Perkin-Elmer Model 580 infrared spectrophotometer coupled with an Interdata 6-16 mini computer can be used for the identification of asbestos. The dust is deposited on a filter material with sufficient transparency in the region required for analysis or a caesium iodide disc containing the bulk material is prepared. The spectra are referred to the diagram shown below for identification of the asbestos type:

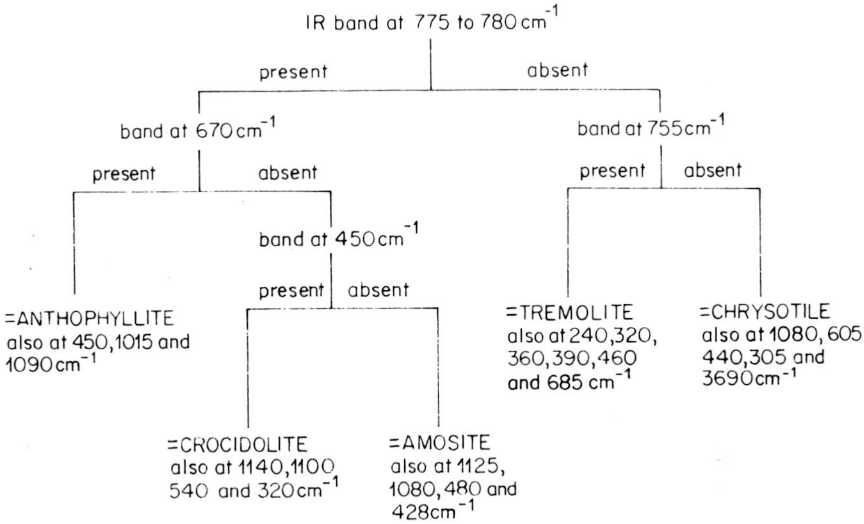

Practical Uniform Co. Ltd.
Northern Sales Office,
Lorenzo Drive,
Liverpool L11 1BH

Telephone: 051-226 1395

The Company supplies a boiler suit that is used on a wide scale by workers engaged with asbestos. The boiler suit is zip-fronted with an outside breast pocket with zip closure: there are no other pockets or side vents. There are elasticated wrists and ankles and an attached hood. The fabric normally favoured is a nominal 4-oz nylon (available in blue or green and small, medium, and large sizes). The garment can also be produced in $7\frac{1}{4}$-oz polyester–cotton (67 : 33) cloth or a ceramic-finish 100% Terylene.

Rotheroe & Mitchell Ltd.
Victoria Road,
South Ruislip,
Middlesex HA4 0LG

Telephone: 01-422 9711

The L2C pump is a pocket-sized sampler designed to sample air close to an employee's face. Model L2SF is available with stabilized flow control. The filter unit has a clip for attachment to jacket lapel, safety helmet, etc., and is attached to the pocket unit (which can be supported with the belt supplied) by a thin PVC tube. The flow-rate can be adjusted from 0.5 to 3.0 l/min. The pump, which weighs 1.14 kg, measures 11.9×9.4×5.7 cm and can be operated for up to 10 h depending on the filter used. A size-selective cyclone unit is available: respirable dust is collected on the filter and non-respirable dust is collected in a removable chamber. The pump can be obtained as an individual item, as can any of the accessories, and is incorporated in the Rotheroe & Mitchell Ltd. Air Sampling Kit, which is suitable for asbestos fibre determination. The kit comprises the L2C pump complete with extension cable and filter holder, a charger for the air sampler, a box of Whatman GF1A filter-papers, two packs of Millipore 0.8-μm filters, a can of Aerosol fixative spray, one pair of tweezers, one screwdriver, and four sample cans, all contained in an 11.9×5.7×9.4cm case.

Shirley Institute
Didsbury,
Manchester M20 8RX

Telephone: 061-445 8141 (W. T. Cowhig)

The Shirley Institute provides a service for the identification of asbestos lagging and other materials, and in particular for showing whether crocidolite (blue) asbestos is present. A bulk sampling method has been developed which avoids disturbing the lagging in order to obtain a representative sample of all the possible layers of material which may be present. A tool rather like a cork-borer is used to penetrate the lagging cleanly to the surface of the pipe or boiler, etc. (Figure 1.16); when the tool is withdrawn, it contains a sample of all layers of the lagging material. The sampling tool containing the asbestos sample is placed in a plastic bag which is returned to

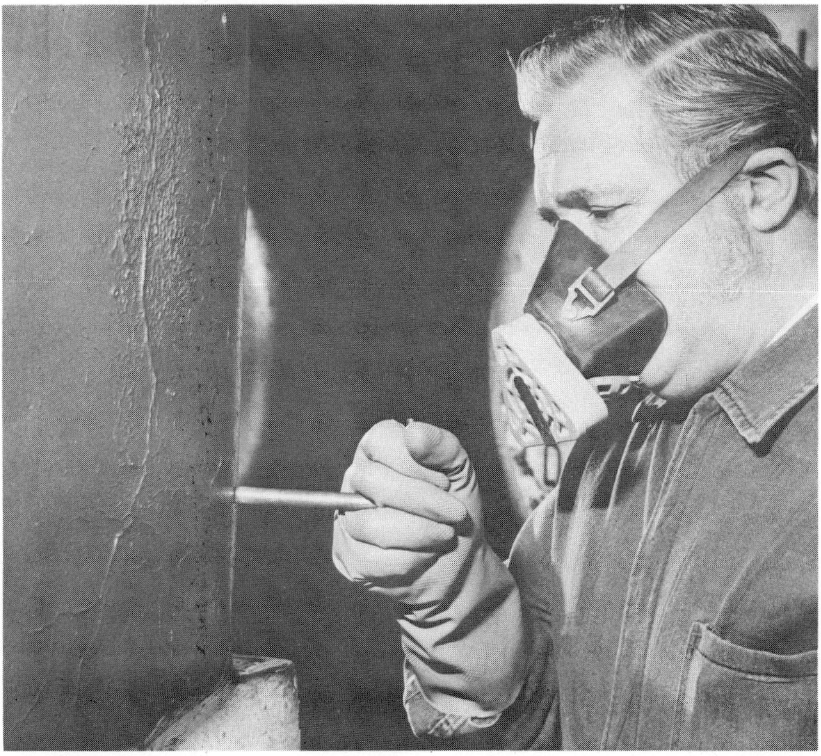

Figure 1.16 Taking a bulk sample of asbestos lagging. (Photograph by courtesy of the Shirley Institute)

the Shirley Institute for analysis (Figure 1.17). The tiny hole in the lagging can be filled with a suitable material and covered over with plaster. An adequate number of tools and plastic bags are supplied on request. Alternatively, a member of the Institute's staff can be sent to collect suitable samples.

When posting asbestos samples, they must be carefully packed in polythene bags and marked 'ASBESTOS' on the outside.

Staysafe and Co. Ltd.,
Yorkshire House,
909 Wolverhampton Road,
Warley,
West Midlands B69 4RR

Telephone: 021-552 5403

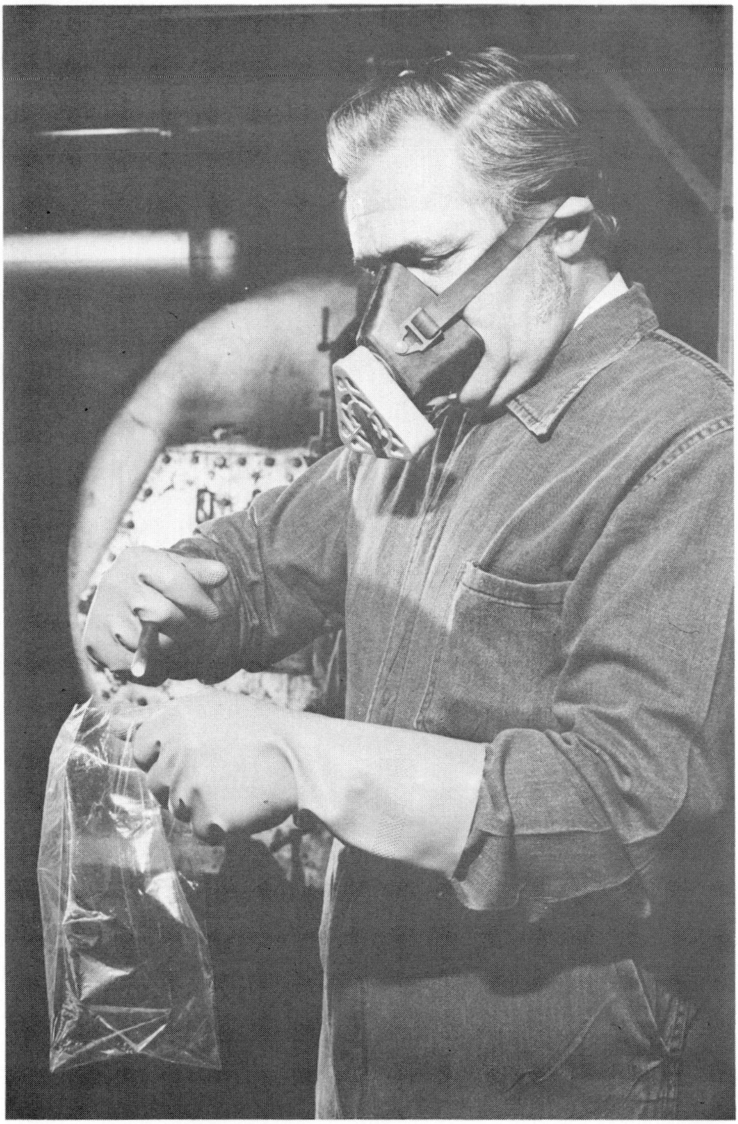

Figure 1.17 Lagging sample being placed in plastic bag prior to being sent away for analysis. (Photograph by courtesy of the Shirley Institute)

Staysafe are manufacturers and suppliers of fire and personal safety equipment and are involved in the manufacture and supply of clothing for use when working with asbestos. The general design of garments available is as follows: all-enveloping boiler suit with permanently attached hood, adjustable by means of a draw-string to enable a close fit to be made round the face, no pockets or access slits, and elasticated at wrist and trouser cuffs to keep the garment in place inside gloves and boots. The hood design allows sufficient room for the wearing of respirators or other breathing equipment, and goggles. The front opening has a fine-tooth zip or flap-over front with Velcro fixing.

The material of choice for the suit is close-weave 100% nylon or Terlyene, although it can be supplied in PolyCotton if conditions necessitate a garment which is warmer to wear.

Thos. W. Ward Ltd.
Albion Works,
Sheffield S4 7UL

Telephone: 0742-26311

The Dismantling Division of Thos. W. Ward has been involved in the removal of asbestos lagging and insultation from power stations, gas works, chemical plants, and other industries for many years. A Special Section, dealing with asbestos stripping under controlled conditions was formed in 1975.

TUC Centenary Institute of Occupational Health,
London School of Hygiene & Tropical Medicine,
Keppel Street (Gower Street),
London WC1E 7HT

Telephone: 01-636 8636 (Professor J. C. McDonald, Director)

Through its Information and Advisory Service, the Institute provides a comprehensive consulting facility for the recognition, evaluation, and control of asbestos hazards at the workplace. The services available include identification of asbestos type by either X-ray diffraction or optical microscopy (dispersion staining method), measurement of airborne dust concentration using the standard techniques, and advice on the control of asbestos dust. The Institute can also provide individual clinical examination of workers suspected of exposure to asbestos dust or carry out epedemiological investigations of work groups.

Welsh Occupational Health Laboratory
Department of Community Medicine,
Welsh National School of Medicine,
Heath Park,
Cardiff CF4 4XN

Telephone: 0222-755944 (Dr. R. L. Kell)

The Laboratory offers an occupational hygiene service which includes the measurement of asbestos fibres in air.

Wolfson Istitute of Occupational Health
Environmental Health Service,
University of Dundee Medical School,
Ninewells,
Dundee

Telephone: 0382-644625

A service in asbestos analysis is offered. Bulk samples are analysed by X-ray diffraction and airborne fibre samples are counted by phase contract microscopy.

(5) REFERENCES

American Water Works Association (1974) A study of the problems of asbestos in water, *Am. Wat. Wks. Ass. J.*, September, 1–22.

Bellhouse, G. (1931) *Report on Conferences between employers and inspectors concerning methods for suppressing dust in asbestos textile factories*, HMSO, London, 35–214.

Buchanan, W. D. (1965) Asbestosis and primary intrathoracic neoplasms, *Biological Effects of Asbestos*, New York Academy of Sciences, New York 507–517.

Cook, P. M., Glass, G. E., and Tucker, J. H. (1974) Asbestiform amphibole minerals: detection and measurement of high concentrations in municipal water supplies, *Science, N.Y.* **185,** 853–855.

Cooke, W. E. (1927) Pulmonary asbestosis, *Brit. Med. J.*, **4,** 1024–1025.

Cunningham, H. M., and Pontefract, R. D. (1971) Asbestos fibres in beverages and drinking water, *Nature, Lond.* **232,** 332–333.

Gilson, J. C. (1971) *Asbestos health hazards.* Some implications of new knowledge to prevention, in *Proceedings of the IVth International Pneumoconiosis Conference, Bucharest, 1971*, Apimondia, Bucharest.

Harries, P. G. (1976) Experience with asbestos disease and its control in Great Britain's naval dockyards, *Envir. Res.*, **11,** 261–267.

HMSO (1970) *Report of the Departmental Committee on Compensation for Industrial Diseases*, Cmd. 3495, 3496, HMSO, London.

Holt, P. F., and Young, D. K. (1973) Asbestos fibres in the air of towns, *Atmos. Envir.*, **7,** 481–483.

Infante, P. F., and Lemen, R. A. (1976) Asbestos in dentistry, *J. Am. Dent. Ass.*, **93,** 221–222.

Merliss, R. R. (1971) Talc and asbestos contaminant of rice, *J. Am. Med. Ass.*, **216,** 2144.

McDonald, S. (1927) Histology of pulmonary asbestosis, *Brit. Med. J.*, **2,** 1025–1026.

Nicholson, W. J., Maggiore, C. J., and Selikoff, I. J. (1972) Asbestos contamination of parenteral drugs, *Science N.Y.*, **177,** 171–173.

Nicholson, W. J., and Pundsack, F. L. (1973) Asbestos in the environment, in *Biological Effects of Asbestos* (Eds. P. Bogovski, J. C. Gilson, V. Timbrell, and J. C. Wagner), IARC Scientific Publications No. 8, International Agency for Research on Cancer, Lyon, 126–130.

Nicholson, P. D., Rohl, A. N., and Weisman, I. (1975) *Asbestos Contamination of the Air in Public Buildings*, EPA Contract number 68-02-1346, U.S. Government Printing Office, Washington, D.C.

Rickards, A. L. (1973) Estimation of submicrogram quantities of chrysotile asbestos by electron microscopy, *Analyt. Chem.*, **45,** 809–811.

Rosato, D. V. (1959) *Asbestos—Its Industrial Applications*, Reinhold, New York.

Sebastien, P., and Bignon, J. (1974) *Contribution à l'Etude de la Pollution Particulaire de l'Air Ambiant de la Ville de Paris par les Microfibrilles d'Amiante*, Rapport Final, Contract No. 206, Ministére de la Qualité de la Vie, Paris.

Sebastien, P., Bignon, J., Gaudichet, A., Dufour, G., and Bonnaud, G., (1976) Les pollutions atmosphériques urbaines par l'asbeste, *Rev. Fr. Mal. Resp.*, **4,** Suppl. 2, 51–62.

Selikoff, I. J., Nicholson, W. J., and Langer, A. M. (1972) Asbestos air pollution, *Arch. Envir. Hlth.*, **25,** 1–13.

Skidmore, J. W., and Jones, J. S. P. (1975) Monitoring an asbestos spray process, *Ann. Occup. Hyg.*, **18,** 151–156.

Wagner, J. C. (1965) The sequelae of exposure to asbestos dust, *Ann. N.Y. Acad. Sci.*, **132,** 691–695.

Wehman, H. J., and Plantholt, B. A. (1974) Asbestos fibrils in beverages. I. Gin. *Bull. Envir. Contam. Toxicol.*, **11,** 267–272.

Willis, R. A. (1960) *Pathology of Tumors*, 3rd Ed., Butterworths, London.

Wolff, A. H., and Oehme, F. W. (1974) Carcinogenic chemicals in food as an environmental health issue, *J. Am. Vet. Med. Ass.*, **164,** 623–629.

Wright, G. D. (1969) Asbestos and health 1969, *Am. Rev. Resp. Dis.*, **2,** 467–479.

Zielhuis, R. L. (Ed.) (1977) *Public Health Risks of Exposure to Asbestos*, European Economic Community-Directorate of Social Affairs, Contract No. 18-6-1975.

CHAPTER 2

The mineralogy of asbestos

Jack Zussman

University of Manchester

(1) INTRODUCTION

The asbestos minerals have an almost unique combination of physical and chemical properties. They take the form of extremely thin and flexible fibres, which at the same time have great thermal stability and tensile strength. In appearance the fibres are often silky, and their flexibility is such as to allow them to be spun into yarn and subsequently made into woven fabric. The most widespread modern uses of asbestos are in 'fireproof' textiles, papers and boards and in brake and clutch linings for many kinds of vehicle and machinery. The low thermal conductivity of asbestos products can be as important as their incombustibility for fire prevention. Composites of asbestos with cement, resins, and plastics are commonly manufactured.

The term asbestos derives from the Greek meaning 'inextinguishable' and was perhaps used by the ancient Greek writers in the sense of 'indestructible' to describe material which was unscathed by flames. The first use of the term in the context of mineralogy came in the middle of the 19th Century as applied to a fibrous amphibole mineral discovered in the Italian Alps.

The three main kinds of asbestos which have had wide commercial exploitation are chrysotile, amosite, and crocidolite. Of these, chrysotile is by far the most abundant and most used. Three others, in decreasing order of importance, are anthophyllite, tremolite and actinolite. The last material, although not uncommon as a mineral, is rare in asbestiform habit.

Of the above six varieties of asbestos, chrysotile alone is a member of the serpentine group of minerals, all of the others belonging to the amphibole group. Chrysotile has yellowish or greenish white fibres which are usually silky in nature, crocidolite is blue and less silky, and amosite has white, grey, pale yellow, or pale brown fibres which are more brittle than those of the former two varieties. Typical specimens are shown in the frontispiece.

Amosite is not strictly a mineral name but is a commercial name derived from the initial letters of Asbestos Mines of South Africa. Amosite usually consists mainly of the amphibole mineral grunerite, but appreciable amounts of other amphibole fibres are present in some specimens.

Silicates other than serpentines and amphiboles, and also some non-silicates, can occur in asbestiform habit, but they are not exploited because of their limited occurrence or inappropriate physicochemical properties.

(2) DEFINITIONS

Some would confine the use of the term 'asbestos' to material which, if present in sufficient amount, would be commercially exploitable because of the special properties mentioned above, but it is probably difficult to exclude specimens which fall slightly short of this because of greater coarseness, brittleness, or weakness of fibre.

The most abundant minerals which could be termed 'asbestos' are chemically hydrous silicates (see below). Those hydrous silicates that are 'asbestos' are composed of particles which are extremely thin and so generally have a very high length to breadth ratio. We are concerned, therefore, with particle morphology and a property known in mineralogy as 'crystal habit'. Figure 2.1 shows a range of crystal habits and their descriptive terms, namely tabular, equant, prismatic, acicular, and asbestiform*.

The limiting length to breadth ratios that differentiate between these terms have not been precisely defined, and it has not in general been thought necessary to do so. Thus, it is a matter of subjective judgment as to whether a crystal is described as acicular (meaning needle-like) or prismatic, but it is doubtful if anyone would use the term acicular for material with a length to breadth ratio much less than 10 : 1.

Since the fundamental fibrils of asbestos have typical widths in the range 20–200 nm, a fibril of any appreciable length will have a very high length to

* There has been a tendency for the term 'asbestiform' to be misused recently as being equivalent to 'asbestos-forming'. Even the later term can be misleading. The mineral tremolite, for example, can be said to be asbestos-forming in the sense that it sometimes occurs as an asbestos variety. One variety may be like asbestos (an aggregate of hair-like fibres) and asbestiform; another may be a single relatively thick crystal and clearly not asbestiform. It does not mean that tremolite *always* forms as asbestos.

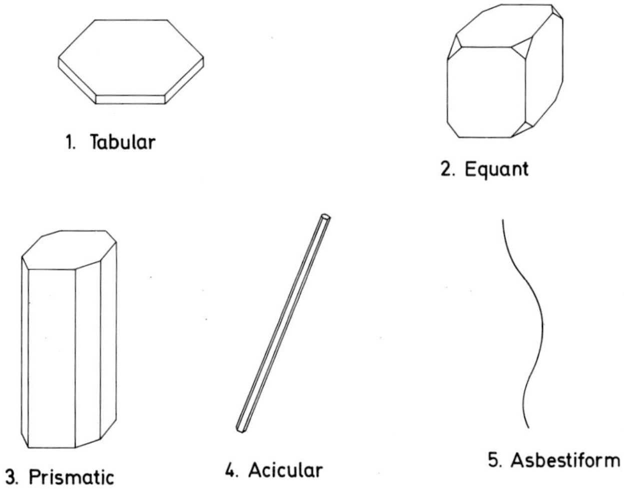

Figure 2.1 Illustration of a range of crystal 'habits' from tabular to asbestiform

breadth ratio. Thus, for example, assuming an average width of 100 nm, a fibril of length 1 mm will have a length to breadth ratio of 10 000 : 1. Even a fibril as short as 10 μm has a length to breadth ratio of 100 : 1, and only a 1-μm length will give a ratio as low as 10 : 1. The characteristic feature of high length to breadth ratio is not, however, applicable on a single fibre basis, but it is more useful if applied statistically. In natural untreated asbestos, although some very short fibres will be present, a high proportion of fibrils with a length to breadth ratio of the order of 100 : 1 or greater can be expected. In samples of milled asbestos (Campbell *et al.*, 1977), most particles were found to have a length to breadth ratio between 5 : 1 and 20 : 1, whereas for a non-asbestos amphibole the majority had a ratio of less than 3 : 1. For milled chrysotile asbestos most particles had a ratio of greater than 50 : 1.

Related to, but not entirely dependent upon, the length to breadth ratio is the property of flexibility. Needle-shaped crystals of some substances will be very brittle, others may flex slightly without breaking and spring back elastically, while others can deform inelastically. The term acicular, meaning needle-shaped, would not normally be applied to crystals that are curved.

A bundle of asbestos fibrils will tend to bend and recover elastically, but when the diameter of the fibre bundle is small so that only few fibrils are involved, then inelastic deformation occurs, giving the curved fibrils common for natural asbestos (Figure 2.10a).

Variations in minerals can be produced not only by the different habits of their component crystals, as described above, but also by the way in which these crystals are aggregated. An important feature of asbestos is that it is an aggregate of hair-like fibrils. These have their lengths approximately parallel to one another but lie in more then one azimuthal orientation. More extensive discussion of definitions pertaining to asbestos is given by Campbell *et al.* (1977).

(3) CHEMICAL COMPOSITION

Chemically, the commercial asbestos minerals are all hydrous silicates. The serpentine minerals, including chrysotile asbestos, have a comparatively simple chemical formula approximating closely to $Mg_3Si_2O_5(OH)_4$.

Because of the widely variable chemical nature of amphiboles, the group as a whole is best considered by means of the general formula $A_{0-1}X_2Y_5Z_8O_{22}(OH, F)_2$, the principal substituents being A = Na, K; X = Na, Ca; Y = Mg, Fe, Al; and Z = Si, Al. Approximate formulae for the amphibole minerals which can occur as asbestos are given in Table 2.1.

The formulae in Table 2.1 are simplified versions showing only the major constituents. It is not uncommon for minor amounts of elements such as Al, Mn, Cr, Ni, and Ti to occur, and it would be surprizing if many other elements did not occur in trace amounts. Examples of actual chemical formulae (Deer *et al.*, 1963) of some asbestos specimens are as follows:

Chrysotile: $Mg_{2.88}Mn_{0.001}Fe^{2+}_{0.003}Fe^{3+}_{0.045}Al_{0.016}Si_{1.95}O_{4.75}(OH)_{4.25}$.

Anthophyllite: $Ca_{0.11}Mg_{5.97}Fe^{2+}_{0.70}Fe^{3+}_{0.12}Mn_{0.006}Al_{0.20}Si_{7.73}O_{21.44}(OH)_{2.56}$.

Grunerite (amosite): $Ca_{0.09}Na_{0.01}K_{0.04}Mn_{0.08}Fe^{2+}_{4.65}Mg_{1.53}Fe^{3+}_{0.49}Ti_{0.03}Al_{0.12}Si_{7.73}O_{21.57}(OH)_{2.43}$.

Crocidolite: $Na_{1.38}Ca_{0.17}K_{0.13}Mg_{3.05}Fe^{3+}_{1.66}Fe^{2+}_{0.48}Al_{0.11}Si_{7.93}O_{21.92}(OH)_{2.08}$.

It should be noted that the chemical analyses from which the above formulae were derived were carried out on minerals separated from rock samples and they cannot be assumed to be entirely free from impurities. Also, it is typical of minerals that they can exhibit wide ranges of composition while still possessing the characteristic structure and general formula.

Table 2.1 Approximate major cation contents of the amphibole asbestos minerals General formula = $X_2Y_5Z_8O_{22}(OH, F)_2$

Mineral	X	Y	Z	Notes
Tremolite	Ca_2	Mg_5	Si_8	
Actinolite	Ca_2	$(Mg, Fe)_5$	Si_8	
Anthophyllite	Mg_2	$(Mg, Fe)_5$	Si_8	$Mg/(Mg+Fe)>0.6$
Grunerite ('amosite')	$(Fe^{2+}, Mg)_2$	$(Fe^{2+}, Mg)_5$	Si_8	Approx. $Fe^{2+}_5Mg_2$
Crocidolite	Na_2	$Fe^{3+}_2Mg_3$	Si_8	

Thus, for example, grunerites from different localities, or even from adjacent areas within one locality, can show different Fe to Mg ratios, and crocidolites may show a range of Mg^{2+}, Fe^{2+} and Fe^{3+} contents.

(4) CRYSTAL STRUCTURES

The chemical constitution of the asbestos minerals and their apparent complexities are more readily understood in terms of their crystal structures.

4(i) Serpentine minerals

The serpentine minerals, of which chrysotile is one variety, have a layered silicate structure. The layers can be regarded as made up of Si atoms, each surrounded by four oxygen atoms at the corners of an almost regular tetrahedron, and Mg atoms each surrounded by six oxygen atoms at the corners of an approximately regular octahedron. Si–O tetrahedra lying on their triangular bases and with apices all pointing in the same direction are linked by sharing all basal oxygen atoms to form a continuous layer. This has approximately 3-fold symmetry but is more conveniently described in terms of a rectangular unit cell which has repeat distances $a \approx 5.3$ Å, $b \approx 9.2$ Å. Gaps in the plane formed by apical oxygen atoms are filled with (OH) ions to form a regular and approximately close-packed array of (O, OH) at this level. Lying above this plane is an array of Mg atoms and above these a plane of (OH) ions such that the Mg atoms are surrounded by octahedra of (O, OH) ions. Each Mg has a triangle of 3 (OH) above it and a triangle below of 2(O) and one (OH), rotated by 60° with respect to the first. Plan and elevation views of this structure are presented in Figure 2.2. The chemical content of a unit cell can be seen to be a multiple of $Mg_3Si_2O_5(OH)_4$.

In building up the three-dimensional serpentine structure, composite layers of the type described are superimposed, and the inter-layer distance is about 7.3 Å. However, one layer may be placed directly above another, or there may be some displacement or rotation between layers. This leads to the possibility of various stacking sequences, different symmetries and multi-layer unit cells. One-, two-, three-, six-, and nine-layer unit cells are known and the symmetry may be trigonal, hexagonal, or monoclinic. Disordered stacking can also occur.

The comparatively simple structure described above is that of the serpentine mineral called lizardite. Lizardites are mostly fine-grained and have platy morphology, but some have lath-like particles with elongation parallel to *a*.

The asbestiform morphology of chrysotile is not obviously reconcilable with a layered crystal structure, and this paradox has been the subject of

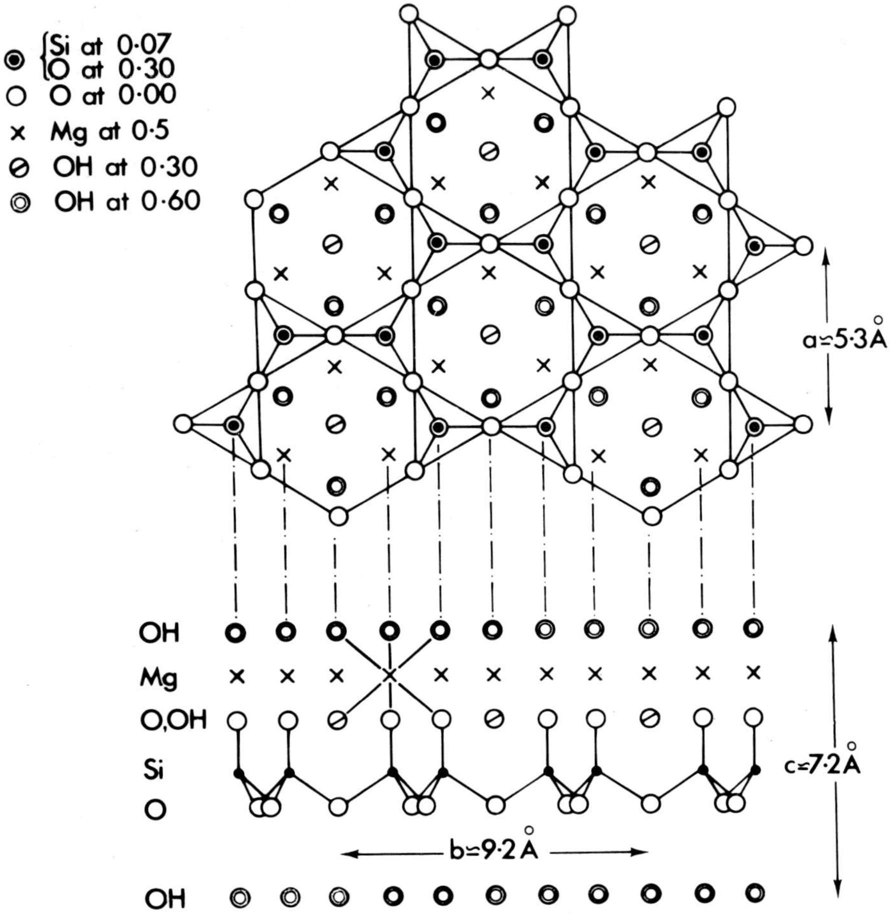

Figure 2.2 Plan and elevation view of idealized structure of serpentine minerals

much painstaking research (e.g. Whittaker, 1956). The ultimate and complete solution came with the direct evidence from high-resolution electron microscopy (Yada, 1967) that in chrysotile the structural layers are curved about the *a* direction to form either scrolls or concentric cylindrical tubes (Figure 2.3). The diameters of such tubes are of the order of 200 Å; thus, a sliver of chrysotile asbestos with cross-section 0.1 mm square contains about 20×10^6 tubular fibrils all in approximately parallel orientation. It is possible, therefore, to strip from an asbestos fibre bundle very fine threads each of which still contains many thousands of fibrils.

A theoretical reason for the curvature of serpentine layers can be found in the composite and polar character of the fundamental sheet. The Si component tends to have a smaller repeat dimension than the Mg component, and

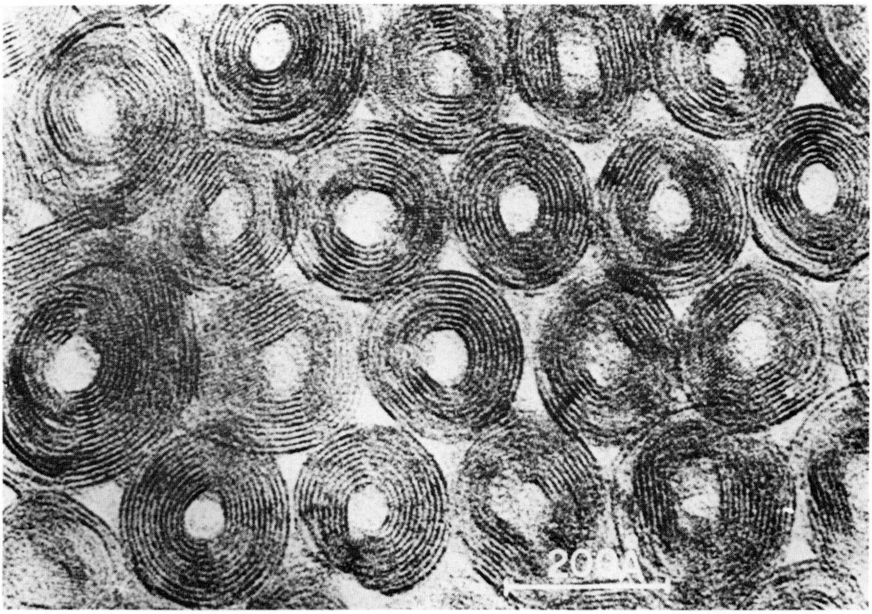

Figure 2.3 High-resolution electron micrograph of transverse section of chrysotile asbestos. (From Yada, 1967)

the mis-match between the two can be overcome by bending, with the Mg octahedra on the outside of the curve. In another variety of serpentine, antigorite, the layers invert at regular intervals thus producing a regularly repeating corrugation. Well formed platy crystals, not tubes, are the result. In lizardites the platy crystals tend to be buckled and disordered, and they are usually limited to very small (sub-microscopic) dimensions.

The growth of very thin fibrils is more readily understood for chrysotile than for amphibole asbestos (discussed later). When chrysotile tubes are formed, a particular radius of curvature may be the most stable, and this may impose an upper limit on the diameters of the tubes.

4(ii) Amphibole minerals

The fundamental unit of the amphibole structure is a chain of SiO_4 tetrahedra linked by sharing corner oxygen atoms in the manner shown in Figure 2.4. The characteristic chain formula is Si_4O_{11} and the repeat distance along its length is approximately 5.3 Å. The chains are four tetrahedra wide and of very great length; fibres of asbestiform amphiboles run parallel to the chain length.

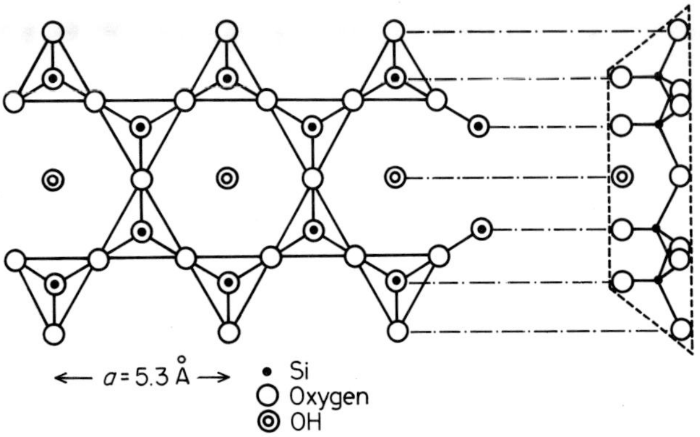

Figure 2.4 Plan and end view of an idealized amphibole chain (Si_4O_{11}) plus additional OH ions

In the amphibole structure the Si_4O_{11} chains are linked laterally by cations as shown in Figure 2.5. In tremolite the cations are Mg and Ca, the Mg ions linking chains by means of strips of Mg (O, OH) octahedra. The oxygen atoms in these strips are the apices of tetrahedra, and the (OH) ions occur as in Figure 2.4. The Ca ions link neighbouring chains across the bases of the tetrahedra and they occur in distorted polyhedra of oxygen atoms. The alternative occupation of Mg (Y) and Ca (X) sites in different amphiboles has been discussed above.

Most amphiboles are monoclinic in crystal symmetry, as a result of the way in which successive chain units are stacked with respect to each other. Anthophyllite, however, is orthorhombic, and the relation of its unit cell to that of a monoclinic amphibole is shown in Figure 2.6. The cell parameters of some amphiboles are given in Table 2.2.

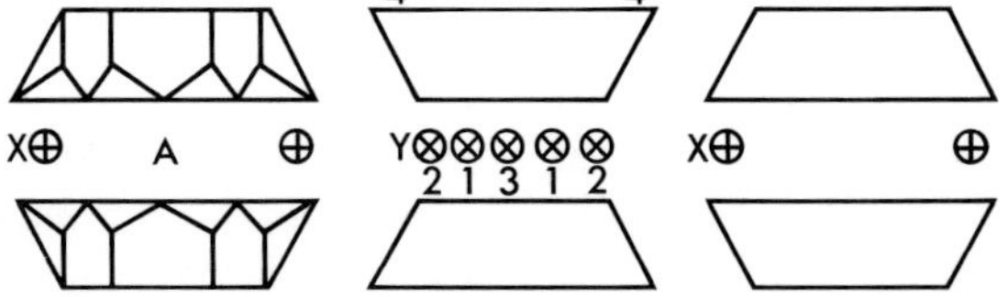

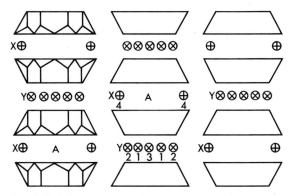

Figure 2.5 Schematic end-view of amphibole chains linked by cations in X and Y sites. In some amphiboles the site A is occupied

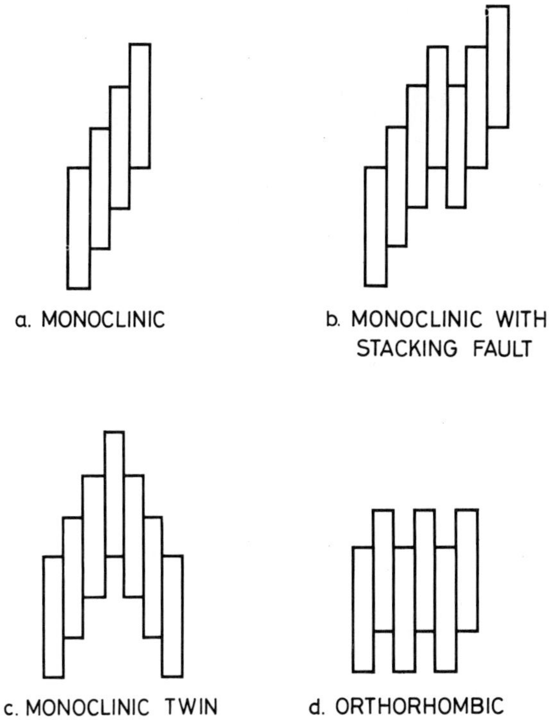

Figure 2.6 Schematic illustration of possible different modes of stacking of neighbouring blocks of amphibole structure

The chain-like structure in amphiboles leads to their possession of good cleavages on (110) planes. If the chains themselves and their linkage across Y sites are regarded as the strongest elements of the structure, paths of weakness can be traced as in Figure 2.7 and an average path can be taken as defining a likely cleavage plane.

Table 2.2 Cell parameters of amphiboles*

Amphibole	*a*(nm)	*b*(nm)	*c*(nm)	β	Reference
Tremolite	0.982	1.805	0.528	104°39′	Papike *et al.* (1969)
Actinolite	0.989	1.820	0.531	104°38′	Mitchell *et al.* (1971)
Grunerite	0.956	1.830	0.535	101°50′	Ghose and Hellner (1959)
Crocidolite	0.974	1.795	0.530	103°54′	Whittaker (1949); Zussman (1955)
Anthophyllite	1.856	1.801	0.528	90°	Finger (1970)

* These values are for particular specimens. Variations in chemical composition can be expected to yield a range of values but usually within 1–2% of those given.

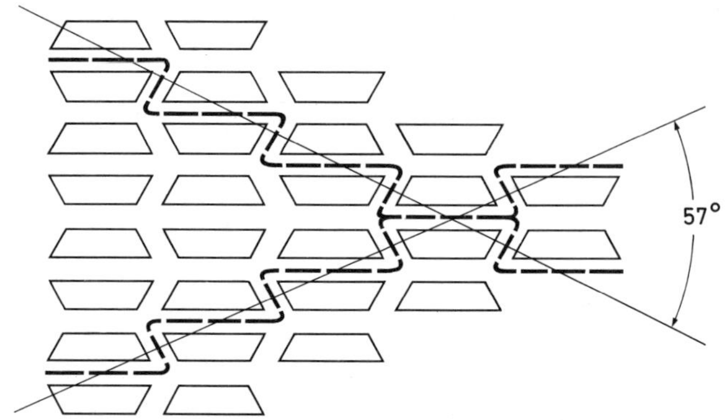

Figure 2.7 Schematic illustration of amphibole structure as seen along the z-axis. Probable paths of weakness are shown, causing cleavages intersecting at approximately 57°

It is tempting to associate the physical nature of asbestos with the chain-like structure and with cleavages, but while these features may be indirectly pertinent they are not fully responsible. Thus, many amphibole specimens are not asbestiform although they possess the chain structure and good cleavage. A distinction should be recognized between the process of cleaving fragments from a single crystal of an amphibole, and that of stripping a fibril or bunch of fibrils from the aggregate that constitutes a specimen of asbestos.

The processes that lead to growth in asbestiform habit are not clearly understood. To form asbestos there presumably has to be multiple nucleation followed by relatively rapid growth along the fibre direction and very limited growth at right-angles to it. Such a process might be influenced by chemical factors, and it is noteworthy that the class of amphibole called hornblendes, in which Al is an important substituent for Si in the structural chains, are not often found with fibrous and still less with asbestiform morphology. Major element chemistry cannot, however, be the only factor, since we find among tremolites some which are and some which are not fibrous or asbestiform. One can only conjecture that other factors such as the temperature regime during crystallization, trace element chemistry, speed of growth, or a combination of these, exert an influence on crystal habit.

The question of variations in nature and properties from one amphibole to another is worthy of further discussion. Although the crystal structures of all amphiboles are broadly the same, these structures are derived (by X-ray diffraction methods) from specimens which, although small, contain about 10^{15} unit cells. They are therefore 'average' structures. Real crystals do not have the ideal exact repetition of unit cells, but contain defects of various kinds, and the abundance and distribution of defects are known to influence mechanical properties, and may also influence morphology and surface properties. Structural detail at this level is not easily detected by X-ray diffraction, but may be investigated more readily by electron microscopy.

An important kind of structural imperfection in amphiboles is the 'stacking fault'. In a perfect crystal of a monoclinic amphibole, slabs of structure parallel to (100) are stacked alongside one another with regular and identical displacements (Figure 2.6a). In a faulted structure, errors in the direction and magnitude of these displacements occur and the frequency of such faults can vary from one specimen to another (Figure 2.6b). When the faults are relatively infrequent, the result can sometimes be described as a twinned crystal (Figures 2.6c and 2.8). When the faults are frequent and regularly repeating, they are no longer really faults but are the regular displacements of a structure with larger unit cell and perhaps different symmetry (Figure 2.6d). This latter condition describes approximately the relationship between monoclinic and orthorhombic amphiboles.

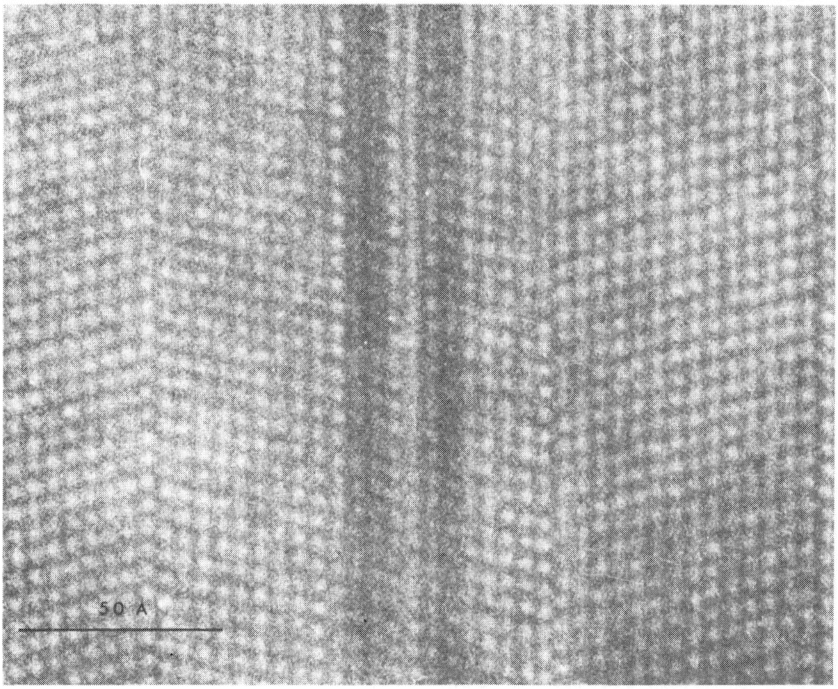

Figure 2.8 High-resolution electron micrograph (beam parallel to *y*) of amosite asbestos showing twinned and otherwise faulted structures. (From Hutchison *et al.*, 1976)

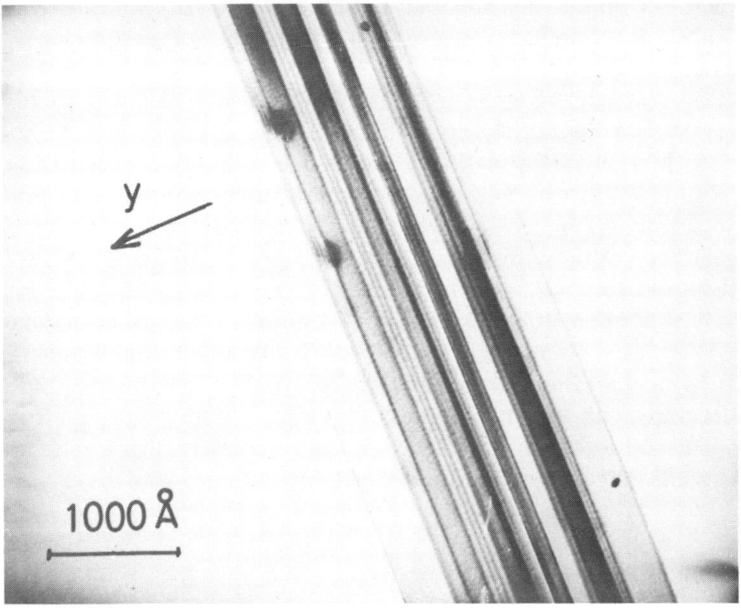

Figure 2.9 Electron micrograph (beam perpendicular to y) of amphibole showing Wadsley defects on (010) planes. (From Chisholm, 1973)

Another kind of structural imperfection is the 'Wadsley defect' or 'crystallographic shear plane'. When these are present the amphibole contains, in addition to the normal double chain of tetrahedra, some single-chain or triple-chain elements of structure. In low-magnification electron micrographs such defects are seen as linear features parallel to (010) (Figure 2.9).

The structures of amphiboles as revealed by X-ray diffraction have been determined mainly from specimens which are not asbestiform because it is much more difficult to obtain the required data from a fibre bundle than from a single crystal, and a single fibril of asbestos is too thin for a practical X-ray experiment. It may be that, even in terms of the 'average' structure, there are subtle but possibly significant differences as between the asbestos and non-asbestos varieties of a particular amphibole.

(5) OCCURRENCES

Serpentinite, the rock which is composed largely of serpentine minerals, is of fairly widespread occurrence, being found for example in Alpine-type settings and as rocks dredged from the ocean bottom. These rocks usually contain some proportion of chrysotile, but either in a low concentration or in the form of disseminated very short fibres. The main commercially exploited occurrences of chrysotile asbestos are in Canada (Quebec, British Columbia, Yukon), the U.S.S.R. (Urals, East Siberia) and Africa (Rhodesia, Swaziland).

In most occurrences serpentinites were originally ultrabasic igneous rocks such as peridotite [mainly olivine, $(Mg, Fe)_2SiO_4$, and pyroxene, $(Mg, Fe)SiO_3$] and dunite (mainly olivine). These became metamorphosed hydrothermally to give serpentinites containing the fine-grained serpentine minerals, lizardite, chrysotile, and antigorite. Veins of long-fibre chrysotile asbestos were probably produced in a further episode of transformation under conditions of relatively low but increasing temperature. An alternative rock from which chrysotile is derived is an impure siliceous dolomite, $CaMg(CO_3)_2$, as found for example in the Transvaal. Among other better known chrysotile occurrences are those of Cyprus and in the U.S.A. (Arizona and Vermont).

Although amphibole minerals occur in a wide variety of igneous and metamorphic rocks, the asbestiform amphiboles have a much more restricted occurrence. The principal amphibole asbestos minerals, amosite and crocidolite, are formed in thermally metamorphosed banded ironstones. The latter are highly siliceous rocks but must also be rich in magnesium and sodium for the formation of amosite and crocidolite, respectively. In each case it is likely that the asbestos fibres are a secondary product of recrystallization from a non-asbestiform vein of the mineral concerned, grunerite for amosite and riebeckite for crocidolite. The main commercially exploited occurrences of crocidolite are in Cape Province, South Africa, and of amosite in the Transvaal. A well known deposit of crocidolite (largely worked-out) is at Wittenoom Gorge, Australia, and another source is Bolivia. The less common anthophyllite asbestos comes mainly from Finland, and tremolite asbestos from Italy.

Mining of asbestos can be either of the open-pit or underground type, and the asbestos fibre is separated by either hand cobbing or machine milling methods. An average rock contains about 10% of fibre.

(6) SYNTHESIS

It has proved possible to synthesize in the laboratory most of the serpentine and amphibole minerals, including some of fibrous morphology (see, for example, Evans *et al.*, 1976; Fedoseev *et al.*, 1968). For serpentines, tubular chrysotile fibres frequently result, but only in very short lengths. For amphiboles, finely fibrous forms and even asbestiform varieties have been synthesized with appreciable fibre length, but many have had rather unusual chemical compositions. Perhaps because of the relative cheapness of natural asbestos, little work has been done on synthesizing longer and stronger asbestos fibres which could have properties adequate for commercial use.

There may be a tendency to look at synthetic inorganic fibres as replacements for asbestos because of the health hazard aspects, but for many uses these do not have as desirable physical properties, and moreover they are not necessarily themselves free from possible adverse health effects.

(7) OPTICAL PROPERTIES

Distinction between the asbestos minerals and others, and between the different kinds of asbestos, can be achieved by simple methods if one is dealing with large fibres or bundles of fibres and if concentrations are not very low. In these circumstances optical immersion methods using a polarizing microscope or X-ray powder diffraction can be diagnostic. Brief accounts of these methods were given in Zussman (1977), and more detailed treatments can be found in standard text-books.

The optical properties of the asbestos minerals are presented in Table 2.3. If one is concerned only with the widely used forms of commercial asbestos, chrysotile, crocidolite, amosite and anthophyllite, the following distinctions are of use.

Crocidolite as a hand specimen has a distinctive blue colour, which is visible also in fine dispersed fibres as viewed under the polarizing microscope. In plane polarized light the fibres show pleochroism (different colours for different light vibration directions), dark blue with the polarizer parallel and a paler grey–blue with the polarizer perpendicular to the fibre length. In addition, crocidolite is unusual in that the refractive index parallel to the fibre length is lower than that perpendicular to it (i.e. in optical terms it is 'length fast'). Amosite is slightly pleochroic, with a darker brown or grey tint parallel to the fibre length. Anthophyllite is colourless and has lower refractive indices than those of the other two amphiboles. Chrysotile is also non-pleochroic and its refractive indices are appreciably lower than those of

Table 2.3 Optical properties of asbestos

Asbestos	R.I. (approx.) α	γ	Optical orientation	Pleochroism
Anthophyllite	1.61	1.63	Length slow	Weak to moderate; γ strongest absorption
Amosite	1.67	1.70	Length slow	γ pale brown; α, β paler brown or yellow
Tremolite	1.60	1.62	Length slow	Non-pleochroic
Actinolite	1.62–1.67	1.64–1.68	Length slow	γ green or green–blue; α, β paler colours
Crocidolite	1.70	1.71	Length fast	Strong pleochroism; γ pale grey–blue; α, β dark blue
Chrysotile	1.54	1.55	Length slow	Non-pleochroic

any amphibole. However, its optical properties are often difficult to determine because of the fineness of its fibres.

Provided that the fibres or fibre bundles are thick enough for resolution in the polarizing microscope (of the order of 1 μm or greater), small numbers of particles can be detected by the optical method, even in a mixture.

When optical studies are made on fibre bundles rather than single fibrils (as is more often the case), the optical properties observed are the integrated effects of the individual units. One property particularly affected in this way is the extinction angle. A single crystal of a monoclinic amphibole (e.g. tremolite, actinolite, riebeckite, or grunerite) will show straight extinction when lying on (100) but inclined extinction on (010) or on a cleavage plane (110). Asbestiform varieties of the above minerals mostly show straight extinction in all orientations about the fibre axis. This can be regarded as the net effect of fibrils in more than one azimuthal orientation. In some asbestos very fine-scale twinning may be responsible, while in others there may be a high degree of randomness of orientation.

The orthorhombic amphibole anthophyllite will show straight extinction for all ($hk0$) planes in single crystals, and also in randomized or twinned aggregates in asbestos.

Chrysotile asbestos, although structurally monoclinic, will show straight extinction because of the randomized orientation produced by its tubular and aggregate nature.

Typical appearances of chrysotile, crocidolite, amosite, and tremolite under the polarizing microscope are shown in Figure 2.10.

(8) X-RAY DIFFRACTION DATA

The various asbestos minerals have distinctive X-ray powder diffraction patterns, that of chrysotile being completely different from those of amphibole asbestos because of its distinctive crystal structure. Among the amphiboles, anthophyllite gives a clearly different pattern because it is orthorhombic and the others are monoclinic. Although the crystal structures and powder patterns, of the monoclinic amphiboles tremolite, crocidolite, and grunerite are broadly similar, the precise positions of peaks and their relative intensities differ.

Indexed powder patterns for crocidolite, amosite, chrysotile, tremolite, and anthophyllite asbestos are given in Table 2.4. The precise d values will vary with chemical variations such as the Fe to (Fe + Mg) ratio in an amosite. The intensities also will be affected by such variations, but they can be appreciably affected by the extent to which fibrils show non-random orientation in the powder specimen.

The sensitivity of the standard X-ray diffraction method to small amounts of amphibole or serpentine in a mixture of compounds is not very great,

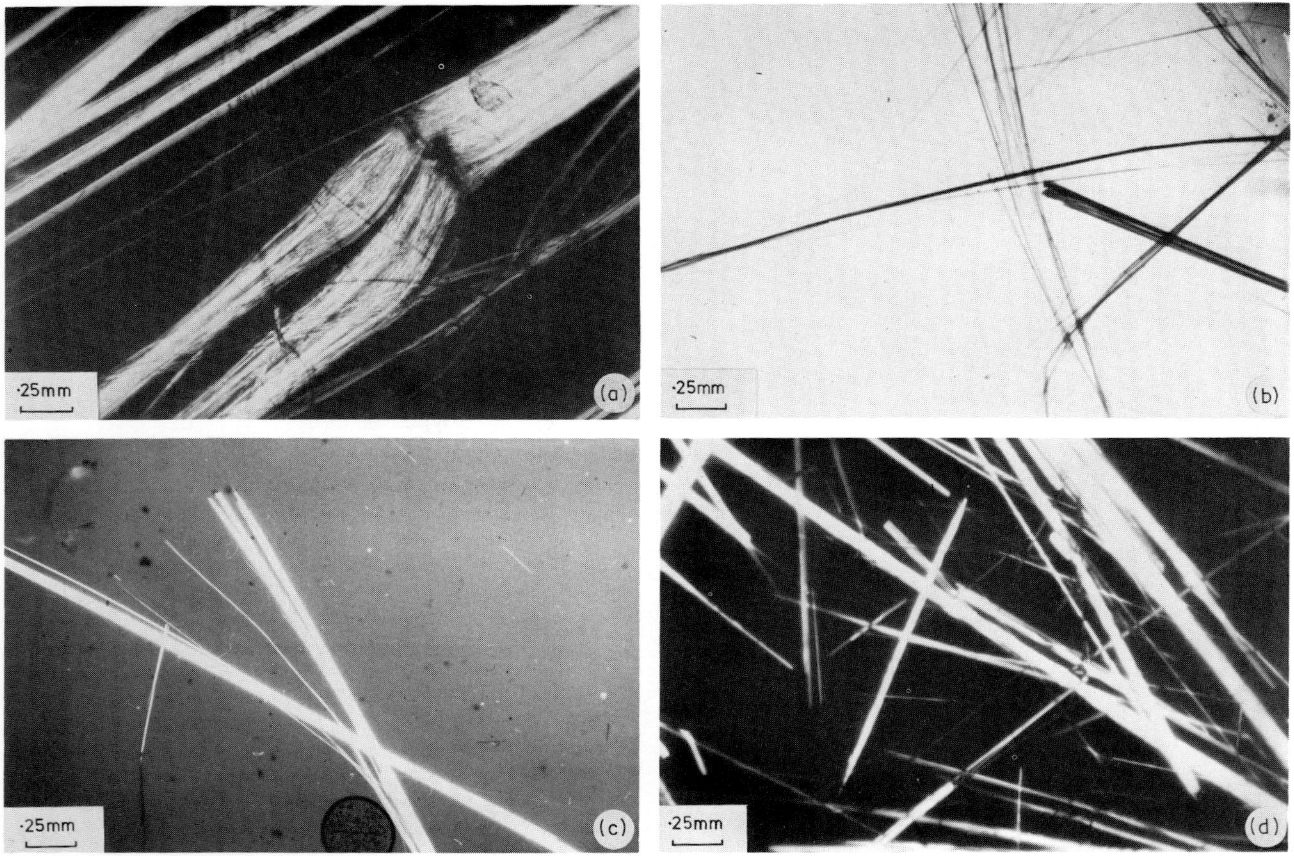

Figure 2.10 Photomicrographs (polarizing microscope) of (a) chrysotile, (b) crocidolite, (c) amosite and (d) tremolite [crossed polars for (a) and (d)]

Crocidolite (Koegas); diffractometer, a 9.74, b 18.06, c 5.31 Å; β 103.73, C 2/m		Amosite (Penge); diffractometer, a 9.51, b 18.30, c 5.33 Å; β 101.06°, C 2/m		Chrysotile (clino-); camera*, a 5.32, b 9.20, c 14.64 Å; β 93.3°		Tremolite (Koluk River, N.W. Alaska); diffractometer, a 9.87, b 18.02, c 5.33 Å, β 104.91, C 2/m				Anthophyllite (UICC); diffractometer, a 18.50, b 17.90, c 5.29 Å; Pnma			
d observed (Å)	hkl	d observed (Å)	hkl	d observed (Å)	hkl	d observed (Å)	hkl	d observed (Å)	hkl	d observed (Å)	hkl	d observed (Å)	hkl
8.42	110†	9.1	020	7.36	002†	9.04	020	2.340	$35\bar{1}$†	9.34	200 + T002†§	2.557	621
4.89	$11\bar{1}$	8.33	110†	4.58	020 b‡	8.44	110†	2.325	$42\bar{1}$	8.90	020	2.536	640
4.50	040†	5.10	130	3.66	004 †	5.09	130†	2.301	$17\bar{1}$	8.23	210†	2.111	561, 152
3.89	$\bar{1}31$	4.68	200	2.66	130 b‡	4.88	$11\bar{1}$	2.213	$24\bar{2}$, 042	4.67	400 + T004†§	2.094	741
3.40	041, 131	4.58	040	2.59	201	4.77	200	2.165	261†	4.56	021 + T020§	2.065	821
3.26	240	4.16	220	2.55	$20\bar{2}$	4.53	040†	2.046	202	4.52	410, 040	2.052	840
3.11	310†	3.27	240†	2.46	202†	4.21	220	2.019	351†	4.41	121	2.023	532
2.79	330†	3.06	310†	2.28	203	3.88	$\bar{1}31$	2.005	370	4.12	420†	1.997	831
2.72	151†	2.79	330	2.21	$20\bar{4}$	3.38	041, 131†	1.967	152, $42\bar{2}$	3.64	231, 430	1.978	062, 751†
2.60	061	2.77	151	2.10	204	3.28	240†	1.897	510†	3.23	421, 440†	1.870	091, 921 + T0.0.10†§
2.54	$20\bar{2}$	2.64	061	1.83	008	3.13	310†	1.868	460	3.12	600 + T006†§	1.844	671†
2.32	$35\bar{1}$†	1.640	$24\bar{3}$	1.75	206	2.94	$15\bar{1}$, 221†	1.621	113	3.05	610, 501†	1.798	581
2.18	261	1.598	$15\bar{3}$	1.536	060†	2.81	330†	1.587	$15\bar{3}$	2.86	521	1.730	812
2.03	351	1.556	600	1.465	0, 0, 10	2.73	$33\bar{1}$†	1.538	422	2.83	450	1.670	690†
1.865	$\bar{4}61$					2.71	151†	1.440	$53\bar{3}$†	2.74	441†	1.616	912
1.631	480, $24\bar{3}$					2.60	$06\bar{1}$			2.72	630	1.561	253
1.619	113					2.536	$20\bar{2}$†			2.692	531		
						2.415	311			2.671	351		
						2.387	350			2.587	061, 112 +T? $13\bar{2}$§		

* When chrysotile is examined by powder diffractometer instead of camera, only the stronger reflections may be detected.
† High intensities.
‡ b = broad.
§ T = talc. The talc present in the UICC anthophyllite sample has particularly strong 002 and 006 reflections.

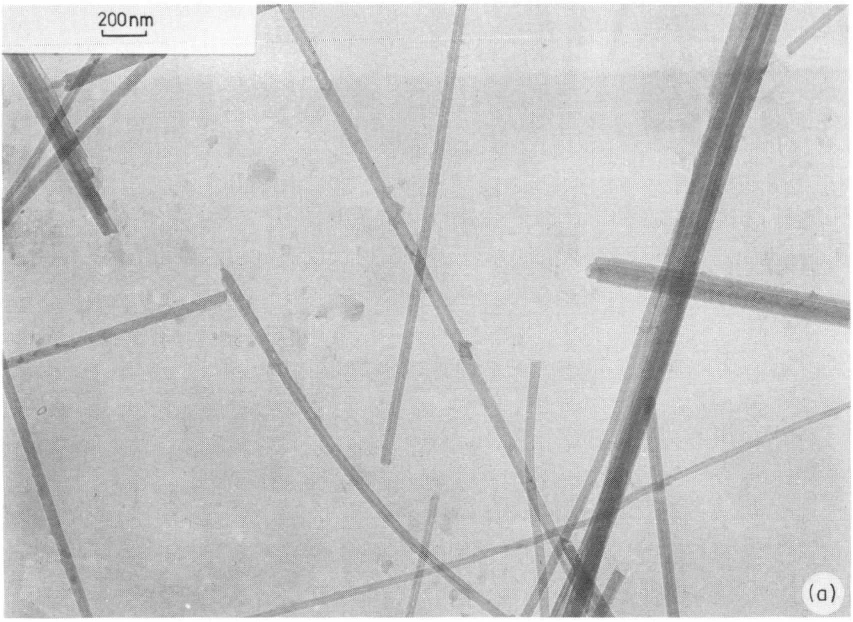
200 nm
(a)

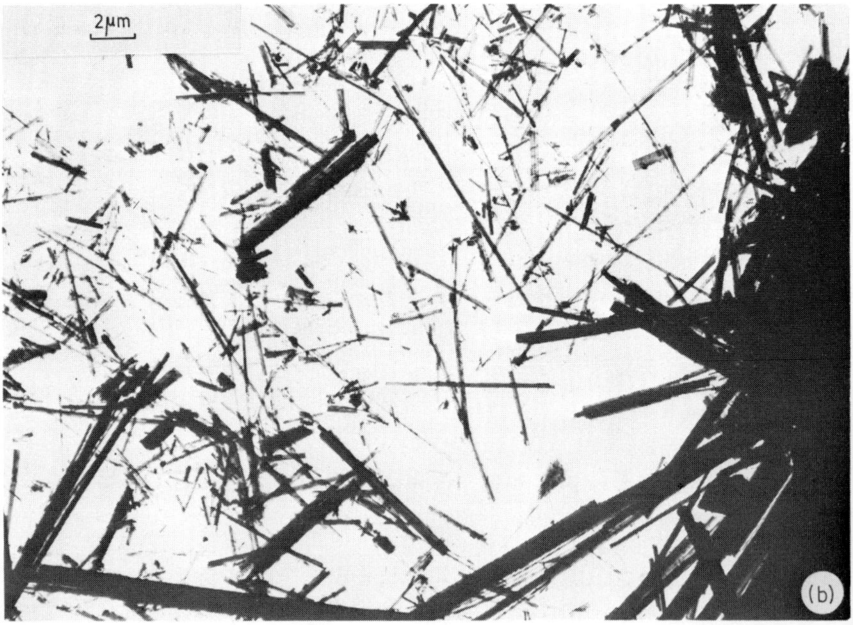
2 μm
(b)

Figure 2.11 Electron micrographs of asbestos fibres: (a) chrysotile (Champness *et al.*, 1976); (b) crocidolite (UICC) (electron micrograph by Dr. P. E. Champness); (c) amosite (UICC)

detection of less than 1–5% (depending on the conditions) being rather uncertain. Mixtures containing more than one amphibole may, of course, also present difficulty.

(9) ELECTRON OPTICAL CHARACTERISTICS

The detection and identification of asbestos particles by electron microscopy and electron diffraction can be time consuming and expensive but it is sometimes essential because of the limitations and ambiguities inherent in other methods. In the electron microscope chrysotile asbestos has a characteristic appearance, usually as long, thin, curved fibrils, and with high enough magnification its tubular nature can be discerned. Its electron diffraction pattern is also different from those of other asbestos minerals. Electron micrographs typical of chrysotile, amosite, and crocidolite are shown in Figure 2.11.

The morphologies and electron diffraction patterns of the amphibole asbestos minerals are similar to one another and require careful study and measurement for differentiation. It should be remembered that the electron diffraction pattern observed depends not only on the crystal structure but also on fibril orientation. Further details are given in Chapter 7.

Additional information about minute particles of asbestos can be obtained by use of analytical attachments to electron microscopes, measuring characteristic X-rays excited by the electron beam. With care, chemical analyses can be obtained by this method which can be useful in resolving ambiguities of identification (see, for example, Champness *et al.*, 1976).

Electron microscopy is essential if it is important to determine numbers and dimensions of particles which are below the limits of optical resolution.

(10) NON-ASBESTIFORM AMPHIBOLE AND SERPENTINE MINERALS

The harmful effects of over-exposure to dusts of commercial asbestos have been intensively investigated for many years. More recently, concern has been expressed over the inhalation or ingestion of particles of other kinds of amphibole and serpentine. In their natural forms the distinction between a large crystal of tremolite and a bundle of tremolite asbestos fibrils is clear but, as explained earlier, the distinction becomes blurred if both are finely comminuted. Little detailed work has been done to compare the properties of the two, but it seems likely that differences do exist that are associated with the different conditions of growth that produced such different natural morphologies. There are indications (Lee *et al.*, 1978), for example, that crystal cleavage fragments show the expected predominant {110} cleavage faces whereas in asbestos particles {100} is the dominant form. A preliminary study (T. Zoltai, personal communication) has shown that the surface properties are significantly different, and it may be that defect and other structural detail differs (see p. 55).

The problem of how best to regard non-asbestos varieties is exacerbated by the accepted definition of a 'fibre', in the context of health hazards, as possessing an aspect ratio of greater than 3:1. Crushed amphibole crystals will have far fewer particles than amphibole asbestos in the >10:1 range but, because of their ready prismatic cleavage, they will have many particles with aspect >3:1.

With a true asbestos a high proportion of particles with ratios >3:1 will in fact have ratios >10:1 or indeed >100:1, and while the concentration of particles >3:1 may be a useful measure of toxicity, it may be the longer particles that are causing the greatest harm. If this is so, a better boundary for dusts from non-asbestos amphiboles might be drawn at aspect ratios of, say, about 10:1.

If it is thought that non-asbestos amphibole dusts, even with their different size populations, and possibly different properties, might be as harmful as asbestos, then in cases where exposure is significant, epidemiological and animal studies would need to be undertaken.

ACKNOWLEDGEMENT

I am grateful to Mrs. M. Dorling for help with photomicrographs and X-ray powder data.

(11) REFERENCES

Campbell, W. J., Blake, R. L., Brown, L. L., Cather, E. E., and Sjoberg, J. J. (1977) '*Selected Silicate Minerals and their Asbestiform Varieties: Mineralogical Definitions and Identification–Characterization*, U.S. Bureau of Mines Information Circular 8751.

Champness, P. E., Cliff, G., and Lorimer, G. W. (1976) The identification of asbestos, *J. Microsc.* **108,** 231–249.

Chisholm, J. E. (1973) Planar defects in fibrous amphiboles, *J. Mater. Sci.*, **8,** 475–483.

Deer, W. A., Howie, R. A., and Zussman, J. (1963) *Rock Forming Minerals*, Vol. 2, *Chain Silicates*, Vol. 3, *Sheet Silicates*. Longmans, London.

Evans, B. W., Johannes, W., Oterdoom, H., and Tromsdorff, V. (1976) Stability of chrysotile and antigorite in the Serpentine multisystem. Schweiz, *Miner. Petrog. Mitt.*, **56,** 79–93.

Fedoseev, A. D., Makarova, T. A., and Kosulina, G. (1968) Synthesis of fibrous richterite under hydrothermal conditions. *Zap. Vses. Min. Obshch.*, **97,** 722–725.

Finger, L. W. (1970) Refinement of the crystal structure of an anthophyllite. *Ann. Rep. Dir. Geophys. Lab., Carnegie Inst. Yr. Bk.*, **68,** 283–288.

Ghose, S., and Hellner, E. (1959) The crystal structure of grunerite and observations on the Mg–Fe distribution. *J. Geol.*, **67,** 691–701.

Hutchison, J. L., Irusteta, M. C., and Whittaker, E. J. W. (1976) High resolution electron microscopy and diffraction studies of fibrous amphiboles. *Acta Crystallogr.*, **A 31,** 794–801.

Lee, R. L., Lally, J. S., and Fisher, R. M. (1977) Important considerations in the identification and counting of mineral fragments, *Workshop on Asbestos*, National Bureau of Standards, Gaithersburg, Maryland, in press.

Mitchell, J. T., Bloss, F. D., and Gibbs, G. V. (1971) Examination of the actinolite structure and four other C 2/m anphiboles in terms of double bonding, *Z. Krist.*, **133,** 273–300.

Papike, J. J., Ross, M., and Clark, J. R. (1969) Crystal-chemical characterisation of clinoamphiboles based on five new structure refinements, in *Pyroxenes and Amphiboles: Crystal Chemistry and Phase Petrology*, Min. Soc. Am. Special Publ. No 2.

Whittaker, E. J. W. (1949) The structure of Bolivian crocidolite, *Acta Crystallogr.*, **2,** 312–317.

Whittaker, E. J. W. (1956) The structure of chrysotile. II. Clinochrysotile, *Acta Crystallogr.*, **9,** 855–862.

Yada, K. (1967) Study of chrysotile asbestos by a high resolution electron microscope, *Acta Crystallogr.*, **23,** 704–707.

Zussman, J. (1955) The crystal structure of an actinolite, *Acta Crystallogr.*, **8,** 301–308.

Zussman, J. (1977) *Physical Methods in Determinative Mineralogy*, 2nd Edn., Academic Press, New York.

CHAPTER 3

Chemistry and physics of asbestos

A. A. Hodgson

Cape Asbestos Fibres Ltd. Middlesex

(1) ASBESTOS: THE RAW MATERIAL

Six species of asbestiform minerals are known, these being derived from two large groups of rock-forming minerals, the serpentines and the amphiboles. Chrysotile or white asbestos is the sole species classified in the serpentine group, but it is by far the most abundant kind of asbestos. The amphibole group includes the asbestiform types, crocidolite, amosite, anthophyllite, tremolite, and actinolite. While the names of the last three types apply equally to the fibrous and crystalline forms of these minerals, the name crocidolite (blue asbestos) is given to the fibrous form of riebeckite and amosite to the fibrous form of grunerite. Table 3.1 shows a classification of these minerals, together with their approximate chemical compositions.

Table 3.1 Classification of varieties of asbestos minerals

Group	Variety	Formula
Amphibole	Anthophyllite	$Mg_7Si_8O_{22}(OH)_2$
	Amosite	$(Fe^{2+}Mg)_7Si_8O_{22}(OH)_2$
	Crocidolite (blue asbestos)	$Na_2Fe_2^{3+}(Fe^{2+}Mg)_3Si_8O_{22}(OH)_2$
	Tremolite	$Ca_2Mg_5Si_8O_{22}(OH)_2$
	Actinolite	$Ca_2(MgFe^{2+})_5Si_8O_{22}(OH)_2$
Serpentine	Chrysotile (white asbestos)	$Mg_3[Si_2O_5](OH)_4$

The bulk of the world's asbestos has developed as cross-fibre seams or veins in their host rocks. The seams contain fibres in an extremely tightly packed parallel formation, the seam width, which of course determines the fibre length, which is generally between 0.5 and 20 mm. Longer fibres are less abundant, but it is not uncommon to find chrysotile, crocidolite, and tremolite fibres up to 100 mm long and amosite fibres up to 250 mm long. Anthophyllite and certain forms of crocidolite in some deposits do not occur in cross-fibre development, but rather as fibrous masses containing randomly orientated cubic blocks of fibre up to 25 mm long. Within each block the fibres lie parallel as in a seam.

1(i) Occurrence and formation

All forms of asbestos are metamorphic minerals that have a close association with their parent rocks. Most chrysotiles have formed in ultrabasic rocks in which olivines and pyroxenes have been altered by hydrothermal action to serpentine. Part of this serpentine has recrystallized as fibrous material in fissures and cracks which developed in the cooling of the original magma (Jenkins, 1960). Fracturing appears to have taken place during the cooling of the original peridotites and dunites, soon after emplacement, and this stage was followed by the intrusion of acid magma, resulting in partial or complete serpentinization of the rocks. Subsequently, a final period of hydrothermal activity gave rise to fissure filling or wall rock-emplacement along the routes of cracks by vein chrysotile. Talc and magnetite are prevailing impurities in many chrysotile deposits, and the magnetite may appear as banded margins to the fibre veins or be disseminated throughout both the wall rock and the fibre.

A secondary source of chrysotile, although small, is mined from serpentinized dolomitic limestones. Such fibre is often of high quality and free of the magnetite associated with deposits where the host rock was of igneous origin.

Chrysotile deposits in the Northern Hemisphere are mid- to late-

Palaeozoic in age. In contrast, the southern African chrysotiles are Precambrian, as are also the amphibole asbestos deposits of South Africa, which latter are estimated to be some 2000 Ma old.

Such serpentine masses which contain chrysotile are usually extensive, both in plan and depth. The chrysotile veins form an irregular lattice work in the host rock and the winning of the fibre necessitates removal of the whole mass of rock for crushing and extraction at the refinery. Consequently, open-cast methods have proved to be the most economical way of working chrysotile deposits and the immense quarries in the asbestos fields of eastern Quebec are typical of this development. There is some underground mining, particularly by block caving methods whereby ore is conveyed from the underside of an ore body while the superimposing rock collapses slowly downwards. There are extensive chrysotile deposits in Russia, China and in southern Africa, particularly in Southern Rhodesia, where most of the exploitation is underground. There the ore bodies are generally tabular in shape with a pronounced dip. Underground development has been carried for several kilometres along the strike and at depths of over 300 m.

The mining of amphibole asbestos is almost completely confined to South Africa (Du Toit, 1945; Keep, 1961) (Figure 3.1). There, crocidolite and amosite asbestos occur in metamorphosed Precambrian sedimentary strata, the banded ironstones, which are included in the Transvaal Super-Group. The Transvaal Super-Group consists of the following:-

Transvaal

(3) *Pretoria Group:* shales quartzites with some intrusives and volcanics
(2) *Chuniespoort Group:*
  (b) *Penge Iron Formation* banded ironstone
  (a) *Malmanie Sub-Group* dolo-dolomite
(1) *Black Reef Formation:* shales, flagstones, quartzites.

North Cape Province

(3) *Koegas Posmasburg Sub-Group:* jaspers, quartzites, some volcanics.
(2) *Ghaap Group:*
  (b) *Asbes Heuwel Sub-Group* banded ironstone
  (a) *Campbell Rand Sub-Group* dolomite
(1) *Smitsdrif Sub-Group:* shales, quartzites, some volcanics

The Transvaal-type rocks are exposed on a broad arc extending about 1000 km from the Orange River in Cape Province northwards to south Botswana and thence east and south into Eastern Transvaal. Within the ironstone formations marker zones and potential asbestos horizons can be traced for hundreds of kilometres.

Despite the consistent nature of the strata there are notable differences in the origins of the two forms of asbestos found here (Cilliers and Genis, 1964). Only crocidolite occurs in the Asbestos Hills of northern Cape Province. Its deposits are lensoid and appear to be dependent on structural

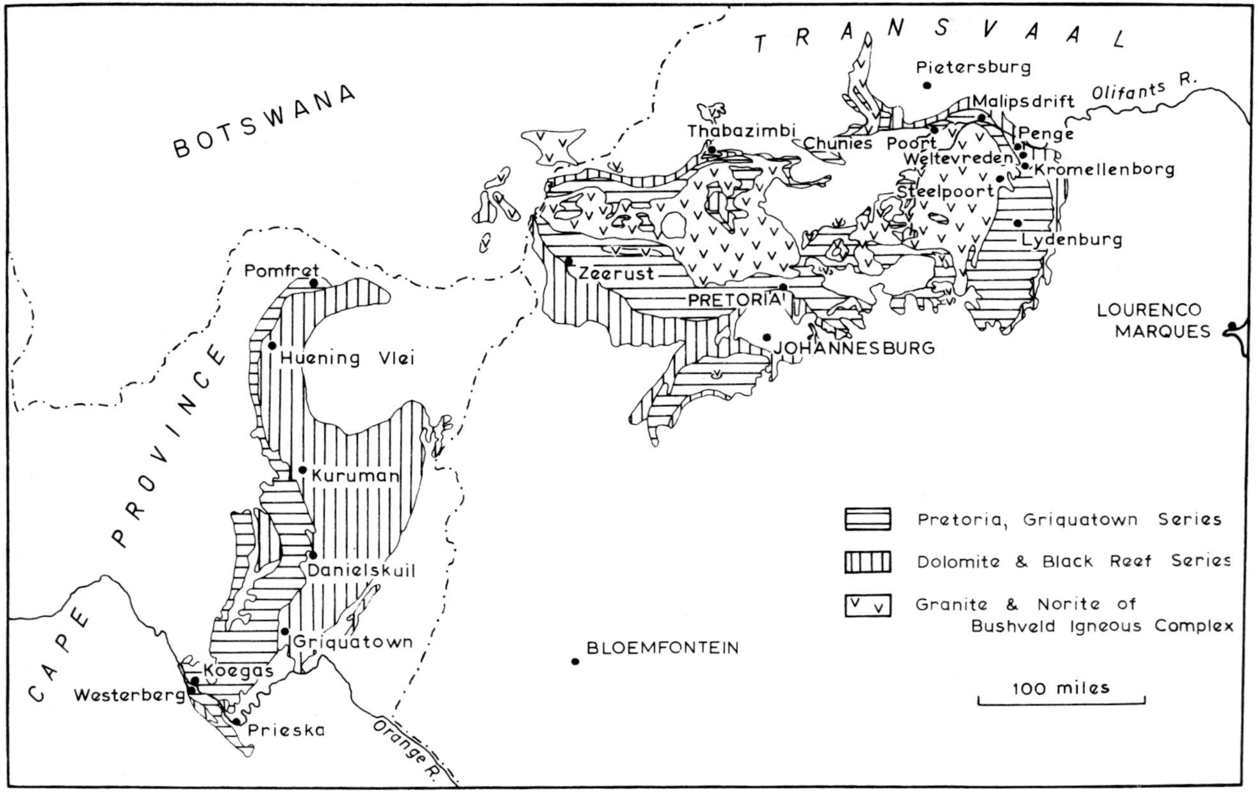

Figure 3.1 Map of deposits in South Africa. (After Du Toit, 1954)

changes in the host rock at intersections of successive folding and cross-folding. It is assumed that the original sediments were the result of reaction between iron hydroxide gels and silica held in solution in an alkaline marine environment and that these consolidated to form the banded ironstones. The banded nature of these rocks is evidence of cyclical changes of climate, temperature and probably atmosphere. Indeed, the tight and thin nature of many of the individual bands is reminiscent of varved sediments and suggests annual changes in deposition. Provided that the chemical composition was right and that subsequent dynamic forces were imposed in the right places, crocidolite asbestos formed as a cross-vein fibre from 0.5 to 100 mm in thickness. There is evidence of seeding of fibre growth from magnetic or ankerite which often form thin layers at the interfaces between fibre viens and host rock. Barren rock in the crocidolite reefs has the same deep blue colour as the fibre and this blue colour is imparted by the dense formation of microscopic crocidolite fibres or riebeckite crystals set in a ground mass of ankerite and silica.

The amosite fields of eastern Transvaal have the same history of deposition at the blue asbestos fields of Cape Province, but there the comparison ends. The host rocks are a dense, dark grey and the fibre seams an ash grey. The fibre seams are generally thicker and more abundant in the asbestos horizons here. There is little development of magnetite, but fibre–rock interfaces often have masses of large grunerite crystals generally disposed parallel to the interface. Such crystals often appear within the fibre seam, but without a preferred orientation. Graphite is an embarrassing contaminant of amosite fibre and occurs in thin seams both in the host rock and adjacent to fibre veins, with occasional graphite screens running haphazardly through the veins.

While dynamic forces have contributed to some extent to the formation of amosite, there is almost no evidence of folding in the rocks in the Penge–Kromellenboog–Weltevreden area, where most of the amosite occurs. The strata is planar and dips at a constant 17° towards the west, and towards the core of the Bushveld Igneous Complex. Indeed, all of the evidence points to amosite being the product of thermal metamorphism, the excessive growth of some fibre veins, masses of grunerite crystals, and graphite being convincing evidence. But why should amosite be formed and not crocidolite? One reason is the lack of sodium. In fact, the strata here generally lacks alkalis except that in some places the host rock has an unusually high potassium content. Another reason is the presence of graphite, which, in conjunction with the heat of the aureole of the Bushveld Igneous Complex, has maintained iron oxides in the reduced state. The iron in amosite is divalent, whereas in crocidolite there are similar proportions of divalent and trivalent iron.

About 80 km northwest of the Penge area, where the exposures of the banded ironstones trend east–west, in the Malipsdrift Pietersburg area, there lies a small elongated area along the Malips River where both crocidolite and amosite occur and from whence the Transvaal crocidolite originates. Crocidolite here appears to have been a product of thermal metamorphism. Fibre seams have characteristic wall formations of large cuboid magnetite crystals, together with some riebeckite. While amosite and crocidolite occur in separate deposits, a curious and inexplicable feature of the fibre horizons is the frequent appearance of veins of both amosite and crocidolite within the same reef. Occasionally crocidolite may pass into amosite in the same vein, with a sharp interface between an upper layer of crocidolite and a lower one of amosite. There is no evidence of graphite or graphitic shale at Malipsdrift, as at Penge, and the chemistry of formation of the amphiboles in this area seems to show a preponderance of ferrous hydroxide gels in the early deposits.

The history of formation of the amphibole asbestos of south Africa has a marked biochemical background. Although the rocks of the Transvaal system are some 2000 Ma old there is evidence in the dolomite series of algal concretions indicating a life form. The banded ironstones in the upper part of the system were deposited in stagnant and anaerobic conditions, in which bacteria appear to have played a large part in determining the ratio of divalent to trivalent iron. Subsequent decomposition and dissipation of these organisms has left a secretion of minor amounts of primitive oils and waxes in the crocidolite fibre of northern Cape Province (Harington, 1964). These do not occur in the Transvaal crocidolite or amosite, but the presence of considerable amounts of graphite in the Penge amosite fields indicates that the residues of primitive life were charred within the aureole of the Bushfeld Igneous Complex shortly after the consolidation of the deposits of the Transvaal system.

Two further occurrences of crocidolite are known, one at Cochabamba in Bolivia and the other at Lusaka in Zambia. The Bolivian deposit produces less than 300 tons of fibre per year, while Lusaka is not mined. Both of these deposits are interesting because fibre formation has an igneous origin, unlike that of South Africa, which has a sedimentary origin.

Chrysotile, crocidolite and amosite are economically and technically the most important forms of asbestos. Of the remaining types of amphibole asbestos, anthophyllite is the best known although its deposits are small. The Finnish anthophyllite of Paakkila, which ceased production in 1970, does not occur in cross-fibre development, but rather as fibrous masses containing randomly orientated cubic blocks of fibre up to 25 mm in length. Within each block the fibres lie parallel as in a seam. The Finnish deposit has a distinct igneous origin. Anthophyllite deposits are not uncommon throughout the world, but their economic importance is small. Similar remarks apply to tremolite, but it is available in Italy, Pakistan and Korea.

Fibrous actinolite is rare, although there is a small production from Taiwan and a reef of such material has been found associated with crocidolite at Koegas, Cape Province, South Africa.

1(ii) World production

Current production of raw asbestos fibre has approached or exceeded 5 Mt per annum in recent years. The production of asbestos has increased dramatically since World War II and has doubled since 1960. In the past 15 years, about 50 Mt of asbestos has been mined, distributed and used in its various applications on a world-wide scale. The annual proportions of chrysotile and the amphiboles have remained consistent, slightly more than 93% of the total being chrysotile and the remainder amosite and crocidolite. Looking to the future, the U.S. Bureau of Mines estimates that known reserves of asbestos are 87 Mt and that total resources may be of the order of 135 Mt. World production in 1977 is shown in Table 3.2.

1(iii) Applications

Of the total asbestos produced annually, about 66% is used in asbestos–cement products. These include flat sheets or sidings, tiles and corrugated sheets for roofing, rain-water pipes and guttering, and pressure pipes capable of working at hydraulic pressures up to 12.5 atm. These products usually contain 10–15% of asbestos fibres, the main function of which is to

Table 3.2 World production of asbestos, 1977 (Source: *QAMA Bull.* Feb. 1978)

	Tonne
Chrysotiles	
U.S.S.R.	2 356 000
*Canada	1 432 000
Southern Africa, inc Rhodesia	412 000
Europe	299 000
China	199 000
U.S.A.	95 000
South America	72 000
Australia	68 000
Other countries	41 000
Amphiboles	
South Africa	
Amosite	67 000
Crocidolite	201 000

* Quebec 1 159 000

act as fibrous reinforcement in the cement. The bulk of asbestos fibre used in the products is of the chrysotile type, but up to 40% of crocidolite and amosite fibres may be incorporated to enhance reinforcement, dispersion of fibre, and the drainage properties of the asbestos–cement mix.

About 75% of crocidolite production from its sole source in South Africa is used in conjunction with chrysotile in the manufacture of asbestos-cement pressure pipes. A small proportion of amosite is frequently used in pressure pipe manufacture as a filter aid, a typical asbestos mix consisting of 60% of chrysotile, 30% of crocidolite and 10% of amosite. Asbestos–cement pipes 600–2000 mm in diameter, despite competition from plastic, concrete and steel pipes, remain one of the most widely used forms of water transportation and the demand for suitable grades for this application has always remained high. The most widely used manufacturing process is based on conventional practice for making board, in which an endless conveyor felt picks up the solid vaccum filter boxes through which excess of water drains away, leaving a thin laminate of the solids on the felt. A making roller or mandrel riding on the felt picks up the laminate, winding it on continuously until a sheet or pipe section of the desired thickness is built up. The green product is air-cured for up to 28 days, or can be cured in autoclaves in less than 24 h, provided that silica is included in the mix of raw materials to complete the reaction. There are alternative methods of manufacture based on asbestos–cement pastes from which excess of water is squeezed out in the forming of the product, and on dry asbestos–cement mixes which are spread out on a conveyor belt and wetted out by a water spray. Chrysotile–cement pastes modified with a cellulose-based extender can be extruded to enable square sections or more complicated shapes to be formed.

Asbestos–cement products have specific gravities in the range 1.6–2.0. They are suitable for exterior panelling and roofing, but are not resistant to thermal shock at temperatures in excess of 800 °C; asbestos–cement products also have a relatively high thermal conductivity. Improved thermal properties can be obtained with materials at with specific gravities of 0.8 or less. These include insulating wall or ceiling boards for interior applications, containing up to 35% of amosite in an autoclaved calcium silicate matrix or in an air-cured matrix of cement with pearlite or kieslguhr. These products are manufactured on similar machinery to that employed in the asbestos–cement industry.

Other thermal insulating materials incorporating amosite fibres include lightweight magnesia, autoclaved calcium silicate products, and materials that contain high proportions of fibre bonded with sodium silicate.

The longer and more valuable grades of chrysotile and blue asbestos are directed to the manufacture of asbestos textiles for inclusion in asbestos clothing, safety curtains, and conveyor belting. Blue asbestos has considerable resistance to chemical attack and finds special application in filtration cloths for use in acidic media, in diaphragms in electrolytic cells, and in fume-exhaust dampers.

Second only to its use in building materials, asbestos has a large range of applications in conjunction with organic resins and similar matrices. Considerable amounts of chrysotiles are incorporated in asphalt floorings and vinyl floor tiles, and various types of asbestos–asphalt mixes have been used for pavings and road surfacings.

Chrysotile asbestos is a crucial, almost irreplaceable raw material in vital automotive products. Combinations of up to 60% of chrysotile with phenolic resins go into the making of brake linings and clutch facings, and also heavy-duty friction materials for other purposes. The frictional properties of these materials are largely dependent on the physical and chemical characteristics of the fibre, although the composition may include small amounts of other minerals and metals to modify the finished product.

Packings, gaskets asbestos papers and millboards, spray insulation and decorative sprays, speciality paints, asphalt felts, acoustic insulations, and corrosion and electrically resistant reinforced plastics are further examples of the varied applications of asbestos.

It should be briefly mentioned that refined asbestos fibre, as produced at the mines for shipment elsewhere, is not necessarily in a form suitable for direct use in many of its applications. Further processing of the fibre may be required and this is usually done at the manufacturing site. This involves milling the raw fibre to increase its degree of fiberization or, in other words, to reduce and split the coarser fibres into finer ones. A variety of methods are used, ranging from edge-runners and rod mills to hammer-mills and attritors. While certain mechanical principles dictate which mills make the most suitable fiberizers of the different types of asbestos, it is also true that particular techniques produce changes in texture and surface properties of specific types of asbestos fibre, and such changes are of profound technical importance.

1(iv) Structure

While recognizing that all asbestos fibres have in common their delineation to fine filaments of immense strength, the chrysotiles on the one hand and the amphiboles on the other are opposed, but complementary to one another in many of their properties. Additionally, the different types of amphiboles possess subtle differences which do not appear to be as well understood as they might.

Projections of the structures of the two major forms of asbestos are shown in Figures 3.2 and 3.3 (Deer *et al.*, 1962, 1963; Yada 1971, 1975). The fine structure of crystalline amphiboles as observed under the microscope is repeated on a superfine scale in fibrous amphiboles and the best possible

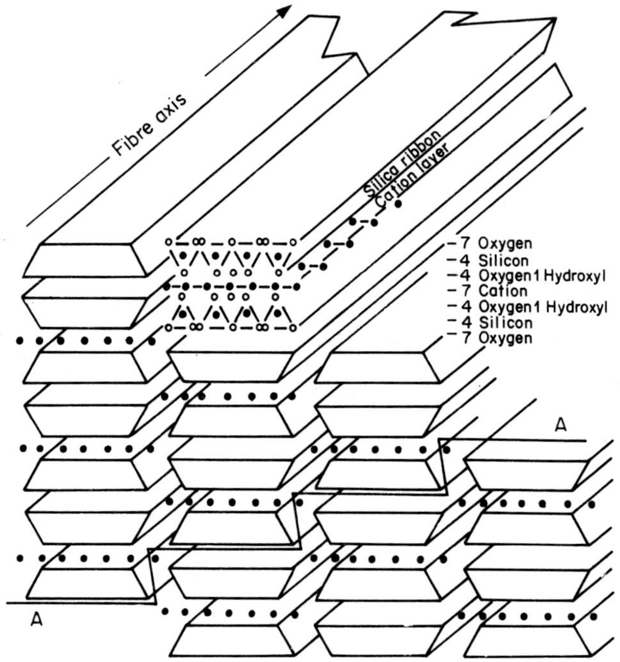

Figure 3.2 Schematic diagram of the crystal structure of an amphibole fibre, indicating the unit cell based on $X_7Si_8O_{22}(OH)_2$. The line A–A represents the edge of the preferred cleavage plane along which the fibres will split to form even smaller fibres

electron microphotographs of the cross-section of a fibre bundle has shown structures akin to the basal cleavage lines in the crystal form. Beyond this point the amphiboles, fibrous or otherwise, differ only in the chemical composition of the cation layer sandwiched between the two silicate ribbons. The external surface of the amphiboles is a silicate layer, having a low surface potential and a degree of hydrophobicity. It will be seen that, except at its edges, the cation layer is well protected.

The fundamental formula for the amphibole group is $X_7Si_8O_{22}(OH)_2$ where X represents cations. The cation sites are labelled in order from edge to edge of the ribbons, viz. $M_4M_2M_1M_3M_1M_2M_4$, and the ordered occupancy of these sites determines the type of amphibole. In the monoclinic amphiboles M_4 sites may be occupied by mono- or di-valent cations with a large radius such as Na^+, K^+, or Ca^{2+}, while the $M_2M_1M_3$ sites are filled di- or trivalent cations with a smaller radius such as Fe^{2+}, Fe^{3+}, or Mg^{2+}. In the orthohombic amphibole anthophyllite, all sites are preferentially filled with Mg^{2+}, with some possible substitution by Fe^{2+} at M_1 and M_3 sites. Transition

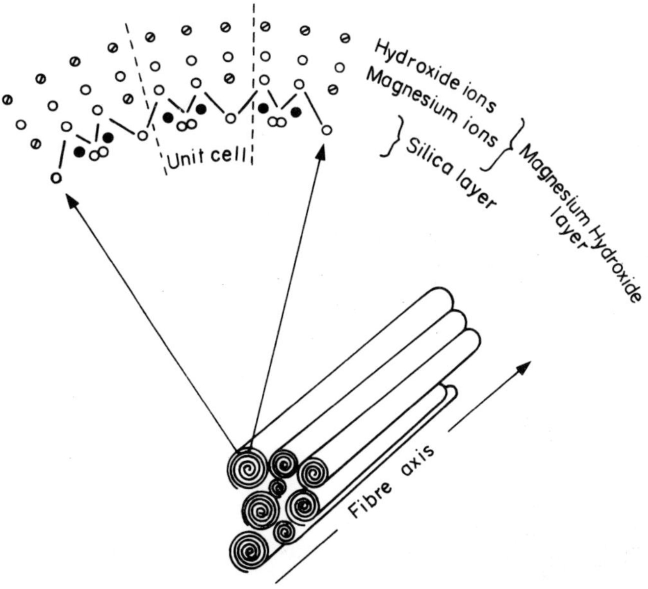

Figure 3.3 Schematic diagram of the structure of a chrysotile fibre formed of several scrolls of individual crystallites. Each scroll is formed from a closely connected double layer having magnesium hydroxide units on its external face and silica units on its inner face. The details of a small section of the scroll show the structure of the double layer and of the unit cell based on $Mg_3(Si_2O_5)(OH)_4$

to monoclinic forms occurs with $4Fe^{2+}$ which, being the larger cations, take precedence over Mg^{2+} in the M_4 and M_2 positions. The order of cation placing in the three most important amphibole fibres is as follows:

	M_4	M_2	M_1	M_3	M_1	M_2	M_4
Anthophyllite:	Mg^{2+}	Mg^{2+}	Mg^{2+}	(Mg^{2+}, Fe^{2+})	Mg^{2+}	Mg^{2+}	Mg^{2+}
Amosite:	Fe^{2+}	Fe^{2+}	(Fe^{2+}, Mg^{2+})	(Fe^{2+}, Mg^{2+})	(Fe^{2+}, Mg^{2+})	Fe^{2+}	Fe^{2+}
Crocidolite:	Na^{+}	Fe^{2+}	Fe^{2+}	(Fe^{2+}, Mg^{2+})	Fe^{2+}	Fe^{3+}	Na^{+}

These differences in cation ordering are connected with differences in the properties of the three types of asbestos. Anthophyllite, which has the slightly smaller unit cell dimensions of the three, is a silky, flexible fibre, with considerable resistance to chemical degradation. Amosite, in contrast, is a harsh, relatively brittle fibre, easily oxidized and having less resistance to chemical degradation than either anthophyllite or crocidolite.

While all chrysotiles have similar chemical compositions, their form bears little apparent relation to the platy elements of lizardite that are the main form of serpentine minerals. The basic structure is similar to that of the kaolinite group, with highly hydroxylated cation layers alternating with silicate layers. In the serpentine, the magnesium hydroxide (brucite) layer has slightly larger dimensions than the silicate (linked SiO_4) layer, resulting in a mis-match, which is resolved in chrysotile into a curvilinear structure of a scroll or coil form. The central axis of this form is the long axis of the fibre. Since the brucite layer forms the external surface of chrysotile, the fibre possesses strongly basic properties, exerts a high surface potential in water and has marked hydrophyllic tendencies. The surface layer may be leached out by water and the magnesium hydroxide content of chrysotile is rapidly removed by acids. The reactivity of the brucite layer is utilized by linkage with polar organic compounds and surfactants (Fripiat *et al.*, 1967, 1971).

1(v) Composition

Although chrysotile occurs in a number of countries, many of the published analyses (Deer, *et al.*, 1962, 1963; Badollet and McGourty, 1958; Jenkins, 1960) of the mineral refer to the source at Theford, Canada: With regard to analyses of amphibole asbestos of South Africa, many of which were gathered together by Du Toit (1945), it is known that before the 1950s the South African amosites and crocidolites were mined from partially weathered zones of strata. Not until 1961 were analyses available (Cilliers *et al.*, 1961) of examples of these kinds of asbestos that could be considered fresh and unaltered by natural weathering.

Table 3.3 contains some analyses of typical chrysotiles from various sources together with their cell contents and Table 3.4 contains similar data from analyses of amphibole fibres. The hydroxyl ions contained in the cell have been determined from thermogravimetric analysis (TGA) data.

Regardless of source, there is a marked uniformity among the compositions of chrysotiles and one has to look hard for differences which might explain variations in other chemical or physical properties. The analyses disclose the presence of two impurities, CO_2 being linked with calcium and magnesium carbonates, while FeO and Fe_2O arise from a magnetite impurity. The magnetite impurity in the Cassiar and Shabani chrysotiles is low and as a result these materials are particularly valued in certain electrical applications. Brucite, indistinguishable by chemical analysis but determined by TGA, is a common impurity in chrysotile, its concentration generally being of the order of 0.5–3% but occasionally being much higher.

Trace amounts of Ni, Cr, Co, and Mn occur in all asbestos minerals, Ni and Cr being significant in chrysotile and Mn in amosite. While some cation substitution may account for these trace elements, their presence is generally due to accessory minerals and partially as free metallic impurities derived from milling machinery (where analyses have applied to graded asbestos fibre) (Cralley *et al.*, 1967, 1968; Monkman, 1975).

Table 3.3 Compositions and cell contents for some typical chrysotiles (Source: Cape Asbestos Fibres Ltd.)

Component	Thetford, King Beaver Mine	British Columbia, Cassiar	Russia, Asbest	Rhodesia, Shabani	Swaziland, Havelock Mine
SiO_2	38.75	40.75	39.00	39.70	39.93
FeO	2.03	0.28	1.53	0.70	0.45
Fe_2O_3	1.59	0.44	0.54	0.27	0.10
Al_2O_3	3.09	3.37	4.66	3.17	3.92
CaO	0.89	0.35	2.03	1.08	1.02
MgO	39.78	41.28	38.22	40.30	40.25
MnO	0.08	0.03	0.11	0.26	0.05
Na_2O	0.10	0.07	0.07	0.04	0.09
K_2O	0.18	0.04	0.07	0.05	0.09
H_2O^+	12.22	12.86	11.37	12.17	12.36
H_2O^-	0.60	0.78	0.77	0.64	0.92
CO_2	0.48	0.44	1.83	2.13	1.04
Total	99.79	100.69	100.20	100.51	100.22
Mg	2.90	2.85	2.80	2.84	2.85
Fe^{2+}, Mn	0.06	0.01	0.06	0.03	0.02
Na, K	0.02	0.01	0.01	0.01	0.01
Al	0.02	0.11	0.13	0.12	0.14
Total	3.00	2.98	3.00	3.00	3.02
Si	1.90	1.93	1.93	1.94	1.92
Al	0.15	0.07	0.14	0.07	0.15
Total	2.05	2.00	2.07	2.01	2.07
O	5.03	4.96	5.25	5.04	5.27
OH	3.97	4.05	3.76	3.96	3.73

Typical trace element levels in chrysotiles, based on mean analytical data from 18 dust-free samples, are MnO, 0.85, Cr_2O_3 0.117, NiO 0.145, CoO 0.009, CuO 0.001, and ZnO 0.008 (Monkman, 1975).

Given a sufficiently wide range of analyses, it is not difficult to distinguish between the various kinds of amphibole asbestos. For example, among the crocidolites the most useful marker elements are Na and Mg (Table 3.4). The mineral from Koegas contains about 6.20% of Na_2O and 1% of MgO, while crocidolite from the Kuruman Hills, about 150 miles north of Koegas, contains 5–5.5% of Na_2O and up to 3% of MgO. Australian crocidolite has a lower total iron content than the Cape Province specimens, although in other respects it is similar to the Pomfret, Cape Province variety. The Malipsdrift and Cochabamba crocidolites are problem minerals, as their analyses immediately indicate. Other methods of investigation by differential thermal analysis and by microscopic examination suggest strongly that these are mixed amphiboles (Hodgson, 1965a, b). Apart from containing a fibrous silica intergrowth (hence the high SiO_2 content by analysis), the Malipsdrift crocidolite is in fact a mixture of crocidolite and amosite. The Bolivian material appears to be a magnesioriebeckite, having close similarities to a crocidolite from Lusaka, Zambia (Drysdall and Newton, 1960).

Amosites have a high iron(II) content, approaching 40% FeO, but within their sole environment there is a variation to types with lower FeO contents, this being associated with diminishing harshness. A variety of amosite known as montasite from Pietersburg contains 31% of FeO and is described as a silky fibre. (Du Toit, 1945). Anthophyllites have high magnesium

Table 3.4 Composition and cell contents of some typical amphibole asbestos varieties (Source: Cape Asbestos Fibres Ltd.)

Component	Crocidolite, Koegas, Cape Province	Crocidolite, Kuruman, Cape Province	Crocidolite, Pomfret, Cape Province	Crocidolite Malipsdrift, Transvaal	Crocidolite, Cochabamba, Bolivia	Crocidolite, Hammers-ley Range, W. Australia
SiO_2	50.90	50.70	52.00	59.41	55.65	52.85
FeO	20.50	17.50	17.65	15.11	3.84	14.94
Fe_2O_3	16.85	18.30	16.05	14.03	13.01	18.55
Al_2O_3	Nil	0.70	Nil	Nil	4.00	0.18
CaO	1.45	1.30	1.20	0.49	1.45	1.07
MgO	1.06	3.05	4.28	3.53	13.09	4.64
MnO	0.05	0.06	Trace	Trace	Trace	Trace
Na_2O	6.20	5.30	6.21	4.63	6.91	5.97
K_2O	0.20	tr.	0.06	0.28	0.39	0.05
H_2O^+	2.37	2.53	2.43	2.07	1.78	2.77
H_2O^-	0.22	0.29	0.26	0.14	Trace	0.22
CO_2	0.20	0.45	0.09	0.09	Trace	0.23
Total	100.0	100.18	100.23	99.78	100.12	101.47
Na, K	1.92	1.59	1.86			1.91
Ca, Mn	0.21	0.13	0.12			0.13
Total $2M_4$	2.13	1.72	1.98			2.04
Fe^{3+}	1.98	2.13	1.85			2.00
Total $2M_2$	1.98	2.13	1.85			2.00
Fe^{2+}	2.68	2.27	2.26			1.90
Mg	0.25	0.69	0.93			1.06
Total $2M_1+M_3$	2.93	2.96	3.19			2.96
Si, Al	7.97	8.00	7.96			7.98
22(O)	22.05	22.03	22.06			21.82
2(OH)	1.95	1.97	1.94			2.18

(Table 3.4 continued)

Component	Amosite, Penge, Transvaal	Amosite, Weltevreden, Transvaal	Anthophyllite, Paakkila, Finland	Tremolite, Pakistan	Prieskaite, Koegas, Cape Province
SiO_2	49.70	51.30	57.20	55.10	53.80
FeO	39.70	35.50	10.12	2.00	25.30
Fe_2O_3	0.03	0.90	0.13	0.32	1.90
Al_2O_3	0.40	Nil	—	1.14	1.20
CaO	1.04	0.95	1.02	11.45	10.20
MgO	6.44	6.90	29.21	25.65	4.30
MnO	0.22	1.76	—	0.10	0.40
Na_2O	0.09	0.05	—	0.14	0.10
K_2O	0.63	0.51	—	0.29	0.40
H_2O^+	1.83	2.31	2.18	3.52	2.60
H_2O^-	0.09	0.05	0.28	0.16	Nil
CO_2	0.09	0.25	—	0.06	0.20
Total	100.26	100.48	100.14	99.93	100.40
Fe^{2+}	1.78	1.28	—	0.23	—
Ca, Mn	0.23	0.33	0.16	1.71	1.71
Mg	—	—	2.08	0.31	—
Na, K	0.07	0.12	—	0.11	0.13
Total $2M_4$	2.08	1.73	2.24	2.36	1.84
$Fe^{2+}(Fe^{3+}, Al)$	1.39	1.35	—	—	1.72
Mg	0.61	0.65	2.00	2.00	—
Total $2M_2$	2.00	2.00	2.00	2.00	1.72
Fe^{2+}	2.09	2.03	1.15	—	2.00
Mg	0.91	0.97	1.85	3.00	1.00
Total $2M_1 + M_3$	3.00	3.00	3.00	3.00	3.00
SI, Al, Fe^{3+}	7.99	8.17	7.79	8.00	8.26
22(O)	21.98	21.96	22.00	22.05	22.46
2(OH)	2.02	2.04	2.00	1.95	1.54
H·	—	—	0.68	—	0.45

contents (30–35% MgO) associated with 5–10% of FeO, and tremolites a similar MgO content, with 10–15% of CaO. The 'prieskaite' discovered by Cilliers in a horizon in the Koegas mine (Cilliers, 1961) may well be described as a ferro-actinolite, having the Ca-Mg-Fe make-up necessary for actinolite, but with excess of iron. It is one of the few known actinolites that is truly fibrous.

Chemical analysis will give some indication of the degree of natural weathering to which an amphibole asbestos has been exposed, a question which particularly applies to amosite and crocidolite. Amosite should contain iron(II) only, and ground surface weathering or sub-surface leaching shows up in the increasing amounts of iron(III) on analysis. Fresh crocidolite has a distinct excess of FeO over Fe_2O_3 in analysis, and otherwise, when the FeO content is equal to or less than that of Fe_2O_3, the crocidolite cannot be considered to be fresh.

The hydroxyl components of the amphiboles have special significance in the deduction of the cell contents. Thermogravimetric and dehydration studies show that only part of the total water liberated at 105 °C by the amphiboles is correctly defined as chemically combined water which can be allotted to the two hydroxylsites.

The cell contents given in Table 3.4 have been compiled from these data. The values are shown in the table in a simplified form, in which cations have been allocated to their positions according to literal interpretation of the substitution rules. In fact, there must be some overlapping, and in detailed breakdown small amounts of appropriate ions must be moved to fill the inner cation sites. Excesses of large ions in M_4 go into the A sites. Anthophyllites and actinolites usually contain more combined water than can be allocated to the two hydroxyl sites, a problem which has been the subject of several mineralogical papers (Zussman, 1955; Francis and Hey, 1956; Hutton, 1956). The current view is that extra hydrogen as hydroxyl may replace some of the oxygen in the Si_4O_{11} bands. In the cell analysis for the anthophyllite and the actinolite, extra hydrogen is shown as H·.

1(vi) Chemical reactions

All forms of asbestos break down progressively to simpler components, such as pyroxenes and silica, when heated to temperatures in the range 600–1000 °C. Reactions of asbestos, as such with other materials at high temperature therefore do not exist. In any reactions involving high-melting-point metals or fused salts, such as the sodium carbonate fusion in the preliminary step of chemical analysis, it can be assumed that primary breakdown of the mineral at the temperature of fusion will be followed by reaction of the products with the fusion material. At lower temperatures, asbestos fibres appear to be resistant to attack by molten materials. Attempts have been made to incorporate these fibres in low-melting-point metals such as tin, zinc, and lead, and there is no apparent reaction with the metal. It is suspected, however, that molten sodium and potassium will attack asbestos fibres through reactions involving both dehydration and extraction of silica.

In connection with the above, asbestos fibres as such do not have melting points: it is the products of decomposition which melt. In general, if the latter contain iron pyroxenes, fusion occurs between 1000 and 1100 °C, which is typical of the high-iron asbestos minerals amosite and crocidolite. The high-magnesium varieties, such as chrysotile and anthophyllite, decompose to give magnesium-pyroxenes in the products which melt at about 1450 °C.

The reactivity of the asbestos minerals towards acids and alkalis is fairly well known. Strong acids rapidly decompose chrysotile with removal of all of its MgO and H_2O content, amounting to about 58% by weight, but amphibole fibres show various degrees of resistance to reaction with acids. All varieties of asbestos will resist prolonged attach by strong alkalis, such as 5 M sodium hydroxide solution, at least up to 100 °C.

Figure 3.4 shows some typical acid-resistance curves for asbestos fibres refluxed in 4 M hydrochloric acid for periods up to 8 h. These differentiate

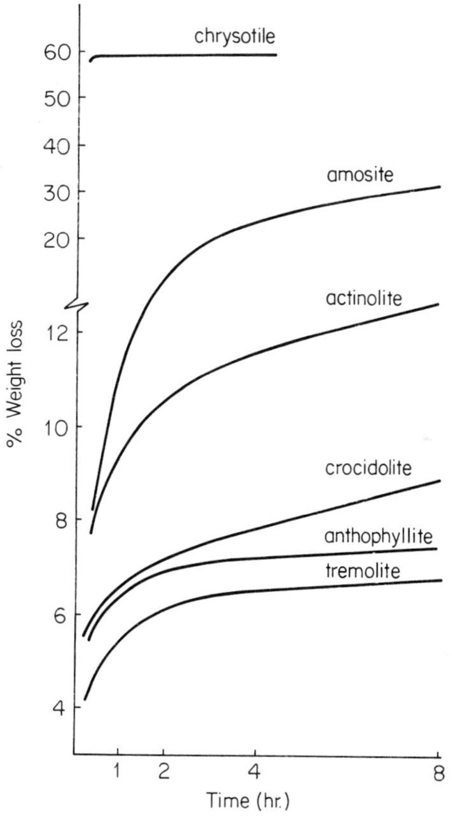

Figure 3.4 Typical acid-resistance curves for asbestos fibres refluxed in 4 M hydrochloric acid

clearly between chrysotile and the amphibole fibres, although the range of reactivity of the latter fibres to hydrochloric acid is wide. Moreover, different types and strengths of acid produced varied results. Hydrofluoric acid will completely decompose all asbestos fibres; both sulphuric and niric acid have much the same effects on amosite and crocidolite regardless of concentration above 0.5 and 1 M, respectively, but orthophosphoric and hydrochloric acids at 4–5 M decompose these two fibres to a far greater extent than the 1 M acids. Chrysotile possesses such a low resistance to acids that it may be decomposed by concentrated acetic acid.

Referring to Figure 3.4, the early stages of acid attack on the amphibole fibres indicate a rapid reaction, due partly to the removal of impurities such as alkali and iron carbonates, but mainly to the decomposition of extremely finely divided fibres and the stripping of imperfect surfaces on coarser fibres. In the later stages of the reactions, the rate of weight loss becomes steady, and is then considered to be indicative of the true acid resistance of the fibre. It should be added that the rate of decomposition is also a direct function of the surface area of the asbestos fibres, and all the curves of Figure 3.4 were produced by fibres of similar surface area to make their comparison valid.

There are obvious structural differences between the two major groups of asbestos fibres, which account for the rapid decomposition of chrysotile by acids and the relatively slow breakdown of the amphiboles. Although the subject has been under experimental review for many years, recent investigations have thrown light on the reaction mechanisms involved (Atkinson *et al.*, 1971; Monkman, 1971; Allen and Smith, 1975). When chrysotile reacts with acids, its outer (MgOH) layers are stripped away and decomposition proceeds by a rate-controlling diffusion process through a layer of siliceous residue. The final residue retains the fibrillar morphology, although without its characteristic tubular structure. Reaction rates are proportional to acid concentration and temperature, and appear to be the same for both monobasic and polybasic acids.

Two possible mechanisms may account for the decomposition of amphiboles in strong acids. Firstly, cations may be continuously extracted from the nucleus, leaving a silica skeleton that has the elements of the original structure. Secondly, the whole nucleus may break down to cations and hydrated silica, which may coat unattacked fibrils and thereby make it more difficult for the acid to reach fresh surfaces. The information in Table 3.5 indicates that all constituent cations are continuously extracted from amosite, crocidolite, and prieskaite, with the exception that little Na^{+} is removed from crocidolite. Some of the Ca^{2+} ion is removed from tremolite in the early stages only, while extraction of Mg^{2+} occurs in the later stages. In anthophyllite, there is little change in the cations removed between $\frac{1}{2}$ and 8 h

Close comparison of the results suggests that certain cations are selectively

Table 3.5 Cations extracted from amphibole fibres by hydrochloric acid after $\frac{1}{2}$-h and 8-h refluxes

Fibre	Cation	Number of cations per unit cell extracted by reflux with 4HCl	
		After $\frac{1}{2}$ h	After 8 h
Anthophyllite	Mg^{2+}	0.46	0.46
	Fe^{2+}	0.15	0.18
Tremolite	Ca^{2+}	0.38	0.38
	Mg^{2+}	0.02	0.41
Prieskaite (actinolite)	Ca^{2+}	0.25	0.40
	Fe^{2+}	0.55	0.92
	Mg^{2+}	0.20	0.32
Crocidolite	Na^{+}	Trace	Trace
	Fe^{3+}	0.32	0.55
	Fe^{2+}	0.33	0.58
	Mg^{2+}	0.01	0.10
Amosite	Fe^{2+}	1.48	2.81
	Mg^{2+}	0.36	1.00

extracted at a greater rate than others, and observation which is not entirely compatible with the breakdown of the whole nucleus.

The rate of acid decomposition of the amphiboles is not entirely reconcilable with the suggested cleavage model based on (110) which exposes the minimum number of cations. Indeed, such cleavage is not compatible with electron microscopical observations, which indicate a (100) or (010) cleavage. (Whittaker, after Hodgson, 1977). Such cleavage will expose all cations to equal attack, although this does not explain the anomaly of low extraction of Na^{+} and Mg^{2+} from crocidolite.

Suspensions of chrysotile in water possess alkalinity in the pH range 10–11. Part of this alkalinity can, of course, be attributed to the solution of brucite impurities, but the continuous extraction of the chrysotile in hot water shows that the fibre itself is decomposed (Holt and Clark, 1960). The extracts contain orthosilicic acid and magnesium, the latter accumulating with time (approximately 1 mg per 100 ml of extract after 5 days). The breakdown of the chrysotile appears to follow a course in which Mg^{2+} ions are leached out, leaving fragments of the Si–O lattice. These fragments are probably of a colloidal form, since they can be hydrolysed to orthosilicic acid by heating the extract after the fibre has been removed.

The activity of asbestos fibres in neutral or buffered solutions at 37 °C has received considerable attention in relation to biochemical investigations. In the continuous leaching of chrysotile there is an initial high release of Mg^{2+}, which subsequently reduces to a steady rate after several days. Correspondingly, there is an initial minimal release of silica, increasing to a much higher steady rate, probably in the form of fragmentary SiO^- units (Chowdhury, 1973).

The leaching of amphibole fibres, as might be expected releases cations only in micromole amounts and at a slowly decreasing rate in continuous 20-day experiments (Chowdhury, 1973). For both amosite and crocidolite the release of Fe is barely detectable. The release of Mg^{2+} is low at commencement and gradually decreases, but in crocidolite the initial release of Na^+ is relatively high, again before decreasing to a very low steady rate. The release of Na^+ can be attributed to the rapid decomposition of imperfect surfaces on exposed fibrils and suggests that cleavage of the cellular units must pass near to or at the edges of cation ribbons.

In connection with its inclusion in asbestos–cement and in autoclaved calcium silicate products, Ball and Taylor (1963a) have investigated the hydrothermal reactions of chrysotile with materials such as magnesium oxide, silica, lime, alumina, iron(III) oxide, tricalcium silicate, and β-dicalcium silicate. All of these materials occur in the chemical and phase compositions of cement. The reactions between chrysotile and MgO, SiO_2, Al_2O_3 or Fe_2O_3 gave products such as forsterite (Mg_2SiO_4), talc ($Mg_3Si_4O_{12}H_2$), and brucite, all with a definite crystallographic orientation relative to the chrysotile. Reactions of this type belong to the category termed topotactic, and occur where there is a three-dimensional similarity between the crystal structures of the starting material and the product. Reactions between chrysotile and Al_2O_3 or Fe_2O_3 gave, in addition, a mica-like product with a high thermal stability. In other reactions, products containing calcium, such as $CaMgSiO_4$ (monticellite), were unorientated.

The hydrothermal conditions used in these investigations require temperatures of 300–570 °C, pressures between 80 and 330 kg/cm^2 and times up to 84 days. These conditions considerably exceed the autoclaving temperatures and pressures used in industrial practice, where 200 °C, 10–12 kg/cm^2 and times of 24 h are considered to be the upper practical and economic limits. Nevertheless, the reactions may occur at these lower temperatures and pressures, although to a much smaller extent. They therefore have a bearing on the type of bond that may be formed between chrysotile and cement and associated compounds. There is no available evidence that amphibole fibres take part in similar reactions, at least in normal manufacturing processes, but at the same time investigations at higher temperatures and pressures have not been attempted. It is believed that the amphiboles, having an inherently better chemical stability than chrysotile, will react to only a very limited extent.

1(vii) Synthesis of the asbestos minerals

Perhaps the earlier work in this subject during the 1920s and 1930s was of academic interest only, but more recently the purpose of synthesis has been sharpened by research in the allied field of metal whiskers. This research has demonstrated that metal-whisker crystals of a strength similar to that predicted by theory can be made. It is supposed that their phenomenal strength arises from their perfection in surface and structural details, and similar materials are being sought among the fibrous silicates in an attempt to improve on the naturally occurring product.

Chrysotile has been synthesized hydrothermally from silica gel and a magnesium salt solution (Ipatieff and Mouromstseff, 1927), from silica gel and magnesium hydroxide solution (Syromyatrukov, 1935), from sodium silicate and magnesium carbonate (Wells, 1929), and as a result of fundamental studies on the $MgO–SiO_2–H_2O$ system (Jander and Wuhrer, 1938), to quote a few earlier attempts. The fibres produced were generally in the form of matted fibrils, about 1 μm in length and recognizable only under the electron microscope. More recently, following further studies on the $MgO–SiO_2–H_2O$ system, Yang (1961) has synthesized chrysotile, mainly of the clino- variety, in bundles of 100-μm lengths, with individual fibres 15–20 μm in length. The synthesis required a mixture of MgO and a bulk silicic acid in a molar ratio of 1.5, and hydrothermal conditions of 300–350 °C and 85–130 kg/cm^2 maintained over 5–10 days. The pH of the system was maintained at 10.3–10.7 by means of sodium carbonate and 1–2% of F^- as ammonium fluoride was added as a mineralizer. The synthetic fibres had the characteristics of the naturally occuring mineral, despite their diminutive size.

The minerals of the amphibole group have been synthesized by fusion of suitable batch compositions in which the OH ion was substituted by fluorine (Bowen and Scheirer, 1935; Comeforo and Kohn, 1954; Saito, 1957). Such fluor-amphiboles, crystallized from melts, form as crystals, needles, and fibres, all of a microscopic size. Apart from the necessity of substituting F for OH, these syntheses have also been limited to the production of soda-amphiboles of the riebeckite–crocidolite and soda–tremolite varieties. Saito (1957) found that by crystallizing a mass of fluor-amphibole from a suitable alkaline melt and then subjecting it to hydrothermal treatment in concentrated sodium hydroxide or sodium carbonate solution at 350 °C for 8–16 h, he could produce fine needles of a fluor-riebeckite up to 15 mm long and 15 μm thick. The kinetics of amphibole formation in both hydrothermal and solid-state reactions indicate a multi-stage process which nears completion at about 900 °C, with subsequent decomposition to pyroxenes and glass at about 1150 °C. The most promising yields are obtained in fluor-riebeckite, fluor-richterite and fluor-arfvedsonite systems, but in non-soda systems fluor-amphiboles are formed in small amounts only. Synthetic homologues containing iron are rarely mentioned and successful synthesis rely mainly on Mg or an unusual combination of Ni and Cr as a substitute for Fe. (Grigoreva *et al.*, 1971; Makarova *et al.*, 1971; Speakman, 1971).

(2) PHYSICAL PROPERTIES OF ASBESTOS FIBRES

2(i) Tensile strength

The physical properties of different varieties of asbestos are summarized in Table 3.6.

This is the most important and most commonly quoted physical property of asbestos fibres. The earliest published data (Badollet, 1951) gave relatively low values for tensile strength, but later work produced much higher results, particularly for crocidolite and chrysotile, the measured maxima for which were 6.03 and 5.68 GN/m^2, respectively (Zukowski and Gaze, 1959). These investigations were important for showing that the observed tensile strengths were dependent on the length of specimens tested, the shortest lengths of fibre in the experimental range 35–2 mm giving the highest strengths. By taking high-speed cine photographs of fibres in the act of breaking, it was possible to see that failure began at a surface flaw and proceeded by the rupture of relatively weak interfibrillar bonds, which in turn exposed fresh surfaces which failed at a different place.

Recent investigations of tensile strength have been based on comparative measurements made on specimens of a standardized minimum length of 3 or 4 mm, compatible with the handlability of the fibres. Also, the technique of measurement has been advanced through the use of the Marsh micro-tensile testing machine (Techne Ltd., Cambridge) which permits the testing of fibres with average diameters in the range 10–20 μm.

The nature of fractures of chrysotile fibres at break point are characterized at one extreme by a number of fibrils protruding from the site of the major break and at the other by a clean transverse break (Bryans and Lincoln, 1971, 1975). Measurements of shear strength and shear modulus of the interfibrillar region have shown that increases in shear strength make the fibre more flaw sensitive and weaker, whereas increases in shear modulus tend to strengthen a fibre. Brittle fibres generally have a higher interfibrillar shear strength than silky fibres. While similar work has not been attempted on amphibole fibres, it is clear that this fracture model extends through to less silky crocidolite fibres and the extremely brittle amosite fibres.

Under the influence of heat in a temperature regime up to 800 °C, the tensile strength of the major types of asbestos fibres increases slightly before diminishing rapidly as structural and decompositional changes take place (Figure 3.5; Burman, 1967). The increase in strength occurs in the region 150–250 °C for amosite and crocidolite and is more marked for chrysotile at 550 °C. Bryans and Lincoln (1975) suggest that these increases in strength in chrysotile are due to increases in shear modulus, probably arising from modifications to the interfibrillar component in the fibre.

The decreases in strength at higher temperatures arise from initial structural changes, such as the formation of oxyamphiboles around 400–500 °C

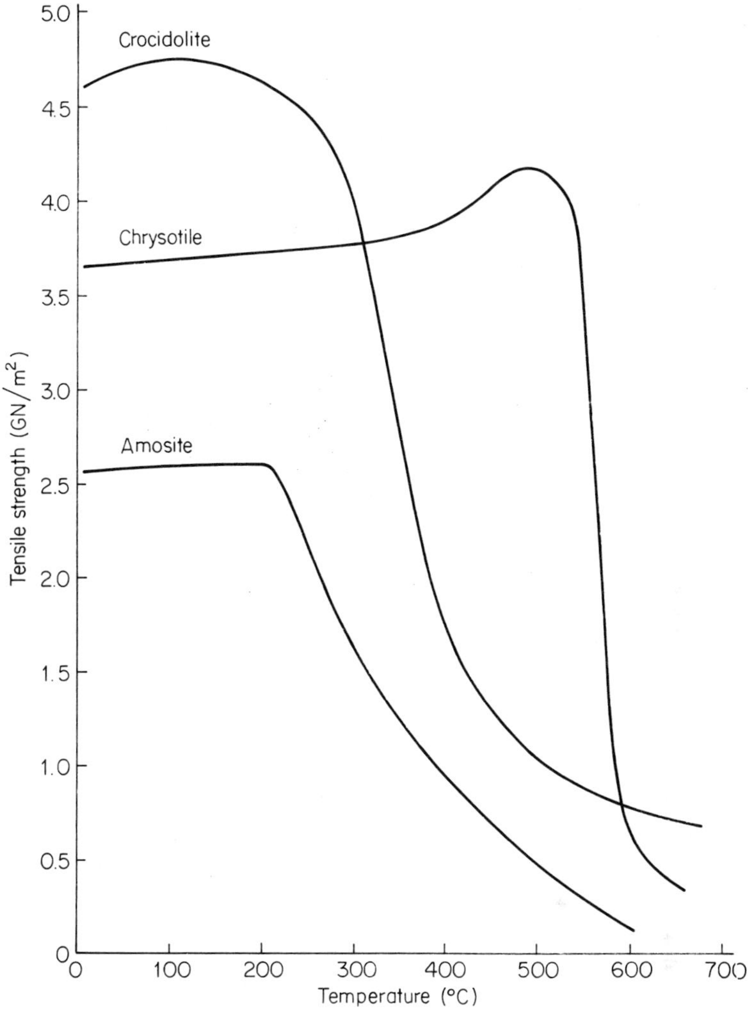

Figure 3.5 Tensile strength changes in heated asbestos fibres

and loss of some structural water in chrysotiles. Above these temperatures, dehydroxylation and structural rearrangements take over and the eventual high-temperature strength of asbestos fibres is only 10–20% of their original strength.

High values above 10 GN/m^2 may be predicted for the tensile strength of silicates by considering the strengths of the stable silicon–oxygen structures which characterize these minerals (Orowan, 1948–49). Strengths approaching 60% of these values have been obtained for very short crocidolite fibres and the general shortfall in terms of average values of tensile strengths requires some explanation. Several factors appear to influence fibre tensile strength in the amphibole series: (a) planar defects which probably contribute to fibrosity; (b) cation composition, which contributes to intrinsic strength; and (c) macroflaws such as physical discontinuities on the fibre axis.

Electron microscope studies of fibrous amphiboles indicate two types of planar defects (Chisholm, 1973), these being (100) twin boundaries or stacking faults and (010) Wadsley defects. The crystallographic features of the latter may be envisaged as structured defects where neighbouring shear planes are separated by either one or three silicate chains instead of two, leading to either talc-like or pyroxene-like regions being present. These structures have also been noted as pyroxene I-beam, amphibole I-beam and triple-chain I-beam (Veblen *et al.*, 1977). Chisholm (1975) believes that the high tensile strength of fibrous amphiboles can be accounted for by the presence of planar defects which inhibit brittle failure. Cracks propagate only until they reach a defect and then proceed by interfibrillar slip, and this is confirmed by observations that fractured ends of the amphibole fibres exhibit a stepped outline.

The several different types of amphibole fibres differ markedly in their mean tensile strengths, and the evidence suggests that these mean figures are representative properties of these fibres. To some extent the paragenesis of asbestos fibres will affect the defects in structure which may determine the strength of the fibre but, in addition, the nature of cation ordering in the amphiboles is important. It is suggested that under strain cation–oxygen bonds will distort and break before silicate chains are ruptured and that the cation composition of the amphiboles directly influences their tensile strength (Prentice, 1977). The most likely plane of rupture is across $M_2M_3M_2$ where there are 18 cation–oxygen bonds, compared with 28 in the $M_4M_1M_1M_4$ plane. The relative strength of the amphibole fibres then hinges on cation preferences in the $M_2M_3M_2$ plane:

	Cation–oxygen bonds	Tensile strength (GN/m^2)
Crocidolite	$12Fe^{3+}$–O	
	$6(MgFe^{2+})$–O	3.43
Amosite	$12(MgFe^{2+})$–O	
	$6Fe^{2+}$–O	2.45
Anthophyllite	$12Mg^{2+}$–O	
	$6Fe^{2+}$–O	1.47

The order of strength of the cation bonding with oxygen is $Fe^{3+} > Fe^{2+} > Mg^{2+}$. High iron contents differentiate the strengths of crocidolite and amosite from that of anthophyllite (and tremolite); and the high Fe^{3+} content of crocidolite together with a high Fe : Mg ratio distinguishes its high tensile strength in comparison with that of amosite.

Macroflaws or regular discontinuities in the genesis of fibre growth probably place physical limitations on realizable tensile strength, as indicated by the relation between fibre length and measured strength. Extrapolation of the observed data would indicate that near-ultimate strengths are possible with fibres a few microns long, if it were practical to test them.

2(ii) Surface area

While tensile strength is the most important intrinsic physical property of asbestos fibres, their surface area is the most important parameter in all their industrial applications. In industrial parlance surface area becomes degree of openness or degree of fiberization. Each process involving asbestos fibre requires a degree of fiberization which, within limits, is critical for that purpose. For example, chrysotile fibres prepared for inclusion in moulded brake linings will have a different degree of fiberization from those prepared for carding in the first stage of textile manufacture. The correct degree of fiberization is achieved by milling the raw asbestos in a hammer mill, pin disc mill, rod mill, or similar types of machine. In multi-stage manufacturing processes, account must also be taken of the further fiberization which may occur through mechanical handling or wet-mixing stages.

The ability of asbestos fibres to open or split into finer fibres through suitable treatment is one of their most valuable properties. Ideally, the best fiberizing technique should open a fibre without damaging it by disintegration into shorter lengths. In practice, this is hardly possible, although laboratory-scale ultrasonic treatment of asbestos fibres shows that ideal fiberization can be achieved.

The term degree of fiberization can be explained by saying that if a cross-sectional area a splits down to n fibres having an average cross-sectional area a/n, then its degree of fiberization has been increased n-fold. If the mean radius of cross-section of the n fibres is r, then it can be shown simply that their absolute surface area is proportional to $1/r$. The degree of fiberization can therefore be expressed directly in terms of the measured surface area of the fibres. Theoretically, of course, the ultimate limit to this value is enormous. The absolute surface area of typical asbestos fibres used in manufacturing processes lies between 30 000 and 90 000 cm^2/g, as measured by the nitrogen absorption method at room temperature. Measurements on the same fibres after out-gassing at higher temperatures (100 and 400 °C) may raise the upper limit of surface area to 140 000 cm^2/g (Addison, 1964). Raw amphibole asbestos as shipped from refineries at the mines may have an absolute surface area of 6000–20 000 cm^2/g and for similar unprocessed chrysotile fibres the value may be up to 30 000 cm^2/g. It will be appreciated from these figures that while the mean surface area of commercial asbestos fibres is several thousand times greater than that of the massive ore, the amount of milling that the fibres receive in manufacturing processes will increase the degree of fiberization by a factor of only perhaps five or six at the most.

In practice, the nitrogen absorption method of measurement of surface area cannot be used and the usual technique on the industrial scale involves an air-permeability measurement based on the method of Rigden (1943). This technique is widely used in the cement and clay-mineral industries, where it gives reasonably accurate results, but it gives low results for asbestos fibres—about three to four times lower than the absolute figures. The theoretical background to the air-permeability method is only true for small spherical particles which can be packed into a volume of low porosity. Asbestos fibres satisfy neither of these conditions and, in fact, an empirical factor must be introduced into calculations to account for their rod-like shape. Nevertheless, the method is adequate for rapid practical purposes. It is usual to state such measurements on asbestos fibres as 'apparent surface area'.

2(iii) Other physical properties

Specific gravities have been determined by weighing in water prepared blocks of fibre approximately 1 cm cube taken from ore specimens. Before weighing the specimens, it is essential that they are out-gassed under water by placing them under high vacuum for several hours. Occluded gases may account for up to 5% of the pore space in fresh specimens and air in weathered specimens may account for up to 20% of the pore space.

The only available electrical data on asbestos concern volume resistivity, which is low for chrysotile and very variable for amphibole fibres (Badollet, 1960).

The magnetic susceptibility of asbestos fibres is generally low (Table 3.6) but there is considerable distinction between the high iron amphiboles and anthophyllite and between the amphiboles and chrysotile, Measured susceptibilities parallel to and at right-angles to the fibre axis show some differences, in the descending order amosite, crocidolite, anthophyllite, chrysotile (National Physical Laboratory, 1972). However, these slight differences are not reflected in the strong reaction of freely suspended asbestos fibres to magnetic fields (Timbrell, 1972, 1975).

Table 3.6 Physical properties of the varieties of asbestos

Property	Source*	Chrysotile	Anthophyllite	Amosite	Crocidolite	Tremolite	Actinolite
Tensile strength (kg/cm^2)	1	31×10^3	24×10^3	25×10^3	35×10^3	$<5 \times 10^3$	$<5 \times 10^3$
Young's modulus (kg/cm^2)	1	1.65×10^6	1.58×10^6	1.65×10^6	1.9×10^6	—	—
Flexibility	—	Good	Fair to brittle	Fair	Good	Brittle	Fair to brittle
Specific gravity	1	2.55	2.85–3.1	3.43	3.37	2.9–3.2	3.0–3.2
Hardness (mohs)	2	2.5–4.0	5.5–6.0	5.5–6.0	4	5.5	6
Specific heat (kcal/g/°C)	2	0.266	0.210	0.193	0.201	0.212	0.217
Volume resistivity (MΩ cm)	2	0.003–0.15	2.5–7.5	Up to 500	0.2–0.5	—	—
Magentic susceptibility mean χ g at $H = 10\ kOe$	3	5.3×10^{-6}	14.3×10^{-6}	78.7×10^{-6}	60.9×10^{-6}	—	—

* 1, Cape Asbestos Fibres Ltd.
2, Badollet (1960).
3, Reproduced with permission of the Director of the National Physical Laboratory.

Under the influence of strong magnetic fields, fibres in aqueous suspension align parallel (P-type) or normal (N-type) to the field. Anthophyllite and crocidolite have P-type fibres only, while amosite has both P and N-type fibres. Tremolite may have only P- or only N-type fibres and chrysotile does not have any preferred orientation. While the method is not absolute, it does provide a means of distinguishing the three common forms of asbestos, chrysotile, crocidolite, and amosite.

2(iv) Optical properties

The simplest and most widely used technique for identifying asbestos fibres relies on their optical properties, in particular their refractive indices (R.I.). The optical data in Tables 3.7 and 3.8 refers to fresh specimens of fibres in the collection of Cape Asbestos Fibres Ltd. Refractive indices correspond to changes in cation composition and as a general rule can be related directly to Fe : Mg ratios. Thus, the lower R.I. for chrysotile and anthophyllite correspond to high Mg contents. South African crocidolite has a higher Fe : Mg ratio than amosite and a correspondingly higher R.I., but Bolivian crocidolite is distinguished by both a lower R.I. and a lower Fe : Mg ratio than its South African counterpart.

(3) THERMAL DECOMPOSITION OF ASBESTOS

It could be said that up to the early 1950s, asbestos fibre was always regarded as a thermally resistant, fire-proof material. Except for statements that the fusion point of asbestos lay above 1000 °C and that the products of decomposition at high temperature were known, no precise limitations were placed on the supposed high degree of heat resistance of asbestos fibres. In

Table 3.7 General optical properties of the varieties of asbestos

Asbestos	Colour	Pleochroism	Birefringence	Orientation	Extinction
Chrysotile	White	Nil	Moderate, 1st order	Length slow	Parallel
Anthophyllite	White	Nil	Moderate, low 2nd order	Length slow	Parallel
Amosite	Ash grey	Nil	Strong, 2nd order	Length slow	Parallel
Crocidolite	Dark blue	Grey blue to dark blue	Weak (masked)	Length fast	Parallel
Tremolite	White	Nil	Moderate, low 2nd order	Length slow	Oblique (max. 20°)
Actinolite	Dark green	Pale green to deep green	Moderate, low 2nd order	Length slow	Parallel

Table 3.8 Refractive indices (20 °C) of the varieties of asbestos

Asbestos	Source	n_α	n_γ
Chrysotile	Canada	1.537–1.554	1.554–1.557
	Russia	1.543–1.550	1.551–1.555
	S. Africa	1.549–1.553	1.553–1.557
	Italy	1.541–1.553	1.547–1.554
Anthophyllite	General	1.578–1.652	1.591–1.676
Amosite	S. Africa (E. Transvaal)	1.670–1.675	1.683–1.694
Crocidolite	S. Africa:		
	N. W. Cape	1.682–1.696	1.686–1.700
	Transvaal	1.681–1.683	1.683–1.687
	Australia	1.680–1.687	1.684–1.691
	Bolivia	1.659	1.673
Tremolite	General	1.599–1.628	1.591–1.676
Actinolite	General	1.600–1.628	1.625–1.655

fact, all varieties of asbestos fibres break down progressively through a series of internal reactions which may begin at temperatures as low as 200 °C. This does not mean that their properties as heat insulators are diminished. Whether compounded partially or wholly into heat-insulating materials, asbestos fibres possess ablative properties in that the products of decomposition at a hot surface from a poorly conductive layer which provides protection to the succeeding layers of insulation. The insulating properties of asbestos depend much more on its fibrous nature and on the poor thermal conductivity of individual fibres than on limitations arising from its thermal decomposition.

The thermal decomposition of asbestos has been investigated in depth by X-ray diffraction, gas absorptiometric and thermal analysis methods in both oxidizing and inert atmospheres. Asbestos lends itself particularly to investigation by thermal analysis and the following paragraphs give typical data and their interpretation.

3(i) Amphibole asbestos

Typical complete thermal analysis curves for crocidolite are shown in detail in Figures 3.6 and 3.7. The differential thermal analysis (DTA) curve of crocidolite in an inert atmosphere has a dehydroxylation endotherm at about 600 °C, which coincides with distinct steps on curves of weight loss and dynamic dehydration. The loss of water from dehydroxylation amounts to about 1.9% in fresh crocidolite and agrees with the theoretical amount. The endothermic–exothermic pattern about 800 °C on the DTA curve corresponds to breakdown of the amphibole anhydride to simpler molecules,

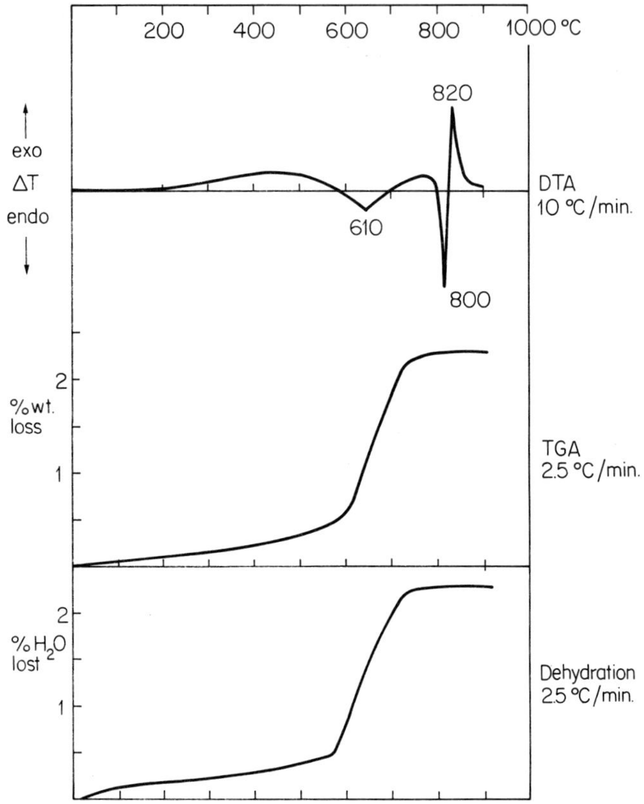

Figure 3.6 Thermal analyses of crocidolite in an inert atmosphere (argon)

in this example a NaFe-pyroxene (probably acmite), magnetite, silica, and a glass phase. These products fuse and melt between 850 and 1000 °C.

Decomposition of crocidolite in an oxidizing atmosphere proceeds somewhat differently. An oxidation process occurs at 400 °C and, although the dehydration curves indicate the loss of water, curves for weight loss do not. Condensation of OH^- groups does not take place, but an oxyamphibole is formed with protons—the hydrogen ions of the hydroxyl groups—migrating to the surface of crystallites to combine with atmospheric oxygen to form water, again 1.9%. There is a corresponding movement of electrons and some Fe^{2+} is converted into Fe^{3+}. Addison et al. (1962) studied the same reaction by gas-absorptiometric methods and posulated the following redox equation for the combined process of dehydrogenation and oxidation:

$$4Fe^{2+} + 4OH^- + O_2 \rightarrow 4Fe^{3+} + 4O^{2-} + 2H_2O$$

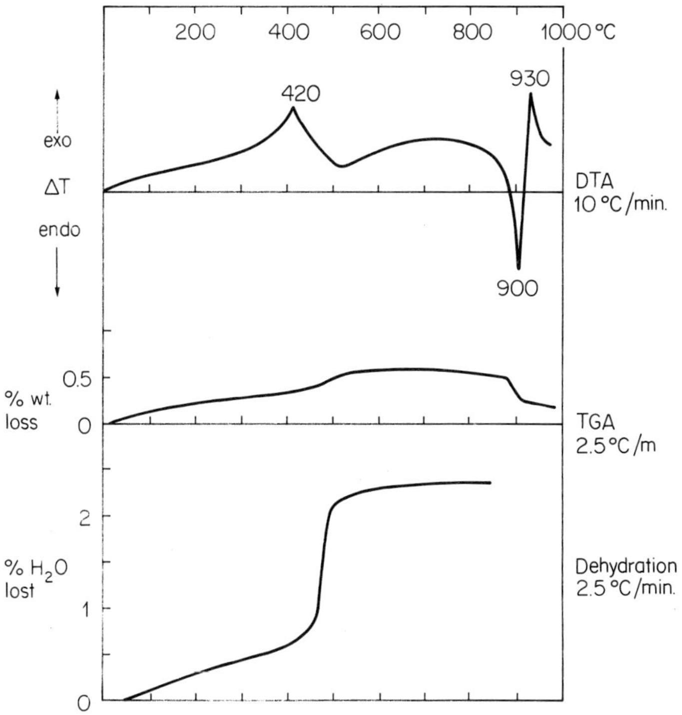

Figure 3.7 Thermal analyses of crocidolite in oxygen

No structural changes accompany this process and none in fact are detected until the temperature of the oxycrocidolite has reached 850 °C. Decomposition takes place completely at 900 °C, giving the pyroxene, magnetite haematite, and silica.

In addition to the true chemically combined water, crocidolite loses about 0.5% or less of water in the range 100–500 °C, independently of the surrounding atmosphere. Similar amounts are attributed to the other amphibole fibres. The sole change with which this loss of water can be associated is in the loss of tensile strength of the fibre, and consequently this particular amount of water is referred to as 'physically combined'. It appears to be associated with the phenomenon of fibrosity in the amphiboles: water molecules form cross-links between individual crystallites, which thus bond into a strong fibrous structure. The distribution in fresh crocidolite amounts to one molecule of water or one cross-link to every four unit cells.

Addison and Sharp (1962) have studied the reduction of fresh dehydroxylated and oxidized crocidolite in a hydrogen atmosphere at temperatures of 450 and 615 °C. The reductions follow different mechanisms–both water and

Table 3.9 Summary of the decomposition reactions of amphiboles

	Oxidising conditions*			Neutral conditions*		
Type	Dehydrogenation	Dehydroxylation	Breakdown	Dehydrogenation	Dehydroxylation	Breakdown
Crocidolite	R 400–600 °C P 420 °C 1.9% H_2O		P 900 °C NaFe-py., haem., sil.	R 550–700 °C P 610 °C 1.9% H_2O		P 800 °C NaFe-py., mag., sil., glass phase
Prieskaite (fibrous actinolite)	R 450–670 °C 1.9% H_2O		R 900–110 °C CaMgFe-py., sil.	R 400–850 °C(2) P 700 °C 1.9% H_2O		P 1040 °C CaMgFe-py., sil.
Amosite	R 600–800 °C 0.9% H_2O	R 800–900 °C 1.0% H_2O	R 600–900 °C FeMg-py., mag., haem., sil.	R 600–800 °C P 780 °C 1.5% H_2O	R 600–800 °C 0.4% H_2	R 600–900 °C FeMg-py., sil.
Anthophyllite	R 600–850 °C 1% H_2O	R 850–1000 °C P 950 °C 2% H_2O	P 950 °C MgFe-py., sil., mag.(?)	R 800–1050 °C P 950 °C 2.7% H_2O	R 800–1050 °C 0.2% H_2	P 950 °C MgFe-py., sil., mag. (?)
Actinolite (semi-fibrous)	R 620–960 °C 1.1% H_2O	R 960–1080 °C P 1040 °C 1.5% H_2O	P 1040 °C CaMgFe-py., sil.	R 620–1080 °C P 1040 °C 2.4% H_2O	R 620–1080 °C 0.2% H_2	P 1040 °C CaMgFe-py., sil.
Tremolite		R 950–1040 °C P 1040 °C 2% H_2O	P 1040 °C CaMgFe-py., sil.	R 950–1050 °C P 1040 °C 2% H_2O		1040 °C CaMg-py., sil.

* R, temperature range; P, typical peak temperature (DTA); (2) indicates two steps in dehydration; py. = pyroxene; haem. = haematite; mag. = magnetite; sil. = silica.

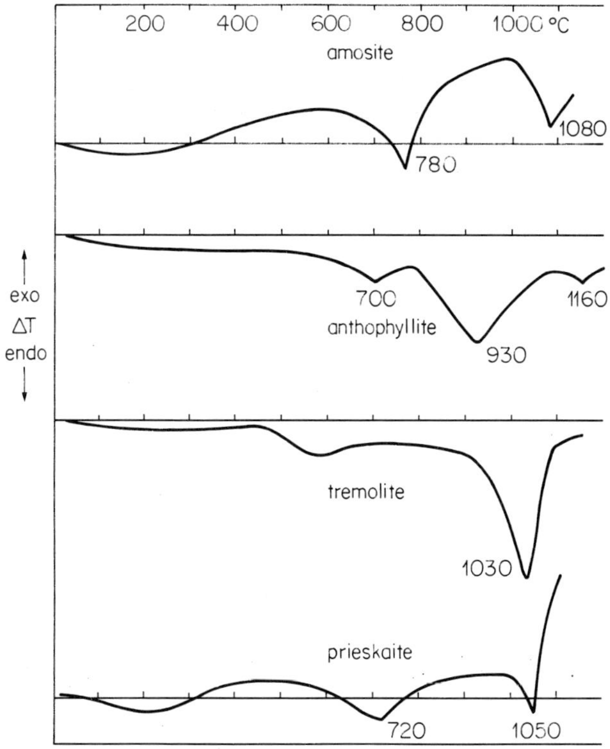

Figure 3.8 Typical DTA curves for amphibole fibres in an inert atmosphere (argon). Prieskaite is a variety of actinolite.

zerovalent iron are formed in each. A pyroxene is formed together with silica at temperatures of 600–850 °C.

The decomposition reactions of the fibrous amphiboles are summarized in Table 3.9, together with typical DTA curves in Figures 3.8 and 3.9 (Freeman, 1962; Hodgson, 1963). The outstanding feature of these reactions is the ability of the amphiboles to lose the elements of chemically combined water by dehydroxylation, dehydrogenation, and oxidation, or by dehydrogenation and expulsion of free hydrogen, depending on the external conditions.

Straightforward dehydroxylation, as for crocidolite in an inert atmosphere, can be expressed by $2OH^- \rightarrow H_2O + O^{2-}$. The dehydroxylation of amosite in an inert atmosphere is, however, complicated by an auto-oxidation of some of the iron(II) content, coupled with the expulsion of an equivalent amount of free hydrogen:

$$4Fe^{2+} + 6OH^- \rightarrow 2Fe_2O_3 + 3H_2$$

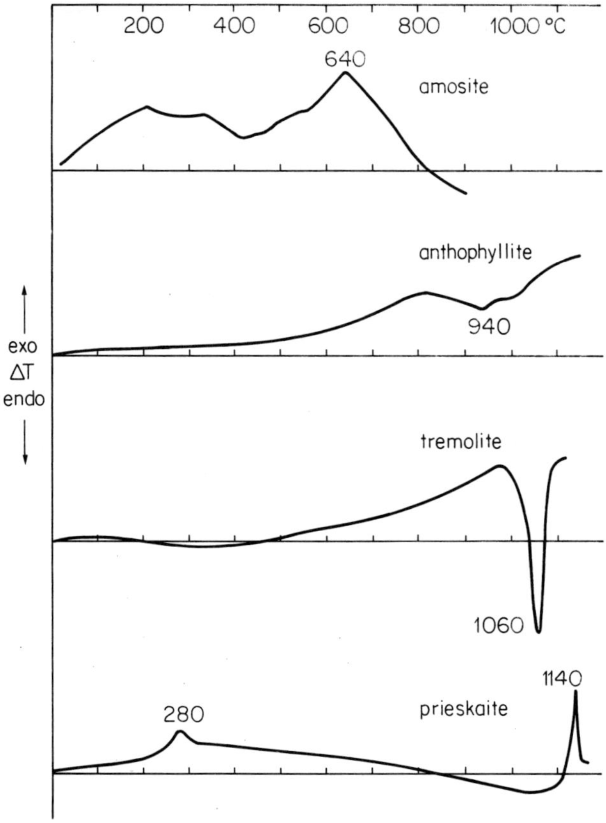

Figure 3.9 Typical DTA curves for amphibole fibres in oxygen. Prieskaite is a variety of actinolite

About one quarter of the total hydrogen content of amosite is lost in this way. The remainder is lost as water in the condensation of OH^- groups. Anthophyllite and non-fibrous actinolite apparently behave similarly, though the amount of free hydrogen expelled will be small, being dependent on the presence of available iron(II). Tremolite loses water solely through a dehydroxylation reaction in both inert and oxidizing atmospheres.

When amosite is heated in oxygen, water forms partly by oxidation of protons and partly by dehydroxylation, in that order, and with some overlapping of the reactions. The oxidation of protons accounts for one quarter to half of the hydrogen present. Anthophyllite and non-fibrous actinolite also lose some water via proton oxidation, although considerably less than does amosite. The fibrous actinolite prieskaite behaves similarly to crocidolite and loses all its proton content to external oxidation.

The amount of water formed from proton oxidation seems to depend on proton mobility. All of the protons in crocidolite (and prieskaite) may be oxidized, up to half of them in amosite, small amounts only in anthophyllite and non-fibrous actinolite, and none in tremolite. The series crocidolite, amosite, anthophyllite (actinolite), tremolite is loosely linked by these several factors: decreasing proton mobility; increasing temperature of dehydroxylation in both inert and oxidizing atmospheres; increasing temperature of structural breakdown; and increasing magnesium content at M_1 and M_3 sites. Addison (1962), in comparing the slow oxidation of amosite with the rapid oxidation of crocidolite, suggested that magnesium blocking at M_1 and M_3 sites must interfere with the redox reaction. This is not the complete explanation because crocidolites with various magnesium contents oxidize at similar temperatures (near 400 °C by DTA) and always indicate complete proton mobility with no tendency for a dehydroxylation reaction. Nevertheless, the magnesium content of the amphiboles appears to be the most important factor in determining their thermal stability.

In summary, the decomposition of these amphiboles consists of three fundamental stages: loss of physically combined water, loss of chemically combined water, and breakdown into simpler products, mainly a pyroxene. The loss of chemically combined water is associated with proton mobility, a phenomenon which cannot be confined solely to reactions in oxidizing atmospheres. While dehydroxylation can be over-simplified as a condensation of OH^- groups, this reaction seems to proceed through migration of protons to reaction zones where they unit with oxygen. In the absence of oxidizing conditions at structural surfaces, these reaction zones must be structural parts. Freeman and Taylor (1960) and Taylor (1962) explain the whole process of decomposition in terms of the formation of donor and acceptor regions in the crystalline structure. The donor region is destroyed and supplies not only oxygen, which will react with protons that migrate towards it, but also cations which migrate to the otherwise undisturbed acceptor regions. The main products of decomposition form in the acceptor regions, and both the products and the starting materials have structures based on nearly close-packed oxygen layers, with cations at interstices. The crystalline products are always formed topotactically, that is, with definite preferred orientations. The smallest changes in cation positions and disturbance of the oxygen layers occur in the formation of the pyroxenes which, in consequence, are always highly orientated.

The way in which these amphiboles lose combined water is of considerable importance in the interpretation of chemical analyses. The question of allocation of the water liberated from the amphiboles has been the subject of several mineralogical papers in the past, particularly with respect to anthophyllite and actinolite where an excess of combined water is found (Zussman, 1955; Francis and Hey, 1956; Hutton, 1956). It is now clear that one cannot assume that water lost above 105–110 °C is chemically combined and that an understanding of the dehydration processes involved is necessary before determinations of combined water can be made. The true content of hydroxyl water in crocidolite (and prieskaite) can be found most conveniently from TGA or dynamic dehydration curves in a neutral atmosphere, but for the other amphiboles dynamic dehydration in an oxidizing atmosphere would be necessary to ensure complete oxidation of protons (Hodgson *et al.*, 1965; Hodgson, 1963). By such methods it is possible to separate physically combined water from chemically combined water, which is normally evolved in a sharp loss over a narrow range of temperature. Fresh amosite, crocidolite, and tremolite give losses which, on calculation of the contents of the cell, agree closely with the $2OH^-$ requirement. Anthophyllite and actinolite may have an excess of OH^- groups and, since not more than two can be allocated to the normal structure, it is assumed that extra hydroxyl groups may replace some of the oxygen atoms in Si_4O_{11} bands.

This general outline of the thermal decomposition of the amphibole fibres applies to the typical minerals. Obviously, a number of variants exist. For example, weathered amosites and crocidolites may have losses of chemically combined water corresponding to $1\frac{1}{2}OH^-$ per unit cell, the loss of up to $\frac{1}{2}OH^-$ being associated with natural oxidation processes. These processes appear to be limited by the disposition of Fe^{2+} ions in M_1 and M_3 sites (Hodgson, 1963). Other than this, the complete decomposition of such materials is not affected by weathering. The detailed pattern of breakdown of crocidolites from the Transvaal, Bolivia, and Australia varies somewhat from the general scheme. In particular, the Transvaal and Bolivian crocidolites behave as mixed amphiboles. The Transvaal material is known to be an intimate physical mixture of crocidolite and amosite, and the Bolivian material has been classified as a magnesio-riebeckite (Drysdall and Newton, 1960) and has characteristics of both anthophyllite and crocidolite on the evidence of thermal decomposition (Hodgson, 1965a, b). The compositions of both tremolites and anthophyllites include pure magnesium varieties as well as ferro-magnesium varieties, and the proportion of iron present will influence the thermal decomposition of these minerals. Usually, the effect of a high iron content is to produce two distinct steps in the loss of chemically combined water and a lowering of the temperature of final decomposition.

Finally in discussing the amphiboles, further reference should be made to the influence of magnesium on their decomposition reactions. In general, the greater is the magnesium content, the higher are the temperatures at which dehydroxylation and breakdown occur. While magnesium blocking at M_1 and M_3 sites can affect the rate of oxidation of certain types of amphiboles, this appears to be subsidiary to the control of ultimate decomposition temperatures. This has been shown to be true for crocidolite itself. By correlating DTA evidence with chemical compositions for 21 crocidolites from South Africa and Australia, it is possible to show that the temperature of breakdown is influenced by the $Fe^{+}:Mg^{2+}$ ratio (Hodgson, 1965a, b). These cations fill the M_1 and M_3 sites. Where this ratio is high, i.e. 8–11, ultimate decomposition occurs at 890–900 °C in oxidizing conditions and at 800 °C in inert conditions. Progressive lowering of the ratio to 2 coincides with a rise in the temperature of breakdown of about 80 °C in both oxidizing and inert conditions. Some similar evidence from tremolites has shown that for a number of specimens having FeO^{-} contents between 7 and 0.3%, the endothermic temperature of decomposition increases from 980 to 1080 °C as the Fe^{2+} content diminishes. All of the evidence points to magnesium as the key to the control of the upper temperature limit of decomposition for these amphiboles and to control of proton mobility, except in crocidolite, where separate redox mechanism operates. In this sense crocidolites are unique among the amphiboles.

3(ii) Chrysotile asbestos

The thermal decomposition of chrysotile follows a two-stage sequence of dehydroxylation and breakdown. The mechanism of decomposition of serpentine, the mother rock of chrysotile, has been studied by a number of workers (Vermaas 1953; Brindley and Zussman, 1957; Ball and Taylor, 1961) and the same mechanism applies to chrysotile (Ball and Taylor, 1963b; Brindley and Hayami, 1965). The dehydroxylation of chrysotile takes place in the range 600–780 °C and at 800–850 C the anhydride breaks down in a sharp transformation to forsterite and silica. The total decomposition corresponds to

$$2Mg_3[Si_2O_5](OH)_4 \rightarrow 3Mg_2SiO_4 + SiO_2 + 4H_2O$$

The forsterite is well crystallized and persists together with the silica up to 1000 °C, but well above this temperature some enstatite is formed:

$$3Mg_2SiO_4 + SiO_2 \rightarrow 2Mg_2SiO_4 + Mg_2Si_2O_6$$

The reactions take place independently of surrounding atmospheres. Typical thermal analysis curves (Figure 3.10) show on DTA a dehydroxylation endotherm and a sharp exotherm corresponding to the forsterite recrystallization, and on TGA a gradual weight loss between 100 and 600 °C and a sharp loss above 600 °C, the total dehydration amounting to about 13% of H_2O.

A 'donor–acceptor' mechanism of decomposition has been applied to the transformation of the serpentine minerals, based initially on a similar mechanism for the dehydration of brucite (Ball and Taylor, 1961). In Ball and Taylor's mechanism donor regions, corresponding to two-ninths of the

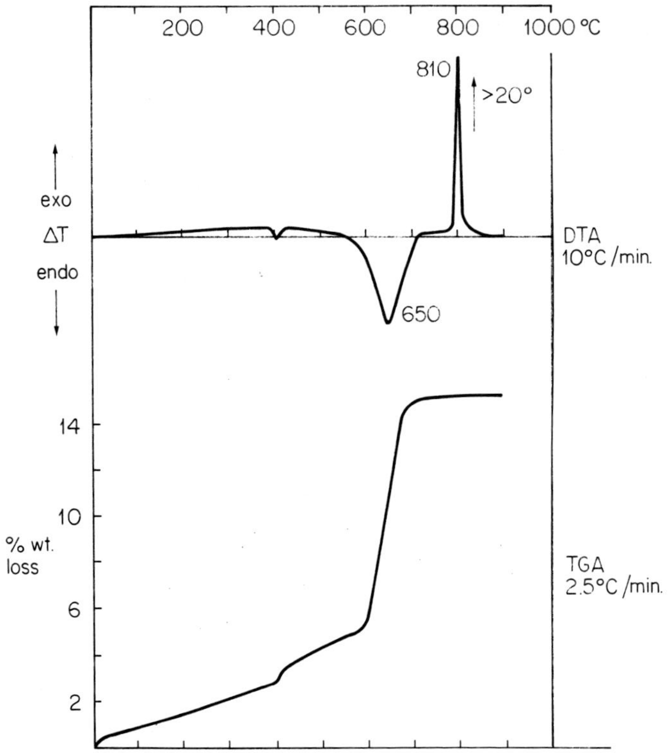

Figure 3.10 Typical thermal analysis of chrysotile

whole, supply cations to acceptor regions and oxygen atoms to react with protons which migrate from acceptor regions and form water. The scheme can be represented as follows:

$$7Mg_3Si_2O_5(OH)_4 \xrightarrow{28H} 2Mg_3Si_2O_5(OH)_4$$

$$600\,^{\circ}C \quad \text{Acceptor region} \xleftarrow[6Mg,\ 4Si]{} \text{Donor region}$$

$$\downarrow \qquad\qquad\qquad \downarrow$$

$$\text{Serpentine anhydride} \qquad 18H_2O$$

$$\downarrow$$

$$800\,^{\circ}C \quad Mg_2SiO_4 + SiO_2$$

The donor regions are completely destroyed and form pores. In Brindley and Hayami's interpretation (Brindley and Hayami, 1965), one-third of the whole becomes a donor region and residual silica remains there after its destruction. A highly disorganized forsterite forms in the acceptor region, which transforms into a well crystallized material at 800 °C. The forsterite always forms topotactically and the energy of its recrystallization is well exemplified by DTA, which can give a sudden recrystallization exotherm of amplitude 50–100 °C. Under static heating conditions the transformation is more gradual, beginning in small nuclei and spreading as cation migrations and packing changes occur. Further information on the dehydration of chrysotile has been obtained by TGA methods. Woodrooffe (1956) examined a number of Canadian chrysotiles using a Chevenard thermobalance operating at a heating rate of less than 1 °C/min. He found three states of dehydration: 100–500 °C, giving up to 1% of H_2O; 500–700 °C, giving 10–13% of H_2O; and 700–900 °C and above, giving 1% of H_2O. The total dehydration ranged from 12 to 15.5% for 16 specimens. The extent and temperature range of the main part of the dehydration agree well with the proposed decomposition mechanism.

However, some investigations of chrysotile by thermal analysis have brought to light a further aspect of these mechanisms. (Hiscock, 1964). In this work 15 specimens of the type of mineral from the Thetford and Asbestos areas of Quebec province were examined, together with a further 15 specimens from world-wide sources. DTA showed that the endothermic peak temperature of dehydroxylation is, with few exceptions, 650 ± 20 °C, and the exothermic peak temperature of recrystallization is almost invariable at 810 ± 10 °C. The TGA and dehydration measurements agreed approximately with those of Woodrooffe, but the main loss of water appears between 600 and 700 °C (± 20 °C) and amounts to 9–11.3% according to the specimen; the loss between 100 and 600 °C is up to 2.5%. The TGA results alone can be misleading since part of the weight loss may be due to the decomposition of brucite and carbonate impurities, but provided that these are accounted for the loss due to water alone agrees well with dehydration measurements.

A theoretical composition of chrysotile based on structural considerations will show that 9.75% of H_2O, equivalent to $3(OH)^-$, rests in the external hydroxyl layer, with 3.25% of H_2O, equivalent to $1(OH)^-$, in the internal hydroxyl layer (see Figure 3.3). There are reasons, therefore, for associating the main dehydration loss of chrysotile with disruption of the external hydroxyl layer, while the sum of losses below 600 °C and above 700 °C appears to be associated with the remaining hydroxyl groups. On this basis, calculation of the disposition of OH^- groups in cell contents gives the results shown in Table 3.10.

The total number of OH^- groups does not often exceed eight, and usually shows deficiencies. Since the measurements commence at 100 °C, it seems unlikely that chrysotile contains interfibril –H–O–H links in the same way that the amphiboles do. All H_2O released above 100 °C can be attributed to chemically combined water. Hydroxyl in the $(OH)_3$ layer approaches a minimum of six, while there are often deficiencies in the $(OH)_1$ layer.

Table 3.10 Disposition of (OH) groups in chrysotile as derived from TGA

Source	$(OH)_6$	$(OH)_2$	Total
Quebec, Thetford (a)	5.9–6.4	1.7–2.0	7.9–8.0
Quebec, Thetford (b)	6.4–6.8	1.6–2.3	7.9–8.7
Quebec, Asbestos	6.3	2.1	8.4
Russia, Asbest (a)	5.9	1.6	7.5
Russia, Asbest (b)	6.7	0.8	7.5
Transvaal, Barberton	6.5	1.4–1.9	7.9–8.4
S. Rhodesia, Shabani	6.5	1.4	7.9
British Columbia, Cassiar	7.1	1.0	8.1
Arizona	6.2	1.2	7.4
Greece	6.2	1.6	7.8
Turkey	6.4	1.5	7.9

The evidence points to proton movement beginning at about 150 °C and to the $(OH)_3$ layer remaining virtually intact up to 600 °C. Initially, proton migration may take place in the $(OH)_1$ layers alone to incipient donor regions where the protons will react with hydroxyl groups to form water. When brucite itself is subjected to thermal decomposition, it dehydrates under static conditions at 250 °C and under dynamic conditions at 350–400 °C. Thus, although a similar mechanism can be used to explain the dehydration of brucite and chrysotile, proton migration in the latter appears to take place in two distinct modes. One of these is delayed until the chrysotile reaches a temperature of 600 °C and then produces a large and rapid dehydration; the other commences at low temperatures and produces a continuous small dehydration which may not be completed up to 900 °C.

The interpretation of weight losses measured by TGA admittedly depends on a measure of technique within experimental operation. Monkman (1967) used slow heating rates of 0.7 °C/min and contained the specimens in a closed crucible. Under such conditions, weight losses in chrysotile between 75 and 325 °C are deemed to be due to loss of loosely bound water, while a two-stage loss form 525 to 725 °C and 775 to 925 °C coincides with the breakdown of the chrysotile lattice and loss of residual compositional (OH) groups. The total loss above 525 °C corresponds to the combined (OH) content of chrysotile and does not point to any differentiation in the mode of loss of (OH) groups. Nevertheless, whichever technique is chosen, the TGA result is repeatable for chrysotiles from different sources and provides a confirmatory means of identifying chrysotiles. Associated minerals such as brucite, magnesite, calcite, and talc can, if present in bulk, be readily distinguished from chrysotile by virtue of the range of their weight losses (Monkman, 1967).

(4) SURFACE PROPERTIES OF ASBESTOS

A consideration of the elements of the chrysotile and amphibole structures leads immediately to the conclusion that the two groups of fibres must have totally different surface properties. Chrysotile has a predominant surface layer of hydroxyl sites connected to an adjacent inner layer of magnesium sites. These alkaline sites would be expected to exert a strong surface activity either by partial solution or by attraction of ions of opposite charge. In contrast, amphibole asbestos has external bands of silica sites which are recognized as having a weakly acidic nature. Such sites, being highly insoluble, would be expected to exert a weak surface charge solely by attraction of suitable ions, comparable to the effect obtained with ordered silica surfaces, such as in quartz.

Fundamental studies of this subject have provided some valuable, though limited, information. The influence of electrolytes on the surface charge of asbestos fibre has been investigated in two ways. Pundsack (1955) determined the isoelectric point of chrysotile by sedimentation methods, by increasing the alkalinity of suspensions of fine fibres until a floc formed and the suspension clarified. Later, Martinez and Zucker (1960) applied a streaming potential technique to the measurement of the zeta potential of chrysotile. They obtained a value of about +100 mV for the zeta potential of chrysotile in neutral solutions, the isoelectric point being at pH 11.8, somewhat higher than the value obtained by Pundsack (1955). Independent studies on amphibole fibres have given values of −10 mV for crocidolite and −20 mV for amosite in neutral solution. Isoelectric points have not been determined, but the zeta potential of these two fibres tends towards zero with increasing alkalinity of the solutions.

There are considerable technical difficulties in the streaming potential method (Ball and Fuerstenau, 1973) and reviews of published zeta potential data for asbestos show little concordance among results. Nevertheless, the technique provides sufficient evidence to confirm theoretical suggestions about the surface charge of asbestos fibres.

The micro-electrophoretic method of measuring zeta potentials has received more attention in recent years and has considerable flexibility in the investigation of surface active phenomena in relation to mineral processing in general and to asbestos fibres in particular.

The effect of pH on the zeta potential of chrysotile and the change in the surface activity of chrysotile due to leaching in both controlled and natural conditions were reported by Chowdhury and Kitchener (1975). Its zeta potential depends not only on its chemical composition but also on the presence of impurities, brucite for example imparting high positive values. Naturally, leached chrysotile gives negative values characteristic of a pure silica surface.

Amphibole asbestos has a resistant structure which undergoes little change when subjected to leaching conditions. Its zeta potential *versus* pH curve approximates to that of quartz and exhibits high negative values at pH 10–11. Amosite asbestos responds to cationic surfactants and its zeta potential at high pH can be reversed from negative to positive values by adsorption of cetyltrimethylammonium bromide (Ralston and Kitchener, 1975). Studies such as these become increasingly important in the interpretation of the function of asbestos in industrial processes such as the manufacture of asbestos–cement and in the investigation of the biological effects of asbestos where biochemical functions are as yet imperfectly understood.

The surface charge of asbestos fibres reflects the sedimentation behaviour of asbestos in aqueous suspension. Measured sedimentation volumes are an important test parameter for asbestos, but the behaviour of fibres in the asbestos–cement–water system is of considerable practical importance. Chrysotiles, having typical small and dense sedimentation volumes, react with cement to form sediments which possess poor permeability to water, while the amphibole fibres under similar conditions form open networks of fibres, giving a relatively high porosity. The hydrophilic nature of chrysotile brings about a further complication in that its hydroxylated surfaces undoubtedly form contiguous phases with the gels formed in the early stages of the hydration of cement particles. The drainage rates of asbestos–cement–water systems based on amphibole asbestos are some three to four times better than those based on chrysotile, a fact which has bearing on the speed of production of most types of asbestos–cement manufacturing machinery. It is a widespread practice to blend amphibole and chrysotile fibres in the manufacture of asbestos–cement products to ameliorate drainage problems.

Numerous attempts have been made to modify the surface properties of chrysotile asbestos, by surface-active agents and other organic absorbates. Generally, these are not as successful as might first appear. The density of hydroxyl groups at the surface of chrysotile is high and no more than a limited number of links with the relatively large molecules of typical surface-active agents, such as quaternary ammonium compounds and sulphonates, can be expected. The residue of unlinked hydroxyl groups must still exert a considerable influence. Polyacrylamides have been suggested as being among the best flocculating agents for asbestos–cement slurries (Nikolaev and Pogoskaya, 1965).

From zeta-potential studies, Martinez (1964) has shown that inorganic polyvalent ions, such as are present in sodium silicate solutions, may be specific in altering the surface charge on chryostile. Sodium phosphate and sodium carbonate solutions bring about a reduction in the positive surface charge of chrysotile through absorption of the anions but do not produce a reversal of polarity. These findings have been applied to the treatment of chrysotile asbestos for manufacturing purposes. By milling the fibre in the presence of limited amounts of dilute sodium silicate solution, the drainage properties of this form of asbestos in asbestos–cement slurries can be improved.

Another obvious means of modifying the surface properties of chrysotile is to subject the fibre to a limited degree of heat treatment so as to drive off part of its chemically combined water. The aim is to reduce the constitutional water content by up to 30%, preferably by flash heating at 500–600 °C (Badollet and Streib, 1955; U.S.P. 2 616 801). The drainage properties of chrysotile in water are improved considerably by this treatment, but these must be offset against losses in tensile strength and reinforcement properties.

Although surface-active agents are not satisfactory modifiers of chrysotile in manufacturing processes, they have a use in another way. A piece of chrysotile ore immersed in a strong household detergent solution, and without any other disturbance, will in the course of time completely fiberize to a mass of extremely finely divided fibres. Amphibole fibres do not respond to this treatment. The effect has been adapted to industrial use where it is desirable to produce aqueous dispersions of chrysotile of a 'fibro-colloidal' character from the relatively coarse commercial fibre. By wet milling the chrysotile fibre with surface-active agents, such as straight sodium soaps, fatty acid sulphonates and more complex derivatives of these, extensive reduction of fibre diameters to values between 200 and 500 Å can be achieved (B.P. 689 692; U.S.P. 2 652 325). Fine fibres such as these have applications in the manufacture of asbestos papers and felts and, of course, in the recent development of wet systems for producing chrysotile asbestos yarns.

Some aspects of the surface chemistry of asbestos fibres have already been discussed. Gas-absorptiometric studies have yielded much information connected with the oxidation and thermal decomposition of asbestos with the surface area of fibres and with their surface chemistry in gaseous systems (Sharp, 1963; Addison *et al.*, 1962). The surface properties of both chrysotile and amphibole asbestos, referred to ammonia or nitrogen atmospheres, show many similarities. Generally, the apparent surface area to ammonia is greater than that to nitrogen at normal temperatures. It is possible that NH_3 links to excess of OH^- sites at fibre surfaces with more than one ammonia molecule per site. At higher temperatures (400 °C) the surface areas determined by either NH_3 or N_2 absorption are similar for each individual sample of fibre. This implies that the surface properties of all asbestos fibres may be modified by out-gassing at higher temperatures, although the mechanisms involved are different. By out-gassing crocidolite in oxygen all surface-absorbed water and chemically combined water (by proton ejection) is removed. Partial oxidation and removal of protons occur in amosite and partial dehydration occurs in chrysotile. Young and Healey

(1954) suggested that hollow tubes of chrysotile contained plugs of water which, when removed at higher temperatures, allowed equal absorption of both NH_3 and N_2. It therefore appears that, regardless of residual hydroxyl sites, profound modification to the surface properties of asbestos fibres can be achieved by the removal of absorbed water at fibre surfaces. Such absorption is not necessarily irreversible.

An interesting outcome of gas-absorptiometric studies on crocidolite concerns its potential as a catalyst. Addison and Sharp (1962) found that after oxidizing crocidolite at 450 or 650 °C it could in turn be reduced with conversion to its Fe^{3+} content to Fe^{2+} or Fe^0 together with the expulsion of water:

$$\text{at } 450\,°\text{C};\ 4Fe^{3+} + 2O^{2-} \rightarrow 4Fe^{2+} + 2H_2O$$

$$\text{at } 615\,°\text{C};\ 2Fe^{3+} + 3H_2 \rightarrow 2Fe^0 + 3H_2O$$

Both of these reactions involve removal of oxygen from the lattice, but if an oxygen atmosphere can be maintained around the crocidolite then the forward oxidation reaction (p. 96) and the reverse reduction can be maintained in equilibrium. Some independent experiments along these lines have shown that carbon monoxide and methane in mixtures with oxygen can be smoothly oxidized at temperatures of 500–600 °C. Crocidolite however, does not offer any advantage over other types of catalysts, such as platinized asbestos or transition metal oxides, mainly because its threshold temperature for oxidation is too high. As a laboratory reagent oxidized crocidolite may be more valuable than the conventional copper(II) oxide as a high-surface-area oxidant. It has been used to detect the evolution of hydrogen from amosite and other amphiboles during their thermal decomposition in inert atmospheres (Hodgson, 1963).

(5) INFRARED SPECTROSCOPIC DATA FOR ASBESTOS

The use of infrared (IR) spectroscopy both for the identification of and structural studies on asbestos is a relatively recent development. Published work on the subject includes that of Parks (1971), Daykin (1971), Farmer (1974), Beckett *et al.* (1975) and Patterson and O'Connor (1966).

All types of asbestos exhibit strong absorptions in a 1200–900 cm^{-1} band attributable to Si–O stretching vibrations, and in a 600–300 cm^{-1} band attributable to Si–O chain and ring vibrations. Additional peaks below 600 cm^{-1} correspond to cation–oxygen vibrations and, in theory, four cation sites for the amphiboles can be distinguished, although in practice it is difficult to detect more than two of these absorption peaks. O–H stretching vibrations give distinctive peaks in the region 3600–3700 cm^{-1}, their amplitude corresponding to the OH content of the sample.

Table 3.11 Principal IR absorption peaks for asbestos minerals (wavenumber, cm^{-1})

Vibrations	Chrysotile	Anthophyllite	Amosite	Crocidolite	Tremolite	Actinolite
O–H stretching	3700	3680	3660	3655	3680	3680
vibrations	3655	3675	3640	3640	3670	3670
			3620	3620		
Si–O stretching	1080	1095	1129	1145	1105	1100
vibrations	1030	1020	1085	1110	1065	1040
	965	980	1000	995	998	998
			890	898	950	953
				880	922	922
Silicate chain		780	775	780	755	758
and ring		755	700	695	685	685
vibrations		712	636	635		
		670				
Cation–oxygen	610		500	545	508	510
stretching	438	495	480	505	461	460
vibrations	410	455	426	450	390	
	305		330	320	360	

Principle absorption peaks in wavenumbers (cm^{-1}) for the different types of asbestos are given in Table 3.11 and the main distinguishing features of each asbestos can be summarized as follows:

Chrysotile: strong O–H peak at 3700 cm^{-1}, coupled with a weaker peak at 3655 cm^{-1}. The ratio of the intensities of these peaks is attributed to the 3:1 ratio of the external and internal hydroxyl groups in chrysotile (Daykin, 1971).

Amosite and crocidolite: weak O–H peaks in a 3600–3660 cm^{-1} band and similar absorption traces throughout. Distinguished by Si–O stretching vibrations in the 1150–850 cm^{-1} band. With highly sensitive instruments the NaO_8 peak at 270 cm^{-1} distinguishes crocidolite.

Tremolite and actinolite: weak O–H peaks, and other absorption peaks in virtually identical positions. Distinguished by difference in peak intensities, tremolite exhibiting five relatively strong peaks in the 900–1150 cm^{-1} band.

Anthophyllite: while similar to other amphiboles, has distinctive peaks at 670 and 450 cm^{-1}.

The preparation of samples for IR spectroscopy is well known and well documented. The use of caesium iodide discs in preference to potassium bromide is advocated for asbestos samples because of the strong absorption of the latter at 400 cm^{-1}. However, absorption peaks above 300 cm^{-1} are not obscured by potassium bromide peaks and the three major types of asbestos can be readily identified by peaks at 305 cm^{-1} for chrysotile, 318–320 cm^{-1} for crocidolite, and 330–332 cm^{-1} for amosite.

The quantitative assessment of asbestos in environmental samples is a new development (Prentice, 1976; Coates, 1977). The technique is rendered complex by interference from other materials but nevertheless it is possible to detect 3% of crocidolite in the presence of amosite. Assessment of airborne samples of asbestos is limited by the minimum detection level of IR radiation at about 0.2 mg of sample and by the fact that such a weight represents a considerable amount of material by currently used airborne sampling techniques.

The decomposition of asbestos by heat, which can be followed by DTA, can also be followed by IR spectroscopy on specimens prepared at different limiting temperatures. The first change which is noted is the diminution and disappearance of O–H absorption peaks, this occurring at 350–500 °C for crocidolite, at 550–600 °C for chrysotile, and at 450–700 °C for amosite. Above 650 °C, the presence of forsterite in heated chrysotile is distinguishable by IR spectroscopy. The amphiboles generally show a shifting of peaks to higher frequencies as the temperature of decomposition is increased and major changes in the spectra occur on complete breakdown of the structures at temperatures above 900 °C.

In such studies, the asbestos specimens were heated in air for periods of 6 h at specific temperatures. The regime is not that of a DTA experiment and consequently recrystallization changes occur at lower temperatures than in DTA. In practice, this is an advantage when it is necessary to identify asbestos in environmental samples which may have been subjected to an elevated temperature regime for a long period. For example, both crocidolite and amosite oxidize to a reddish brown colour at 450–600 °C and are visibly indistinguishable, but can be detected separately by IR data from the experimental technique outlined above.

ACKNOWLEDGEMENTS

The author is indebted to the Directors of Cape Asbestos Fibres Ltd. for permission to make use of unpublished work in the Company's records and to the staff of the Research Laboratory for their assistance in the preparation of this chapter.

(7) REFERENCES

Addison, C. C., Addison, W. E., Neal, G. H., and Sharp, J. H. (1962) *J. Chem. Soc.*, **1962,** 1468.

Addison, W. E., and Sharp, J. H. (1962) *J. Chem. Soc.*, **1962,** 3693.

Addison, W. E. (1964) Nottingham University personal communication.

Allen, M. P., and Smith, R. W. (1975) *3rd International Conference on Asbestos Minerals, Quebec.*

Atkinson, A. W., Gettins, R. B., and Rickards, A. L. (1971) *2nd International Conference on Asbestos Minerals, Louvain.*

Badollet, M., and McGourty, J. P. (1958) *Trans. Can. Inst. Min. Metall.*, **61,** 168.

Badollet, M., (1951) *Trans. Can. Inst. Min. Metall.*, **54,** 151.
Badollet, M. (1960) *Modern Plastics Encyclopedia*, McGraw-Hill, New York.
Badollet, M., and Streib, W. C. (1955) *Trans. Can. Inst. Min. Metall.*, **58,** 33.
Ball, B., and Fuerstenau, D. W. (1973) *Miner. Sci. Engng*, **5,** (4,) 267.
Ball, M. C., and Taylor, H. F. W. (1961) *Mineral. Mag.*, **32,** 754.
Ball, M. C., and Taylor, H. F. W. (1963a) *J. Appl. Chem., Lond.* **13,** 145.
Ball, M. C., and Taylor, H. F. W. (1963b) *Mineral. Mag.*, **33,** 467.
Beckett, S. T., Middleton, A. P., and Dodgson, J. (1975) *Ann. Occup. Hyg.*, **18,** 313.
Bowen, N. L., and Scheirer, J. F. (1935) *Am. Miner.*, **20,** 543.
Brindley, G. W., and Zussman, J. (1957) *Am. Miner.*, **42,** 461.
Brindley, G. W., and Hayami, R. (1965) *Mineral. Mag.*, **35,** 189.
Bryans, R. G., and Lincoln, B. (1971) *2nd International Conference on Asbestos Minerals, Louvain.*
Bryans, R. G., and Lincoln, B. (1975) *3rd International Conference on Asbestos Minerals, Quebec.*
Burman, D. R. (1967) *1st International Conference on Asbestos Minerals, Oxford*
Chisholm, J. E. (1973) *J. Mater. Sci.*, **8,** 475.
Chisholm, J. E. (1975) *3rd International Conference on Asbestos Minerals, Quebec.*
Chowdhury, S. (1973) *Ph.D. Thesis*, University of London.
Chowdhury, S., and Kitchener, J. A. (1975) *Int. J. Miner. Process.*, **2,** 277.
Cilliers, J. J., Freeman, A. G., Hodgson, A. A., and Taylor, H. F. W. (1961), *Econ. Geol.*, **56,** 1421.
Cilliers, J. J. (1961) *Ph.D. Thesis*, Pretoria University.
Cilliers, J. J., and Genis, J. H. (1964) *Crocidolite Asbestos in Cape Province. The Geology of some Ore Deposits in Southern Africa*, Vol. III., Geological Society of South Africa Johannesburg.
Coates, J. P. (1977) *The Infra Red Analysis of Quartz and Asbestos*, Perkin-Elmer Ltd., Beaconsfield, Bucks.
Comeforo, J. E., and Kohn, J. A. (1954) *Am. Miner.*, **39,** 537.
Cralley, L. J., Keenan, R. G., and Lynch, J. R. (1967) *Am. Ind. Hyg. J.*, **28,** 452.
Cralley, L. J., Keenan, R. G. Kupel, R. E., Kinser. R. E., and Lynch, J. R. (1968) *Am. Ind. Hyg. J.*, **29,** 529.
Daykin, C. W. (1971) *2nd International Conference on Asbestos Minerals, Louvain.*
Deer, W. A., Howie, R. A., and Zussman, J. (1962, 1963) *Rock Forming Minerals*, Vols. 2 and 3, Longmans, London.
Drysdall, A. R., and Newton, A. A. (1960) *Amer. Miner.*, **45,** 53.
Du Toit, A. (1945) *Trans. Geol. Soc. S. Afr.*, **48,** 161.
Du Toit, A. (1954) *The Geology of South Africa*, Oliver and Boyd, Edinburgh.
Farmer, V. C. (1974) *The Infra Red Spectra of Minerals.* Mineralogical Society, London.
Francis, G. H., and Hey, M. H. (1956) *Mineral. Mag.*, **31,** 173.
Freeman, A. G., and Taylor, H. F. W. (1960) *Silikattechnik*, **11,** 390.
Freeman, A. G. (1962) *Ph.D. Thesis*, Aberdeen University.
Fripiat, J. J., Mendelovice, E., and De Kimpe, C., 1967. *1st International Conference on Asbestos Minerals, Oxford.*
Fripiat, J. J., Zapata, L., Van Meerbeek, A., della Faille, M., van Russelt, M., and Mercier, J. P., (1971) *2nd International Conference on Asbestos Minerals, Louvain.*
Grigoreva, L. F., Chigareva, O. G. Mikirticheva, G. A., and Krupenikova, Z. V. (1971) *2nd International Conference on Asbestos Minerals, Louvain.*
Harington, J. S., and Smith, M., (1964) *South African J. Sci.*, **60,** (9), 283.
Hiscock, D. G., (1964) Internal paper, Cape Asbestos Fibres Ltd., Research Laboratory, Barking.

Hodgson, A. A. (1963) *Ph.D. Thesis*, University of London.
Hodgson, A. A. (1965a) *Fibrous Silicates*, Lecture Series No. 4, Royal Institute of Chemistry, London.
Hodgson, A. A. (1965b) *Mineral. Mag.*, **35,** 391.
Hodgson, A. A., Freeman, A. G., and Taylor, H. F. W. (1965) *Mineral. Mag.*, **35,** 445.
Holt, P. F., and Clark, S. G. (1960) *Nature, Lond* **185,** 237.
Hutton, C. O., (1956) *Acta Crystallogr.*, **9,** 231.
Ipatieff, W., and Mouromsteff, D. (1927) *Kokl. Akad. Nauk SSSR*, **85,** 647.
Jander, W., and Wuhrer, J. (1938) *Z. Anorg. Allg. Chem.*, **235,** 273.
Jenkins, G. F. (1960) *Asbestos. Industrial Minerals and Rocks*, 3rd Edition, American Institute of Mining, Metallurgical and Petroleum Engineers U.S.A.
Keep, F. E. (1961) *Amphibole Asbestos in the Union of South Africa*, 7th Commonwealth Mining and Metallurgical Congress, Johannesburg.
Makarova, T. A., Korytkova, E. N., and Nesterchuk, N. L. (1971) *2nd International Conference on Asbestos Minerals, Louvain.*
Monkman, L. (1967) *1st International Conference on Asbestos Minerals, Oxford.*
Monkman, L. (1971) *2nd International Conference on Asbestos Minerals, Louvain.*
Monkman, L. (1975) *3rd International Conference on Asbestos Minerals, Quebec.*
Martinez, E., and Zucker, G. L. (1960) *J. Phys. Chem., Ithaca*, **64,** 924.
Martinez, E. (1964) *Can. Min. Metall. Bull.*, **630,** 104.
National Physical Laboratory (1972) Report commissioned by Cape Asbestos Fibres Ltd.
Nikolaev, K. N., and Pogoskaya, T. I. (1965) *Stroit. Mater.*, **11,** 8.
Orowan, E. (1948–49) *Rep. Prog. Phys.*, **12,** 186.
Parks, L. F., (1971) *2nd International Conference on Asbestos Minerals, Louvain.*
Patterson, J. H., and O'Connor D. J. (1966) *Aust. J. Chem.*, **19,** 1155.
Prentice, F. J. (1976, 1977) Internal papers, Cape Asbestos Fibres Ltd. Research Laboratory
Pundsack, F. L. (1955) *J. Phys. Chem., Ithaca*, **59,** 892.
Ralston, J., and Kitchener, J. A. (1975) *J. Colloid Interface Sci.*, **50,** 242.
Rigden, P. J. (1941) *J. Soc. Chem. Ind., Lond.* **62,** 1.
Saito, H. (1957) *Synthetic Asbestos Research*, Yamanashi University, Kofu-Shi, Japan.
Sharp, J. H. (1963) Ph.D. Thesis, Nottingham.
Speakman, K. (1971) *2nd International Conference on Asbestos Minerals, Louvain.*
Syromyatrukov, F. V. (1935) *Econ. Geol.*, **1935,** 89.
Taylor, H. F. W. (1962) *Clay Miner. Bull.*, **5,** 45.
Timbrell, V. (1972) *Microscope*, **20,** 365.
Timbrell, V. (1975) *3rd International Conference on Asbestos Minerals, Quebec.*
Vermaas, F. H. S. (1953) *J. Chem. Metall. Min. Soc. S. Afr.*, **53,** 191.
Veblen., D. R. Buseck, P. R., and Burnham, C. W. (1977) *Science, N.Y.*, **198,** 4315.
Wells, F. W., (1929) *Am. J. Sci.*, **18,** 35.
Whittaker, E. J. W., in discussion on Hodgson, A. A. (1977) *Phil. Trans. R. Soc. Lond.*, **A286,** 611.
Woodrooffe, H. M. (1956) *Trans. Can. Inst. Min. Metall.*, **59,** 363.
Yada, K. (1971) *2nd International Conference on Asbestos, Louvain.*
Yada, K. (1975). *3rd International Conference on Asbestos, Quebec.*
Yang, J. C. (1961) *Am. Miner.*, **46,** 748.
Young. G. J., and Healey, F. W. (1954) *J. Phys. Chem., Ithaca*, **58,** 881.
Zukowski, R., and Gaze R. (1959) *Nature, Lond.* **183,** 35.
Zussman, J. (1955) *Acta Crystallogr.*, **8,** 301.

CHAPTER 4

Attitudes to asbestos

Seymour S. Chissick

King's College, London

Over the years, the medical and scientific evidence concerning the increased health hazards associated with occupational exposure to asbestos has become such that the United States Department of Labor Occupational Safety and Health Administration designated asbestos as a human carcinogen in 1975. Laboratory experiments with hamsters, mice, rabbits, and rats have shown that all commercial forms of asbestos are carcinogenic and the evidence concerning the occupational exposure of man indicates that this is also true for humans. There are indications that in addition to those employed in the mining and fabrication of asbestos, individuals living in the neighbourhood of asbestos factories and crocidolite mines and in households having contact with asbestos workers are at risk. The general population may also be exposed to asbestos fibres in various ways, via air, food, drink, medicinal preparations, and consumer asbestos-containing items.

At present, it is not possible to assess whether there is a level of exposure to asbestos for humans below which an increased risk of cancer would not occur (International Agency for Research on Cancer, 1977). Because of the very widespread usage of asbestos and the possible risks, different attitudes to the material are adopted at this time by different groups involved, i.e. governments, employers (mine operators, manufacturers, and users), employees (including trades unions), and the general public.

(1) THE ATTITUDES OF GOVERNMENTS

When the various effects of exposure to asbestos dust became apparent, in the United Kingdom Her Majesty's Government introduced the Asbestos Industry Regulations 1931. These regulations [designed to reduce the dust hazard in those parts of the industry, largely involving asbestos textile manufacture (but also including asbestos–cement, brake linings and insulation material and a total of some 300 factories were registered), where asbestosis appeared to occur] required the adoption of certain precautions aimed at reducing the exposure of asbestos workers to the dust. The general effect of the regulations was to require the use of exhaust ventilation systems in such a way as to ensure that no asbestos dust was allowed to escape into the air of any room where people worked, and to require that breathing apparatus and special clothing should be provided for and used by workers where such dust was unavoidable.

The application of the required preventive measures ensured a significant reduction in the incidence of asbestosis in those processes to which the Regulations applied, but there were a number of processes, such as the application and removal of asbestos insulation and lagging, to which the Regulations did not apply. Also, the absolute ban on the emission of asbestos dust in those activities which were covered by the Regulations proved unworkable and unforceable in practice. Because of the long interval between first exposure to asbestos dust and the development of asbestosis, it took 25 years before it was apparent that while there was an obvious improvement in those parts of the asbestos industry where the 1931 Regulations applied, asbestosis was continuing to appear in all areas of the industry. To rectify this situation, the Asbestos Regulations 1969 came into operation in May 1970, and these apply to asbestos factories, construction sites (including building and demolition), engineering construction, electrical stations, and ships under construction and repair, etc. The 1969 Regulations thus have a wider application than the 1931 Regulations and they contain more positive standards of dust control. More recently, the Health and Safety at Work etc. Act 1974, Sections 2, 3, and 4 (concerning the general duties imposed on employers, the self-employed, and persons having control of non-domestic premises in connection with which a trade, business or other undertaking is carried on) relating to health, safety, and welfare at work may be applied to processes and places not covered by the Asbestos Regulations 1969, with the effect that similar standards to those required under the Act need to be adopted.

The Asbestos Regulations 1969 are designed to protect workers from industrial disease which might arise from the inhalation of asbestos dust (defined in the Regulations as dust consisting of or containing asbestos to such an extent as is liable to cause danger to the health of employed persons) and this protection is a statutory obligation wherever asbestos materials are used in such a way as to give rise to the emission of dust dangerous to the health of employees. The Regulations apply to every process (in the United Kingdom) involving asbestos or any article or compound wholly or partly of asbestos, except a process in connection with which asbestos dust cannot be given off. The major requirements of the Regulations are as follows:

(i) The process must be carried out in a safe way, i.e. under an exhaust draught, which prevents the entry of asbestos dust into the air of the workplace or, if this is not practicable, protective clothing and respiratory equipment must be provided for employees who must be fully instructed in its proper use.

(ii) The premises and plant must be kept clean, i.e. as far as practicable cleaning should be carried out so that asbestos dust neither escapes nor is discharged into the air of any workplace. This may be accomplished by means of vacuum cleaning equipment fitted with high-efficiency filters on the exhaust outlet and is facilitated by ensuring that all inner surfaces of walls are smooth and that rooms are constructed in such a manner that ledges and other surfaces where dust may settle are avoided. For premises coming into use for more than 8 h per week after May 14th, 1970, where asbestos dust may be found, the construction must be in accordance with the Regulations, i.e. there must be smooth, impervious interior surfaces and as few surfaces as practicable on which asbestos dust can settle, and a central vacuum cleaning system must be provided, piped through the building and no asbestos dust may escape or be discharged from it into the air of any workplace.

(iii) Storage and distribution of asbestos must be carried out such that no dust may escape, e.g. by the use of suitable closed receptacles.

(iv) Suitable accommodation must be provided for protective clothing and respirators, so that they can be put on, taken off, and stored conveniently; the protective clothing must be cleaned at suitable intervals.

(v) No young persons may be employed in designated activities (any work requiring respirators/other protective equipment, e.g. sweeping up dust).

While the Asbestos Regulations 1969 refer to a level of asbestos dust liable to cause danger to the health of employed persons, this level is not specified in the Regulations. Guidance as to what level is considered by H.M. Government as liable to cause such danger is contained in the Health and Safety Executive Guidance Note E10, *Asbestos-Hygiene Standards and Measurement of Airborne Dust Concentrations*, which replaces the earlier Department of Employment Technical Date Note 13, *Hygiene Standards for Airborne Asbestos Dust Concentrations for Use with the Asbestos Regulations 1969*. The Guidance Note is intended to provide interim guidance pending further recommendations from the Advisory Committee on Asbestos. It is emphasized in the Note that the criteria are provisional and that only practical guidance is given, it being left to the Courts to give a binding decision on the interpretation of the law. Indeed, the current standard may well not provide effective protection against asbestosis. In 1966 X-rays of asbestos workers, who entered the industry after the 1931 Regulations came into effect, indicated few signs of clinical disease. However, X-rays of the workforce then employed taken in 1970 showed that many employees now showed abnormal findings either in the lung or in the covering of the pleurae (Lewinsohn, 1972). There is thus a difference between the prevalence of abnormal X-ray findings among the workers X-rayed in 1966 and those X-rayed in 1970. In addition, among over 200 family contacts of former asbestos industry workers, nearly 40% have been reported as having X-ray changes (lung scarring) characteristic of asbestos exposure (Anderson *et al.* 1978), even though the individual levels of exposure must have been much lower than those of occupational circumstances.

However, until the Advisory Committee on Asbestos provides its recommendations, the Health and Safety Executive will use the following criteria in assessing whether or not there is a compliance with the Asbestos Regulations and the Health and Safety at Work etc. Act 1974:

> the occupational exposure to asbestos dust should never exceed 0.2 fibres of crocidolite per cm^3 of air when measured over any 10-min period or 2 fibres per cm^3 of air of other types of asbestos when measurements are averaged over a 4-h period, and short-time exposure should not exceed 12 fibres per cm^3 of air when measured over any 10-min period.

The approach of the American Government has been to lay down by law the maximum allowable exposure of workers to asbestos, above which protection is deemed to be necessary. The 1970 Occupational Safety and Health Act emphasized the need for standards to protect the health of workers exposed to potential hazards at their place of work and a standard for occupational exposure to asbestos was included in the Occupational Health and Safety Act Standard published in 1971. The Standard, based on a 1969 Federal Standard issued under the Walsh-Healey Public Contracts Act, set an exposure limit of 12 asbestos fibres (of length greater than 5 μm) per cm^3 of air. Within 7 months, as a result of a petition, an emergency temporary standard for occupational exposure was published (effective from July 1972) which stated that:

> the 8-h time-weighted average concentration of asbestos dust to which employees are exposed shall not exceed 5 fibres per cm^3

[greater than 5 μm in length and as determined by the membrane filter method at 400–450× magnification (4-mm objective) phase-contrast illumination]. Concentrations above 5 fibres per cm^3 of air but not exceeding 10 fibres per cm^3 of air may be permitted up to a total of 15 min in an hour for up to 5 h in an 8-h day.

In October 1975, the Occupational Health and Safety Administration proposed a revision of the United States standard for occupational exposure to asbestos to apply to all employments covered by the Act, excluding the construction industry, for which a separate revision was to be applicable. The proposed rules were to have the following effects:

(i) The maximum permissible exposure to asbestos to be 0.5 fibres per cm^3 of air for an 8-h time-weighted average (TWA) with a ceiling exposure of 10 times this concentration for any period not exceeding 15 min.

(ii) Work areas where a person may be exposed to above-average concentrations of asbestos fibres in excess of either of the two limits to be designated Regulated Areas and only authorized persons to be allowed to enter such areas. The authorized persons to be provided with protective clothing and receive information, instruction, and training relevant to the nature of their work. Danger signs to be displayed at each regulated area and danger labels afixed to all asbestos containers where the contents may release asbestos fibres in excess of the exposure limits.

(iii) The retention period for monitoring and medical records to be the duration of employment plus 20 years or 40 years, whichever is the greater.

(iv) Employee exposure to asbestos fibres to be controlled to or below the permissible exposure limits by engineering controls, work practices, and personal protection.

The Occupational Health and Safety Administration realized that the stated TWA exposure may well not be a safe level of exposure: ideally, for known carcinogens there should be no detectable concentrations. However, the Occupational Safety and Health Act requires that feasible standards be established, reflecting both technological and economic factors and the Occupational Safety and Health Administration deemed the stated figures to be appropriate. However, because of technological and economic problems, the effective date for the proposed exposure levels will be delayed. The current (effective from July 1976) 8-h TWA airborne concentration of asbestos fibres to which any employee in the U.S.A. may be exposed is a maximum of 2 fibres longer than 5 μm per cm^3 of air, with a ceiling concentration of 10 such fibres.

It was proposed that the concentration of airborne asbestos fibres to which a single unprotected worker may be exposed in a shift in the work area concerned be determined in the working location likely to contain the highest concentration of asbestos fibres. The result could then be deemed to apply to all of the employees in the work area and to other shifts if it could reasonably be considered that there was the same level of exposure. The employees or their representatives could observe all steps in the measuring/monitoring procedure and were entitled to receive an explanation of the procedures and details of the results. Monitoring is carried out by the membrane filter method (Leidel *et al.*, in press) using a battery-powered personal sampling pump. Air is drawn through a cellulose ester membrane filter and the resulting sample is transformed into a transparent, optically homogeneous gel, and fibres are sized and counted by phase-contrast microscopy at 400–450× magnification.

Where the level of exposure may be above the permitted levels, it was proposed that employers be required to establish and implement a written programme to reduce the exposure to or below the permitted exposure limits using engineering and work practice controls only, e.g. general mechanical ventilation, local exhaust ventilation systems (on saws, abrasive wheels, drills, etc.), and handling, cutting, removing, etc., asbestos in a sufficiently wet state to prevent the emission of airborne asbestos fibres in excess of the exposure limits. Respirators, which must be approved by the National Institute for Occupational Safety and Health, were not to be used in order to achieve the permissible exposure levels under normal working conditions. Respirators were to be used only during the time necessary in order to install emergency controls or implement work practice controls designed to reduce exposure to or below the permitted levels, or in work situations in which engineering controls and work practice controls were insufficient to reduce exposure to below the permitted levels, or in an emergency. Respiratory protection against asbestos fibres is summarized in Table 4.1.

Employees assigned to regulated areas were to be provided with a training programme (details of which were to be made available to the Occupational Safety and Health Administration on request) by the employer, at the time of initial assignment and at least annually thereafter. The employees' training programme would include details of the following: the specific nature of the operations which could result in exposure to asbestos fibres and the necessary protective steps; the engineering controls and the work practices associated with an employee's job assignment; the purpose, proper use and limitations of respiratory protection equipment; the purpose of the medical surveillance programme and a description of the programme; and a review of the standard for occupational exposure to asbestos. The standard and its appendices to be made readily available to all employees working in regulated areas.

One purpose for the proposed establishment of Regulated Areas was to

Table 4.1 Respiratory protection against asbestos fibres (U.S.A.)

Over 2000 times the applicable exposure limit prescribed.	Self-contained breathing apparatus with a full facepiece in pressure-demand (positive-pressure) mode.
Up to 2000 times the applicable exposure limit prescribed.	A type C supplied air respirator with a full facepiece operated in pressure-demand or other positive-pressure mode or with a full facepiece, hood, or helmet operated in continuous flow mode.
Up to 1000 times the applicable exposure limit prescribed.	(A) A type C supplied air respirators operated in pressure-demand or other positive-pressure or continuous-flow mode; or (B) a powered air-purifying respirator with a high efficiency particulate filter:* or (C) self-contained breathing apparatus in pressure-demand (positive-pressure) mode.
Up to 50 times the applicable exposure limit prescribed.	(A) A high-efficiency particulate filter respirator with a full facepiece; or (B) any supplied air respirator with a full facepiece; or (C) any self-contained breathing apparatus with a full facepiece.
Up to 10 times the applicable exposure limit prescribed.	(A) Any air-purifying respirator with replaceable particulate filter; or (B) any single use respirator with or without valve; or (C) any supplied air respirator; or (D) any self-contained breathing apparatus.

* High-efficiency filter: 99.97% efficient against 0.3 μm size dioctyl phthalate.

limit the exposure to high levels of asbestos fibres to as few persons as possible. A roster of all persons entering a Regulated Area was to be made daily and maintained for at least 40 years or the period of employment plus 20 years, whichever is the longer. Clean, dry, whole-body protective clothing, head covering, gloves, and foot covering were to be provided by the employer at least once per day and used by the employee. Special provision was to be made for changing, showering, and toilets, etc.

The extended retention period for records was to assist with the development of essential data on the causes of asbestos-related diseases, some of which are known to have a latency period of up to 40 years. The resulting epidemiological and diagnostic investigations could be used to determine, for example, dose–response (exposure–pathology) relationships in asbestos-related diseases. Company records, which were to be made available to the Occupational Safety and Health Administration on request, were to include specific items of data regarding employee exposure monitoring and medical examinations. The employee exposure measurement records were also to be made available to the individuals concerned and their designated representatives for examination and copying. Considerable modifications have been introduced into the proposals following the detailed comments of the U.S. asbestos industry (see p. 135).

The format in which the records can be present can vary and Appendices 4.1–4.6 are intended to show the type of information and detail required.

Because of the widespread public anxiety and uncertainty about the health risks from asbestos, the British Health and Safety Commission, in agreement with the H.M. Ministers concerned, set up an Advisory Committee on Asbestos in 1976, to undertake a wide-ranging review of health risks to workers and members of the public which may arise from exposure to asbestos. The Committee, under the Chairmanship of Mr. William (Bill) Simpson (who is also Chairman of the Health and Safety Commission) has the following terms of reference:

> 'To review the risks to health arising from exposure to asbestos or products containing asbestos including:
> persons exposed at work;
> members of the public exposed to asbestos generated from work activities;
> members of the public exposed to asbestos from consumer products and from asbestos waste.
> To make recommendations as to whether any further protection is required'.

The Committee has published an interim statement [*Asbestos Health Hazards and Precautions* (HMSO, London, 1977)] designed to put the risks from asbestos in perspective, in the light of present knowledge. In the statement the Committee recommends that great care be taken in the use of asbestos and materials containing it, particularly:

(*a*) exposure to all forms of asbestos dust should be reduced to the minimum reasonably practicable;

(*b*) occupational exposure to asbestos dust should never exceed the following: for crocidolite, 0.2 fibres per cm^3 when measured over a 10-min period; for other types of asbestos, 2 fibres per cm^3 when measurements are averaged over a 4-h period; short-term exposure should not exceed 12 fibres per cm^3 when measured over any 10-min period.

The Committee reccomends the following control measures, which are reproduced by permission of Her Majesty's Stationery Office:

(*i*) *Asbestos in the workplace and in construction and shipbuilding operations*

The following principles of control should be observed in order to minimize the risks:

(a) The creation of dust containing asbestos should be avoided. When it is impossible to avoid creating asbestos dust, it should be controlled by means of enclosure or local exhaust ventilation. Where these methods are impracticable, respiratory protective equipment and protective clothing should be provided and used. The protective clothing should be laundered under special arrangements and should not be taken home. In any event, strict regard should be paid by both employers and employees to the necessary precautionary measures and the relevant legislation.

(b) These principles should also be applied to work with asbestos-based products in construction operations and shipbuilding. However, the Committee recognizes that some processes will inevitably create some dust and where the work is of a short-term nature, the provision of local exhaust ventilation may be impracticable. In such circumstances dust concentrations can be significantly reduced by the use of hand tools instead of power tools.

(c) A very high standard of precautions should be observed in the removal of thermal insulation containing asbestos (de-lagging) and sprayed asbestos coatings (as, for example, during demolition and shipbreaking operations). These precautions should include complete segregation of the working area and thorough soaking of the insulation with water or wet steam. In addition, workers should be provided with, and wear, full protective equipment.

(d) Safe methods of disposal should be used. Dust and waste should be put into impermeable containers which are then tightly sealed. Guidance on the precautions to be taken when asbestos waste is produced or disposed of is given in the Code of Practice for Handling and Disposal of Asbestos Waste Materials produced by the Asbestosis Research Council.

(e) The risk of getting lung cancer from inhaling asbestos is many times more serious if one smokes cigarettes. Smoking should be avoided by anyone who is exposed to asbestos dust regularly.

(*ii*) *Asbestos in existing buildings*

Present evidence suggests that dangers from asbestos in buildings are likely to arise only when products containing asbestos are damaged, either accidentally or during maintenance or repair, and the asbestos fibres are released and dispersed in the air. Where friable materials, e.g. sprayed asbestos insulation, have become or could become damaged, they should be either removed or protected by a suitable coating or covering. The disadvantage of the second solution is that the problem can recur if the coating or covering becomes damaged or deteriorates. On the other hand, during the period of removal of asbestos materials incorporated in buildings, dust levels will usually be generated higher than those that will occur if the asbestos materials are left undisturbed, so in many cases the second solution will be preferable. The balance of advantage will vary from building to building.

(iii) Asbestos in domestic products

The main non-commercial use of asbestos is in do-it-yourself building materials. There are also some domestic products which contain asbestos, such as some pieces of electrical equipment. There is little risk of fibres being dispersed from domestic products in normal use provided they are in good condition. In using do-it-yourself materials containing asbestos the following principles should be observed to minimize the risk:

(a) Avoid creating dust. Use hand tools instead of power tools. Keep the work damp as far as possible.

(b) If it is impossible to avoid creating dust, use a damp cloth to clean up so that the dust is not inhaled. If possible, do the work outside or in a well ventilated space, and work upwind of the source of the dust and away from bystanders.

(c) Cloths used for cleaning up should not be simply left to dry out. Wash them thoroughly or (better still) dispose of them properly while still damp, by sealing in plastic bags.

It is also wise to seek advice before disturbing or fixing asbestos roof insulation material or asbestos in central heating systems.

In addition to the work of the Asbestos Advisory Committee, the Government-operated Employment Medical Advisory Service has been involved in a continuing survey of asbestos workers, in conjunction with the Occupational Medicine and Hygiene Laboratories at Cricklewood, since 1971. The main purpose of the survey is to test the effectiveness of the 1969 Asbestos Regulations (particularly in dust control methods) and to learn as much as possible about the medical effects of exposure to asbestos of different types and of varying, but measured, doses of dust. Over 1200 workers were involved in phase 1 of the survey, which is based on a biennial brief clinical medical examination, completion of a standard respiratory symptom and smoking questionnaire, and a chest X-ray. The results (clinical and radiological changes) are related to the dust estimations carried out at the work places, obtained by sampling the atmosphere over a 4-h period by personal samplers using one man in ten of those who are potentially exposed to asbestos dust.

In phase 2 of the survey, which embraced the larger manufacturers of asbestos products, 5000 workers were medically examined and 700 personal air samples taken; 92.6% of the dust counts were below 2 fibres per cm^3. Phase 3 of the survey started in 1973 and is meant to cover all remaining workers covered by the Asbestos Regulations.

The survey is aimed to cover some 24 000 workers in the asbestos industry and so far X-ray examinations of nearly 1000 of the workers have revealed early signs of asbestosis and they have been referred to their doctors. In a BBC news broadcast (December, 1977) it was alleged by Mr. Max Maddon, MP, that, in the case of a further 627 of these workers, the X-rays showed some discrepancy which could indicate the first signs of asbestosis and that the Employment Medical Advisory Service had decided not to inform these people, thereby withholding information from possible risk people when passing it on could help safeguard their health. The matter had been taken up with the Department of Employment and there was grudgingly implied confirmation of the situation. In the same radio broadcast, the Government action was defended by Dr. Kenneth Duncan of the Health and Safety Executive, on the basis that there is no evidence to suggest that anything can be done clinically for the parties concerned. They come into category 01 and the Health and Safety Executive is only interested in parties in category 1—diagnosable asbestosis—where something useful can be done for the patient. Category 01 patients include those with chronic bronchitis and those with doubtful X-rays. There is no reason to believe that large numbers of these workers in category 01 do in fact have early diagnosable asbestosis, although some may have.

The category 01 asbestos workers who stay in the industry are seen after a further 2-year period, when the doubt may have disappeared or some change may have occurred to bring them into category 1. The most useful thing a category 01 worker could do would be to give up smoking; leaving the industry makes little difference at this stage. All asbestos workers are advized to give up smoking in any case, but little attention is paid to the advice. The Employment Medical Advisory Service does not inform category 01 patients because they do not wish, nor do they need, to change their life styles at this stage, and the claim is made that nothing is being withheld from these people because there is nothing useful to tell them—there is no point in appearing to be terribly concerned by putting into a situation something that is not there.

Among actions taken by other governments, the Swiss safety authority has in practice specifically barred the use of asbestos construction products, including those used for building insulation and sound-proofing, and is committed to a decision on the use of asbestos for lagging, pipes, roofing, and fire insulation. The use of any new asbestos-containing product must be notified by Swiss employers to the Schweizerische Unfallversicherungsanstalt (SUVA—Swiss National Institute for Insurance Against Accidents) and to satisfy the SUVA that there is no 'suitable' alternative available. The SUVA arranges for tests to be carried out in workshops and on construction sites to make sure that the MAK standard of 1 mg of asbestos per m^3 (equivalent to 2 fibres per cm^3) of air during 1 week of 8–9 working hours per day is not exceeded. If the level is exceeded the workforce must wear filter masks. Since January 1972, Denmark has had a regulation prohibiting the use of asbestos (but not asbestos-cement products) for insulation against noise, etc. The Dutch Safety Institute has banned crocidolite asbestos in the Netherlands from April 1st, 1978, and allowable concentrations of other kinds of asbestos dust in the workplace are reduced to the same levels as in the U.K. The spraying of asbestos is forbidden. The Netherlands currently uses 38 ktons of asbestos per annum (50% of which is crocidolite) in the manufacture of asbestos–cement and for insulation and reinforcement. Rules are being introduced to control disposal of asbestos waste and for the notification of the Factory Inspectorate when new processes are started or existing processes modified.

(2) THE ATTITUDES OF THE ASBESTOS INDUSTRIES

In order to appreciate the attitudes and actions of the British asbestos industry at various times it is necessary to realize that for many years neither the industry (which has the prime responsibility for making the arrangements to protect workers from dangers to health from asbestos) nor H.M. Factory Inspectorate (who, until 1974, took the responsibility for enforcing the legislative health and safety requirements—a role which was taken over in 1974 by the Health and Safety Inspectorate) were fully aware of the full extent of the dangers from the dust produced by processing asbestos or of the means of controlling the dust. The 1930 Report (Merewether and Price, 1930) *Effects of Asbestos Dust on the Lungs and Dust Suppression in the Asbestos Industry*, the so-called Merewether Report, stated (p. 12):

> *. . . that in order to prevent the full development of the disease (asbestosis) amongst asbestos workers within the space of an average working lifetime, it is necessary to reduce the concentration of dust in the air of the workrooms to a figure below that pertaining to spinning at the time over which these cases (i.e. those dealt with in the Merewether, 1930 Report) were exposed . . .*

This was taken to imply that the conditions arising from fibre spinning carried on without exhaust draught may be productive of so little dust that exhaust ventilation for its suppression was not (in the light of the existing knowledge) necessary, and thus the dust produced from fibre spinning under good conditions (but without exhaust ventilation) was regarded as the 'dust datum'. Hence, no special recommendations were made concerning dust suppression in those areas of the industry where the amount of dust normally evolved was regarded as *safe* with reference to the datum level. As a result, there was a general acceptance during a period of over 30 years—between the publication of the Merewether Report and the development in the 1960s of the scientific knowledge that the dangers were greater than had originally been appreciated—of dust conditions which it has subsequently been established should not have been allowed. Thus, the accepted conditions existing in the asbestos industry between 1931 and 1964 arose out of the imperfect understanding of the dangers of asbestos dust and the tendency to accept that the implementation of the recommendations of the Merewether Report would suffice to deal with the dangers as they were then understood. There was also the problem of enforcing the 1931 Regulations during this time: it was not practicable to comply with the statutory requirements to exclude asbestos dust absolutely from the workplace; there was no fully reliable means of accurately assessing how much asbestos dust was present in the atmosphere of the workplace; in order to effect a successful prosecution, it was necessary to prove injury to health and the state of medical knowledge was such that the Medical Inspectorate were reluctant to give firm evidence of a link between a case of asbestosis and the specific conditions in a factory at a particular moment of time.

During the 1960s, it became apparent that asbestos dust was more dangerous than had been assumed and that the standards of dust prevention which had earlier been believed to be adequate to keep the risks to health within tolerable limits were not, in fact, adequate. This led to reviews of the legislative and medical aspects of the problem and the development of new safety standards.

The chronology of some of the major events in the Industry in the U.K. over about the last, 80 years, with particular reference to the (*i*) declared and (*ii*) actual attitudes of the Industry at the time, and the prevailing detailed knowledge of the health hazards associated with asbestos and the scientific ability to measure meaningfully the concentration of asbestos dust in air, is traced on p. 6–9. It is clear that in spite of the expansion of asbestos textile manufacture in the U.K. during the first quarter of the 20th Century, it was only towards the end of the 1920s that the special biological effects of asbestos started to become generally recognized. The hazards were first appreciated in textile production because the processes were largely dry and readily produced dust, and many of the workers formed part of a close-knit group whose employment records went back for a considerable period (providing a large sample of people exposed to dust over varying periods on which investigations into the hazard could be carried out). The year 1931 is regarded in the British asbestos industry as the turning point. Prior to this time workers were exposed to excessive and dangerous concentrations of asbestos dust through ignorance on the part of all concerned (employers, Government Inspectors, and the medical profession). However, following the Merewether Report and the pooling of information and experience by the manufacturers, there were rapid and significant improvements in methods of dust control—enclosing machinery, damping the work, extracting dust from the breathing zone of operators, and greater use of respirators. These activities on the part of that part of the industry covered by the Asbestos Regulations produced a significant reduction in the incidence of asbestos disease in the factories. However, the Regulations did not apply to work outside the factories, some of which, e.g. work with thermal insulation, involved relatively high levels of asbestos dust. Because asbestosis can take years to develop (the average period is about 17 years), many years were to pass before it was realized that the earlier generally accepted practices and levels of exposure were not adequate for the complete elimination of the disease. This realization led to the introduction of the 1969 Asbestos Regulations, imposing very stringent standards on virtually all parts of the manufacturing industry. However, because of the time factor in the development of asbestosis, it will not be until the late 1980s that the full effect of these Regulations can be evaluated. The Asbestos industry is optimistic that the current application of the 1969 Regulations will solve the asbestosis problem and reduce the risks to a negligible level—with the reservation that absolute safety can no more be guaranteed for asbestos handling than it can for any other industrial process.

The Asbestos industry in the U.K. has imposed a voluntary ban on the import of blue asbestos fibre since 1970 and accepted resonsibility for its workers who contract asbestos-related disease, by the provision of medical care, compensation, and the transfer of affected employees to lighter, less exposed work. The Industry believes that the penalties for infringement of the Asbestos Regulations should be severe. The following extract from the Policy Manual of one of the U.K.'s largest manufacturers of asbestos-containing materials illustrates the present standards of medical control practised in the industry (reproduced by kind permission of the Asbestos Information Committee).

MEDICAL EXAMINATIONS: Establishments Handling Asbestos

1 This instruction outlines the Group Policy on the medical examination of those employed within establishments manufacturing asbestos products, or handling asbestos products in any way. It also explains the procedure for initial and periodic X-Rays.

2 Regulations

2.1 So far as the Asbestos industry is concerned, a person working in any of the following processes, unless carried on occasionally, is liable for medical examinations:
2.1.1 breaking, crushing, disintegrating, opening or grinding of asbestos and the mixing or sieving of asbestos, or any admixture of asbestos and all processes involving manipulation of asbestos incidental thereto;
2.1.2 all processes in the manufacture of asbestos textiles including preparatory and finishing processes;
2.1.3 the making of mattresses composed wholly or partly of asbestos and processes incidental thereto.
2.2 The National Insurance (Industrial Injuries) (Prescribed Diseases) Regulations 1959, which should be studied by all management concerned, lay down that persons in such employment must be medically examined *initially* before the end of the second month of employment and *periodically* thereafter every 2 years, or as directed by a pneumoconiosis medical board.

3 Group Policy

3.1 It is Group Policy to require a regular medical examination of all employees whose work involves regular contact with asbestos or products containing asbestos.
3.2 In factories or work situations where asbestos is not used, medical examinations and X-Rays will only be carried out as determined by the Senior Director or Manager on site on the advice of the Medical Officer and then only in individual cases as needed.

4 Situations where asbestos is handled or products containing asbestos are handled

4.1 The full medical examinations and X-ray procedures will apply and the necessary records must be kept in all situations where asbestos is handled or products containing asbestos are handled.
4.2 Categories of employees which must be covered are:
4.2.1 All hourly paid works employees—irrespective of where they work.
4.2.2 Staff employed in the following categories:
Works managers
Works supervisors
Technical, R & D and laboratory staff
Work study engineers
Quality control staff
Factory, clerical and stores personnel
4.2.3 Any other staff who regularly enter Production Departments, Laboratories or Stores.
4.3 Other staff located on the premises are not included in these requirements, but they should be offered medical examinations and X-rays on a voluntary basis.

5 Medical examinations and X-rays

5.1 All employees defined in the Category 4.2 will be medically examined and X-rayed on or after recruitment (see 6.3.1) or when changing their occupation, unless they have already been examined and passed fit within the last 6 months.
5.2 X-rays will continue at 2-yearly intervals or more frequently, if in the opinion of the Medical Officer this is necessary in an individual case.
5.3 Medical examinations will continue at the descretion of the Medical Officer.
5.4 Employees, including staff grades, who have at any time been in occupations defined in 4.2 should continue to be X-rayed at least every 2 years, and examined at the discretion of the Medical Officer. The Factory Doctor will notify the Plant Manager or Personnel Department if this frequency needs to be increased in the case of any particular individual, and also when this precaution becomes unnecessary.
5.5 It is also most important to ensure that individuals who no longer work in asbestos situations continue to be medically examined and X-rayed at the requisite intervals throughout the remainder of their service with the Company.

6 Recruitment

6.1 The greatest possible care must be taken to ensure that persons with any health risk, particularly a chest complaint, are not recruited, and whenever possible initial examinations should be arranged before the person terminates his previous employment. Where this is difficult, employment should be made conditional upon satisfactorily passing a medical examination (including X-ray where appropriate) subsequently. This should be arranged through a local chest physician if it is not possible for the person to attend a company-sponsored examination.
6.2 Contracts of employment should be suitably worded to include the requirement for initial and subsequent medical examinations.

7 Records

7.1 Suitable records will be kept showing the department where each man has worked—with the exact dates of entering and leaving each department. These records must be complete—over the full period whilst the individual works for the Company and must not—under any circumstances—be destroyed.
7.2 A suitable cross-reference system must also operate, so that the individual records and dust count records are capable of being compared, e.g. names and geographical boundaries of departments must be identical.
7.3 The dust count records will be maintained as instructed separately.
7.4 A medical record card must be maintained by the medical staff on a strictly confidential basis for each individual. This must show the date and details of all medical examinations and X-rays. (For administrative reasons this should also be duplicated on the Personnel records.)

8 Responsibilities

8.1 It is the responsibility of the Director or Manager responsible for each location to ensure that the legal requirements, and the requirements of this instruction, are carried out, and to ensure that proper records are kept.

Asbestos Regulations 1969

Instructions for Action

1 Introduction

These instructions replace and bring up to date those issued in November 1970. The objective continues to be to reduce health risks to employees by ensuring a uniform standard of control of asbestos dust, and a consistent interpretation of the Asbestos Regulations within the Group factories.

2 Responsibilities

Factory managers or managers responsible for site operations must be fully aware of their responsibilities for the health and safety of employed persons. These responsibilities are comprehensive, involving employees, sub-contractors and plant, and are clearly set out in Regulation 5.

3 Interpretation of the Regulations concerning dust concentrations

3.1 H.M. Factory Inspectorate (HMFI) regard the achievement of a maximum average concentration of 2 fibres per cm^3 over a 4-hour working period as their ultimate objective, with peak concentrations in any 10-min period not exceeding 12 fibres per cm^3
3.2 On the basis of this interpretation, the following categories of concentrations have been established. The action to be taken is also indicated.

3.2.1 HIGH DUST CONCENTRATION

Where the concentration exceeds 12 fibres per cm^3 in any two consecutive 10-min samples, the area must be clearly identified by a notice (as specified in Section 5). Respirators of the approved type and protective equipment must be provided for and worn by each person employed in that area or entering that area while operations are in progress. HMFI will press for improvement in dust extraction equipment where this is practicable, or other forms of control.

3.2.2 MEDIUM DUST CONCENTRATION

Where concentrations are above 2 fibres per cm^3 but below 12 fibres per cm^3 in successive 10-min samples, action to be taken will be based on the average concentration over a

4-h continuous period of work (normally a half shift). Improvement in dust control will be required if the average concentration exceeds 2 fibres per cm^3. Some Inspectors may stipulate the wearing of protective equipment in such cases; if this happens the advice of the Group Consultant for Environmental Control should be obtained. If the average concentration exceeds 2 fibres per cm^3 and it is impracticable to provide dust extraction equipment, respiratory protection must be provided.

3.2.3. LOW DUST CONCENTRATION

Where concentrations are so controlled by exhaust ventilation that they do not exceed 2 fibres per cm^3, no further action will be required other than observance of the general requirements of the Regulations as to maintenance of exhaust equipment and high standards of cleanliness.

3.2.4 DUST-FREE AREAS

Where the nature of the process or operation is such that asbestos dust concentrations are below 2 fibres per cm^3, the substantive provisions of the Regulations, particularly those concerned with the provision of dust extraction equipment or protective equipment, will not be enforced.

4 Monitoring and identification of dusty areas or operations

4.1 Monitoring will normally be undertaken in order to provide information concerning one or more of the following:

4.1.1 MACHINE OPERATION

Tests will be taken at locations or stations situated not more than 1 m from the machine or part of the machine on or with which the process or operation is taking place and at that stage in the process when dust emission is likely to be at its peak.

The test equipment will normally be mounted in a position corresponding to the breathing zone of an operative. Where different types or grades of material are processed, measurements should be made when working with the material likely to produce most dust.

The results are identified as 'machine counts'.

4.1.2 GENERAL ENVIRONMENT

Tests are taken at fixed positions within an area or department, these being typical of the concentration encountered by any person entering that area.

The results are identified as 'background counts'.

4.1.3 OPERATOR'S BREATHING

Tests are taken in personal positions or positions changing with the movement of the operative, representing the atmosphere which he breathes during his working day. A sample will be taken in the 'operator's breathing zone' and the results identified as such. Casella personal samplers, wherever possible worn by the operative, should be used.

4.2 The basic method of measurement to be employed is described in the Asbestos Technical Committee (ARC) Technical Notes 1 and 2. It is desirable that the location and sampling period of measurements should correspond as closely as possible with those which may be used by HMFI. For all categories, the standard sampling time should be 10 min, and wherever possible at least two samples should be taken. It is only necessary to sample over a longer period (up to 4 h) when the 10-min sample indicates dust levels above 2 fibres per cm^3 and in order to test whether, on a time-weighted average, the level will be acceptable. It is not essential for a full 4-hour period to be measured. If a shorter period, normally not less than 30 min, shows that the average level is below 2 fibres per cm^3, no further measurement is necessary.

4.3 It is necessary to identify all areas, operations and processes affected by the Regulations and to determine what further action, if any, is necessary. The overall objective is to reduce the dust concentration in all areas below 2 fibres per cm^3.
4.4 A plan of the factory upon which 'areas or sites of high dust concentration' are clearly marked should be kept up to date and be available for inspection. Any area or operation, however limited, in which an operative may be exposed to asbestos dust concentrations above 12 fibres per cm^3 should be identified in this way.
4.5 All positions within the factory where asbestos or asbestos products are processed or handled must be monitored regularly and permanent records kept of the results obtained. They will be identified on factory plans and sufficient information recorded to ensure consistency of repeat sampling. A summary of the results of monitoring should be prepared monthly and a copy sent to the Group Consultant for Environmental Control.
4.6 The frequency of measurement on a routine (as distinct from an experimental or exploratory) basis should be as follows:
4.6.1 Areas or operations where asbestos dust may be present but where experience has shown the dust levels to be below 2 fibres per cm^3—not less than once every 3 months.
4.6.2 Areas or operations of low dust concentration—not less than once every 3 months.
4.6.3 Areas or operations of medium or high dust concentration—not less than once per month.

N.B., Where operations producing medium or high concentrations are the subject of planned modification, they should be designated as such in the monthly report, with a note of the date when the modification is to be completed. Relevant dust counts need only then be reported quarterly until the planned completion date is reached.
4.6.4 Areas where high dust concentrations may occur but which are not working areas (e.g. dust collecting chambers, or enclosed areas containing automatically operated processes) should be monitored at 3-monthly intervals.
4.6.5 The category of dust concentrations of an area or operation should not be lowered, so reducing the frequency of monitoring, unless there is firm evidence that the improvement in dust levels can be consistently maintained.

5 Factory notices

5.1 A copy of the statutory extract of the Asbestos Regulations 1969 (Form F. 2358) must be exhibited on the works notice board.
5.2 Factory notices to be displayed in areas of high dust concentration should read as follows:

> PROTECTED AREA
>
> Respirators and protective overalls must be worn by all persons entering or working in this area while work is in progress

The words 'while work is in progress' should be omitted if the area is an enclosure to which entry is forbidden while machinery is running, and where loose fibre or dust is liable to be encountered when entry is made for maintenance or other purposes. In other cases, the word 'area' may be inappropriate; in this event a more suitable description should be used.

6 Exhaust ventilation

6.1 Unless it is impracticable, exhaust ventilation must be provided, maintained and used in all areas where the dust concentration regularly exceeds 2 fibres per cm^3.
6.2 The equipment must be inspected at least once in every 7 days (this can be done by the Department Supervisor) and a check list retained by the factory management.
6.3 The installation must be thoroughly examined and tested by a competent person at least

once in every 14 months. A signed report of the result must be made within 14 days and attached to the general register.

7 Respirators and protective clothing

7.1 Respirators and protective clothing must be worn at all times by all persons present in areas of high dust concentration.
7.2 The types of respirators approved for use within factories are listed in the ARC Control and Safety Guide No. 1 (Appendix A). The familiar dust mask (ori-nasal type) is suitable where concentrations do not exceed 40 fibres per cm^3. Pressure-type respirators must be used if concentrations are above 40 fibres per cm^3 and air line respirators where concentrations exceed 800 fibres per cm^3. Managers must have a supply of these special types available or know where they can be obtained at short notice.
7.3 Protective clothing must be such as to prevent asbestos dust becoming deposited on the personal clothing of the employee, i.e. it must completely cover such clothing. Cuffs should be close-fitting and short sleeves, unless worn with bare arms, are not suitable. Head gear must also be provided. Synthetic fabrics, or mixtures of synthetic and cotton, are preferred, since asbestos fibre adheres to cotton. Sources of supply for suitable clothing are given in the ARC Control and Safety Guide No. 1 (Appendix C).

8 Issue and care of respirators and protective clothing

8.1 Each employee who regularly works in concentrations where protective equipment must be worn shall be issued with two respirators and duplicate sets of protective clothing.
8.2 Each employee required to wear respiratory protective equipment must be properly instructed in its use.
8.3 The employee will be required to sign a receipt for the equipment and an acknowledgement that he understands how and when it must be worn. A specimen form for this purpose is given in Control and Safety Guide No. 1 (Appendix B).
8.4 Arrangements must be made for the regular cleaning of respirators and for the replacement of filters. The frequency of such attention will depend on local circumstances. The normal life of the filter will be determined by examination. A properly trained person must be made responsible for respirator cleaning and maintenance. A maintenance record should be kept and a suitable form for this purpose is given in Control and Safety Guide No. 1 (Appendix B).
8.5 All protective clothing must be cleaned at suitable intervals; this can be carried out in the factory or by an outside contractor. In the latter case, all clothing must be packed in suitably sealed containers (polythene bags can be used) and the containers must be clearly and boldly marked 'Asbestos-contaminated clothing'. Arrangements for such packing should be discussed with the laundry before the contract is concluded.

9 Accommodation for clothing and respirators

9.1 Accommodation for clothing not worn during working hours (provided in accordance with Section 59 of the Factories Act) must be such or so situated that asbestos dust is not deposited on it. Lockers provided within departments where asbestos dust occurs must therefore be dust proof, or cloakroom accommodation must be provided elsewhere, in a clean room.
9.2 Where protective clothing has to be provided in accordance with the Regulations, accommodation must be provided in a conveniently accessible position:
(a) for putting on/taking off protective equipment and clothing;
(b) for the storage of such equipment and clothing when not in use;
(c) the equipment and clothing when not in use must be kept in the storage accommodation provided except when removed for cleaning and maintenance.

10 Employment of young persons

No-one under 18 years of age may be employed in any situation where it is necessary to wear respiratory protective equipment.

11 Factory cleaning

11.1 All areas and equipment must be cleaned regularly to ensure that asbestos waste and dust does not accumulate.
11.2 If vacuum cleaning or some equally dustless method is not possible, the area being cleaned must be treated as being of high dust concentration, warning notices should be set up (see Regulation 5), and all employees present must wear respirators and protective clothing.
11.3 Where large areas are involved, management should consider the organization of expert cleaning teams with special access equipment, etc. The teams would operate at night or weekend and, although receiving premium rates of pay, could possibly be more economical and certainly more likely to achieve the necessary standard.

12 Storage and handling of asbestos waste

12.1 The methods to be followed for the storage and distribution of asbestos fibre is included in the ARC Guide, *Handling Asbestos Fibre.* An impermeable bag, securely tied, is a suitable form of closed receptacle. On no account may non-impermeable bags (e.g. hessian sacking) be used for storing or collecting loose asbestos, mixtures of asbestos with other materials, or asbestos waste.
12.2 Loose asbestos, waste, or mixtures of asbestos with other materials must not be transported about the factory except in receptables closed so that dust cannot escape. Open trucks, skips, drums, etc., must not be used.
12.3 Disposal of waste must be organized in strict compliance with the ARC Code of Practice. This has been issued by the Ministry of Housing and Local Government to local authorities who may be expected to insist on these methods. In particular, contracts for collection and disposal of waste must be drawn up in compliance with and with reference to these provisions.
12.4 These instructions may require some amendment in the light of the Poisonous Waste Act to be enacted during 1972 (*Ed.*—this Act was repealed in 1974 with the introduction of the Control of Pollution Act).

13 Meal and other breaks

The Factories Act (1961) prohibits the taking of breaks or meals in the room where processes are going on (other than intervals allowed in the course of spells of continuous employment). Protective clothing, issued under Regulation 8, must not be worn in premises provided for taking meals or breaks.

14 New buildings

Special consideration must be given to the requirements of the Regulations (Regulation 13) when buildings are being designed or converted.

15 Crocidolite

The Group Consultant for Environmental Control should be consulted regarding the special precautions necessary if it is proposed that crocidolite, or products containing crocidolite, be used or stored for any reason.

In America, the progress of regulations concerning asbestos usage followed a somewhat different course and it is interesting to review the attitude adopted by a member of the U.S. Asbestos Industry to the Occupational Health and Safety Administration's proposed rules concerning occupational exposure to asbestos (*Federal Register*, October 9th, 1975).

Occupational Health and Safety Administration Proposals	*Proposed Revision by Johns-Manville Corporation**
(1) The scope of the rules to apply to every place of employment (but construction work, etc., is excluded) where asbestos or a product containing asbestos is manufactured, processed, packaged, stored, applied, used, or otherwise handled.	The scope of the rules to apply to every place of employment (but construction work, etc., is excluded) where asbestos or a product containing asbestos is manufactured, processed, packed, applied, or used.
(2) 'Asbestos' includes chrysostile, amosite, crocidolite, tremolite, anthrophyllite, and actinolite, and every product containing any of these minerals.	'Asbestos' includes chrysolite, amosite, crociodolite, and members of the tremolite, anthrophyllite, and actinolite mineral group when they occur in fibrous habit.
(3) 'Emergency' means an unforeseeable and unexpected occurrence likely to release airborne concentrations of asbestos fibres in excess of 5 fibres per cm^3 of air.	'Emergency' means an unforeseeable and unexpected occurrence likely to release airborne concentration of asbestos fibres in excess of 10 fibres per cm^3 of air.
(4) Permissible exposure to airborne concentrations of asbestos fibres:	Permissible Exposure Limits for airborne concentrations of asbestos fibres:
(i) 8-hour time-weighted average concentration. No employees may be exposed to an 8-hour time-weighted average airborne concentration of asbestos fibres in excess of 0.5 fibres per cm^3 of air, as determined on the basis of a 40-hour work week and as described by the method prescribed.	(i) 8-hour time-weighted average concentration. No employees may be exposed to an 8-hour time-weighted average airborne concentration of asbestos fibres in excess of 2 fibres per cm^3 of air, as determined by the method prescribed.
(ii) Ceiling concentration. No employee may be exposed to airborne concentrations of asbestos fibres in excess of 5 fibres per cm^3 of air, as determined over a period up to 15 minutes, by the method prescribed.	(ii) Ceiling concentration. No employee may be exposed to airborne concentrations of asbestos fibres in excess of 10 fibres per cm^3 of air, as determined over a single sampling period of 15 minutes, by the method prescribed.
(5) Regulated Areas. Any work area where a person may be exposed to airborne concentrations of asbestos fibres in excess of either of the limits imposed shall be designated a regulated area. Only authorized persons may be allowed to enter such an area. A daily rosta of all persons entering a regulated area shall be made and maintained.	Regulated Areas. A regulated area shall be established for those particular operations where asbestos concentrations are in excess of the Permissible Exposure Limits. Access to regulated areas shall be limited to authorized persons. A record of all employees assigned to work in a regulated area shall be made and maintained.
(6) The purpose of all monitoring required is to measure accurately the airborne concentrations of asbestos fibres in a workplace to which employees would be exposed if they worked in the area without the use of personal protective equipment, such as respirators.	The purpose of all monitoring required by this paragraph is to measure the airborne concentrations of asbestos fibres in a workplace to which employees would be exposed if they worked in the area without the use of personal protective equipment, such as respirators. Similarly, it may not be necessary to

* The Johns-Manville Corporation is the largest producer of asbestos fibre in the Western world and the largest manufacturer of asbestos-containing products in the U.S.A. The proposed revisions are reproduced by kind permission of the Corporation.

continuously monitor an employee for an 8-hour period. An employee may be monitored for a sufficient period of time so as to be representative of his 8-hour exposure to asbestos fibres.

(7) Initial Monitoring. Every employer shall cause every place of employment where asbestos fibres may be released to be monitored in such a manner as to determine whether employees are exposed to concentration of asbestos fibres in excess of either of the two limits prescribed in paragraph (4). If either limit is exceeded, the employer shall immediately undertake a compliance programme in accordance with paragraph (10).

Initial Monitoring. A programme of initial monitoring and measurement shall be undertaken in each establishment to determine concentrations of asbestos fibres, without regard to the use of respirators, in excess of the Permissible Exposure Limits. However, no monitoring is required where asbestos fibres have been modified by a bonding agent, coating, binder, or other material so that during reasonably foreseeable use, handling, storage, disposal, processing, or transportation, no airborne concentrations of asbestos fibres in excess of the Permissible Exposure Limits will be released.

(8) Frequency of Monitoring. If monitoring shows that an employee's exposure is above either limit prescribed:

Frequency of Monitoring. Where a determination shows any employee exposures without regard to the use of respirators, in excess of the Permissible Exposure Limits, a programme for monitoring employee exposures to asbestos fibres shall be established. Such a programme:

(i) The monitoring shall be repeated every month, except as otherwise provided. If monitoring shows that an employee's exposure is below both limits prescribed, the monitoring shall be repeated every three months, except as otherwise provided.

(i) Shall be repeated not less than once during each 6-month period where any employee is exposed, without regard to the use of respirators, in excess of the Permissible Exposure Limits.

(ii) If two consecutive monitorings made at least 5 days, but not more than 3 months, apart, show that an employee's exposure is below both limits, monitoring need not be repeated, except as otherwise provided.

(ii) May be discontinued for a representative employee only when at least two consecutive monitoring determinations made not less than 5 working days apart, show exposures for that employee at or below the Permissible Exposure Limits.

(iii) Whenever an employer has reason to believe that an employee's level of exposure has changed because of a change in production, process, controls, or other relevant factors, the employee shall be monitored as soon as practicable.

(iii) Whenever there has been a production, process, or control change which may result in an increase in employee exposure to asbestos fibres or whenever the employer has any other reason to suspect an increase in exposure levels, a new determination under this paragraph shall be made.

(9) Employee Notification.

(i) Within five (5) working days after the receipt of the measurement results, the employer shall notify the employee in

Employee Notification.

(i) Within five (5) working days after the receipt of the measurement results, the employer shall notify the employees in writing of writing of the results concerning the employee's exposure.

(ii) Where the results reveal an employee's exposure to be above either of the Permissible Exposure Limits, such notification shall also include a statement of the corrective action being taken to reduce exposure to or below the Permissible Exposure Limits.

the results. Such notification need not be individually disseminated.

(ii) Such notification shall also include a statement of the corrective action being taken to reduce exposure to or below the Permissible Exposure Limits.

(10) Methods of Compliance. Employee exposure to asbestos fibres shall be controlled to or below the permissible limits by engineering controls, work practices, and personal protection controls:

(i) Engineering and work practice controls. Engineering controls shall be instituted immediately to reduce employee exposure to or below the permissible exposure limits, except to the extent that such controls are not feasible. Where engineering controls which can be instituted immediately are not sufficient to reduce exposure to or below the permissible exposure limits, they shall nonetheless be used to reduce exposure to the lowest practicable level, and shall be supplemented by the use of work practice controls. Where engineering and work practice controls are not sufficient to reduce employee exposure to or below the permissible exposure limits, they shall nonetheless be used to reduce exposure to the lowest possible level, and shall be supplemented by the use of respirators, in accordance with paragraph (14) of this section.

Methods of Compliance. Employee exposure to asbestos fibres shall be controlled to or below the Permissible Exposure Limits by engineering controls, work practices, and personal protection controls as follows:

(i) Engineering, work practice, and personal protection controls. Engineering and/or work practice controls shall be instituted immediately to reduce employee exposure to or below the Permissible Exposure Limits, except to the extent that such controls are not feasible. Wherever feasible engineering and work practice controls which can be instituted immediately are not sufficient to reduce exposures to or below the Permissible Exposure Limits, they shall nonetheless be used to reduce exposure to the lowest practicable level, and shall be supplemented by the use of respiratory protection in accordance with paragraphs (14) of this section.

(ii) Particular tools. All hand-operated and power-operated tools which may produce or release asbestos fibres in excess of the exposure limits prescribed in paragraph (14) of this section, such as, but not limited to, saws, scorers, abrasive wheels, and drills, shall be provided with local exhaust ventilation systems.

(ii) Particular tools. Power-operated tools which may produce or release asbestos fibres in excess of the Permissible Exposure Limits prescribed in paragraph (14) of this section, such as, but not limited to, saws, abrasive wheels and drills, shall be provided with local exhaust ventilation systems.

(11) Local Exhaust Ventilation. Local exhaust ventilation and dust collection systems shall be designed, constructed, installed, and maintained in accordance with the American National Standard Fundamentals Governing the Design and Opera-

Local Exhaust Ventilation. Local exhaust ventilation and dust collection systems shall be designed, constructed, installed and maintained in accordance with recognized engineering practices.

tion of Local Exhaust Systems, ANSI Z9.2-1971.

(12) Mechanical Ventilation. When mechanical ventilation is used to control exposure, measurements which demonstrate the effectiveness of the system to control the exposure, such as capture velocity, duct velocity, or static pressure, shall be made at least every 3 months. Measurements of the system's effectiveness to control exposure shall also be made within 5 days of any change in production, process or control which might result in any change in employee exposure.

(13) Compliance Programme.

(i) Every employer shall establish and implement a written programme to reduce exposures to or below the permissible exposure limits solely by means of engineering and work practice controls.

(ii) The written programme shall include:

(A) A description of each exposed operation, e.g. crew size, operating procedures, and maintenance practices.

(B) Engineering plans and studies used to determine the controls for the operation.

(C) A report of the technology considered in meeting the permissible exposure limits

(D) Monitoring date.

(E) A detailed schedule for implementation of the engineering controls and work practices that cannot be implemented immediately, as well as for the development and implementation of any additional engineering and work practices necessary to meet the permissible exposure limits.

(F) Other relevant information.

(iii) Written plans for compliance programmes shall be submitted, upon request to the Assistant Secretary and Director, and shall be available at the work site for examination and copying by the Assistant Secretary and the Director. Such written plans shall be revised and updated at least every six months to reflect the current status of the programme.

Mechanical Ventilation. When mechanical ventilation is used to control exposure, measurements which demonstrate the capture velocity, duct velocity, or static pressure, shall be made as frequently as required in accordance with sound engineering practice. Measurements of the system's effectiveness to control exposure shall also be made within 30 days of any change in production, process, or control which will likely result in an increase in airborne concentrations of asbestos fibres to which employees will be exposed.

Compliance Programme. A programme shall be established and implemented to reduce exposures to or below the Permissible Exposure Limits, or to the greatest extent feasible, solely by means of engineering and work practice controls, as soon as practicable. Written plans for such a programme shall be developed and furnished upon request for examination and copying to authorized representatives of the Assistant Secretary and the Director. Such plans shall be updated at least every six months, to reflect the current status of such a programme.

(14) Respiratory Protection.	Respiratory Protection.
(*1*) Use. Respirators shall be used where required under this section. Compliance with the permissible exposure limits may not be achieved by the use of respirators except:	(*1*) Use. Where respiratory protection is required under this section, compliance with the Permissible Exposure Limits may not be achieved by the use of respirators, except:
(i) during the time period necessary to install engineering or work practice controls, or	(i) during the time period necessary to install engineering or work practice controls, or
(ii) in work situations in which engineering controls and supplemental work practice controls are insufficient to reduce exposure to or below the permissible exposure limits, or	(ii) in work operations in which engineering and work practice controls are not feasible, or
(iii) in emergencies,	(iii) in work operations in which engineering and work practice controls are insufficient to reduce exposure to or below the Permissible Exposure Limits, or
	(iv) in work operations in which engineering and work practice controls are not practicable because these operations are infrequently performed. In such cases, respirators may be used by an employee only for a period of time not to exceed 2 hours during any work day and 1 hour during any 4-hour period of time, or
	(v) in emergencies.
(*2*) Selection. Where respirators are permitted by this section, the employer shall select them from among those approved by the National Institute for Occupational Safety and Health, U.S. Department of Health, Education and Welfare.	(*2*) Selection. Where respiratory protection is required under this section, the employer shall select them from among those approved by the National Institute for Occupational Safety and Health, U.S. Department of Health, Education and Welfare.
(15) Respirator Programme.	Respirator programme.
(i) Employees who wear respirators shall be allowed to leave work areas to wash their face and respirator facepiece to prevent skin irritation due to respirator use.	
(ii) No employee shall be assigned to tasks requiring the use of a respirator if, based upon his most recent examination, an examining physician has determined that the employee would be unable to function normally while wearing a respirator, or that the safety or health of the employee or other employee would be impaired by his use of a respirator. To the maximum extent possible, such	(ii) No employee shall be assigned to tasks requiring the use of a respirator if, based upon his most recent examination, the examining physician has determined that the employee will be unable to function normally while wearing a respirator, or that the safety or health of the employee or other employees will be impaired by his use of a respirator. To the extent permitted by any applicable collective bargaining agreement, such employee

employee shall be rotated to another job, or given the opportunity to transfer to a different position, whose duties he is able to perform, with the same employer, in the same geographical area and with the same seniority, status and rate of pay he had just prior to such transfer.	shall be rotated to another job, or given the opportunity to transfer to a different position, whose duties he is able to perform, with the same employer, in the same geographical area and with the same seniority, status and rate of pay he had just prior to such transfer, if such a different position is available.
(16) Personal Protective Clothing.	Personal Protective Clothing.
(i) The employer shall provide, and require the use of, personal protective clothing, such as coveralls or similar whole body clothing, head coverings, gloves and foot coverings, for any employee exposed to airborne concentrations of asbestos fibres which exceed either of the limits prescribed in paragraph (4) of this section.	(i) Where employees are assigned to work in regulated areas, the employer shall provide, and require such employees to wear personal protective clothing, such as coveralls and foot coverings to prevent contamination of the employee's street clothing.
(ii) Clean and dry protective clothing and equipment shall be provided to each affected employee at least daily.	(ii) Clean and dry protective clothing and equipment shall be provided to each affected employee at least weekly, except where circumstances require more frequent replacement.
(iii) The employer shall ensure that all protective clothing and equipment is removed only in designated change rooms.	(iii) The employer shall require that all protective clothing and equipment required by this section be removed only in the designated change rooms.
(iv) The employer shall ensure that no employee removes contaminated protective clothing and equipment from the change room, except for the purpose of cleaning, laundering, maintenance or disposal.	(iv) The employer shall require employees not to remove used protective clothing and equipment required by this section from the change room, except for the purpose of cleaning, laundering, maintenance or disposal. However, this shall not restrict re-use of protective clothing and equipment in accordance with paragraph (ii) of this section.
(v) Contaminated protective clothing and equipment shall be placed in impermeable closed containers. (vi) The employer shall inform any person who launders or cleans protective clothing and equipment contaminated with asbestos of the potentially harmful effects of exposure to asbestos fibres.	(v) Protective clothing and equipment required by this section and ready for cleaning, laundering, maintenance or disposal shall be placed in dust–proof closed containers or sealed dust-proof bags, and such containers or bags shall bear suitable labels.
(17) Hygiene Facilities and Practices.	Hygiene Facilities and Practices.
(i) Change rooms. Where employees wear protective clothing and equipment, clean change rooms equipped with storage facilities for street clothes and separate storage facilities for protective clothing and equipment shall be provided.	(i) Change rooms. Where employees are required to wear protective clothing and equipment in accordance with paragraph (16) of this section, clean change rooms equipped with storage facilities for street clothes and separate storage facilities for protective clothing and equipment shall be provided.

	(ii) Showers. Employees working in regulated areas shall be required to shower before leaving at the end of the work shift. The employer shall provide shower facilities in accordance with 1910.141(d).	(ii) Showers. Employees working in regulated areas shall be provided with shower facilities and requested to shower before leaving at the end of the work shift. The employer shall provide shower facilities in accordance with 1910.141(d).
	(iii) Lavatories. Employees working in regulated areas shall be required to wash face, hands and forearms prior to eating, drinking, or smoking. The employer shall provide an adequate number of lavatories for this purpose which shall meet the requirements of 1910.141(d).	(iii) Lavatories. Employees working in regulated areas shall be requested to wash hands, face and forearms prior to drinking, eating, or smoking.
	(iv) Arrangement of shower facilities. Clothes lockers and shower facilities shall be arranged so as to separate regulated areas and uncontaminated areas.	(iv) Arrangement of shower facilities. Clothes lockers and shower facilities shall be arranged so as to separate regulated areas and uncontaminated areas.
	(v) Arrangement of lavatory facilities. Lavatory and toilet facilities which are located in regulated areas shall be arranged so that no access is available from them to uncontaminated areas.	(v) Arrangement of lavatory facilities. To the extent practicable lavatory and toilet facilities which are located in regulated areas shall be arranged so that no access is available from them to non-regulated areas.
	(vi) Prohibition of activities in regulated areas. The presence or consumption of food or beverages, the presence or use of smoking or non-food chewing shall be prohibited in regulated areas.	(vi) Prohibition of activities in regulated areas. The employer shall advise employees that they should not possess or consume food or beverages, or possess or use smoking or non-food chewing products in regulated areas.
(18)	Medical Surveillance.	Medical Surveillance.
	(i) General. Every employer shall provide or make available at his cost medical examinations relative to exposure to asbestos, as required by this paragraph. If an employee refuses a medical examination provided in accordance with this paragraph, the employer shall inform the employee of the possible health consequences of such refusals, and shall obtain a signed statement from the employee stating that such employee has been informed of the consequences and refuses to be examined.	(i) A programme of medical surveillance shall be instituted for each employee assigned to work in a regulated area. The programme shall provide each employee with an opportunity for examinations and tests in accordance with this paragraph.
	(ii) All medical examinations and procedures shall be performed by or under the supervision of a licensed physician selected by the employer, and shall be provided without cost to the employee.	(ii) All medical examinations and procedures shall be performed by or under the supervision of a licensed physician selected by the employer, and shall be provided without cost to the employee.
	(iii) The employer shall inform employees that this medical examination is required pursuant to this standard. The	(iii) The employer shall inform employees that this medical examination is required pursuant to this standard. The employer shall

employer shall request each employee who refuses any required medical examination to sign a statement indicating that the employee refuses to be examined.

(iv) Preplacement. The employer shall provide or make available to each of his employees, within 30 calendar days following his first employment in an area exposed to airborne concentrations of asbestos fibres, a comprehensive medical examination. Such examinations shall include, as a minimum, a chest roentgenogram (posterior–anterior 14×17 inches), a history to elicit symptoms of respiratory disease, pulmonary function tests to include forced vital capacity (FVC) and forced expiratory volume at 1 second (FEV_1), and for employees with 10 or more years of exposure to airborne concentrations of asbestos fibres or who are 45 years of age or older, a sputum cytology examination.

(v) Annual. Every employer shall provide or make available comprehensive medical examinations to each of his employees exposed to airborne concentrations of asbestos fibres at least annually. Such examination shall include, as a minimum, a chest roentgenogram (posterior–anterior 14×17 inches), a history to elicit symptoms of respiratory disease, pulmonary function tests to include forced vital capacity (FVC and forced expiratory volume at 1 second (FEV_1) and for employees with 10 or more years of exposure to airborne concentrations of asbestos fibres or who are 45 years of age or older, a sputum cytology examination.

(vi) No medical examination is required of any employee if adequate records show that the employee has been examined in accordance with this paragraph within the past 1-year period.

(vii) Physician's written opinion

(*1*) With respect to each examination required by this paragraph, the employer shall obtain a written opinion from the examining physician, containing the

request each employee who refuses any required medical examination to sign a statement indicating that the employee refuses to be examined.

(iv) At the time of the initial assignment in a regulated area, or upon institution of medical surveillance, a comprehensive medical examination shall be performed or made available. Such examination shall include, as a minimum, a chest roentgenogram (posterior–anterior 14×17 inches), a history to elicit symptoms of respiratory disease, pulmonary function tests to include forced vital capacity (FVC) and forced expiratory volume at 1 second (FEV_1).

(v) The examination provided for above in this section shall be repeated or made available at the following intervals after the initial examination: fourth year, sixth year, eighth year, tenth year and annually thereafter, for each employee assigned to work in a regulated area.

(vi) No medical examination is required of any employee if adequate records show that the employee has been examined in accordance with this paragraph within the past 4-, 2- or 1-year period, as the case may be.

(vii)

(*1*) With respect to each examination required by this paragraph, the employer shall obtain a written statement from the examining physician, containing the following:

following:

(A) the physician's opinion as to whether the examined employee has any medical conditions which would place the employee at increased risk of material impairment of his or her health from exposure to asbestos fibres, or which would directly or indirectly, be aggravated by such exposure;

(B) any recommended limitations upon the employer's exposure to asbestos fibres, or upon the use of protective clothing and equipment, such as respirators; and

(C) a statement that the employee has been informed by the physician of any medical conditions which require further examination or treatment.

(*2*) The written opinion shall not reveal specific findings or diagnoses unrelated to occupational exposure to asbestos fibres.

(*3*) A copy of the written opinion shall be provided to the affected employee;

(viii) Termination of employment. The employer shall provide or make available within 30 calendar days before or after termination of employment of any employee exposed to airborne concentrations of asbestos fibres, a comprehensive medical eximination. Such examination shall include, as a minimum, a chest roentgenogram (posterior–anterior 14 × 17 inches), a history to elicit symptoms of respiratory disease, pulmonary function tests to include forced vital capacity (FVC) and forced expiratory volume at 1 second (FEV_1) and, for employees with 10 or more years of exposure to airborne concentrations of asbestos fibres or who are 45 years of age or older, a sputum cytology examination;

(ix) Withdrawal from exposure. No employee shall be exposed to asbestos fibres in such a way as would put the employee at increased risk of material impairment of his or her health from such exposure. This determination may be based on the physician's written opinion.

(A) the physician's specific findings or diagnosis related to the employee's occupational exposure to asbestos;

(B) any recommended limitations upon the employee's use of respirators.

(*3*) A copy of the physician's written statement shall be provided to the affected employee.

(19)	Danger labels. Effective date. Labels meeting the required specifications of the section shall not be required until such time as the employer has exhausted the current supply of labels and packages containing such labels meeting the specifications of paragraph (g)(2)(ii) of section 1910.1001, as promulgated on June 7, 1972.
(20) Waste disposal. Asbestos waste, scrap, debris, bags, containers, equipment, and asbestos-contaminated clothing, consigned for disposal, which may produce in any reasonably foreseeable use, handling, storage, processing, disposal, or transportation, airborne concentrations of asbestos fibres in excess of the exposure limits prescribed in paragraph (4) of this section shall be collected and disposed of in sealed impermeable bags, or other closed, impermeable containers.	Waste disposal. Asbestos waste, scrap, debris, bags, and containers, consigned for disposal, which may produce in any reasonably foreseeable use, handling, storage, processing, disposal, or transportation airborne concentrations of asbestos fibres in excess of the Permissible Exposure Limits shall be collected and disposed of in sealed dust-proof bags, or other closed, dust-proof containers.
(21) Retention of exposure, medical, and assignment records. (i) Exposure records: each record of an employee's exposure shall be maintained for at least 40 years, or for the duration of the employee's employment plus 20 years, whichever period is the longer. (ii) Medical records: each record shall be maintained for at least 40 years, or for the duration of the employee's employment plus 20 years, whichever period is longer. (iii) Mechanical ventilation measurements: each record shall be maintained for at least 3 years. (iv) Employee training: each record shall be maintained for at least 3 years. (v) Rosters: each roster required shall be maintained for at least 40 years or for the duration of the personnel's employment plus 20 years, whichever period is longer.	Retention of exposure, medical, and assignment records. (A) Within a three-year period following the effective date of this section, the Assistant Secretary shall furnish written notification to certain employers of the protocol for and commencement of a long term epidemiological study with regard to human exposure to asbestos. At those specific places of employment selected for such study by the Assistant Secretary, the following records shall be maintained for at least 40 years, or for the duration of the employee's employment plus 20 years, whichever period is longer: (i) employee exposure records as required; (ii) employee medical records as required; and (iii) employee assignment records as required. (B) At all other places of employment, such records shall be maintained for a period of at least 3 years. However, adequate summaries of such records shall be maintained for at least 40 years, or for the duration of employment plus 20 years, whichever period is longer. Retention of mechanical ventilation measurement and employee training records. The

	following records shall be maintained for at least 6 months: (*1*) mechanical ventilation measurement records as required; (*2*) employee training records as required.
(22) Transfer of Records. In the event the employer ceases to do business and there is no successor to receive and retain his records for the prescribed period, these records shall be transmitted by mail to the Director of OHSA, and each employee and former employee shall be individually notified in writing of this transfer.	Transfer of Records. In the event the employer ceased to do business and there is no successor to receive and retain his records for the prescribed period, these records shall be transmitted by mail to the Director of OHSA.
(23) Observation of Monitoring. (i) Employee observation: the employer shall give employees or their representatives an opportunity to observe any measuring or monitoring of their exposure to asbestos fibres conducted pursuant to this section. (ii) Observation procedures: without interfering with the measurement, observers shall be entitled to receive an explanation of the monitoring or measurement procedures, observe all steps related to the measurements and record the results obtained.	Observation of Monitoring. (i) Employee observation: the employer shall give affected employees or their designated representatives a reasonable opportunity to observe any measuring or monitoring of their exposure to asbestos fibres conducted pursuant to this section. (ii) Observation procedures: without interfering with the measurement, observers shall be entitled to observe all steps related to monitoring and measurements as required and record the results obtained.

(3) THE ATTITUDES OF THE EMPLOYEES IN THE ASBESTOS INDUSTRY AND OF THE TRADES UNIONS

In the United States an estimated 50 000 people are involved in the manufacture of asbestos-containing products from asbestos raw materials; this figure does not include the secondary manufacture of asbestos-containing products, such as electrical or thermal insulation, or products which include previously manufactured components that contain asbestos. There are about 40 000 field insulation workers in the United States who are exposed to asbestos dust and the activities of these field workers is estimated to cause secondary exposure to some three to five million other workers.

In the United Kingdom it has been estimated (Asbestos Information Centre, personal communication) that there are 24 000 workers involved in the asbestos-consuming industries.

With the very large number of workers exposed to asbestos dust and its related hazards, it is hardly surprizing that the trades unions have viewed with growing concern the mounting evidence of the dangerous nature of asbestos. The dust exposure of an individual worker is very variable, as is the individual susceptibility to cancer from asbestos exposure. However, there are a number of circumstances where the risks to workers and their families are particularly high: building and construction industries; thermal insulation and lagging industry (including removal, repairs, etc.); manufacture and repair of asbestos in brakes and clutches; the machining, cutting, drilling, fixing, etc., of asbestos-based products involved in building construction, maintenance, and repair work; the handling of asbestos gland packings, gaskets, and protective clothing; the replacement of refactory linings in foundry induction furnaces; and the knitting and laying in of asbestos yarn in asbestos knitted fabrics. Despite improvements in the safety regulations and the control of the factory environment in the asbestos industry, deaths from cancer due to inhalation of asbestos dust are expected to continue to rise during the 1980s as the effects of earlier exposure of workers under less strictly controlled conditions become realized, e.g. of about 2700 workers who were employed at a large London asbestos factory between 1913 and 1968 and now still under medical supervision, up to 10% of the male employees and up to 12% of the female employees are expected to die of mesothelioma.

In the United Kingdom, the Trades Union Congress (TUC) has expressed the views of the trades union movement as a whole in its evidence to the Advisory Committee on Asbestos. The evidence was compiled and approved by the General Council after full consultation with all affiliated organizations, and the 1977 Congress substantially endorsed its principle recommendations. Given that all forms of asbestos are carcinogenic to man, the TUC feels that the importation into, and use of, all absestos in the U.K. should eventually be banned. The importation should be stopped when safe and effective substitutes are developed and, pending this action, the use and importation of asbestos should be severely restricted, except for the importation of crocidolite-containing products which should be banned forthwith (so far as can be ascertained from industrial sources, no crocidolite asbestos as such has been imported into the U.K. since 1970). The TUC does not accept that it is possible to lay down a threshold limit value (TLV) for a known carcinogen such as asbestos. This is because (a) medical authorities disagree as to whether a safe level of exposure to carcinogens exists and, if such a level does exist, what level it should be for asbestos fibres; (b) while in theory the asbestos dust levels in manufacturing and handling processes can be controlled down to very low levels (about 0.1 fibres per cm^3), some activities involving asbestos give rise to higher levels of dust (e.g. construction, building, and demolition) which would be very difficult to reduce even using wetting and bagging techniques, and there is little point in setting a

TLV which it would be difficult and impracticable to adhere to; and (c) it is technically difficult to measure low concentrations of asbestos dust quickly and accurately, e.g. sampling over a 4-h period is required in most circumstances when measuring asbestos dust at the 0.2 fibres per cm^3 level, and further time is involved in the counting process. Thus, effective enforcement of asbestos dust concentrations at very low levels is difficult and time consuming, and may expose people to dangerous levels of asbestos dust during the measurement period.

The TUC has suggested an interim maximum allowable concentration (MAC) of 0.2 asbestos fibre per cm^3 of contaminated air, to apply to all forms of asbestos dust, on the grounds that it will be exceedingly difficult to (i) carry out certain asbestos-involving processes at concentrations below this value for any length of time, and (ii) enforce standards lower than this level. The TUC's recommendation is not meant to imply that concentrations at or below this level are safe. It merely recognizes what is feasible at the present time and the purpose is to indicate to employers, workers, and Health and Safety Executive inspectors that exposure to asbestos dust will be allowed so long as the concentration over any 10-min sampling period is less than 0.2 fibre per cm^3. In addition, the TUC recommends that the Health and Safety Commission should draw up and publish guidance to safety representatives/asbestos workers concerning the risks and severe dangers inherent in working with asbestos, particularly with regard to the stripping of asbestos-containing insulation. The TUC also supports the idea that thermal insulation contractors should be licensed under the provisions of S(3) of the Health and Safety at Work etc. Act 1974, and that only fully trained and licensed workers should be permitted to strip asbestos-containing insulation.

Dissatisfaction has been expressed by the TUC over the low level of fines imposed by magistrates for breaches of the existing Asbestos Regulations and over the cynical disregard of employers and magistrates towards the purpose of safety statutes and their Regulations. Following from this, the TUC has recommended that the Government introduce a *minimum* fine of £400 to follow conviction for any breach of the forthcoming Asbestos Regulations.

Other recommendations by the TUC include:

(1) the production of an annual list of all asbestos-containing products and processes;

(2) manufacturers, importers, suppliers, and designers to be required:

(*i*) to give written notification to the Health and Safety Executive (HSE) concerning the importation/planned importation of any raw asbestos into the U.K. (the raw material to be imported only in strong, secure containers which are not liable to puncture or burst and which are prominently labelled: 'DANGER—RAW ASBESTOS', together with information on the hazards and precautions to be taken in connection with exposure to asbestos dust. The importation of raw asbestos and of asbestos products, consisting of or containing fibres longer than 10 μm and thinner than 1 μm to be prohibited;

(*ii*) to inform the HSE of all products and materials which they manufacture, import, etc., and which contain any asbestos;

(*iii*) to affix labels to all asbestos products and that for products in Class A of the asbestos list (see Table 4.2) to contain a suitable warning and safety information;

(3) registers to be prepared by employers listing all workers who may be regularly exposed to asbestos dust, and the contents of the registers to be made available to EMAS and HSE officers and relevant safety representatives at the place of employment;

(4) employers to state their policy towards continued used retention and future purchases of asbestos products/processes in their written safety policy statements;

(5) the continuing Employment Medical Advisory Service study into the morbidity and mortality of asbestos workers to be extended to all 20 000 workers in asbestos risk industries and regular interim reports of this study to be made available;

(6) all asbestos workers to be given pre-employment and regular 6-monthly examinations and the records to be kept for 50 years or until the death of the worker concerned;

(7) that lung cancer and laryngeal cancers that develop in workers exposed to asbestos dust be prescribed under the industrial injuries scheme;

(8) that glass fibre (thicker than 3 μm) substitutes be used in place of asbestos for the time being, unless safer and practicable substitutes exist;

Table 4.2 Information recommended to appear on the annual list of asbestos-containing products/processes, and to be widely distributed to all employers, unions, and safety representatives. Class A processes/products involve or may involve the risk of exposure to levels of asbestos fibres above 0.2 fibres per cm^3. Class B processes/products are those for which safe and practicable substitutes exist. Products and processes in both Class A and Class B should not be used.

Process	Type of asbestos involved	Trade name(s) of product	Identifying characteristics	Class A	Class B

(9) the use of full air line respiratory equipment during the stripping of both blue and white asbestos insulation;

(10) that the Health and Safety Executive devise new methods for the determination of concentration and nature of fibre samples taken from contaminated workplaces—the new methods to trap and count all fibres and fibrils longer than 10 μm and thinner than 1 μm.

In addition to the TUC, individual trades unions, both inside and outside the overall asbestos industrial area, have declared attitudes on the use of asbestos and asbestos-containing materials:

(*i*) The National Federation of Building Trades Employees, while acknowledging that the use of asbestos creates a certain health hazard, is opposed to a ban on asbestos and believes that it should continue to be used for most of the purposes for which it is used at present under proper control, so that the risks are reduced to small, acceptable levels (National Federation of Building Trade Employees, 1976).

(*ii*) The National Union of Teachers has expressed concern over the potential risks to the health of children and staff arising from the use of asbestos and asbestos-containing products in schools, and draws attention to the need to make financial provision to deal with the problem. The Union is of the view (Evidence submitted to the Advisory Committee on Asbestos 1976/77) that: all sprayed asbestos furfaces should be covered with a suitable sealant or cladding, that regular checks on the condition of the sealant/cladding should be carried out, and that any change discovered should be put right immediately; air samples should be taken to ascertain the level of airborne asbestos dust and the results made known to the recognized Trades Unions in a given school; all Local Authority employees who may have cause, in the course of their duties, to work with asbestos should be fully appraised to the risks involved both to themselves and others in the vicinity and should be given appropriate guidance as to the necessary precautions to be taken; where alternative materials are reasonably practicable, they should be used in preference to asbestos. The Union also expresses concern at the possible asbestos content of such commonly used school equipment and commodities as ironing boards, oven gloves, and laboratory and workshop boards.

(4) THE ATTITUDES OF PRESSURE GROUPS, THE MEDIA, AND THE GENERAL PUBLIC

While asbestos has been widely used on a large scale for many years and the health hazards to asbestos workers have been well recognized and documented during this period, it was not until the 1970s that considerable and sustained public attention has been focused on asbestos. Lay members of the public have no real knowledge of what asbestos is, what its effects on health may be, the conditions under which it may be considered to be safe or unsafe, etc., and interested groups have taken it upon themselves to try to mould public opinion and attitudes. The viewpoints of the asbestos industry and of some of those who have taken it upon themselves to be the spokesmen of the general public are diagonally opposed and the conflicting viewpoints have generated fears in the minds of people about something they do not understand. The widespread conflicting publicity and expert advice over the health risks associated with exposure to asbestos *dust* has led to a position of confusion and the voice of moderation attempting to get the problem into a realistic context has often failed to be heard in the face of the massive efforts mounted by the asbestos industry and its opponents.

Among those who have taken an extreme view on the subject in the U.K. and who have attempted to influence and mould public opinion against the use of asbestos are the following: *Socialist Worker*—a paper of the International Socialists—in their publication '*Asbestos—the Dust that Kills in the Name of Profit*; Pat Kinnersley in his book *The Hazards of Work: How to Fight Them*"; and Mrs Nancy Tait in her pamphlet *Asbestos Kills.* In America, Paul Brodeur has dealt with the subject in his book *Expendable Americans.*

By manipulation of facts, by the presentation of statements of opinion as if they are statements of fact, and by evidence presented or omitted, etc., a powerful and sustained anti-asbestos lobby has grown up during the 1970s with considerable influence on workers and public opinion. It is interesting to consider, for example, some of the statements in *Asbestos Kills*, and to look at the corresponding response by the Asbestos Information Committee (Table 4.3).

Nancy Tait has also challenged the Government's action in not telling some asbestos workers about the X-ray discrepancies which might indicate that they are sufferers from untreatable asbestos related disease (see p. 125) and the failure of the Government to ensure that they avoid exposure to asbestos in the hope that progression of the disease will be slowed. She believes that it is indefensible for workers threatened by these diseases to be left in ignorance and the category 01 workers should be warned at this stage and given the opportunity to change employment, with re-training as necessary. Also, smoking withdrawal clinics should be set up in the asbestos companies.

Other groups who have appointed themselves the spokesmen of the public in the U.K. are as follows:

1. Asbestos Action, which, according to the evidence submitted to the Advisory Committee on Asbestos, is an independent and voluntary group of individuals and representatives (Ed: of what?) who came together 'because of their concern at the inadequate public safeguards and public information about the risks to health from the use of asbestos in its many applications'.

While accepting that asbestos offers important and useful benefits, Asbestos Action has attempted to identify areas of public concern about the use and dangers of asbestos, to pose questions and to suggest possible courses of action. Asbestos Action claims that the health risks arising directly and indirectly from the use of asbestos have never been pursued with sufficient, independent, public vigour; that the risks have never been brought in any sustained manner to the attention of those who come into contact with asbestos directly or indirectly; and that the community has never been allowed to consider these dangers in any considered and informed fashion, let alone allowed to undertake any collective value judgement as to whether they wish to accept the health risks which flow from the use of asbestos.

In the evidence presented to the Advisory Committee on Asbestos (HMSO. 1977), Asbestos Action appears to have overlooked the current legal situation and the true facts in some cases. Two examples from the evidence will suffice:

(*i*) '*Asbestos dust on workmen's clothing serves as a ready source of contamination of the employee's home. Such exposure has contributed in the death of wives of asbestos workers. Others in the family can face exposure if asbestos-laden overalls are washed in the family wash, launderette, or dry cleaned.*

These family exposures can be virtually eliminated by proper use of changing rooms, showers, and laundry facilities which ought to be readily available to those working with asbestos'.

The Asbestos Regulations 1968 are quite specific on the matter of working with asbestos under such conditions that it is impracticable to prevent the entry into the atmosphere of asbestos dust—which could then find its way on to workmen's clothing. Protective equipment must be provided by the employer and worn and special provision has to be made for the putting on and taking off of protective clothing (overalls and headgear which will, when worn, exclude asbestos) and for its laundering. The overalls worn by asbestos workers should exclude and not readily hold asbestos dust. Asbestos fibres are known to adhere readily to some types of overall material and material consisting of 60% polyester fibre and 40% cotton has a relatively low tendency to become dusty, and materials made wholly from synthetic fibres are even better in this respect. The overalls should be close fitting at the neck, ankles, and wrists and caps should be provided. In certain instances plastic gloves, rubber boots, and plastic aprons may be necessary.

Suitable accommodation should be provided for storage of protective clothing, separate from that for outdoor clothing. The cleaning of protective clothing provided in accordance with the Regulations is the responsibility of the occupier of the factory and it should either be cleaned at the factory or despatched for cleaning in a suitable container clearly and boldly marked 'asbestos-contaminated clothing'. Asbestos workers should not take overalls

Table 4.3 Claims and counter claims on the hazards of asbestos

Claim: Nancy Tait, *Asbestos Kills*	Counter-claim: Asbestos Information Committee, *Asbestos Kills?*	Comments
1. Only in 1973/74 did the Health and Safety Executive take measurements in factories *other than* those operated by the large asbestos manufacturers, and 24.4% of 400 samples were then found to be above the Hygiene Standard of 2 fibres per cm^3.	In 1973 the Factory Inspectorate found in their sampling of 700 workers *among* the larger asbestos manufacturers that 92.6% were below the 2 fibres per cm^3 standard. In 1974 they found as a result of extending their sampling to a wider range of asbestos users at 38 factories that 75.8% were below 2 fibres per cm^3 and 92% were below 4 fibres per cm^3. A factory inspector is empowered to ask for an improvement in the standard of control if he finds concentrations between 2 and 12 fibres per cm^3. By 1974 92% of 400 samples taken in British Industry were at or below 2 fibres per cm^3.	Assuming the reliability of the figures quoted, both parties are emphasizing different situations, Nancy Tait concentrating on the smaller organizations and the Asbestos Information Committee on the larger organizations. The total number of workers involved in each category would have been statistically more meaningful. Nevertheless, the large number of factories above the 2 fibres per cm^3 standard is of importance only if the personnel concerned were *not* provided with and seen to use respiratory protective equipment. Provided that the workers were adequately protected, etc., the actual level of dust is not of major significance—it only becomes important for unprotected workers and neither party make mention of personal protective devices at the sites concerned.
2. 'It has, however, been possible to show that mesothelioma can be produced by slight exposures, and on present knowledge we must assume that no amount of exposure is completely free from risk'—synopsis of a 1968 report (made available by the DHSS in 1976) of the Standing Medical Advisory Committee's Standing Sub-Committee on Cancer.	The report was written in 1968. The medical paper on which it is based did not say 'slight exposures' but 'exposures of short duration'. The growing evidence of dose–response relationships for mesothelioma (Newhouse, 1973) indicates that exposure of short duration would also have to be relatively heavy to cause mesothelioma.	The World Health Organization (WHO) accepts that the general urban population is exposed to low levels of asbestos; the WHO further accepts (World Health Organization, 1977) that at present it is not possible to assess whether there is a level of exposure to humans below which an increased risk of cancer would not occur.
3. A member of the Municipal and General Workers' Union has asbestosis. His only link with asbestos was that he lived as a child in a pre-fab made of corrugated asbestos sheeting (Tait, 1976).	There are no cases whatever of asbestosis from non-occupational cause in any of the medical literature.	The actual statement, taken from *Asbestos—the Dust that Kills*, is 'The General & Municipal Workers' Union has a member with asbestosis whose only link with asbestos, *it is thought*, is a childhood spent in a pre-fab made of corrugated asbestos sheeting'. A lot depends on what is

		meant by 'asbestosis' and by the 'medical literature'. Kiviluoto (1965) refers to several hundred cases of endemic asbestosis in Finland and East Germany, involving pleural plaques similar to those seen in asbestos workers yet having no subjective symptoms of respiratory disease, in a publication entitled *Biological Effects of Asbestos*.
4. How can a solution to the general cancer problem ever be found when no attempt is made to control the use of a powerful carcinogen?	This completely ignores the existence of the Asbestos Regulations 1969, the Factories Act, the Asbestos Research Council Codes of Control and Safety Guides, the work of the Health and Safety Executive, the work of the Safety Officers and Safety Committees throughout industry and the work of occupational hygienists throughout industry. Asbestos is only a powerful carcinogen when inhaled in excessive amounts.	There are in general three major agents which are considered at this time to contribute to the cause of cancer: chemicals, ionizing radiations and viruses. While there is considerable speculation on the mechanism by means of which the members of the different classes induce cancer in man, the precise mechanism is unknown at this time. However, because of the different nature of the causative agents, it is reasonable to consider there may be a minimum of three distinct problems rather than one general problem. Fibres of many materials (asbestos, glass, aluminium oxide) in the size range 10–80 $\mu m \times 2.5\ \mu m$ (of the order of magnitude of viruses) have been shown to be carcinogenic, while non-fibrous crocidolite and chrysotile did not produce tumours (Stanton, 1973). Thus, chrysotile both bulk and powdered asbestos are not powerful carcinogens.
5. Counting and measuring of asbestos fibres using the electron microscope must be the method of choice if an accurate count is required.	One must distinguish between optical microscopy, which is a convenient tool for routine monitoring work situations, and electron microscopy, which is the best tool for research purposes. Electron microscopy is unsuitable for practical routine monitoring of the work environment.	There are a number of techniques for the examination of airborne asbestos dust samples as collected by the membrane filter method (see Chapter 6).

home to be washed. Asbestos workers not covered by the Regulations will certainly be covered by the Health and Safety at Work etc. Act 1974, which will require similar facilities under Sections 1 (1b) (protecting persons other than persons at work against risks to their health and safety arising out of or in connection with the activities of persons at work) and 3(1) (concerning the duty of an employer to conduct his undertaking in such a way as to ensure, so far as is reasonably practicable, that persons not in his employment who may be affected thereby are not exposed to risks to their health or safety).

Any employer who does not provide the necessary clothing, changing facilities, laundry facilities, etc., is in breach of the law and can be reported to the Health and Safety Inspectorate, who can be expected to issue the appropriate improvement or prohibition notice. Thus, what Asbestos Action infers is taking place is illegal, and the mechanism for dealing with such matters (including the provision for unlimited fines and/or for imprisonment for up to 2 years) exists already.

(*ii*) '*Asbestos exposures can result from fibre contamination of other products. Talc often co-exists with varieties of asbestos and the use of cosmetic talcum powders can produce a significant asbestos exposure. Substantial amounts of anthophyllite asbestos are contained in some "talcum powders". Industrial talcs, especially, are liable to be contaminated as a major source of such talc co-exists with extensive deposits of asbestos*'.

A minimum detection level of 1% by weight of chrysotile has been reported in talc (Shelz, 1974) and, according to the World Health Organization, this level of contamination has rarely been observed in natural talcs. It is interesting to note that the Chairman of Asbestos Action is Mr. Max Madden, Labour MP for Sowerby, in whose constituency the Cape Industry Ltd. Acre Mill factory was located. Mr. Madden was one of the Members of Parliament who had called for an independent inquiry into asbestos rather than the Health and Safety Commission's investigation.

2. Socialist Worker. The evidence submitted by the *Socialist Worker* to the Advisory Committee on Asbestos is full of emotive language and distorted facts. Phraseology such as *murder for profit, murderous health hazard, connivance of the state agencies in evasion of the law, plague of asbestos-induced cancer, and a situation where scores of workers still go on being slaughtered in the future*, together with allegations of the use of slave labour (in Africa) and inferences that polyvinyl chloride and fibre-glass are dangerous, can do little to induce sensible debate and much to inflame public opinion. For example, in the evidence reference is made to '*the terrible scandal of Hebden Bridge*', i.e. implying the conditions at the Cape Asbestos Company (now Cape Industries Ltd.) Acre Mill Factory which resulted in 70 certified cases of asbestosis. Yet the District Inspector of Factories visited the factory several times during the period 1957–61 and is on record as stating that in general the factory had created a much better

impression on him, with regard to dust control, than others he had experience of before and that the Acre Mill Factory was regarded as a good example of how the Asbestos Regulations should be complied with. Indeed, the factory was often used as one to which trainee inspectors were taken to show them what was required (Marne, 1976).

Moreover, the *Socialist Worker* does not appear to distinguish between 'dying from asbestosis' and 'dying with asbestosis', nor does it appear to recognize that cases of asbestosis now must result from conditions that existed some considerable time ago.

3. The National Press has also presented a distorted case, selecting items out of context and presenting statements of opinion in the guise of statements of fact. The announcement that the Department of Trade and Consumer Protection had agreed with asbestos manufacturers and users that a voluntary labelling scheme would be introduced for asbestos products was presented in *The Sundary Times* (August 4th, 1976) under the heading *How asbestos can turn killer around the home. The Daily Mail* (April 28th, 1976) article *Is this killer dust swept under the mat*? contains erroneous sentences. The Department of Education and Science is attributed with telling the reporter *killer asbestos could have been used in about 8700 of the 13 000 new schools built between 1945 and 1975.* Since the article deals specifically with blue asbestos, by implication blue asbestos is referred to in the above quotation, although no blue asbestos as such has been imported into the U.K. since 1969 and would not have been used as an insulation material by any reputable organization since that date.

The article also claims that since 1969 the price of blue asbestos has made it virtually too expensive to be used. In practice, it has been the voluntary ban on the importation of blue asbestos by the asbestos industry, coupled with the low permitted worker exposure levels, which have been the reason for the non-use of blue asbestos. Indeed, it is reported in *Cape Review* (Spring 1976) that the production of blue asbestos is now higher than ever—this would not be so if it were too expensive to use (in other countries). It is reported that there are schools (un-named) which have blue asbestos dust 2–3 in deep in ducts and that a mouse runs along the ducts stirring up the dust. A few moments reflection will show that this situation is very difficult to visualize: 2–3 in of blue asbestos dust could only result from the almost total disintegration of the duct lining, and even a mouse might be expected to have difficulty in running along this type of surface!

The Science Editor of *The Financial Times* (April 28th, 1976) announced the launching of Nancy Tait's *Asbestos Kills* under the heading *Asbestos bit like 'Russian roulette' victims widow says*, while on the same day and on the same subject the *Morning Star* heading was *Asbestosis: fact-hiding to blame for suffering*, and that of *The Sun* was *The deadly asbestos bullet—it takes only one shot of dust to kill a man.* It is interesting to look at some of the phraseology used by *The Sun's* reporter during the examination of *the frightening truth about a hidden menace that threatens all of us*—

> Representative articles were taken to Sussex University to be checked (for the presence of asbestos) on the University's giant electron microscope (*N.B.*, the microscope in question is not of abnormal size as this phrase implies). Asbestos *bullets* showed up in samples from a scouring pad, wall filler and brake linings (a bullet normally means a ball or cone of lead used in small firearms: neither the shape nor the material can be said to apply to asbestos). The filter horrified University chemist and physicist Mrs. Patricia Bills [Mrs. Bills is (1976), in fact, an Experimental Officer in the Department of Chemical Physics. Whilst a graduate (B.Sc., London), her status in the academic hierarchy of the University of Surrey is not very high]. The University's specialist in electron microscopy is attributed to have said '*You can never say that anything with asbestos in it is safe*' (ignoring the whole range of encapsulated asbestos products which are no less safe than many other commonly used products of our industrial society).

4. The British Society for Social Responsibility in Science, in its evidence to the Advisory Committee on Asbestos, claims that several million (i.e. >3 million) workers in Britain are probably exposed to asbestos dust at their workplace. The most important aspect of the evidence, not substantially covered in other published submissions, is the criticism of the British Occupational Hygiene Society and the *safe* level of asbestos dust declared by the Society and accepted as a world-wide standard. The British Association for Social Responsibility in Science is of the opinion that:

(i) the British Occupational Hygiene Society Sub-Committee on Asbestos is virtually a front for the asbestos industry, where members make up over 50% (5 out of 9 members) of the Sub-Committee;

(ii) the *safe* standard, set by this 'independent' committee, relates only to fibrogenic effects of asbestos and not to the carcinogenic effects for which no standards exist at the present time;

(iii) the *safe* standard level was based on information that the Association regards as unreliable (dust level estimates) and suspect (company records) and this information was obtained from the textile industry, where the incidences of asbestos disease are lower than in other industries, e.g. insulation;

(iv) optical microscopy is not suitable for the determination of the whole size range of biologically significant asbestos dust and that the future standard should be based on the use of the electron microscope.

The Association believes:

(i) there is no safe level of exposure to asbestos dust;

(ii) asbestos should be replaced with proven safe substitutes and all synthetic mineral fibres should be manufactured so that they do not break down in use to respirable fibres (i.e. diameter $\geqslant 3.5\,\mu m$);

(iii) where asbestos is worked the dust must be reduced to the minimum level possibly by engineering control methods which must be continuously improved;

(iv) people working with, or near, asbestos need detailed instructions on the best methods available for reducing the formation of dust to a minimum and that such methods should be continuously developed and published;

(v) personal respirators are uncomfortable to wear constantly and render speech difficult and, while there are certain areas of work where there is no alternative, it is unrealistic to use personal respirators as the basis of a safe system of work. Available respirators should be tested for dust removing efficiency and comfort and the results published;

(vi) evidence suggests that peak asbestos dust exposure values are important because the body's defence mechanism may be overwhelmed at these high levels, so that any sampling procedure must record the peak values and these must be taken into account when evaluating the efficiency of extraction equipment etc.

5. The Consumers' Association state in their evidence to the Advisory Committee on Asbestos that, where there may be a danger to health from the production process and products of an industry, people need, and have a right, to know:

(a) what precautions they should take;

(b) the current state of knowledge as to the nature of the danger;

(c) the situations at which, and places where the danger is like to be found.

To these ends the Consumers' Association proposes that:

(i) *all* asbestos containing materials to carry warning labels and, if appropriate, be accompanied by an instruction leaflet;

(ii) motor vehicle handbooks to include instructions on how to work safely with brake linings;

(iii) domestic installers of asbestos-containing products should be required to warn the householder that asbestos is being used, that the asbestos should be permanently marked to facilitate future identification, and all work with asbestos should be done in accordance with the best recommended practices;

(iv) the Advisory Committee on Asbestos inquires into and reports on, with recommendations, the problems of asbestos in the home, water supplies, the atmosphere and general environment, and the transport and disposal of asbestos and the problems asbestos creates for local authorities, etc.;

(v) Environmental Health authorities to offer an advisory service on the disposal of unwanted asbestos and asbestos waste in general and to make the availability of the service known;

(vi) the Advisory Committee on Asbestos to consider whether all cases of mesothelioma are being identified, registered, and the exposure history fully explored;

(vii) people working regularly with asbestos be strongly urged to stop smoking;

(viii) the import of crocidolite and crocidolite-containing products be banned and that products/processes that are most likely to release dust (e.g. powdered wall plugging compounds, spraying of asbestos) should be banned;

(ix) substitute materials be developed and used.

APPENDIX 4.1. ASBESTOS MONITORING EXPOSURE RECORD

Name and Address of Organization

To be kept for a minimum of three years and made available on request to the Assistant Secretary of Labour for Occupational Safety and Health, the Director of the National Institute for Occupational Safety and Health, and to authorized representatives of either.

Exposure levels of all employees monitored				
Name	Social Security No.	Job title	Exposure level	If exposure level in excess of prescribed limit, date of written personal notification

Date of sampling . . . .
Date of analysis . . . .
Location . . . .
Details of operation involving exposure to asbestos fibres . . . .
. . . .
. . . .
Details of ventilation system (if any) . . . .
. . . .
Details of protective devices worn (if any) . . . .
. . . .

Sampling method used ....
....
Sampled by ....
Analytical method used ....
....
Analysed by ....

Signed ....
Status ....
Date ....

N.B. Individual employee exposure records must be made available to the individuals concerned and their designated representatives for examination and copying.

APPENDIX 4.2. PRE-EMPLOYMENT ASSESSMENT EXAMINATION

Name and Address of Organization

Name of Employee ....
Address of Employee ....
Date and place of birth ....
Year of first employment in the asbestos industry ....

	Personal history			Family history		
Rheumatoid arthritis Lupus erythematus Nasal obstruction Severely impaired respiratory function Impaired cardio-vascular function	None	Weak	Strong	None	Weak	Strong

↑ ↑

Advise not to work with asbestos

APPENDIX 4.3

Initial Medical Examination: Details for Employees who consent to the Medical Examination

(To be carried out within 30 calendar days of first employment in an area exposed to airborne concentrations of asbestos fibres. No medical examination is required in the case of an employee for whom there are adequate records to show that he/she has been examined as required during the past 12-month period.)

(1) Date of Examination . . . .

					Sputum cytology examination required		
Age	41	42	43	44	45.		
Years of employment to date in the asbestos industry	6	7	8	9	10	11	12 . . . .

Height . . . . Pulse rate . . . .
Weight . . . . Temperature . . . .
Chest circumference (relaxed) . . . .
Chest circumference (expanded) . . . .
Pets/livestock, etc . . . .
Athletic activities . . . .

(2) History of Respiratory Disease

(a) Smoking habits . . . .
(b) Past exposure to asbestos or other dusts . . . .
. . . .
(c)

	Present	Absent
Pulmonary symptoms		
Cardio-vascular symptoms		
Gastro-intestinal symptoms		

(d) Physical examination with special attention to pulmonary rales, clubbing of fingers, and other signs related to cardio-pulmonary symptoms.

(3) Pulmonary Function Tests =
(a) Vital capacity (VC)

(b) Forced expiratory volume at 1 sec (FEV_1) =
(c) Predicted normal VC (PNVC) =
(d) Corrected VC (CVC) =
(e) Forced VC (FVC) =

$$\frac{CVC}{PNVC} \times 100 = \qquad \text{(Should} = 100\%\text{)}$$

$$\frac{FEV_1}{FVC} = \qquad \text{(Should} = 75\%\text{)}$$

N.B. Smokers with FEV_1/FVC in range 70–75% may be encouraged to reduce/give up smoking using the figures. Repeat measurements after 3 months' abstinence should show significant improvement.

(4) Posterior–anterior chest X-ray (14 × 17 inches) to be attached.
(5) I examined............... on.......................... and
(delete as necessary and initial)
(i) In my opinion he/she has no medical condition which will place him/her at an increased risk of material impairment of his/her health from exposure to asbestos fibres, or which will be aggravated directly or indirectly by such exposure;
(ii) I recommend that:
(a) his/her exposure to asbestos fibres be limited to a maximum of ....
(b) he/she wear a respirator or other protective equipment (specify) ....
(iii) I have informed .... of any medical conditions related to occupational exposure to asbestos that require further treatment (copy of written opinion as sent to employer to be attached).

Signed ....
Status ....
Date ....

(The above information could usefully be printed on a large envelope which could also house the 14 × 17 inch chest X-ray records.)

N.B. This record is to be retained for at least twenty years and made available to the Assistant Secretary of Labor for Occupational Safety and Health, the Director of NIOSH, to authorized physicians and medical consultants of either of them, and, upon request of an employee or former employee, to his physician.

APPENDIX 4.4

Annual Medical Examination: Details for employees who consent to the medical examination.

(1) Date of Examination............Annual examination number........

						Sputum cytology examination required		
Age	41	42	43	44		45......................		
Years of employment to date in the asbestos industry	6	7	8	9		10	11	12..........

Height .... Pulse rate ....
Weight .... Temperature ....
Chest circumference (relaxed) ....
Chest circumference (expanded) ....
Pets/livestock, etc ....
Athletic activities ....

(2) History of Respiratory Disease

(a) Smoking habits ....
(b) Past exposure to asbestos or other dusts ....
....

(c)

	Present	Absent
Pulmonary symptoms Cardio-vascular symptoms Gastro-intestinal symptoms		

(d) Physical examination with special attention to pulmonary rales, clubbing of fingers, and other signs related to cardio-pulmonary symptoms.

(e) Employee's medical complaints related to exposure to asbestos.

(3) Pulmonary Function Tests

(a) Vital capacity (VC) =

(b) Forced expiratory volume at 1 sec (FEV_1) =

(c) Predicted normal VC (PNVC) =

(d) Corrected VC (CVC) =

(e) Forced VC (FVC) =

$$\frac{CVC}{PNVC} \times 100 = \qquad \text{(Should} = 100\%\text{)}$$

$$\frac{FEV_1}{FVC} = \qquad \text{(Should} = 75\%\text{)}$$

N.B. Smokers with FEV_1/FVC in range 70–75% may be encouraged to reduce/give up smoking using the figures. Repeat measurements after 3 months' abstinence should show significant improvement.

(4) Posterior–anterior chest X-ray (14 × 17 inches) to be attached.

(5) Blood Examination (to be carried out in cases where there is a continued decline in health):

Packed cell volume =

Oxygen =

Carbon dioxide =

pH =

(6) I examined........................... on.. and
(delete as necessary and initial)

(i) In my opinion he/she has no medical condition which will place him/her at an increased risk of material impairment of his/her health from exposure to asbestos fibres, or which will be aggravated directly or indirectly by such exposure;

(ii) I recommend that:

(a) his/her exposure to asbestos fibres be limited to a maximum of..

(b) he/she wear a respirator or other protective equipment (specify)..

(iii) I have informed .... of any medical conditions related to occupational exposure to asbestos that require further treatment (copy of written opinion as sent to employer to be attached).

Signed.....................
Status ....
Date ....

(The above information could usefully be printed on a large envelope which could also house the 14 × 17 inch chest X-ray records.)

N.B. This record is to be retained for at least twenty years and made available to the Assistant Secretary of Labor for Occupational Safety and Health, the Director of NIOSH, to authorized physicians or medical consultants of either of them, and, upon request of any employee or former employee, to his physician.

APPENDIX 4.5

Termination of Employment Examination Details for employees who consent to the medical examination

(To be carried out within 30 calendar days before or after termination of employment involving exposure to airborne concentrations of asbestos fibres.)

(1) Date of Examination..
Expected date of termination of employment.................................
Actual date of termination of employment.....................................

					Sputum cytology examination required		
Age	41	42	43	44	45.......................		
Years of employment to date in the asbestos industry	6	7	8	9	10	11	12..........

Height .... Pulse rate ....
Weight .... Temperature ....
Chest circumference (relaxed) ....
Chest circumference (expanded) ....
Pets/livestock, etc ....
Athletic activities ....

(2) History of Respiratory Disease
 (a) Smoking habits ....
 (b) Past exposure to asbestos or other dusts ....
 (c)

	Present	Absent
Pulmonary symptoms Cardio-vascular symptoms Gastro-intestinal symptoms		

 (d) Physical examination with special attention to pulmonary rales, clubbing of fingers, and other signs related to cardio-pulmonary symptoms.

(3) Pulmonary Function Tests
 (a) Vital capacity (VC) =
 (b) Forced expiratory volume at 1 sec (FEV_1) =
 (c) Predicted normal VC (PNVC) =
 (d) Corrected VC (CVC) =
 (e) Forced VC (FVC) =

$$\frac{CVC}{PNVC} \times 100 = \qquad \text{(Should} = 100\%)$$

$$\frac{FEV_1}{FVC} = \qquad \text{(Should} = 75\%)$$

N.B. Smokers with FEV_1/FVC in range 70–75% may be encouraged to reduce/give up smoking.

(4) Posterior–anterior chest X-ray (14×17 inches) to be attached.
(5) Blood Examination (to be carried out in cases where there is a continued decline in health):
 Packed cell volume =
 Oxygen =
 Carbon dioxide =
 pH =

(6) I examined.........................on... and I have informed..of any medical conditions

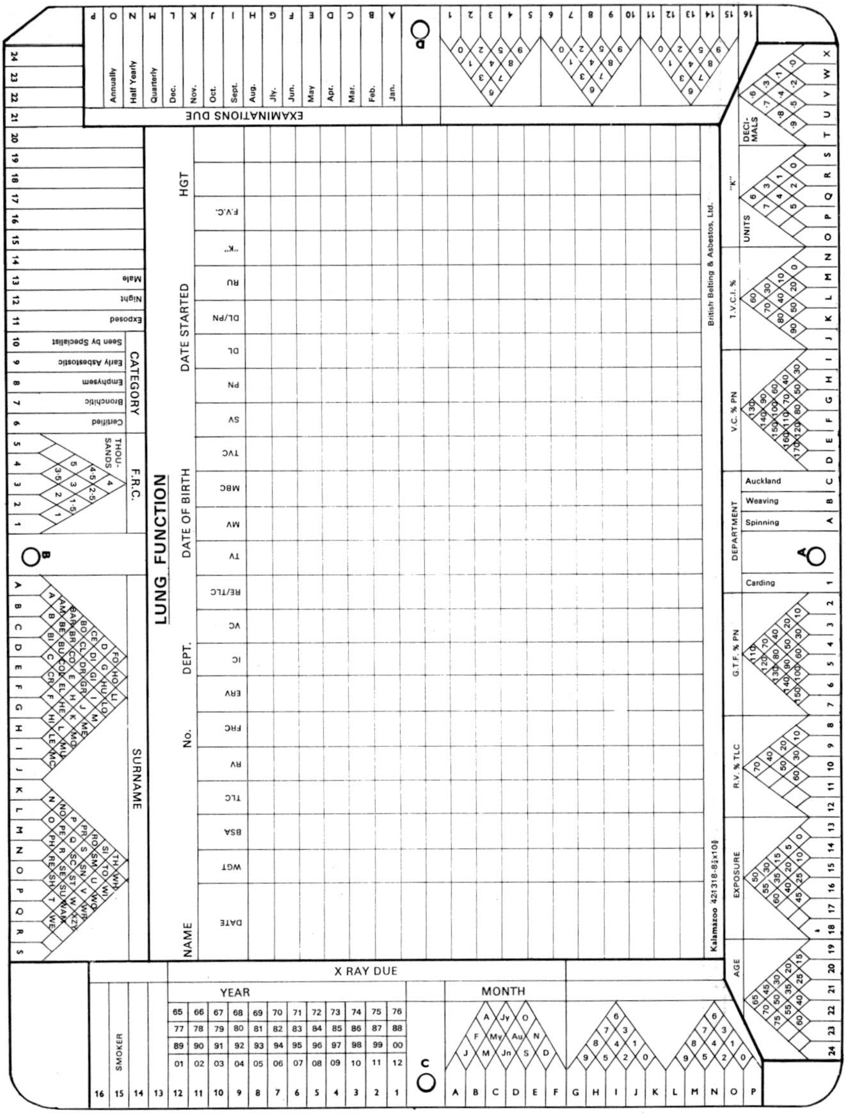

Figure to Appendix 4.6a

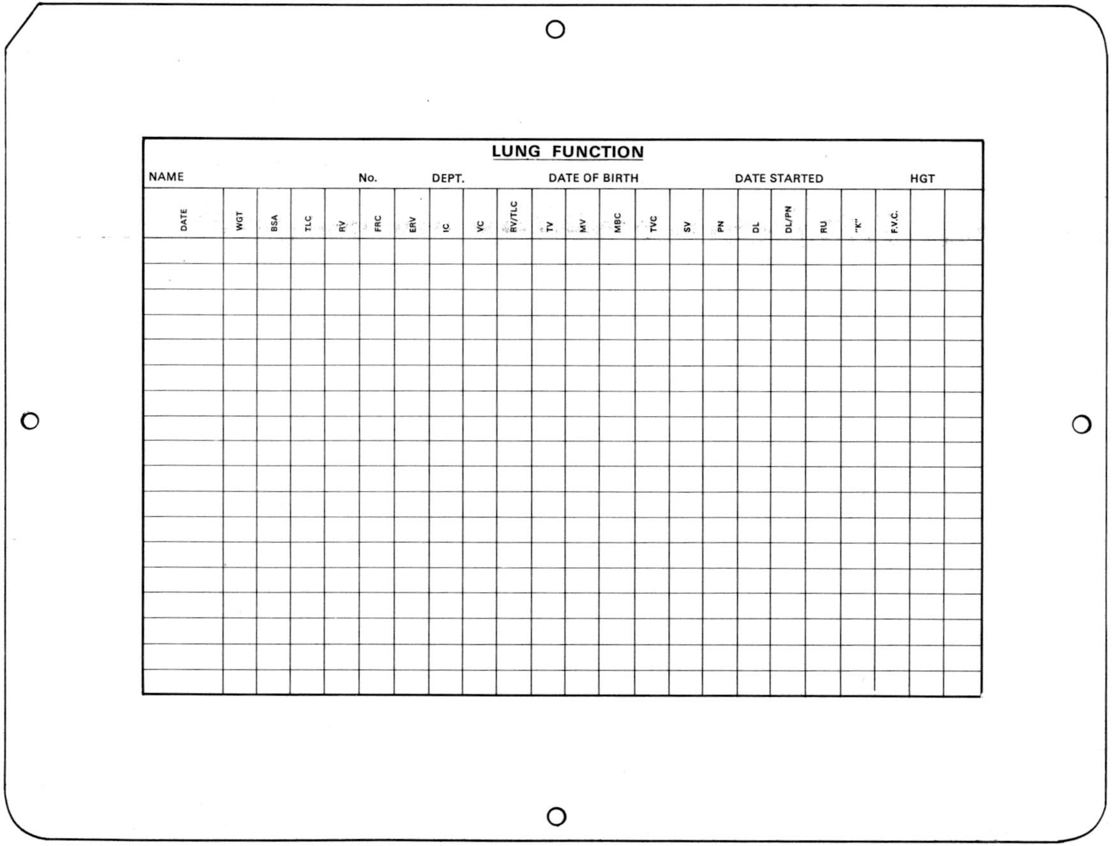

LUNG FUNCTION

NAME No. DEPT. DATE OF BIRTH DATE STARTED HGT

DATE	WGT	BSA	TLC	RV	FRC	ERV	IC	VC	RV/TLC	TV	MV	MBC	TVC	SV	PN	DL	DL/PN	RU	"K"	F.V.C.		

Figure to Appendix 4.6b

related to occupational exposure to asbestos that require further treatment (copy of written opinion as sent to employer to be attached).

Signed....................
Status.....................
Date.......................

(The above information could usefully be printed on a large envelope which could also house the 14×17 inch chest X-ray records.)

N.B. 1. Because of the insidious nature of the disease, medical supervision must continue after exposure to asbestos dust has ceased. Apart from the continuing need to make a diagnosis at the earliest possible stage in the interests of the work concerned, it is also important for compensation and epidemological purposes.

This record is to be retained for at least twenty years and made available to the Assistant Secretary of Labor for Occupational Safety and Health, the Director of NIOSH, to authorized physicians and medical consultants of either of them, and, upon request of an employee or former employee, to his physician.

APPENDIX 4.6a

Following medical examination, the results can usefully be transferred to punch cards (Example of card reproduced by courtesy of Scandura Ltd. A BBA Group Company).

APPENDIX 4.6b

The obverse of the card shown in Appendix 4.6a.

(11) REFERENCES

Anderson, H. A., Selikoff, I. J., Lilis, H., and Daum, S. (1978) *Ann. N.Y. Acad. Sci.*, in press.
H.M.S.O. (1977) *Selected Written Evidence submitted to the Advisory Committee on Asbestos* 1976–77, HMSO, London, 88–90.
International Agency for Research on Cancer (1977) *Evaluation of the Carcinogenic Risk of Chemicals to Man, Vol.* 14, *Asbestos*, International Agency for Research on Cancer, Lyon, 81.

Kiviluoto, R. (1965) Pleural plaques and asbestosis: further observations on endemic and other non-occupational asbestosis, in *Biological Effects of Asbestos, Ann. N.Y. Acad. Sci.*, **132,** 235.

Leidel, N. A., Bayer, S. G. and Zummualde (in press) *USPHS/NIOSH Membrane Filter Method for Evaluating Airborne Asbestos Fibres*, U.S. Department of Health, Education and Welfare, Ohio, U.S.A.

Lewinsohn, H. C. (1972) The medical surveillance of asbestos workers, *R. Soc. Hlth. J.*, **92** (2), 69–77.

Marne, A. (1976) *Report of the Parliamentary Commissoner for Administration to Mr. Max Madden, M.P.* (unpublished).

Merewether, E. R. A. and Price, C. E., (1930) *Effects of Asbestos Dust on the Lungs and Dust Suppression in the Asbestos Industry*, HMSO, London, 34–206.

National Federation of Buildings Trades Employees (1976) *Selected Written Evidence Submitted to the Advisory Committee on Asbestos, 1976–77*, 93.

Newhouse, M. L. (1973) Asbestos in the workplace and the community, *Ann. occup. Hyg.*, **16,** 97–107.

Shelz, J. P. (1974) The detection of chrysotile asbestos at low levels in talc by differential thermal analysis, *Thermochim. Acta*, **8,** 197–204.

Stanton, F. M., (1973) in *Biological Effects of Asbestos*, IARC Scientific Publications No. 8, International Agency for Research on Cancer, Lyon, 289–294.

Tait, N. (1976) in *Asbestos Kills*, The Silburg Fund, London, p. 4.

World Health Organization (1977) *Evaluation of the Carcinogenic Risk of Chemicals to Man, Vol.* 14, *Asbestos*, International Agency for Research on Cancer, Lyon.

CHAPTER 5

Non-occupational asbestos emissions and exposures*

Benjamin E. Suta

and

Richard J. Levine†

Center for Resource and Environmental Systems Studies,
SRI International,
Menlo Park, California

SYNOPSIS

This chapter reviews the sources of asbestos emissions and the levels of asbestos exposure outside the workplace. The various sections are entitled:

* Based on material included in *Asbestos: An Informational Resource*, Ed. by Richard J. Levine, U.S. Dept. Health, Education, and Welfare DHEW Publication No. (NIH) 78–1681, May 1978, and supported under contract number NO1-CN-55176.

† Present address: Chemical Industry Institute of Toxicology, Research Triangle Park, North Carolina 27709.

The most significant exposures of humans to asbestos occur in the workplace. However, persons not employed in asbestos-related occupations are also exposed to asbestos fibres that originate from natural sources or from man-created sources such as the manufacture and use of asbestos products. Such asbestos may be inhaled—as, for example, in an office building in which the air is contaminated by asbestos insulation—or it may be ingested with water, food, and drugs (or, inadvertently, parenterally inoculated). These exposures, termed 'non-occupational', are the subject of this chapter.

(1) ASBESTOS EMISSIONS FROM NATURAL SOURCES

Of greatest concern for natural environmental contamination are those areas where non-commercially exploited asbestos is contained in rock that might be distrubed by natural means, such as weathering or landslides, or by human intervention, such as mining, road building, construction, and tilling of the soil. In such cases, free asbestos fibre may be deposited on to soil or enter air and water (National Academy of Sciences, 1971), thereby contributing to levels of contamination in the ambient air and in water as discussed in this chapter.

The geographic distribution of ultrabasic and metamorphic rock formations in the United States that could possibly contain asbestos is shown in Figure 5.1. It can be seen that the primary areas of source rock are Minnesota, New England, and many of the Western and Southeastern states (especially in the vicinity of the Appalachian Mountains). Because of high population density, the most critical areas for emissions from natural sources appear to be Eastern Pennsylvania, Southeastern New York, Southwestern Connecticut, and greater Los Angeles and San Francisco.

(2) ASBESTOS EMISSIONS FROM HUMAN-CREATED SOURCES

Human-created sources of non-occupational exposures to asbestos include the mining and milling of asbestos, the transportation of asbestos materials and products, the maufacture, installation, use, and demolition of asbestos products, and the disposal of wastes.

Estimated annual emissions in the United States from asbestos mining and milling, manufacturing, use of asbestos products, and disposal of wastes are shown in Figure 5.2. Although these estimates are uncertain, by at least an

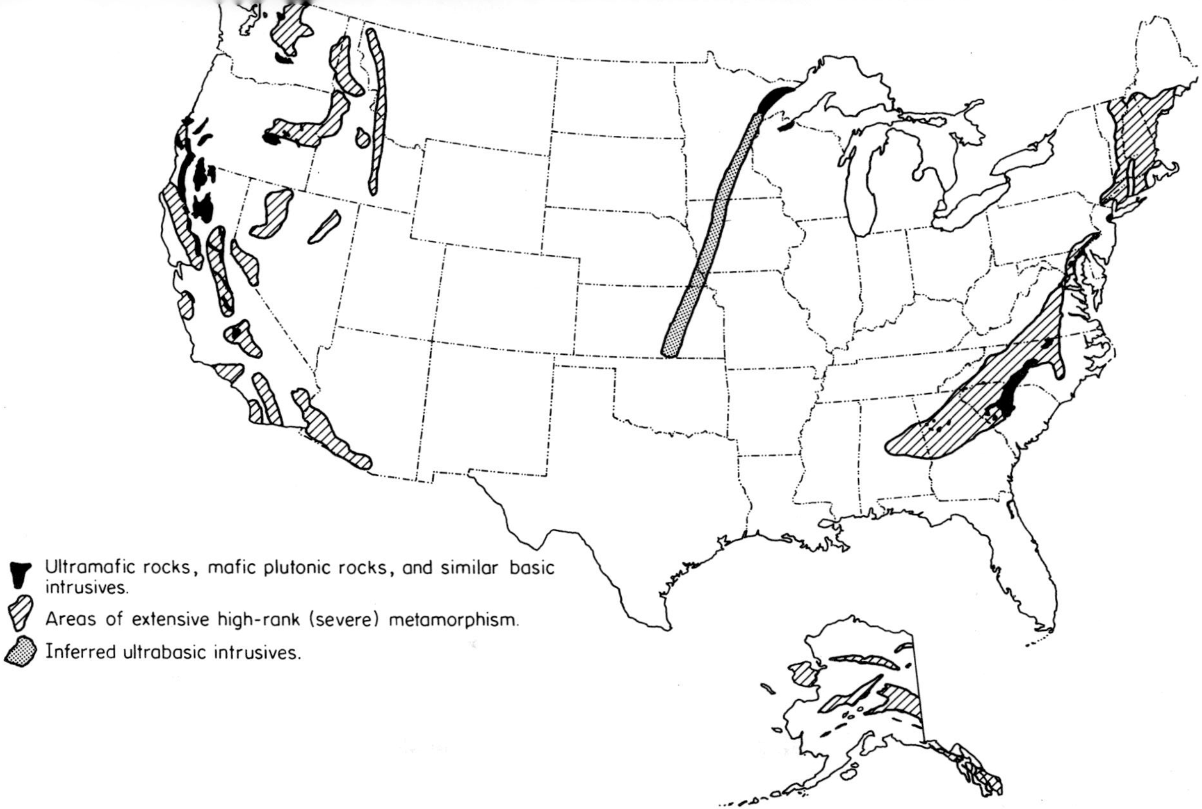

Figure 5.1 Distribution of ultrabasic and metamorphic rock formations in the United States. *Note:* in Hawaii, the type of mineral alteration that could lead to asbestos formation is restricted (to the vicinity of Koolau and Molokai volcanoes on the island of Oahu). (Source: *Tectonic Map of North America*, U.S. Geological Survey; Levine, 1978)

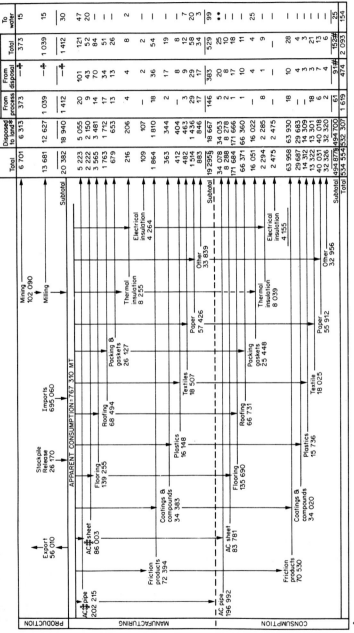

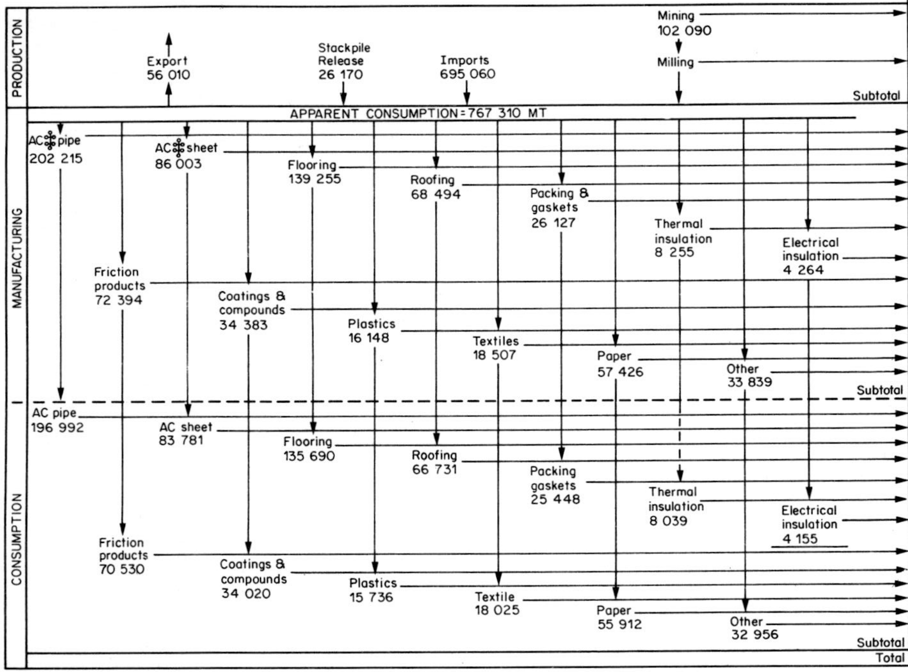

	ASBESTOS DISPOSALS AND EMISSIONS					
			Emissions			
			To air			
	Total	Disposed to land*	From process	From disposal	Total	To water
	6 701	6 313	373	—♣	373	15
	13 681	12 627	1 039	—♣	1 039	15
Subtotal	20 382	18 940	1 412	—♣	1 412	30
	5 223	5 055	20	101	121	47
	2 222	2 150	9	43	52	20
	3 565	3 481	14	70	84	—
	1 763	1 712	17	34	51	—
	679	653	13	13	26	—
	216	206	4	4	8	2
	109	107	—	2	2	—
	1 864	1 810	18	36	54	—
	363	344	2	17	19	—
	412	404	—	8	8	—
	482	463	3	9	12	7
	1 514	1 436	29	29	58	20
	883	846	17	17	34	3
Subtotal	19 295§	18 667	146	383	529	99
	34 078	34 053	5	20	25	**
	8 288	8 278	2	8	10	—
	171 684	171 666	1	17	18	—
	66 371	66 360	1	10	11	—
	16 051	16 022	—	4	4	25
	2 294	2 285	8	1	9	—
	2 475	2 475	—	—	—	—
	63 958	63 930	18	10	28	—
	29 687	29 683	—	4	4	—
	14 312	14 309	—	3	3	—
	13 322	13 301	18	3	21	—
	40 031	40 018	6	7	13	—
	32 326	32 320	2	4	6	—
Subtotal	494 877§	494 700	61	91#	152#	25
Total	534 554	532 307	1 619	474	2 093	154

✤ Asbestos cement.

* i.e. amounts of asbestos refuse that reach the land: emissions from disposal do occur (and are accounted for in this chart under Emissions).

♣ All emissions considered to be 'process' emissions.

§ Subtracting the manufacturing and consumption disposals and emissions (514 172 tons) from total apparent consumption yields a figure of 253 138 tons as 'net addition to asbestos products in use' for the year.

Includes an estimated 15 tons emitted from incineration of solid waste.

Figure 5.2 Disposals and emissions of asbestos from asbestos production, manufacturing, and consumptions in the United States—1974 (metric tons)

order of magnitude, several important conclusions can be drawn:

(a) asbestos is preponderantly disposed to land, least to water;

(b) most asbestos disposed to land is consumer waste which is more likely to be disposed at uncontrolled waste dumps and handled by persons unaware of the hazards;

(c) disposal to land is an important source of atmospheric asbestos and, because of proximity to urban populations, may be even more significant to health than the emissions to air that come from mining and milling.

(3) REDISTRIBUTION AND FATE OF ASBESTOS IN THE ENVIRONMENT

Because asbestos is exceptionally resistant to thermal and chemical degradation, it persists in the environment and can be widely redistributed by both natural forces and human means. The magnitude of this redistribution is governed by an extraordinarily complex set of factors which include the height of the emission source, the rates of air and water flow, fibre diameter, rain, thermal air inversions, electrostatic forces, agglomeration of particles, and the density of vehicular traffic on asbestos-containing landfill, to name only a few.

3(i) Redistribution by air

If, for example, asbestos is emitted to air as part of a 'large' agglomerated particle, it will settle to earth relatively quickly and thereby have a limited potential for environmental contamination; thus, concern over the relatively large amounts of asbestos emitted to air from mines is somewhat attenuated by knowledge that the mining processes tend to produce relatively large particles. At the same time, however, an appreciable fraction of the large mass of asbestos discharged by mills is in the form of free fibres that may remain in the atmosphere for long periods of time, travel great distances, and result in exposure of many people. Studies of atmospheric pollution in the vicinity of asbestos mines and mills in Finland showed small amounts of asbestos dust as far away as 27 km (Laamanen *et al.*, 1965).

A simplified calculation of 'drift distance' for two sizes of asbestos fibres has been made. Fibres were presumed to be injected at a height of 50 ft (15.2 m) into a constant crosswind of 10 miles/h (4.5 m/s) with no net effect of turbulence. The locality was assumed to be rural with a 'roughness height' equivalent to that of a wheat field. A small fibre, one with a diameter of 0.1 μm and a length of 100.0 μm, under such conditions would drift 1120 km; a large fibre 1.0 μm in diameter and 50 μm long would drift 13.3 km (Levine, 1978).

3(ii) Redistribution by water

Asbestos borne by water can also travel considerable distances. Studies of Lake Superior, reported in 1974, indicate that asbestos particles can move several hundred miles or more (United States of America *vs.* Reserve Mining Company, 1974). A later report shows that high river flows in surrounding regions have resulted in high fibre counts in the Philadelphia and Atlanta water supplies (Stewart, 1976).

Water samples taken from wells located in areas containing asbestos rock have shown elevated concentrations of asbestos. A well at Malvern, Pennsylvania, drilled in a belt of serpentine rock, had an asbestos content of up to 0.157 μg/l., in contrast to a well at Glendale, Arizona, in an area known not to have such rock, which had an asbestos content of 0.023 μg/l. or less (American Water Works Association, 1974).

3(iii) Ultimate fate of asbestos fibres

Very little has been reported about the ultimate fate of asbestos fibres once they are released to the environment. While it is known that fibres can be readily subdivided by mechanical means into fibrils of submicron diameter, it has not been established if fibres are subdivided by natural means. It does seem likely, however, that natural forces such as erosion, grinding, abrasion, moisture, and temperature gradients would cause their eventual subdivision.

All types of asbestos resist prolonged attack by strong alkalis. However, it has long been known that hydroxyl groups of chrysotile, in contrast with other asbestos types, will react with weak acids and even water, causing magnesium and silicon to be released from the crystal lattice (Harrington *et al.*, 1975; Clark and Holt, 1961; Reimschussel, 1974). Generally despite some degradation, it is felt that the fibrous morphology is retained (Harrington *et al.*, 1975). Thus, to a limited extent, chrysotile may undergo decomposition through reaction with water and acid present in the environment.

The temperatures required for thermal decomposition of asbestos are seldom attained in the natural environment. With chrysotile, dehydration occurs at about 100 °C, followed by dehydroxylation, which may begin at temperatures at low as 200–250 °C. At 800 °C, full dehydroxylation is achieved, leaving behind an amorphous residue (Harrington *et al.*, 1975). The thermal decomposition of amphibole asbestos occurs at similar temperatures.

(4) EXPOSURE TO AIRBORNE ASBESTOS

As one would expect, airborne asbestos can be found in the vicinity of asbestos mines, mills, manufacturing facilities, and wast dumps. However, elevated levels may also be found near concentrations of braking vehicles, in buildings in which asbestos spray products have been used, and in the cars and homes of asbestos workers who have contaminated them with dust brought from the work area on clothing, body, or equipment. Asbestos may be inhaled by persons who install their own asbestos roofing or flooring, or who repair such items as automobile brakes and clutches, home heating and plumbing systems, wires for toasters and waffle irons, or the walls of their homes.

Other situations with possible exposures to airborne asbestos include the use of roads and driveways surfaced with asbestos-bearing gravel or paving, humidifiers charged with water containing high levels of asbestos, talcum powders, paints containing asbestos, and cigarettes with asbestos filters (reportedly no longer used in American cigarettes). Even powdered papier mâché mixes, which are widely used in elementary schools, have been found to contain 50% or more asbestos (Aaronson and Kohl, 1972).

These exposures may be of a once-only or intermittent nature, but, because of the cumulative permanence of a small portion of inhaled asbestos, they contribute to a person's total risk. One has only to look at the range of asbestos products (upwards of 3000) to visualize the opportunities for inhalation of fibres (see Table 5.1).

Asbestos is also found as a contaminant in the ambient air, and this source is treated first in the discussion of exposure sources that follows.

4(i) Exposure from ambient air

Most persons do not live in areas where the atmospheric asbestos level is elevated owing to proximity to construction sites or to asbestos manufacturing facilities, mines, or mills. Neither are most persons engaged in part-time installation or repair of asbestos products. Ambient air, therefore, constitutes the greatest source of atmospheric exposure for the population at large.

There is a paucity of atmospheric asbestos concentration data, due, in part to the cost and difficulty of obtaining it. Those observations that are available are considerably imprecise, perhaps by several orders of magnitude. Moreover, sometimes data have been reported in terms of mass of asbestos per volume of air and at other times as concentration of fibres. Usually only chrysotile asbestos has been measured.

The scant data available for ambient levels of asbestos in rural air include a reported range of 0.01–0.1 ng/m^3 (Thompson and Morgan, 1971) and levels of 40–100 electron-microscope-visible fibres/m^3 found in a remote area of California (Murchio *et al.*, 1973).

Average reported concentrations of asbestos in urban air vary between 0.09 and 70 ng/m^3 (Table 5.2) and between 0 and 2400 electron-microscope-visible fibres/m^3 (Murchio *et al.*, 1973). Fibre readings for

Table 5.1 Selected asbestos products and their end uses*

Floor tile	Gaskets and packings	Friction products	Paints, coatings, and sealants	Asbestos-reinforced plastics	Asbestos–cement pipe
Office floors	Valve components	Clutch/transmission components	Automotive/truck body coatings	Electric motor components	Chemical process piping
Commercial floors	Flange components	Brake components	Roof coatings and patching compounds	Moulded product compounds for high-strength weight uses	Water supply piping
Residence floors	Pump components	Industrial friction materials			Conduits for electric wires
	Tank sealing components				

Asbestos textiles	Asbestos paper	Asbestos–cement sheet
Packing components	Gas vapour ducts for corrosive compounds	Hoods, vents for corrosive chemicals
Gasket components	Fireproof absorbent papers	Chemical tanks and vessel manufacturing
Roofing	Table pads and heat-protective mats	Portable construction buildings
Commercial/industrial dryer felts	Heat/fire protection components	Electrical switchboards and components
Heat/fire protective clothing	Molten glass handling equipment	Residential building materials
Clutch/transmission components	Insulation products	Molten metal handling equipment
Electrical wire and pipe insulation	Gasket components	Industrial building materials
Theatre curtains and fireproof draperies	Underlayment for sheet flooring	Fire protection
	Electric wire insulation	Insulation products
	Filters for beverages	Small appliance components
	Appliance insulation	Electric motor components
	Roofing materials	Laboratory furniture
		Cooling tower components

* Source: Asbestos Information Association/North America.

Table 5.2 Atmospheric asbestos concentration data for urban areas (ng/m^3)

Place	1970 population	Sample size	Concentration (ng/m^3)		Reference
			Average	Range	
Berkeley, Calif.	116 716	4	6.8	2.1–12	Nicholson *et al.* (1975)
Boston, Mass.	641 071	1	5.0	—	Nicholson *et al.* (1975)
Chicago, Ill.	3 366 957	1	24	9.5–200	Nicholson *et al.* (1975)
Dayton, Ohio	243 601	7	6	0.4–11.0	Heffelfinger *et al.* (1972)
Frankfort, Ky.	21 356	3	0.08	0.02–0.15	Heffelfinger *et al.* (1972)
Houston, Texas	1 232 802	4	5	4–6	Heffelfinger *et al.* (1972)
Los Angeles (freeway)	2 816 061		27†		Murchio *et al.* (1973)
Los Angeles (control)			43†		Murchio *et al.* (1973)
New York City, N.Y.	7 867 760	10	13.2	8.2–41	Nicholson *et al.* (1975)
Manhattan, N.Y.	1 524 541*	7	30†	8–65	Nicholson and Pundsack (1973)
Brooklyn, N.Y.	2 601 852*	3	19†	6–39	Nicholson and Pundsack (1973)
Bronx, N.Y.	1 472 216*	4	12†	2–25	Nicholson and Pundsack (1973)
Queens, N.Y.	1 973 708*	4	9†	3–18	Nicholson and Pundsack (1973)
Staten Island, N.Y.	295 443*	4	8†	5–14	Nicholson and Pundsack (1973)
Pittsburgh, Pa.	520 117	3	4	2–8	Heffelfinger *et al.* (1972)
Philadelphia, Pa.	1 948 609		70†	45–100	Selikoff *et al.* (1972)
Port Allegany, Pa.	2 703		15†	10–20	Selikoff *et al.* (1972)
Ridgewood, Pa.	27 547		20†	—	Selikoff *et al.* (1972)
San Francisco, Calif.	715 674	2	25	8.7–68	Nicholson *et al.* (1975)
Washington, D.C.	756 510	4	21	1.6–40	Heffelfinger *et al.* (1972)

* Also included with New York City.
† Identified as chrysotile asbestos.

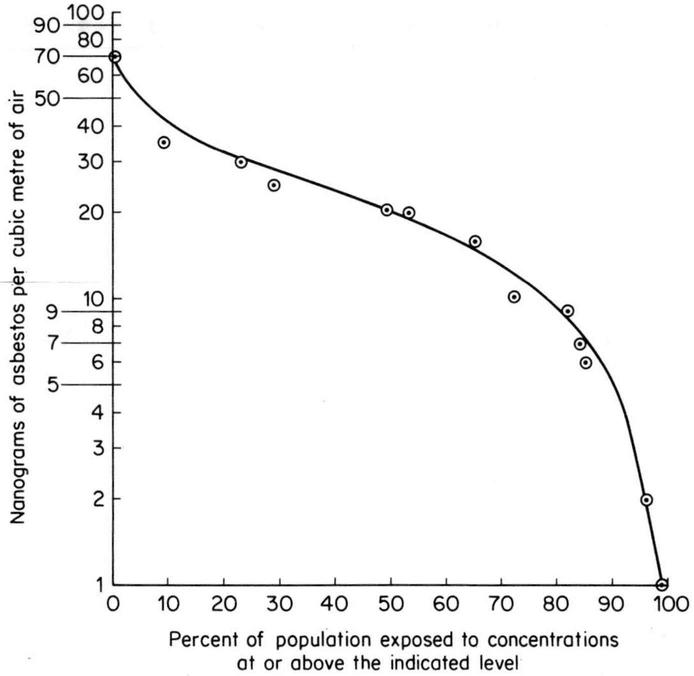

Figure 5.3 Percentage of urban population that breathe various levels of asbestos mass in ambient air

Silver Bay, Minnesota, which is not typical since it is near a taconite milling operation, ranged up to 150 000 electron-microscope-visible fibres/m^3 (Great Lakes Research Advisory Board, 1975).

Using concentration data expressed in terms of mass from Table 5.2 (representing 20.3 million people), the percentage of the urban population at risk to various atmospheric levels of asbestos has been calculated and is displayed in Figure 5.3. Assuming that a standard man breathes 15 m^3 of air daily (U.S. Public Health Service, 1960), one can make crude estimates of the annual amounts of asbestos inhaled from the ambient atmosphere for a U.S. urban population (Table 5.3)*. It can be seen that 50% of the urban population inhales 110 μg or more of asbestos per year, and 10% inhales 230 μg or more, compared with a range of 0.05–0.5 μg annually for rural populations.

4(ii) Exposure from air near asbestos industrial facilities

The possible situations of non-occupational exposure during mining and milling, transportation, and manufacture of products containing asbestos are

* These estimates do not pertain to persons living in the vicinity of asbestos mines, mills, or manufacturing facilities.

Table 5.3 Inhalation of asbestos from ambient air

Proportion of urban population consuming indicated level or more (%)	Atmospheric concentration		Annual inhalation	
	Fibres*/m^3	ng/m^3	Millions of fibres*	μg
50	5 000	20	27.4	110
25	7 500	28	41.1	158
10	10 500	40	57.5	220
5	12 500	48	68.4	268
1	16 000	64	87.6	350
Rural populations	—	0.01–0.1	—	0.05–0.5

* Electron-microscope-visible fibres, assuming 250 fibres per nanogram of asbestos.

reviewed briefly below. This review is followed by estimates of asbestos concentrations near asbestos industrial facilities and amounts inhaled annually by populations residing nearby.

4(ii).1 *Asbestos mining, milling, and product manufacture*

Asbestos fibres are released in mining and milling during removal of overburden and preparation of ores bodies for strip and open-pit mining, as well as during drilling and blasting to remove ore. Other emissions occur from ore piles and waste dumps that are exposed to wind and to distrubance by bulldozers. Drying, crushing, grinding, and screening of the ore result in the release of fibres. In this connection, the large volume of air required for air-aspiration milling (7–10 tons of process air for every ton of fibre produced) (Johns-Manville Corporation, 1976), together with the length of time that asbestos fibres can remain suspended in air, generates a significant potential for emissions.

Baghouses are the predominant engineering control measure used to remove airborne dust in mills and manufacturing plants from effluent air streams. A study in 1974 showed that baghouses using specified filter materials, and when properly designed, operated, and maintained, have a collection efficiency of over 99.99% for fibres greater than 1.5 μm in length (Harwood *et al.*, 1974). However, the collection efficiency for fibres less than 1.5 μm in length was approximately 98%. Since generally there is an enormous number of short fibres, a considerable amount of fibres can be released to the atmosphere even after passage through the best available baghouses. Fibres removed by ventilation and filtering that are not reintroduced into the production process must be disposed of outside the plant and may result in emissions to air.

4(ii).2 *Transportation of materials containing asbestos*

Movement of asbestos ore from mine to mill in open trucks, often over roads paved with mill tailings, may contribute to the overall contamination of the environment. However, three out of five mills operating in the United States are located at the mines, and the remaining two are separated by short, rural, distances (32 and 55 miles).

Shipment of milled asbestos fibre, usually in bags, can result in emissions when bags are broken, but such emissions are minimized by pressure packing and unitization. If bags are re-used, either in the asbestos industry or elsewhere, they may become a source of contamination (Brodeur, 1973).

Emissions of fibres could occur during the shipment of manufactured products, but they would be negligible, since most manufactured products contain asbestos tightly bound in a matrix. A perhaps more important emission source would be the transporting of asbestos-containing solid wastes in open vehicles through urban areas. Also, the transportation of other asbestos-bearing ores, such as talc and taconite, and their products may result in environmental emissions.

4(ii).3 *Estimated atmospheric concentrations and exposures*

There have been few measurements of asbestos concentrations in air near asbestos industrial facilities, and an accurate assessment of emissions from these facilities is lacking. Concentrations have therefore been estimated for U.S. factories using mathematically derived dispersion curves* of assumed plant emissions (see Appendix). Emissions have been determined for 1974 and take into account the modifying effects of airpollution control technology (Levine, 1978).

In the case of mines and mills, which are located in rural areas, a radius of 30 km was used to define a population living 'near-by'†; whereas for asbestos-product manufacturing plants, which are generally in urban areas, a radius of 5 km was used. Results obtained from the model at distances greater than the selected radii approached the levels of ambient air.

Estimated atmospheric asbestos concentrations and amounts of asbestos inhaled annually for populations living in the vicinity of U.S. asbestos industrial facilities are given in Table 5.4. Keeping in mind the imprecision of the estimates, it can be seen that except for the vicinity of asbestos mines and mills, atmospheric exposures to asbestos are not much greater than (less than twice) the median ambient urban exposure of 20 ng/m^3 (Table 5.3).

* Based on the binormal continous plume dispersion model (Turner, 1970).

† The finding that air in the vicinity of asbestos mines and mills in Finland contained elevated quantities of asbestos are far away as 27 km supports the selection of this radius (Laamanen *et al.*, 1965).

Table 5.4 Atmospheric exposure to asbestos in the vicinity of asbestos industrial facilities

Type of facility	Plants analyzed: Number	Plants analyzed: Estimated portion of total plants (%)	Population at risk*	Background	Atmospheric asbestos concentration† (ng/m^3 or thousands of $fibres/m^3$)‡	Annual amount of asbestos inhaled† (μg or millions of fibres§
Mines and mills	5	100	94 000	Rural	400.0(6000)‖	2,190(32 900)‖
Friction products	51	90	4 200 000	Urban	23.0(44)	125(240)
				Rural	3.4(24)	19(130)
Gaskets packing, or insulation	39	65	2 940 000	Urban	24.0(40)	130(220)
				Rural	4.4(20)	24(110)
Textiles	24	95	2 500 000	Urban	21.0(28)	115(155)
				Rural	1.3(8)	7(45)
Cement	35	50	1 600 000	Urban	27.0(120)	148(660)
				Rural	7.2(100)	39(550)
Flooring products	23	90	1 450 000	Urban	21.0(32)	115(175)
				Rural	1.0(12)	5(66)
Roofing products	26	70	1 400 000	Urban	30.0(73)	164(400)
				Rural	9.5(53)	52(290)
Paper	18	80	955 000	Urban	33.0(84)	180(460)
				Rural	13.0(64)	71(350)

* Population residing within 30 km of mines and mills and within 5 km of manufacturing plants.
† Includes nominal atmospheric concentration background.
‡ Electron-microscope-visible fibre data were calculated from mass data assuming 1000 fibres/ng. Actual conversion may be 100–10 000 fibres/ng; hence, reported figures may be one order of magnitude high or low.
§ Consumption assumes a daily inhalation of 15 m^3.
‖ Median population exposure for indicated 'at risk' population followed by upper 10% population exposure in parentheses.

Persons employed in the asbestos industries and exposed to the maximum permissible asbestos concentrations (2 optical-microscope-visible fibres greater than 5 μm long per cubic centimetre) over a 40-h work week, it is estimated would inhale 220 000 μg (or 220 000 million electron-microscope-visible fibres) each year*. Of the populations living near asbestos industrial facilities, only the upper one percentile of persons residing in the vicinity of asbestos mines and mills would inhale annually an amount of asbestos equivalent to that inhaled by persons working in the asbestos industries.

4(iii) Exposure from asbestos manufactured products

Most asbestos is incorporated into finished products where the fibres are bound in a matrix (e.g. asbestos-cement pipe and sheet, flooring and roofing products, and friction products), and this reduces the possibilities for contamination. Yet, by the application of sufficient energy, fibres may be dislodged from even tightly bound materials; automobile brake linings are an example.

Clearly, there are opportunities for human non-occupational atmospheric exposure during installation, use, and repair of asbestos products. However, since there are so many products that use asbestos or materials that may be contaminated with asbestos, it would be almost impossible to estimate human exposure for each product type. In the paragraphs below, some data and information are presented for automotive friction materials and spray asbestos.

4(iii).1 *Automotive friction materials*

Friction materials used in automotive brake linings, disk pads, and clutch facings contain an average concentration of 50% by weight of chrysotile asbestos (Harwood, 1972). It has been estimated that about 118 million pounds of asbestos are used annually to produce brakes in the U.S.A. and, assuming a 15% grinding and milling loss, approximately 103 million pounds of asbestos are actually incorporated into brakes. Similarly, it has been estimated that 4.5 million pounds are incorporated annually into automotive clutch facings (Jacko *et al.*, 1973).

However, since brakes and clutches are usually repaired before they are completely worn out, and since some working automobiles are scrapped, not all of this automotive asbestos will be released to the atmosphere. The estimate of how much actually is worn away annually is 74 million pounds of

* Assuming (1) 50 electron-microscope-visible fibres for every optical-microscope-visible fibre greater than 5 μm long, and (2) 1000 electron-microscope-visible fibres per nanogram.

asbestos, but only a small amount of this is released as fibrous material (Jacko *et al.*, 1973).

Tests performed on brake linings have indicated that under conditions of normal usage, considerable alteration of the asbestos occurs. One study has reported that most of the dust collected from brake drums is non-fibrous and is similar in appearance to thermally degraded asbestos. It is suggested that temperature at the points of contact of brake linings and drum actually reach levels at which thermal degradation of asbestos can occur (Harwood, 1972).

Three research studies of asbestos emissions from brake linings give estimated percentages of free fibre at 1% or less, 0.3%, and less than 0.02% (Jacko *et al.*, 1973; Lynch, 1968; Anderson, 1973). In the first of these studies, it was estimated that annually in the U.S.A. there are 239 340 lb of asbestos fibre emissions from cars, buses, and trucks, and that, of this amount, 204 952 lb drop out on the roadway, 7655 lb become airborne, and 26 733 lb are retained within the brake and clutch housings (Jacko *et al.*, 1973)*. These atmospheric emissions are of greatest concern in urban areas near traffic routes with high volumes of braking vehicles.

Electron and light microscopy were used in a recent study to analyse the number and size of asbestos fibres collected from air at four Los Angeles freeway loop sites and from upwind ambient air controls within 200 ft of the freeway (Murchio *et al.*, 1973). Concentrations of chrysotile asbestos at the four freeway sites were low (generally in the range 0–12 000 electron-microscope-visible fibres/m^3, and they did not differ significantly from concentrations of chrysotile in the matched upwind ambient air samples (0–9000 fibres/m^3). Concentrations of amphibole asbestos did not differ between freeway or controls (1200 fibres/m^3). There was no correlation of asbestos fibre concentrations in the freeway samples with number or speed of motor vehicles passing by during the sampling periods, nor was there a correlation with wind direction or velocity.

Measurements were also made of chrysotile and amphibole asbestos at the San Fransisco Bay Bridge toll plaza, and the concentrations of each there were found to be 1400 electron-microscope-visible fibres/m^3. This compared with an average San Francisco Bay Area atmospheric chrysotile concentration of 500 fibres/m^3 (Murchio *et al.*, 1973).

In another study made in 1973, asbestos air levels adjacent to a New York City toll booth were found to be three to five times background levels (Anderson, 1973).

* There is apparently an error in the data used in this study from *Brake Emissions: Emission Measurements from Brake and Clutch Linings from Selected Mobile Sources*, March 1973, EPA (NTIS 68-04-0020). Total emissions should have been 239 340 lb and, therefore, the other emission figures have been scaled up accordingly in this chapter.

4(iii).2 *Spray asbestos*

From 1958 to 1973, spray materials containing 10–30% of asbestos were used extensively to fireproof girders, spandrels, and decking of high-rise office buildings, and use of spray asbestos for decorative and acoustical purposes dates from the mid-1930s (Selikoff *et al.*, 1972; Nicholson *et al.*, 1973). Erosion of such spray materials alone may cause asbestos fibres to enter building air, but the materials might also be damaged and dislodged, for example by workmen repairing fixtures inside the space between a ceiling and the floor above. In large office buildings, air is often returned to the ventilation system through these spaces.

The results of a study of high-rise office buildings are given in Table 5.5.

Table 5.5 Concentrations of asbestos in the air of office buildings*

	Average asbestos concentration (ng/m^3)		Type of spray		
Location	Building air	Outside air	Cementitious	Fibrous	Decorative acoustic
New York City, N.Y.					
Turin House	8.2	18			×
Steinman Building	41	33			×
EXXON Building	29	8.0		×	
McGraw-Hill	9.2	2.0		×†	
Hippodrome Building	11	14		×	
1133 Avenue of Americas	29	9.2		×	
888 7th Avenue	77	12		×	
1700 Broadway	12	9.2	×		
TWA Terminal	17	—			×
Buddhist Temple	27	—			×
Boston, Mass.					
JFK Building	2.5	5.0	×		
Chicago, Ill.					
U.S. Gypsum	9.5	24	×		
CNA Plaza	200	—		×	
Berkeley, Calif.					
Department of Health	12	18	×		
Great Western	7.0	5.0	×		
UC Cafeteria	2.1	4.3			×
Harmon Gym.	7.5	0			
San Francisco, Calif.					
Wells Fargo	8.7	3.9		×	
Metropolitan Life	68	46	×		

* Source: Nicholson *et al.* (1975).
† Non-asbestos spray was used in this building.

Table 5.6 Summary of asbestos concentrations in the air of office buildings*

Type of spray	Number of samples	Asbestos concentration (ng/m^3) Average	Range	Average building elevated concentration†
Cementitious	6	18.5	2.5–68	+5.4
Fibrous	6(5)‡	59.1	8.7–200	+21.5
Decorative acoustic	5(3)	19.1	2.1–41	−1.3
No asbestos	2	8.4	7.5–9.2	+7.4
All outside atmosphere	16	13.2	0.0–46	—

* Source: Nicholson *et al.* (1975).
† Average of building asbestos concentrations minus their corresponding outside concentrations.
‡ Number of outside air samples.

Buildings treated with fibrous, as opposed to cementitious, asbestos sprays had average asbestos concentrations that were about twice the level of concentrations in the outside atmosphere (Nicholson *et al.*, 1973).

The inhalation of asbestos by office personnel working in buildings treated with fibrous asbestos spray can be estimated by using the concentration data given in Table 5.6. It is assumed that persons work 8 h per day for 240 days per year, and that, on working days, one third of asbestos air exposure is as at the level inside office buildings and two thirds is at ambient urban levels (20 ng/m^3, using the 50% population value from Table 5.3). Thus, it is calculated that a person working in a building insulated with fibrous spray asbestos inhales about 160 μg of asbestos per year, and level approximately 45% above the median value for the total urban population (110 μg).

Flaking of sprayed asbestos has been reported inside schools, libraries, dormitories, and warehouses (*New York Times*, 1977; *Engng. News Rec.*, 1974; Sawyer, 1977; Lumley *et al.*, 1971). Air concentrations may range from 0.02 optical-microscope-visible fibre/ml under quiet conditions to 4.0 fibres/ml during dry dusting (Sawyer, 1977).

The application of spray asbestos presents a hazard not only to the persons spraying, but also to nearby construction workers, passers-by, and neighbourhood residents. There is now a U.S. Government Standard which stipulates that spray-on materials used to insulate or fireproof structures, pipes, and conduits must contain less than 1% of asbestos on a dry-weight basis. For the spray application of material containing more than 1% of asbestos used to insulate or fireproof machinery or equipment, no visible emissions are permitted (*Federal Register*, 1973). Spray-on paints, decorative materials, and weatherproofing, however, are not regulated.

Prior to implementation of these regulations, data were obtained (see

Table 5.7 Concentrations of asbestos in the air near spray fireproofing sites*

Sampling locations (miles)	Number of samples	Asbestos concentration (ng/m^3) Average	Range
1/8–1/4	11	60	9–375
1/4–1/2	6	25	8–54
1/2–1	5	18	3.5–36

* Source: Nicholson and Pundsack (1973).

Table 5.7) at various building sites in lower Manhattan where asbestos-containing fireproofing materials were being sprayed (Nicholson and Pundsack, 1973). Generally, average atmospheric concentrations within $\frac{1}{4}$ mile of a construction site were at least twice the back-ground level. The use of current spray materials that contain 1% of asbestos or less may be expected to result in elevated asbestos air concentrations no greater than one tenth of those levels occurring previously.

4(iv) Exposures from disposal of asbestos products and wastes*

Solid wastes produced from the manufacture and use of asbestos-containing products and from demolition† can be emission sources, and in the past these waste materials were often disposed of without regard to their emission potential. Moreover, disposal may result in the co-mingling of asbestos-containing wastes with municipal wastes in open dumps, thus creating a long-term emission source.

Industrial asbestos wastes include process wastes such as dust, slurries, waste from overspraying, and mill tailings; waste collected by air control equipment (e.g. the dust from sawing, grinding, drilling, etc., that is vented to control devices); scrap; and emptied asbestos shipping bags. Exposures during waste disposal are minimized when waste dust is wetted, slurried, transported in a closed container, or pelletized. Scrap asbestos wastes can be handled safely if fibres are bound in a matrix so that atmospheric emissions

* Except where otherwise indicated, the material summarized here has been based on U.S. Environmental Protection Agency (1974).

† For many years asbestos has been incorporated into materials such as insulation, cement sheet, roofing, and floor tiles and used in constructing industrial and commercial buildings and ships. When such buildings and ships are demolished, areas of loosened asbestos, especially from insulating materials, are open to the ambient air and can emit fibres. Obviously, demolition will continue to be a source of emissions in the future, requiring control measures. (For the most part, single-family residences contain only small amounts of asbestos insulation.)

are unlikely (e.g. many wastes from fabrication of cement pipe and boards, friction products, floor tile, paper products containing appropriate binders, and numerous gasket materials).

An asbestos-containing waste common to almost all manufacturers of asbestos products is the shipping bags in which milled fibre is received. If the bags are not shredded and incorporated directly into the product mix, they are incinerated or disposed of in landfills, sometimes sealed in plastic bags. Occasionally, they may be treated like non-asbestos wastes and may result in exposures to unknowing handlers, which may increase the risk of disease. (Brodeur, 1973).

Water may become polluted with asbestos fibres in manufacturing, particularly in the paper and cement product industries, and during use in wet cyclones for cleaning exhaust gases from factories. The slurry waste from such processes may be directed either into settling ponds and the water recirculated (the dried waste being disposed of in landfills), or it may be dumped directly into convenient sewers, rivers, or lakes. In either case, but especially in the latter, it can contribute to contamination of the environment with asbestos.

The results of a survey of waste disposal methods used by asbestos product manufacturers are given in Table 5.8 (Harwood *et al.*, 1974). They show that 37% of the plants surveyed use dumps for wastes and that another 13.4% use landfills. The remainder re-use, sell, store or wet-slurry their wastes. Re-use is most common in asbestos product industries where milled fibre is a primary ingredient, e.g. manufacturers of asbestos–cement pipe and paper. Of greatest concern from the standpoint of emissions are those wastes that are disposed of in uncovered dumps.

Asbestos mills generate vast amounts of waste. Whereas a large manufacturing waste disposal site may have a surface area of 12 000 m^2 (about 3

Table 5.8 Methods of waste disposal used by asbestos product manufacturers*

Method	No. of plants using method	Proportion of total (%)
Re-use	38	39.2
Dump	36	37.0
Landfill	13	13.4
Sell	5	5.2
Store	3	3.1
Wet slurry†	2	2.1
Total	97	100.0

* Source: Harwood *et al.* (1974).
† Ultimate method of disposal not specified.

acres), a large mill tailings disposal site may occupy 400 000 m^2 (about 100 acres). Mill waste may contain from less than 1% of asbestos by weight, as in the case of the Vermont mill (which disposes of over 10^6 tonnes of asbestos tailings annually), to over 30% of asbestos in some California operations.

Obviously, there are opportunities for asbestos emissions from the disposal of non-industrial wastes, which constitute the majority of asbestos wastes disposed to land (Figure 5.1). Most notable of these are the renovation and demolition of ships and buildings, which may contain large amounts of friable asbestos insulation as well as many other asbestos products.

Three studies of asbestos concentrations in air near asbestos waste dumps, all conducted prior to the establishment of U.S. environmental standards in 1975, have been published (Murchio *et al*, 1973; U.S. Environmental Protection Agency, 1974; Harwood and Blaszak, 1974). It is apparent from these studies that atmospheric asbestos concentrations in the vicinity of waste disposal sites (often in urban areas) are considerably higher than background concentrations—perhaps 10 to 1000 times higher—and may even approach occupational levels. (The revised U.S. environmental asbestos standard (*Federal Register*, 1975) that went into effect after these data were recorded may help to reduce emissions.)

4(v) Exposures of asbestos workers' families

Families of persons employed in the asbestos industry may be subjected to asbestos contamination that augments their exposures from other sources. Workers may bring asbestos fibres home on their skin or clothing or on equipment such as lunch boxes and automobiles. Atmospheric concentrations of asbestos in the homes of asbestos workmen have been reported to be 100–5000 mg/m^3 (Nicholson *et al.*, 1973), similar to estimated concentrations in the vicinity of asbestos mines and mills and much higher than the 0.09–70 ng/m^3 reported for U.S. cities (see Table 5.2)*.

(5) EXPOSURE TO ASBESTOS IN DRINKING WATER

Drinking water is one of the possible routes by which humans are exposed to asbestos. Comtamination of drinking water may be due partly to erosion from natural deposits of serpentine and other asbestos-containing materials, as noted in Section 3. Substantial contamination may also result from improper disposal of asbestos wastes. These wastes may be effluents that are directly discharged into water systems, or they may be released to the

* It is known that occupants of households of asbestos workers have elevated rates of asbestosi and mesothelioma (Newhouse and Thompson, 1965).

atmosphere or disposed of on the land, and subsequently join the water system.

Another potential contaminator of drinking water is the piping and pumping of municipal water distribution systems. About 200 000 miles of asbestos–cement pipes are used to carry water to U.S. consumers, and the pipes provide a source of asbestos fibres from leaching and errosion (American Water Works Association, 1974; Millette, 1976).* Gaskets and insulation used in treatment and pumping mechanisms are other possible contaminators (Stewart, 1976).

5(i) Asbestos content of drinking water supplies

There are major difficulties in determining the asbestos content of water. It is not uncommon for results from duplicate samples analysed by competent workers in independent laboratories to vary by more than an order of magnitude (Levine, 1978). Further, most of the data that are available are for 'grab' samples, i.e. samples of a few litres of water that are taken from one source at one time.

The degree to which grab samples represent the characteristics of an entire municipality's water supply over location and time could be questioned—some municipalities receive their water from several sources, and seasonal and climatic variations can change the asbestos content of water. In San Francisco, water from other sources is supplemented by water from a reservoir that is located in a chrysotile rock area. (Recall also the study mentioned previously which showed higher asbestos fibre counts in Philadelphia and Atlanta water supplies during periods of high river flows in the surrounding regions.) Further, a study published in 1974 found that the effect of climatic conditions on the amount and mineralogical nature of the suspended solids in the Duluth water supply is most evident when heavy rainfalls are followed by an increase in the amount of suspended solids resulting from river runoff and shore erosion (Cook *et al.*, 1974); obviously, if these solids contain asbestos, the asbestos concentrations of the water will increase.

5(ii) Elevated asbestos levels

With all the problems inherent in the available data, they still, when taken in their entirety, provide an indication of the levels of asbestos in drinking water supplies. Data on the concentration of asbestos in U.S. drinking water were extracted from nine different sources, representing 105 water supplies for some 32 million people (Table 5.9). An effort was made to select data

* However, a recent study found no significant release of chrysotile asbestos from asbestos–cement pipe exposed to the action of moderately agressive water (Hallenbeck *et al.*, 1978).

Table 5.9 Asbestos contents of U.S. drinking waters

Place	1970 population	10^6 fibres/l.			μg/l.*			Reference
		N	Mean	Range	N	Mean	Range	
Birmingham, Ala.	300,910		BDL					U.S. Environmental Protection Agency (1975)
Montgomery, Ala.	133 386			BDL–0.12(C)				U.S. Environmental Protection Agency (1975)
Tuscaloosa, Ala.,	65 773		0.45(C)					U.S. Environmental Protection Agency (1975)
Fairbanks, Alaska	14 771		BDL					U.S. Environmental Protection Agency (1975)
Anchorage, Alaska	52 500		0.07(A)					U.S. Environmental Protection Agency (1975)
Glendale, Ariz.	36 227				34	0.010	0–0.075	American Water Works Assoc. (1974)
Globe, Ariz.	7 333				1	0.114		American Water Works Assoc. (1974)
Yuma, Ariz.	29 007		0.12(C)					U.S. Environmental Protection Agency (1975)
Jonesboro, Ark.	28 300		NSS					U.S. Environmental Protection Agency (1975)
Little Rock, Ark.	132 483		0.27(C)					U.S. Environmental Protection Agency (1975)
Van Buren, Ark.	8 373		40					Tobin (1976)
Alameda County, Calif.	—	2	0					Cooper and Murchio (1974)
Burlingame, Calif.	27 320	1	0					Cooper and Murchio (1974)
Contra Costa County, Calif.	—	1	0					Cooper and Murchio (1974)
Long Beach, Calif.	358 633				1	0.618		American Water Works Assoc. (1974)
Livermore, Calif.	37 703	1	0					Cooper and Murchio (1974)
Milbrae, Calif.	20 781	1	0					Cooper and Murchio (1974)
Pittsburgh, Calif.	20 651		NSS					U.S. Environmental Protection Agency (1975)
Redding, Calif.	16 900		0.5					Tobin (1976)
Redwood City, Calif.	55 686	1	0.0					Cooper and Murchio (1974)
Sacramento, Calif.	254 413		NSS					U.S. Environmental Protection Agency (1975)
San Diego, Calif.	696 769				1	0.835		American Water Works Assoc. (1974)
San Francisco, Calif.	715 674	8	0		8	0		Stewart (1976)
San Francisco, Calif.			1.54(C)					U.S. Environmental Protection Agency (1975)
San Francisco, Calif.		2	0.6(C)	0.2–1.0(C)				Cooper and Murchio (1974)
San Jose, Calif.	445 779	1	0					Cooper and Murchio (1974)
Weaverville, Calif.	1 489		4.5					Tobin (1976)
Boulder, Colo.	66 870		BDL					U.S. Environmental Protection Agency (1975)

Denver, Colo.	514 678	8	0.007	0–0.036	8	0.042	0–0.333	Stewart (1976)
New Haven, Conn.	137 707		NSS					U.S. Environmental Protection Agency (1975)
Stratford, Conn.	51 100		0.38					Tobin (1976)
Wilmington, Del.	80 386		0.29(C)					U.S. Environmental Protection Agency (1975)
Washington, D.C.	756 510		NSS					U.S. Environmental Protection Agency (1975)
Ft. Lauderdale, Flo.	139 590		NSS					U.S. Environmental Protection Agency (1975)
Melbourne, Flo.	40 236		0.30(C)					U.S. Environmental Protection Agency (1975)
Miami, Flo.	334 859		BDL					U.S. Environmental Protection Agency (1975)
Atlanta, Ga.	496 973	4	5.75	0–12	4	0.192	0–0.574	Stewart (1976)
Cairo, Ill.	5 600		NSS					U.S. Environmental Protection Agency (1975)
Chicago, Ill.	3 366 957	1	0		1	0		Stewart (1976)
Kankahee, Ill.	31 500		1.6					Tobin (1976)
Indianapolis, Ind.	744 624		0.18(C)					U.S. Environmental Protection Agency (1975)
Kansas City, Kan. & Mo.	675 300	3	0		3	0		Stewart (1976)
Topeka, Kan.	125 011		NSS					U.S. Environmental Protection Agency (1975)
Witchita, Kan.	276 554				2	1.006	0.417–1.593	American Water Works Assoc. (1974)
Ashland, Ky.	29 245		BDL					U.S. Environmental Protection Agency (1975)
New Orleans, La.	578 000		0.88					Tobin (1976)
Boston, Mass.	641 071	6	3.98	0–10	6	19.55	0–35.7	Stewart (1976)
Bay City, Mich.	49 449		1.2			0.03(C)		Beaman and File (1976)
Eagle Harbor, Mich.	20		0.13(A)			2		Cook (1975)
Marquette, Mich.	22 300		0.16(A)			1		Cook (1975)
Midland, Mich.	35 176		0.6			0.001		Beaman and File (1976)
Ontonago, Mich.	2 432		0.24(A)			3(A)		Cook (1975)
Saginaw, Mich.	91 849					0.0013		American Water Works Assoc. (1974)
Beaver Bay, Minn.	362	1	3(A)					Cooper and Murchio (1974)
Beaver Bay, Minn.			12.4(A)			420		Cook (1975)
Cloquet, Minn.	8 699		1.0(A)			60		Cook (1975)
Duluth, Minn.	100 578	9	39.33(A)	17–74.(A)	7	13.3(A)	2.7–27(A)	Nicholson (1974)
Duluth, Minn.			1.62(A)			110		Cook (1975)
Duluth, Minn.			24	10–35				Beaman and File (1975)
Duluth, Minn.		4	2.3(A)	1–5(A)				Cooper and Murchio (1974)
				BDL–0.4(C)				
Duluth, Minn.				1.1–120(A)				U.S. Environmental Protection Agency (1975)
Duluth, Minn.						190	30–800	Cook *et al.* (1974)

Table 5.9 (*Continued*)

Place	1970 population	10^6 fibres/l. N	Mean	Range	μg/l.* N	Mean	Range	Reference
Grand Marias, Minn.	197 649		0.2(A)			5		Cook (1975)
Grand Marias, Minn.			0			0		Nicholson (1974)
Silver Bay, Minn.	3 504		0.26(A)			50		Cook (1975)
Silver Bay, Minn.		1	2.(A)					Cooper and Murchio (1974)
Two Harbors, Minn.	4 437		1.95(A)			140		Cook (1975)
Two Harbors, Minn.		1	2.5(A)					Cooper and Murchio (1974)
Jackson, Mass.	153 968		NSS					U.S. Environmental Protection Agency (1975)
Independence, Mo.	111 662			0.36–0.58(C)				U.S. Environmental Protection Agency (1975)
Kansas City, Mo.	507 087		0.07(C)					U.S. Environmental Protection Agency (1975)
Springfield, Mo.	120 096		0.30(C)					U.S. Environmental Protection Agency (1975)
St. Louis, Mo.	622 236		NSS					U.S. Environmental Protection Agency (1975)
Elizabeth, N.J.	112 654		BDL					U.S. Environmental Protection Agency (1975)
Jersey City, N.J.	260 545		0.016(C)					U.S. Environmental Protection Agency (1975)
Manville, N.J.	13 029		0.82					Tobin (1976)
Buffalo, N.Y.	462 768		0.13(C)					U.S. Environmental Protection Agency (1975)
Elmira, N.Y.	39 945		NSS					U.S. Environmental Protection Agency (1975)
Glen Falls, N.Y.	17 222		BDL					U.S. Environmental Protection Agency (1975)
New York, N.Y.	7 867 769	6	0		6	0		Stewart (1976)
New York, N.Y.			0					Nicholson (1974)
Niagara Falls, N.Y.	85 615		NSS					U.S. Environmental Protection Agency (1975)
Rochester, N.Y.	296 233		BDL					U.S. Environmental Protection Agency (1975)
Cincinnati, Ohio	425 524		NSS					U.S. Environmental Protection Agency (1975)
Dayton, Ohio	243 601		NSS					U.S. Environmental Protection Agency (1975)
Muskogee, Okla.	37 331		BDL					U.S. Environmental Protection Agency (1975)
Tulsa, Okla.	331 638		BDL					U.S. Environmental Protection Agency (1975)
Bethlehem, Pa.	72 686		NSS					U.S. Environmental Protection Agency (1975)
Erie, Pa.	129 231		0.07(C)					U.S. Environmental Protection Agency (1975)
Malvern, Pa.	2 583				24	0.119	0.001–0.977	American Water Works Assoc. (1974)
Philadelphia, Pa.	1 948 609	16	16.95	0–130	16	0.143	0–0.588	Stewart (1976)
South Pittsburgh, Pa.	520 117		0.21(C)					U.S. Environmental Protection Agency (1975)

Newport, R.I.	34 562			0.4–1.0(C)				U.S. Environmental Protection Agency (1975)
Providence, R.I.	179 213				3	0.409	0.267–0.579	American Water Works Assoc. (1974)
Columbia, S.C.	113 542		0.13(C)					U.S. Environmental Protection Agency (1975)
Greenville, S.C.	61 208		NSS					U.S. Environmental Protection Agency (1975)
Chattanooga, Tenn.	119 082		0.13(C)					U.S. Environmental Protection Agency (1975)
Chattanooga, Tenn.			4.7					Tobin (1976)
Clarksville, Tenn.	31 719		0.90(C)					U.S. Environmental Protection Agency (1975)
Memphis, Tenn.	623 530				1	1.696		American Water Works Assoc. (1974)
Nashville, Tenn.	447 877			0.43–0.80(C)				U.S. Environmental Protection Agency (1975)
Abilene, Texas	89 653		BDL					U.S. Environmental Protection Agency (1975)
Amarillo, Texas	127 010		0.09(A)					U.S. Environmental Protection Agency (1975)
Dallas, Texas	844 401	1	0		1	0		Stewart (1976)
Houston, Texas	1 232 802	5	0					Cooper and Murchio (1974)
Brattleboro, Vt.	12 800		0.11(C)					U.S. Environmental Protection Agency (1975)
Crystal Springs, Vt.	—		NSS					U.S. Environmental Protection Agency (1975)
Eden, Vt.	100		0.08(C)					U.S. Environmental Protection Agency (1975)
Enosburg, Vt.	1 266		0.05(C)					U.S. Environmental Protection Agency (1975)
Jericho-Underhill, Vt	725		NSS					U.S. Environmental Protection Agency (1975)
North Troy, Vt.	774			0.98–2.2(C)				U.S. Environmental Protection Agency (1975)
Quarry Hill, Vt.	—		NSS					U.S. Environmental Protection Agency (1975)
Richmond-Harrington, Vt.	935		NSS					U.S. Environmental Protection Agency (1975)
Charlottesville, Va.	38 883		NSS					U.S. Environmental Protection Agency (1975)
Seattle, Wash.	530 831	4	0.850	0–1.9	4	0.303	0–1.21	Stewart (1976)
Seattle, Wash.				BDL–1.812(A) NSS–2.464(C)				U.S. Environmental Protection Agency (1975)
Ashland, Wisc.	9 615		0.31(A)			20		Cook (1975)
Superior, Wisc.	32 237		4.(A)			1.4(A)		Nicholson (1974)
Superior (Wells), Wisc.			0.03(A)			0.8(A)		Cook (1975)
Cheyenne, Wyo.	40 914		NSS					U.S. Environmental Protection Agency (1975)
San Juan, Puerto Rico	452 749		NSS					U.S. Environmental Protection Agency (1975)

* BDL = below detection limit (less than 50 000 fibres/l.);
NSS = not significant (less than 5 fibres in 20 fields);
C = chrysotile;
A = amphibole.

only for finished drinking water. Some of the samples were taken from taps in the water supply system; others were taken at municipal water stations (Levine, 1978).

Chrysotile fibres, amphibole fibres, or both, were found to be present in 56 of the 105 water supplies. Fibre concentrations for individual samples varied between 0 (below detectable limits) to 130 million electron-microscope-visible fibres/l. Mass asbestos concentrations for individual samples were found to vary between 0 (below detectable limits) to 800 mg/l*.

A number of cities that take their drinking water from Lake Superior have elevated levels of asbestos. These elevated levels may be due partly to erosion of the local asbestos-containing minerals. They may also result from local taconite mining operations that have discharged into the lake, over the past 20 years, millions of tons of tailings containing 38–44% amphibole (United States of America *vs.* Reserve mining Company, 1974) (note: this source was not included in Figure 5.1).

4(iii) Estimated asbestos consumption from water

The distribution of asbestos fibre and mass concentrations by percentages of the consuming population as derived from Table 5.9 is shown in Figures 5.4 and 5.5, and Table 5.10. Since all authors did not report data in terms of both fibre and mass concentrations, the two types of data (fibre and mass) are not based on identical consuming populations.

The population-at-risk is truly representative only of the places for which asbestos concentration data were available. The cities sampled were not sampled at random; indeed, some were selected because asbestos was suspected to be in their drinking water. Hence, it might be argued that Table 5.10 overestimates asbestos intake in water for the U.S. population as a whole.

It can be seen in Table 5.10 that 50% of the population studied consumes at least 6.9 million electron-microscope-visible fibres, or 19.6 μg, per year, and 10% consumes at least 2000 million fibres, or 710 μg, per year. The significance of these levels of consumption on health cannot be ascertained at present, and there are currently no regulations on the asbestos content of drinking water.

* Some authors reported their data as fibres per litre, some reported micrograms per litre, and some reported both. Attempts made to convert between micrograms and fibres, based on data expressed in terms of both, were largely unsuccessful because of widely varying results. It may be that fibre-size distributions vary so significantly from place to place that no overall correlation is possible.

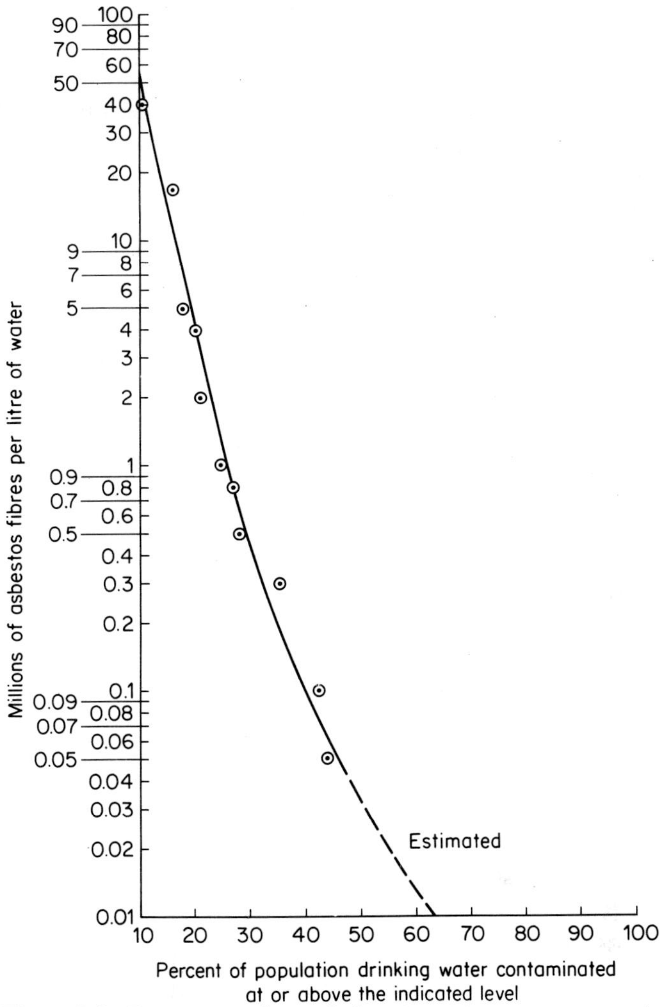

Figure 5.4 Percentage of population that drink water contaminated by various levels of asbestos fibres

(6) EXPOSURE TO ASBESTOS IN FOODS AND DRUGS

Asbestos contents of food and drugs have not to date, been, well established, and the U.S. has no regulations concerning the content of asbestos in foods and non-parenteral drugs.

Foods may be contaminated during their agricultural phase from asbestos in air and soil and from asbestos impurities in talc used as a pesticide vehicle. Root uptake of contaminated water, and deposition of such water

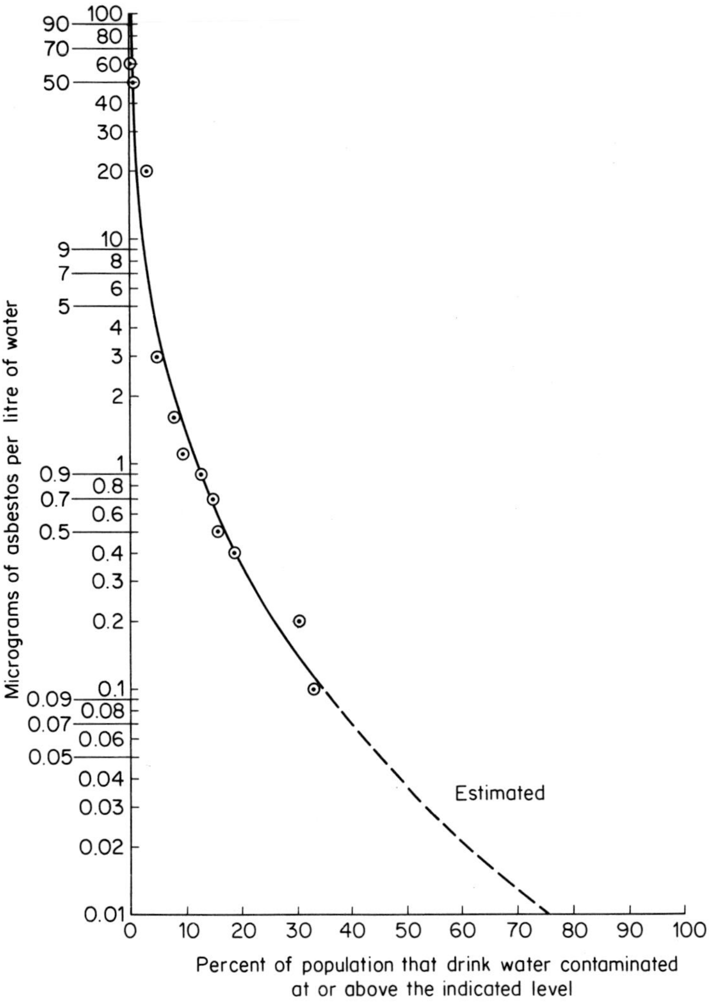

Figure 5.5 Percentage of population that drink water contaminated by various levels of asbestos mass

directly on to leafy surfaces by sprinkler irrigation, might also be possible mechanisms of asbestos contamination. However, no published literature has been found to substantiate any of these possibilities. One instance of accidental contamination of food in the course of transportation (shipping) has been documented (U.S. Food and Drug Administration, 1973).

Processed foods may become contaminated with asbestos either from

Table 5.10 Consumption of asbestos from drinking water

Proportion of population consuming indicated level or more (%)	Assumed concentrations		Annual consumption†	
	10^6 f/l.	μg/l.	10^6 f*/year	μg/year
50	0.0125	0.035	6.9	19.6
25	0.2	0.21	110.0	115.0
10	4.0	1.3	2 190.0	710.0
5	17.0	4.0	9 310.0	2 190.0
1	37.0	30.0	20 260.0	16.425.0

* Electron-microscope-visible fibres.
† Assumes a person consumes 1.5 l. of water per day.

water used in their preparation or from asbestos filters, adhesives, rubber, and resins used in processing and packaging (Code of Federal Regulations, Title 21, Section 121). Foods in which asbestos filters are used may include beer, wine, liquors, fruit juices, sugar, lard, and vegetable oil (Wolff and Oehme, 1974). Asbestos filters have also been employed to process cider, condiments, drinking water, mouthwashes, syrups, tonics, and vinegar (U.S. National Institute for Occupational Safety and Health, 1972). Talc, which may contain asbestos as an impurity, may be added to processed foods directly as an ingredient or indirectly, as a form-release agent or in packaging materials (see the details p. 200 on talc).

The authors of one study found between 1.1 and 6.6 million electron-microscope-visible fibres/l. in U.S. and Canadian beer, and between 1.7 and 12.2 million fibres/l. in Canadian soft drinks. The fibre content of the water used in the beer and soft drink samples is not known. They also reported between 1.8 and 11.7 million fibres/l. in wines from various parts of the world (Cunningham and Potefract, 1973)*. In another study, between 13.1 and 24.0 million fibres/l. were found in one manufacturer's gin, and the water used in processing of the gin contained between 3.3 and 8.7 million fibres/l. (Wehman and Plantholf, 1974).

Asbestos filters also have been used in the processing of drugs, including antibiotics, and blood plasma. However, since April 1975, the U.S. Food and Drug Administration has disapproved their use in preparing parenteral drugs and biologics (Code of Federal Regulations, Title 21, Section 133). Asbestos filters may be used, however, in the manufacture of non-parenteral drugs and their ingredients.

* During preparation of certain French wines, powdered asbestos may be mixed directly with the wine (Conseil Superieur d'Hygiene Publique de France, 1976).

Talc

Mineralogical analysis of 51 commercial talcs showed that none were pure talc (Schultz and Williams, 1942), and a study of New York State talc deposits indicated that the asbestiform minerals tremolite and anthophyllite may constitute major fractions of commercial talc (Kleinfeld *et al.*, 1973). In another study, in which 22 available consumer talcum products were examined, it was found that fibrous constituents were present in all of the samples (Cralley *et al.*, 1968). More recently, of 21 consumer talcums and powders tested, ten were found to contain up to 14.5% of asbestos (generally tremolite and anthophyllite, but sometimes including small amounts of chrysotile) (Rohl *et al.*, 1976).

A considerable amount of talc and soapstone is used in pesticides (33 677 short tons in the U.S. for 1970) (Brobst and Pratt, 1973). Most of this material is applied by air dissemination, either dry or as a droplet spray, and so constitutes a potential health hazard not only to farm labourers and to persons spraying pesticide, but also to the general population residing nearby. Some proportion of the mineral carrier will be deposited on to crops and ultimately may be ingested in food. Some talc may find its way into drinking water supplies.

Talc may be used as a free flow agent, anti-caking agent, form-release agent, polishing agent, dispersant, and filler in the preparation of many processed foods and food additives. It has been used as a lubricant in dried spices and flavouring mixes and as a vehicle and dispersant in enrichment formulae for a variety of foods. It may be used to polish ground nuts and coated chewing gums as a gum base, as a form-release agent for moulded stick gum, candies, and foods, and as an anti-caking agent in certain foods including salami, ground nuts, chewing gum, and candy (Wolff and Oehme, 1974; Mickus, 1976; Miller, 1976; Blejer and Arlon, 1973; Eisenberg, 1974; Pitco, 1976). Talc-impregnated paper and other materials have been used to package such foods as meat, salt, macaroni, rice, cornflakes, and dry milk. Talc has also be used in the clarification of liquids (*Federal Register*, 1975b).

Talc has been extensively employed in processing rice, where it has functioned as an abrasive in polishing, a cosmetic imparting a white sheen, a lubricant to prevent the oil-coated hyperenriched fraction of wash-resistant enriched rice from sticking, a vehicle and dispersing agent in enrichment formulae for non-wash-resistant rice (marketed with directions not to rinse or drain after cooking), and a protective agent against hydration, oxidation spoilage, and insect infestation (Mickus, 1976).

One third of the rice consumed in the U.S.A. and its territories is non-enriched talc-and-glucose-coated rice. Most coated rice is marketed to oriental and Puerto Rican populations in California, Guam, Hawaii, and Puerto Rico. Coated rice marketed in Puerto Rico is enriched (Miller, 1976). Instructions to rinse before cooking, which would wash away a considerable portion of the vitamins added in enrichment dusting, accompany non-enriched coated rice, presumably to limit ingestion of talc. However, even after repeated washing, asbestos fibres may be shown to contaminate talc-coated rice (Blejer and Arlon, 1973).

Talc (which may contain asbestos) has been used in the manufacture of capsules and tablets. Other uses of asbestos and talc in medicine and dentistry include allergy syringes with asbestos-packed pistons (Bernstein and Moteff, 1976), the preparation of periodontal packings using asbestos powder (Council on Dental Therapeutics, 1976), and the dusting of surgical gloves with talc†. 'Poudrage', or the injection of talc, has been used surgically to sclerose the pleural cavity.

† A recent survey revealed that talc is still being used in the U.S. as a dusting powder and form-release agent in the manufacture of some surgical gloves (SRI International, 1976).

(7) APPENDIX. CALCULATION OF ATMOSPHERIC ASBESTOS CONCENTRATIONS IN THE VICINITY OF MAJOR U.S. ASBESTOS INDUSTRIAL FACILITIES

Production figures for asbestos mines and mills were readily available (Table 5.11), but consumption of asbestos at individual asbestos-product factories had to be estimated as follows. Annual consumption of the particular asbestos product sector (U.S. Bureau of Mines, 1974) was multiplied by the fraction of sector sales accounted for by the factories under study (Margolin and Igwe, 1975), and this by the estimated factory fraction of total sector employment (Margolin and Igwe, 1975).

Annual source emissions were then determined by multiplying 1974 asbestos production or consumption by an emission factor (Tables 5.11 and 5.12). The emission factors used were based on recent emissions data and took into account improved air pollution controls (Fowler, 1977). Source emission rates were expressed in terms of nanograms per second.

Downwind dispersion curves (plots of distance *versus* atmospheric asbestos concentration at any point divided by source emission rate) were constructed for four different wind conditions (Harwood *et al.*, 1974), using the Binormal Continuous Plume Dispersion Model (Turner, 1970). Because the curves for the four different weather conditions showed a consistent pattern, a single dispersion curve was constructed from the geometric mean of the four downwind projections. It was assumed that the wind would be equally likely to blow in any direction and that at any point from the source the mean atmospheric asbestos concentration would be one quarter of that determined from the downwind dispersion curve.

Populations living in the vicinity of asbestos mines and mills or manufacturing plants were defined as living within a radius of 30 or 5 km, respectively. Results obtained from the model at distances greater than the

Table 5.11 Asbestos emissions to air from mining and milling*

Mine	Annual production (metric tons)	Emission factors (Kg/metric tons)			Total emissions (metric tons)		
		Mining	Milling	Total	Mining	Milling	Total
Vermont	36 138	4	9	13	145	325	470
Coalinga	27 104	4	18	22	108	488	596
Union Carbide	31 621	3	5	8	95	158	253
Copperopolis	4 517	4	9	13	18	41	59
Jacquay	2 710	2.5	10	12.5	7	27	54
Total	102 090				373	1039	1412

* Source: Fowler (1977).

Table 5.12 Air emission factors for asbestos product manufacture*

Product category	Emission factor (kg/metric ton)	Product category	Emission factor (kg/metric ton)
Asbestos–cement pipe	0.1	Thermal insulation	0.5
Asbestos–cement sheet	0.1	Electrical insulation	0.1
Flooring	0.1	Friction products	0.25
	0.25	Textiles	0.15
Packing and gaskets	0.5	Paper	0.5

* Source: Fowler (1977).

selected radii approached the levels of ambient air. In the vicinity of mines and mills, the total population of towns (1970 U.S. Census) was assigned the exposure level determined for the town geographic midpoint. A series of concentric circles at 1-km distances was plotted around manufacturing plants, and the nearby population was assumed to be uniformly distributed within townships or census tracts. The asbestos air concentration determined for the midpoint of each circular segment was used to represent the level of exposure for the estimated population contained within.

(8) REFERENCES

Aaronson, T., and Kohl, G. (1972) Papier mâché, *Environ*, **14**(10), 25–26.

American Water Works Association (1974) A study of the problem of asbestos in water, *Am. Wat. Wks. Ass. J.* **66**(9, Pt. 2), 1–22.

Anderson, D. (1973) *Emission Factors for Trace Substances*, EPA Publication No. 650/2-74-087, Environmental Protection Agency, Washington, D.C.

Anderson, H. A., Lilis, A., Fischbein, A., Daum, S., and Selikoff, I. J. (1974) *Asbestos disease resulting from household exposure to occupational dusts*, *Chest*, **66,** 318–319.

Berstein, I. L., and Moteff. J. (1976) *Possible asbestos hazards in clinical allergy*, *J. Allergy Clin. Immunol.*, **57,** 489–492.

Beaman, D. R., and File, D. M. (1976) *Quantitative determination of asbestos fiber concentrations*, *Analyt. Chem.*, **48,** 101–110.

Blejer, H. P., and Arlon, R. (1973) Talc: a possible occupational and environmental carcinogen, *J. Occup. Med.*, **15,** 92–97.

Brobst, D. A., and Pratt, W. P. (1973) *United States Mineral Resources*, U.S. Geological Survey Professional Paper No. 820.

Brodeur, P. (1973) *The Expendable Americans*, Viking Press, New York.

Castleman, B. C., Camarota, L. A., Fritsch, A. J., Mazzocchi, S., and Crawley, R. G. (1975) The hazards of asbestos for brake mechanics, *Publ. Hlth. Rep.*, **90,** 254–256.

Chem. Mark. Rep. (1975) Asbestos hazard seen in patching, spackling, and repair of brakes, Aug. 18.

Clark, S. G., and Holt, P. F. (1961) Studies on the chemical properties of chrysotile in relation to asbestosis, *Ann. Occup. Hyg.*, **3,** 22–29.

Code of Federal Regulations, Title 21, Section 121, Parts 2520, 2562, 2576, and 2587.

Code of Federal Regulations, Title 21, Section 133, Parts 133.8 and 133.9.
Conseil Superieur D'Hygiene Publique de France. (1976) *Amiante et Alimentation. Rapport d'Enquetes Effectués en Octobre 1976 par l'Inspection de la Repression des Fraudes et du Controle de la Qualité Chez les Productués et Negociants en Vins, Séance du* 18 *Octobre.*
Cook, P. M., Glass, G. E., and Tucker, J. H. (1974) Asbestiform amphibole minerals: detection and measurement of high concentrations in municipal water supplies, *Science, N.Y.*, **85,** 853–855.
Cook, P. M. (1975) Semiquantitative determination of asbestiform amphibole mineral concentrations in western Lake Superior water samples, *Adv. X-ray Analy.*, **1975,** 18.
Cooper, R. C., and Murchio, J. C. (1974) *Preliminary Studies of Asbestiform Fibers in Domestic Water Supplies,* University of California, Berkeley, AMRL-TR-74-125, Paper No. 5.
Council on Dental Therapeutics (1976) Council on Dental Materials and Devices: Hazards of asbestos in dentistry. *J. Am. Dent. Ass.*, **92,** 777–778.
Cralley, L. J., Key, M. M., Groth, D. H., Lainhart, W. S., and Ligo, R. M. (1968) Fibrous and mineral content of cosmetic talcum products, *Am. Ind. Hyg. Ass. J.*, **29,** 350–354.
Crowder, J. V., and Wood, G. H. (1972) *Control Techniques for Asbestos Air Pollutants,* EPA Publication No. AP-117, Environmental Protection Agency, Research Triangle Parke, North Carolina.
Cunningham, H. M., and Pontefract, R. D. (1973) Symposium on industrial chemicals as food contaminants, *J. Ass. Off. Analyt. Chem.*, **56,** 976–981.
Diederichsen, R. Personal communication, SRI International, Menlo Park, California, 1976.
Eisenberg, W. V. (1974) Inorganic particle content of food and drugs, *Envir. Hlth. Perspect.*, **9,** 183–191.
Engng. News Rec. (1974) Asbestos fallout, *Engng News Rec.*, **193**(13), 41.
Federal Register (1973) **38**(66), 8829–8830, April 6.
Federal Register (1975a) **39**(208), 48292–48311, October 14.
Federal Register (1975b) **40**(51), 11866–11869, March 14.
Fowler, D. P. (1977) *Disposals and Emission, of Asbestos in the United States,* SRI International, Menlo Park, Calif.
Great Lakes Research Advisory Board (1975) *Asbestos in the Great Lakes Basin,* International Joint Commission.
Hallenbeck, W. H., Chen, E. H., Hesse, C. E., Patel-Mandti, K. K., and Wolff, A. H. (1978) Is chrysotile asbestos released from asbestos–cement pipe into drinking water, *J. Am. Wat. Wks. Ass.*, **70,** 97–102.
Harington, J. S., Allison, A. C., and Badami, D. V. (1975) Mineral fibers: Chemical, physiochemical, and biological properties, *Adv. Pharmac. Chemother.*, **12,** 291–402.
Harwood, C. F. (1971) *Asbestos Air Pollution Control,* Illinois Institute for Environmental Quality, PB-205208.
Harwood, C. F. (1972) Asbestos air pollution from the wear of brake linings, *IITRI,* Chicago, April.
Harwood, C. F., Siebert, P., and Blaszak, T. P. (1974) Assessment of particle control technology for enclosed asbestos sources, *IITRI,* EPA-650/2-74-088.
Harwood, C. F., and Blaszak, T. P. (1974) Characterization and control of asbestos emissions from open sources, *IITRI,* EPA-650/2-74-090, September.
Heffelfinger, R. E., Melton, C. W., and Kiefer, D. L. (1972) *Development of a Rapid*

Survey Method of Sampling and Analysis for Asbestos in Ambient Air, Battelle Laboratories, Columbus, Ohio, Contract No. CPA 22-69-110.

Jacko, M. G., Du Charme, R. T., and Somers, J. H. (1973) How much asbestos do vehicles emit?, *Auto Eng.*, **81** (6), 38–40.

Johns-Manville Corporation (1976) Personal communication.

Kleinfeld, M., Messite, J., and Langer, A. M. (1973) A study of workers exposed to asbestiform minerals in commercial talc manufacture, *Envir. Res.* **6,** 132–143.

Laamanen, A., Noro, L., and Raunio, V. (1965) Observations on atmospheric air pollution caused by asbestos, *Ann N.Y. Acad. Sci.*, **132,** 240–254.

Levine, R. J., (Monograph Manager) (1978) *Asbestos: An Informational Resource*, U.S. Dept. of Health, Education, and Welfare, National Cancer Institute, Division of Cancer Control and Rehabilitation, Prevention Branch.

Lumley, K. P. S, Harries, P. G., and O'Kelley, F. J. (1971). Buildings insulated with sprayed asbestos: a potential hazard, *Ann. Occup. Hyg.*, **14,** 255–257.

Lynch, J. R. (1968) Brake lining decomposition products. *J. Air. Pollut. Contr. Ass.*, **18,** 824–826.

Margolin, S. V., and Igwe, B. U. N. (1975) *Economic Analysis of Effluent Guidelines: The Textiles, Friction, and Sealing Materials Segment of the Asbestos Manufacturing Industry*, EPA Publication EPA 230/2-74/030, A. D. Little Inc. and U.S. Environmental Protection Agency.

Mickus, R. (1976) Personal communication, Rice Growers Association.

Miller, D. (1976) Personal communication, U.S. Food and Drug Administration.

Millette, J. R. (1976) Analyzing for asbestos in drinking water, *News Environ. Res. Cincinnati*, Environmental Protection Agency, Washington D.C., Jan. 16.

Murchio, J. C., Cooper, W. C., and De Leon, A. (1973) *Asbestos Fibers in Ambient Air of California*, California Air Resources Board, Berkeley, California.

Murphy, R. L., Levine, B. W., Bazzaz, F. J., Lynch, J. J., and Burgess, W. A. (1971) Floor tile installation as a source of asbestos exposure, *Am. Rev. Resp. Dis.*, **104,** 576–580.

National Academy of Sciences (1971) Asbestos: The Need for and Feasibility of Air Pollution Controls, National Academy of Sciences, Washington, D.C.

New York Times (1977) Jan. 4, p. 31.

Newhouse, M. L., and Thompson, H. (1965) Epidemiology of mesothelial tumors in the London area, *Ann. N.Y. Acad. Sci.*, **132,** 579.

Nicholson, W. J., and Pundsack, F. L. (1973) Asbestos in the environment, in *Biological Effects of Asbestos*, IARC Scientific Publications No. 8, International Agency for Research on Cancer, Lyon, pp. 126–130.

Nicholson, W. J., Rohl, A. N., and Weisman, I. (1973) *Asbestos Contamination of the Air in Public Buildings*, Mt. Sinai School of Medicine, EPA-450/3-76-001.

Nicholson, W. J. (1974) Analysis of amphibole asbestiform fibers in municipal water supplies, *Envir. Hlth. Perspect.*, **9,** 165–172.

Nicholson, W. J. Rohl, A. N., Fischbein, S. A., and Selikoff, I. J. (1975) Occupational and community asbestos exposure from wallboard finishing compounds, *Bull. N.Y. Acad. Med.*, **51,** 1180–1181.

Pitco (1976) Personal communication, Ford Gum and Machine Company.

Reimschussel (1974) *Asbestos in the Environment*, (Ed. J. R. Kramer, O. Mudroch, and S. Tihor), McMaster University, for IJC Great Lakes Research Advisory Board.

Rohl, A. N., and Langer, A. M. (1974) Identification and quantification of asbestos in talc, *Envir. Hlth. Perspect.*, **9,** 95–109.

Rohl, A. N., Langer, A. M., Selikoff, I. J., and Nicholson, W. J. (1975) Exposure to asbestos in the use of consumer spackling, patching, and taping compounds, *Science, N.Y.* **189,** 551–553.

Rohl, A. N., Langer, A. M., Selikoff, I. J., Tordini, A., and Klimentidis, R. (1976) Consumer talcums and powders: mineral and chemical characterization, *J. Toxic. Envir. Hlth.*, **2,** 255–284.

Sawyer, R. N. (1977) Asbestos exposure in a Yale building, *Envir. Res.*, **13,** 146–169.

Schultz, R. Z., and Williams, C. R. (1942). Commercial talc—animal and mineralogical studies, *J. Ind. Hyg.*, **24,** 75–79.

Selikoff, I. J., Nicholson, W. J., and Langer, A. M. (1972) Asbestos air pollution, *Arch. Envir. Hlth.*, **25,** 1–12.

Stewart, I. M. (1976) *Asbestos in the Water Supplies of Ten Regional Cities*, EPA Publication No. 560/6-76-017, Environmental Protection Agency, Washington, D.C.

Thompson, R. J., and Morgan, G. B. (1971) *Determination of Asbestos in Ambient Air*, Presented at Identification and Measurement of Environmental Pollutants Symposium, Ottawa, Ontario, Canada, June 14–17.

Tobin, P. (1976) Personal communications about data developed under contract with McCrone Associates, U.S. Environmental Protection Agency.

Turner, D. B. (1970) *Workbook of Atmospheric Dispersion Estimates*, U.S. Public Health Service Publication No. 999-AP. 26.

U.S. Bureau of Mines (1974) *Mineral Yearbook* (preprint).

U.S. Environmental Protection Agency (1974) *Background Information on National Emissions Standards for Hazardous Pollutants—Proposed Amendments to Standards for Asbestos and Mercury*, EPA Publication No. 450/2-74-009a, Environmental Protection Agency, Washington, D.C.

U.S. Environmental Protection Agency (1975) *Preliminary Assessment of Suspected Carcinogens in Drinking Water (Appendices)*, Interim Report to Congress, Environmental Protection Agency, Washington, D.C.

U.S. Food and Drug Administration (1973) FDA detains lima beans contaminated with asbestos after accident, *Fd Chem. News*, September, 24.

U.S. National Institute for Occupational Safety and Health (1972) *Report of Review Committee*, NIOSH Pub. No. HSM-72-10267.

U.S. Public Health Service (1960) *Radiological Health Handbook*, U.S. Public Health Service, Washington, D.C., 193.

United States of America *vs.* Reserve Mining Company (1974) Court Proceedings No. 5–72, Civil 19, United States Court of Appeals, 8th Circuit, June.

Wehman, H. J., and Plantholf, B. A. (1974). Asbestos fibrils in berverages: PTI gin, *Bull. Envir. Contam. Toxic.* **11,** 267–272.

Wolff, A. H., and Oehme, F. W. (1974). Carcinogenic chemicals in food as an environmental health issue, *J. Am. Vet. Med. Ass.*, **164,** 623–629.

CHAPTER 6

Monitoring and identification of airborne asbestos

Stephen Thomas Beckett

Institute of Occupational Medicine, Edinburgh

(1) SYNOPSIS

The presence of asbestos dust has been associated with asbestosis, bronchial carcinoma, and mesothelioma. The monitoring of airborne fibre is therefore of great importance, as only fibre which is inhaled is linked with these diseases. The introduction to this chapter considers the background to asbestos monitoring and considers the different types of samples which can be taken, and also the relative advantages and disadvantages of fibre mass as opposed to fibre number concentration measurements. The membrane filter method is used by most occupational hygiene laboratories to monitor the industrial environment, and this is detailed in Section (3). Methods of detecting very low asbestos concentrations, direct-reading dust monitoring equipment, lesser used instruments and methods of historical interest are then described. Finally, some of the techniques used to identify the type of fibres present in airborne dust samples are reviewed.

(2) INTRODUCTION

Airborne asbestos dust is usually monitored for one of three reasons. Firstly, large numbers of samples are taken to check compliance with legislation. As part of the standards issued by most controlling authorities recommended methods are described by which this monitoring should be done (for example, Department of Employment and Productivity, 1970). Secondly, within the asbestos industry regular sampling is carried out to determine the efficiency of dust suppression equipment. Here it is frequently necessary to know only the relative amount of dust present, and direct-reading dust monitoring instruments play a key role. Finally, an increasing number of samples are taken for epidemiological purposes. For this it is essential that standard methods be used which can be related to one another, and which remain constant over many years. This work includes monitoring the exposure of people outside the asbestos industry and may involve measuring extremely small amounts of asbestos.

In practice, it is not possible to obtain an absolute measure of the dust inhaled and retained in the lung. With any sampling method that is adopted, therefore, there will be inherent errors that must be understood and allowed for in the interpretation of the sample results. Care must be taken to ensure that the samples obtained are representative of the airborne dust at the sampling point, and are sufficient in number so that variations of concentration with time and space can be allowed for. Initially, dust sampling instruments were either too heavy to be easily portable (e.g. the thermal precipitator) or only took very short duration samples (e.g. konimeter, Draeger pump). The long-period sampling instruments are normally used to monitor the environmental air, and take what are commonly called 'static' samples. The short-period or 'snap' samplers operate most satisfactorily in high dust concentrations, where a measurable amount of asbestos can be captured. Neither 'static' nor 'snap' samples may be representative of the airborne dust breathed by exposed personnel, however. Hygiene standards normally refer to Threshold Limit Values (TLVs). These are the levels at which it is believed that nearly all workers may be repeatedly exposed throughout a 40-h week without adverse effect. Many asbestos workers do a series of different jobs throughout a shift and may be exposed to dust for only relatively short periods. In such instances, and indeed in the majority of situations, a sample taken with the instrument attached to the person being monitored has distinct advantages. This provides what is known as a 'personal' sample. In this case the instrument should be completely portable and the air sampled in the breathing zone of the person concerned. Normally this is taken to be anywhere within 30 cm of the ori-nasal region.

In order for the results of the monitoring to have any meaning, it is necessary that they provide a consistent measure of the dust of pathogenic significance. When evaluating a potentially hazardous dust either its mass or number concentration is normally measured. The mass can be either that of all the airborne particles (total dust), or that of the proportion thought to be capable of reaching the lung alveoli, commonly referred to as respirable dust. For coal dust the work of Jacobsen *et al.* (1970) has shown that the mass of respirable dust present in the air corresponds much more closely with radiological change than does particle number. This work was based on a long-term study in British coalmines, but unfortunately similar epidemiological data are not available for asbestos exposure, and both fibre mass and number concentration measurements are therefore used. The total mass of airborne dust is monitored in the U.S.S.R. provided that the asbestos content is greater than 10%. This, however, creates a difficulty when interpreting the results, because whereas in an asbestos textile factory most of the dust will be asbestos, in mining or the asbestos–cement industry the majority of the dust may be other material. A more satisfactory method of evaluation is to determine the weight of asbestos in this total dust sample using chemical or physical methods of analysis. This, however, will monitor all asbestos fibres, including those too large to penetrate into the alveoli. One of these fibres could have a mass of many times that of the potentially harmful dust in the sample, and so this method in fact only provides an upper limit to the amount of asbestos present. Alternatively, instruments containing particle size selectors, known as elutriators, can be used for sampling the dust. These were designed, however, using the aerodynamic properties of spherical particles (Breuer, 1964; Fuchs, 1964; Walton, 1954), and rely on the larger particles separating out by virtue of their higher falling speed. At present, however, the sampling characteristics of these instruments with fibrous dusts are not fully understood and an additional degree of uncertainty is introduced when they are used. One of the alternative German Federal Republic standards for asbestos, however, requires the use of one of these instruments (Deutsche Forschungsgemeinschaft: Senatskommission zur Prüfung Gesundheitsschadlicher Arbeitsstoffe, 1976). Once again physical or chemical analysis can be used to determine the amount of asbestos present.

The measurement of the airborne fibre number concentration is normally preferred in most countries. This avoids some of the difficulties and uncertainties of mass monotoring by using a microscopical method to separate out only those particles thought to be potentially harmful from a sample of all the dust from the air. The results obtained, however, depend upon the microscopical magnification and technique used and also upon any restriction placed upon the type of particle counted. Some early attempts to monitor the exposure of asbestos workers recorded all the particles thought to be respirable, but most present methods involve counting only those fibres within a limited size and shape range. One advantage of this method is that, unlike the mass standard, the 2 fibres/cm^3 British Standard for chrysotile asbestos (which has been adopted by several other countries), is based upon an epidemiological study (British Occupational Hygiene Society Committee on Hygiene Standards, 1968, 1973). This type of measurement is also likely to provide the basis of any revised standard within the foreseeable future.

(3) THE MEMBRANE FILTER METHOD

This is now the most widely used method for monitoring asbestos dust in industry. In 1972 a report to the World Health Organization International Agency for Research on Cancer recommended that inter-laboratory trials be carried out, so that this method could be standardized (Advisory Committee on Asbestos Cancers, 1973).

3(i) Outline of technique

Measurements are taken by drawing a known volume of air through a membrane filter. This filter is then made transparent, and the number of fibres fitting a standard definition of size and shape which are in the deposit are counted using a phase-contrast microscope. The mean fibre concentration during the sampling period can then be calculated. Where fibre identification is needed, different types of sampling filters and analytical techniques may be required.

It is also possible to use an entirely different technique, with a green or black membrane, which is fixed with Perspex but not cleared. This is then mounted without a coverglass and examined by incident light using a 4-mm objective and a suitable vertical illuminator (such as the Cooke Universal Illuminator). Tests seemed to indicate that this method gives similar results to the usual phase-contrast method (Addingley, 1966), but in general it has not found wide-scale usage for asbestos monitoring, probably because of the increased difficulty in setting up and operating the microscope.

3(ii) Definition of the Fibres which are Evaluated

The membrane filter method was developed by the British Asbestosis Research Council (Holmes, 1965) in order to try to monitor only those fibres thought to be capable of causing lung damage. At the time of its inception, asbestos or ferruginous bodies were thought to play a major role in the development of asbestosis. These bodies are fibres surrounded by protein and iron frequently found in the sputum of asbestos workers. These fibres are normally longer than 10 μm (Vorwald *et al.*, 1951; Beattie, 1961; Beattie and Knox, 1961) and it was therefore concluded that it was necessary to evaluate only the longer fibres. Initially it was decided to monotor those fibres between 5 and 100 μm in length, although this upper limit is no longer applied in most countries, except South Africa (Walton *et al.*, 1976). As only a relatively small proportion of airborne fibres are longer than 100 μm, however, the difference is insignificant. In addition, at the time of inception of the membrane filter method, it was thought that the number of longer fibres was a constant proportion of the total number (Addingley, 1966), but it has since been found that some processes, notably carding, preferentially produce longer fibres. More recent experimental evidence, however, has shown that the longer fibres (>10 μm) are in fact the most potentially dangerous, in particular in the development of cancer (see, for example, Stanton, 1973), showing that the limitation of monitoring to the larger fibres was probably justified.

An added advantage of monitoring the longer fibres was that this can be done by optical microscopy, thereby avoiding the sophisticated electron optical equipment that is required in order to detect the smaller fibres. It was also found that by restricting the type of microscope and range of magnification used, good agreement could be obtained between different laboratories evaluating the same samples. A magnification of approximately 500× was chosen in the U.K. (Asbestosis Research Council, 1971a). With increasing magnification more fibres are 'seen', but the level of inter-laboratory agreement depends more and more upon the quality of the microscope and the skill of the observer. The lack of ability to see all fibres does not invalidate the method, however, as it is only required to produce an index of the hazard, which may in fact not be proportional to all the fibres present. In order to distinguish a fibre from the other dust in the sample, the definition of a fibre having a length to diameter ratio (i.e. aspect ratio) of at least 3:1 was chosen. This is now used internationally by occupational hygienists when monitoring fibrous dusts, whereas 10:1 is more commonly used by engineers and fibre technologists.

In order to cause disease, the airborne particles must be capable of penetrating into the lung. Many, however, are caught in the nose and larger airways by sedimentation or impaction. For spherical particles a generalization of experimental measurements of the proportion of airborne particles of different sizes able to reach the alveoli produced a plot known as the Johannesburg curve (Orenstein, 1960). Here the size of a particle is expressed in terms of its aerodynamic diameter, that is the diameter of a sphere of density 1 g/cm^3 with the same falling speed in air as the particle. According to this convention, all particles larger than 7.1 μm aerodynamic diameter and about 50% of the 5 μm aerodynamic diameter particles are removed before reaching the lungs. Fibres, however, behave very differently aerodynamically, and it has been shown that those with high aspect ratios fall through the air at a rate which is proportional to their diameters, but independent of their lengths, and that asbestos fibres have aerodynamic diameters approximately three times their actual diameters (Timbrell, 1965, 1973). Asbestos fibres with high aspect ratios and with diameters greater than approximately 2.5 μm are therefore unable to reach the alveoli. When monitoring asbestos, the diameter limit is made greater than this so as to ensure that all of the respirable fibres are evaluated. In Australia, Belgium, France, Sweden and the U.K., only fibres with diameters less than 3 μm are evaluated, whereas a 5 μm limit is used in South Africa and Finland. Canada, Denmark and the U.S.A., however, do not place any restriction on the diameter of the fibres counted, provided that it is less than one third of its length. A summary of the definitions of a countable fibre used in different countries, together with the standards to which they apply, is given in Table 6.1.

In general, the requirements for a fibre to be evaluated are as follows:

1. a length at least three times its diameter;
2. a length greater than 5 μm;
3. a diameter less than 3 μm (some countries only);
4. visible under a phase-contrast microscope at a magnification of approximately 500×.

A major advantage of using the microscope to select a definite size range of fibres is that this permits the operator to sort by size and exclude fibres not considered to be respirable. All fibres can therefore be collected on the filter and there is no necessity for a pre-selector such as a cyclone or horizontal elutriator, as used for spherical dust. This is important as a satisfactory air elutriation method for asbestos fibres has not yet been reported (Asbestosis Research Council, 1971b).

A further constraint placed by some codes of practice for evaluating asbestos is that the fibres fitting the above classification shall be asbestos (see Section 7). This can lead to difficulties as some non-asbestos fibres have the same morphology as asbestos fibres, for example organic fibres and chrysotile, and gypsum fibres and amphibole asbestos (Middleton, 1978). In a comparison of the counting procedures in nine countries (Walton *et al.*, 1976), it was noted that many laboratories counted all fibres fitting the size

Table 6.1 Comparison of the airborne asbestos dust standards and fibre definitions applied in different countries

Country	Standard (chrysotile only)	Fibre definition Length (μm)	Diameter (μm)	Aspect ratio
Australia	4 fibres/cm^3	>5	<3	>3:1
Belgium	2 fibres/cm^3	>5	<3	>3:1
Canada	2 fibres/cm^3	>5	No limit	>3:1
Denmark	2 fibres/cm^3	>5	No limit	>3:1
Finland	5 fibres/cm^3, to be lowered to 2 fibres/cm^3	>5	<5	>3:1
France	2 fibres/cm^3	>5	<3	
Federal Republic of Germany	0.10 mg/m^3	Respirable asbestos only		
	4.0 mg/m^3	Asbestos in mixed respirable dust		
	Since 1976 an alternative number standard has been applied			
	2 fibres/cm^3	>5	<3	>3:1
Norway	5 fibres/cm^3, to be lowered to 2 fibres/cm^3	>5	<3	>3:1
South Africa	2 fibres/cm^3	5–100	<5	>3:1
Sweden	1 fibre/cm^3	>5	<3	>3:1
U.K.	2 fibres/cm^3	>5	<3	>3:1
U.S.A.	5 fibres/cm^3, to be lowered	>5	No limit	>3:1
U.S.S.R.	2 mg/m^3	Total dust if asbestos content >10%		

definition, whereas others limited themselves to those fibres whose morphology appeared to be that of asbestos.

3(iii) The membrane filter

Cellulose ester filters are normally used for asbestos fibre monitoring. The Asbestosis Research Council (1971a) in the U.K. recommends the use of 0.8–5.0 μm pore size filters, whereas the Australian code of practice (National Health and Medical Research Council, 1976) suggests that only a pore size of 0.8 μm should be used. Although many asbestos fibres have diameters much less than this, they are in fact captured by the filter and the optical fibre count is not affected by penetration. Smaller pore sizes (e.g. 0.2 μm) may be used when it is essential for all fibres to be retained on the filter surface, for example when coating for transmission electron microscopy. These samples are normally restricted to special situations, as a large pump is required to overcome the pressure drop across the filter. When it is necessary to evaluate a sample by scanning electron microscopy, Nuclepore filters (Nuclepore Corp., Commerce Circle, Pleasanton, U.S.A.) are frequently used (Beckett, 1973). These are manufactured by etching the tracks of high-energy particles through polycarbonates, and have a very smooth surface, giving a uniform background on which the fibres can be easily identified. These filters, however, have different filtration characteristics from the cellulose ester membranes, and very significant fibre penetration occurs when sampling at pore sizes of 5 μm and above (Spurny *et al.*, 1976b).

Filters of diameter 25 and 37 mm are most commonly used, and sampling filter holders are available commercially. Smaller 13 mm filters, however, are being developed for some uses (Winters, 1976), in particular where there is very little dust or where the sampling volume is limited. These, however, present difficulties at high sampling rates where larger pore size filters may be required to overcome the increased pressure drop. Some laboratories prefer cellulose ester membrane filters with a gridded pattern printed on the surface. This enables the microscopist to find more easily the plane of the dust deposit, which may be difficult for sparse samples.

3(iv) Sampling

The filter is placed in a holder, where it is supported by a gauze or thick pad, which helps in controlling the distribution of air through the filter. The 25 mm Gelman holder (Figure 6.1) is normally used with the filter surface completely exposed. The 37-mm Millipore holder (Figure 6.2) can also be used in this way, or alternatively the face cap may be left in place and the small plug removed. The latter reduces the risk of damage to the filter by

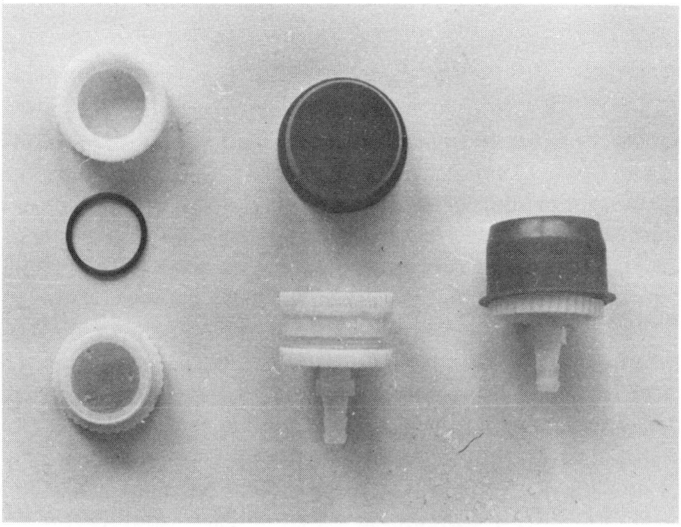

Figure 6.1 Gelman 25-mm filter holder and dust cap

Figure 6.2 Millipore 37-mm filter holder

large high-velocity particles, but has the disadvantage that if any large particles are present, a small portion of the centre of the filter may be obscured by them (Edwards and Lynch, 1968). This should not alter the count, but the effect can be overcome by drilling six additional 4-mm holes in some face caps and fitting these to the holders during sampling. The addition of a $1\frac{3}{4}$-in. plenum between the air inlet and the filter has been reported to give a more uniform deposit with this holder (Bartosiewicz, 1973). Breslin and Stein (1975) measured the collection efficiency of the Millipore holder in still air and found that it was good for respirable sizes of spherical particles. In wind velocities of 2 m/s, however, its sampling characteristics were less satisfactory (Ogden and Wood, 1975).

When monitoring asbestos in moving air streams, for example in air ducts of filtration or exhaust systems, isokinetic sampling is required. This involves matching the velocity of the air entering the sampling equipment to that flowing in the duct. A sampling head must therefore be designed so that when it is inserted in the air stream it does not alter the flow pattern (British Standards Institution, 1971).

Personal samples require the filter to be placed in a holder within the operator's breathing zone. Various methods of wearing the holder have been evaluated, e.g. using a head harness, plastic jacket, shoulder harness, or lapel filter holder, but no significant difference was found (Harness, 1973). Normally a shoulder harness or a lapel filter holder is used with the sampling surface facing downwards or vertically (Figure 6.3). Upwards-facing filters are avoided because of the high risk of contamination. In addition, the filter may be protected by a cap, or by leaving the end on the Millipore holder, in order to prevent the contamination from extraneous sources such as dusty overalls. The volume of air sampled is determined by measuring the flow-rate of the air through the filter. This can be carried out connecting a variable-gap flow meter or bubble meter on the sampling side

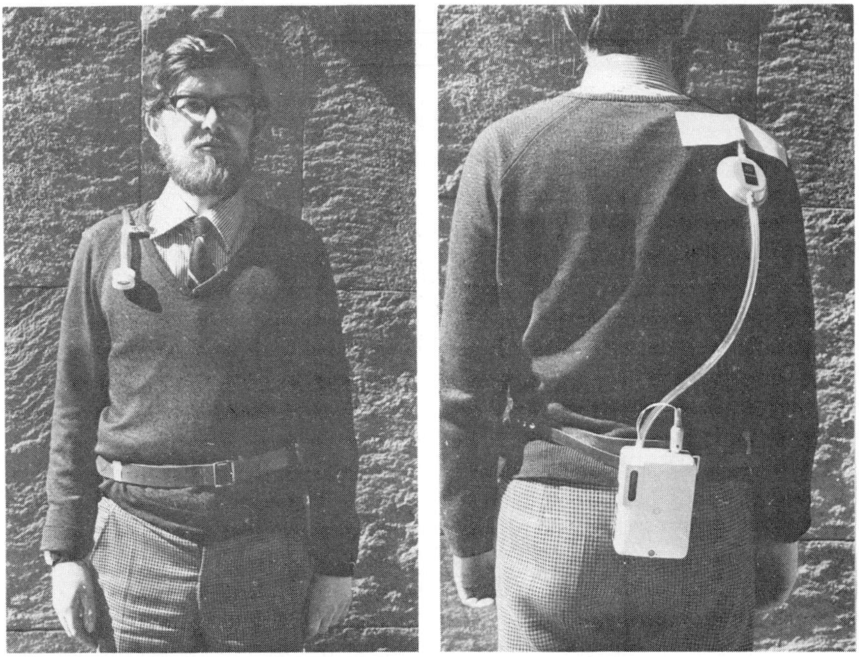

Figure 6.3 Operator wearing personal sampling equipment

of the filter holder. Flow meters are usually calibrated at atmospheric pressure and, where they are incorporated between the pump and filter head, allowance must be made for the reduced pressure in which they are operating. Flow meters incorporated in the sampling pumps are liable to error in that they record all of the air drawn through the pump, including any from leaks or bleed systems. The flow-rate should be checked at regular intervals and at the end of sampling, so that any changes due to filter blockage or pump malfunctions can be noted. Some personal sampling pumps are, in fact, able to compensate for small increases in pressure drop across the filter. A comparative study of the pumps at present available on the U.K. market has recently been completed (Wood, 1977).

The flow-rate used and the sampling period are varied according to the conditions being monitored and the purpose of the sample. According to the British Department of Employment Technical Data Note 13 (Department of Employment and Productivity, 1970), the prime samples should be of 10 min duration and, if the resultant dust levels fall between 2 and 12 fibres/cm^3, 4-h samples are required in order to assess the mean dust levels. The pumping rates of many commercially available instruments, e.g. Casella or Rotheroe and Mitchell, can be varied from approximately 0.5 to 2 l./min. This is suitable for the collection of the shorter samples, but in some

Figure 6.4 Draeger hand pump for taking 'snap' samples

industrial environments this is too great for a 4-h sample. During the longer period, a dense deposit of non-asbestos dust would be collected, which obscures the fibres during the microscopical evaluation. A number of representative samples must therefore be collected during the 4-h period and time-weighted means calculated. An alternative method is to take a series of cumulative snap samples using one membrane filter only. The total volume of air sampled, to give results comparable with a continuous 10-min sample, should not however be less than 2 l. and the separate portions of the sample should be taken at convenient and equal intervals. Draeger hand pumps (Figure 6.4) or similar devices are used to take this type of sample (Asbestosis Research Council, 1971b). More recently, increased emphasis has been placed on the 4-h sample (Health and Safety Executive, 1976) and pumps have been developed to sample at flow-rates of the order of 100 cm^3/min or less (Winters, 1976) to enable long-period samples to be taken on a single filter.

3(v) Transportation of filters

Once the sample has been taken, the open end of the filter holder should be covered with a cap to prevent contamination, and the filter removed and mounted in a clean atmosphere away from the sampling environment. Some codes of practice (Asbestosis Research Council, 1971a) suggest that the dust deposit should be fixed on the membrane surface while the filter is in the sampling head. Two methods are described. One involves applying several drops of polymethylmethacrylate (Perspex) solution (0.025% in chloroform) to the membrane while clean air is being drawn through it by means of a low-velocity pump. The alternative method is to spray the filter surface with a cytological fixative from an aerosol dispenser. Great care must be taken to avoid disturbing the dust deposit. Tests carried out by Harness (1973) to compare fixed and non-fixed filters carried in their holders by road vehicle did not reveal any significant difference. It is therefore very doubtful whether this procedure is required, and it is omitted from some codes pf practice (National Health and Medical Research Council, 1976).

3(vi) Mounting of the filter

The filter must now be made transparent so as to enable the sample to be examined by transmission optical microscopy. Most filter samples are mounted on standard 25 × 76 mm microscope slides. The 25-mm filters can be mounted whole, but the larger 37-mm diameter filters must be divided into sectors.

There are many liquids which will make the filter transparent, e.g. cyclohexanone, dioxan + 5% water, and cynamaldehyde. Some only clear certain makes of filters, while others are used for specific purposes, such as fibre identification. Three methods, however, are widely used, namely triacetin, dimethyl phthalate and diethyl oxalate, and techniques based upon acetone vapour. With all of these methods the contamination during sample preparation may be significant when assessing filters that contain relatively few asbestos fibres. The effect of such contamination should be investigated by the preparation of control filters. The refractive indices of the media vary between approximately 1.4 and 1.6. A comparison of the counts from different parts of the same filter, mounted by different methods, did not show any significant difference between triacetin and dimethyl phthalate/diethyl oxalate (Walton *et al.*, 1976), although acetone clearing was thought to increase the count by up to one third above these two methods. The cause of this, and other effects due to the mountant, have not yet been fully investigated.

Triacetin (glycerol triacetate) (Asbestosis Research Council, 1971a) gives samples which normally remain countable for more than a year. Apart from the U.K., it is commonly used in France and Belgium. The main problem associated with this mounting technique is the disturbance of the dust deposit collected on the filter. To avoid this, the triacetin must first be spread out on the microscope slide into a pool a little less than the filter diameter. The filter is then placed dust side uppermost on to the triacetin and left for approximately 2 min, before being covered with the slip. When possible this should be about the same size as the filter. Ideally, a ring of undissolved filter is left around its perimeter. The use of too much triacetin may lead to migration of the deposit.

Dimethyl phthalate with diethyl oxalate (Asbestos Textile Institute, 1971) is widely used in the U.S.A., Scandinavia and South Africa. The chief disadvantage of this method is the relatively short lifetime of the samples. It is recommended that they be evaluated not more than 1 month after preparation, apparent loss being detected after about 46 days (Edwards and Lynch, 1968). This was thought to be due to migration of the fibres over a wide area, producing a reduced overall density. The mounting solution used in this method is a 1:1 mixture of dimethyl phthalate and diethyl oxalate. Normally 0.1–0.05 g of dissolved membrane filter material per cubic centimetre of solution is also incorporated. The purpose of the dissolved material is to provide a solution with as high a viscosity as possible without being difficult to handle. The highly viscous solution delays the migration of particles outward from the centre of the sample. The filter will clear without this additional material, but in this case the sample should be evaluated within 24-h.

When 37 mm filters are used, a sector is cut from the filter using a clean scalpel. This sector may be of any size to fit on to the glass slide, but wedge-shaped pieces with about 1×2 cm are recommended by some workers (Edwards and Lynch, 1968). A drop of the above solution is smeared on the slide into a shape corresponding to the sample section. The filter is then placed dust side up on top of this, and then covered with a clear cover-slip, which is pressed down lightly to remove air bubbles. The filter is usually completely dissolved in 15 min, with any background granularity disappearing within 1 or 2 days.

When samples are required to remain permanent over a period of several years, acetone mounting techniques are normally used. All involve the use of acetone vapour to clear the filter and produce samples on which the fibres tend to lie more in one plane than do the other methods. It is essential that the filter on the slide be placed very quickly into saturated vapour either from a jet, or by placing it inside a flask of boiling acetone. The use of liquid acetone or low-concentration vapour usually results in movement of the deposit or distortion of the filter. The cleared filter sets very hard, and several methods are used to attach the cover-glass. In the technique used chiefly in the U.S.A., the sample complete with cover-glass is replaced in the acetone vapour. An alternative method developed in Australia (National Health and Medical Research Council, 1976) involves the addition of a drop of triacetin on the cover-glass in order to obtain improved optical properties. Immersion liquids of matching refractive index can also be used to interface the filter and cover-glass.

3(vii) Microscopical Evaluation

Although the filter is now transparent, many of the fibres themselves cannot be seen when using a normal transmission optical microscope, because the refractive index of the background medium is very close to that of the asbestos. The effect is therefore equivalent to looking for a glass needle in a tank of water, and very few fibres are visible. The light which passes through the fibre, however, has a small change of phase relative to that which goes through the background alone. The human eye cannot see differences in phase, but only changes of amplitude or wavelength of the light. Microscopes have therefore been developed which can change small differences in phase into differences in amplitude, and thereby make the fibres more easily visible. Those which do this are known as phase-contrast (Bennett *et al.*, 1951) or interference-contrast microscopes (Vickers Instruments Ltd., 1957). The former are the most widely used as they are less expensive. This microscope operates by absorbing a large proportion of the light from the background and changing its phase by 90° relative to that which passes through the particle. Normal phase contrast absorbs 65–70% of the light, but the British code of practice (Asbestosis Research Council, 1971a) recommends 90–95% absorption for asbestos counting to increase the visibility of the fibre and reduce observer eye fatigue. Depending on whether the background light is advanced or retarded in phase, the microscope is known as positive or negative phase. In general, positive phase is used for asbestos counting (Walton *et al.*, 1976). The interference contrast uses a combination of polarizing filters and a quarter-wave plate to obtain the same effect. Although this type of microscope is not in general use, it can have advantages over phase-contrast equipment for dust containing mixed fibres, when the polarizing analyser can sometimes be used to identify non-asbestos fibres.

As has been mentioned previously, the number of fibres evaluated depends on the magnification of the microscope used. The different codes of practice normally specify a value within the range 400–700×. At this magnification, a 40×(4-mm) objective is frequently used, together with 10× eyepieces. Additional magnification may be obtained by an additional lens between the objective and eyepiece.

Before using the microscope for counting, its phase-contrast or interference optics must be correctly aligned, as failure to do so may result in a count reduced by up to 50% (Harness, 1973). This is especially important for chrysotile fibres, which have a significantly lower contrast than amphibole fibres (e.g. amosite, crocidolite), because of the smaller difference between the refractive indices of the fibres and the mountants. For phase contrast, the phase ring in the object lens must be in alignment with that below the substage condenser. These rings can be observed by using an auxiliary

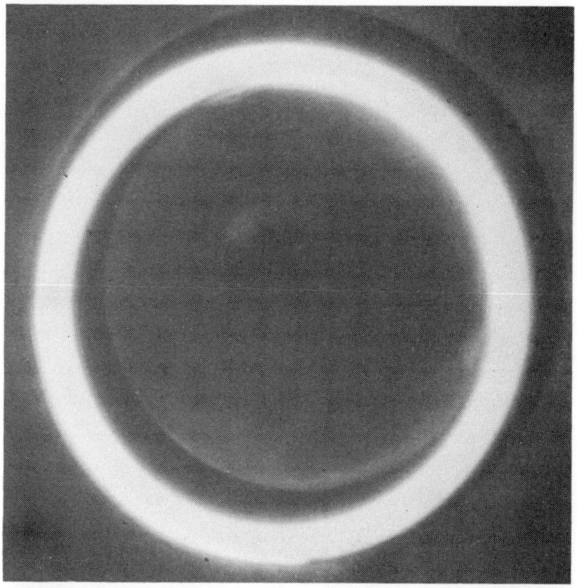

Figure 6.5 Phase rings in microscope in alignment

viewing telescope in one of the eyepieces, or with a focusable Bertrand lens, incorporated in many modern microscopes. Correct alignment exists when the central bright ring has a dark edge on both the inner and outer sides (Figure 6.5) and no small chinks of brighter light are visible (Figure 6.6). The alignment should be checked periodically during routine use, and following any adjustment to the objective lens or movement of the complete microscope.

The correct illumination is also very important, as too high or too low an intensity can result in fibres becoming less easily visible. The use of a green or blue filter, as recommended in most codes of practice, is also desirable to reduce eyestrain.

All methods of optically evaluating membrane filter samples require an eyepiece graticule (or reticule) to be used. This enables markings of known size and shape to be viewed in the plane of focus of the microscope. Several designs are used, for example the Porton $\sqrt{2}$, BS 3625, and Patterson Globe and Circle. The sizes of the markings on the graticule are calibrated against a stage micrometer. This need only be done once, unless the eyepieces or objectives are changed, the magnification is altered in any way, or the microscope has been re-assembled following cleaning. The graticules usually consist of a series of circles of different diameters, surrounding a central area, normally a rectangle, which is itself divided into smaller areas (see Figure 6.7). This main area is known as the graticule grid. In the techniques

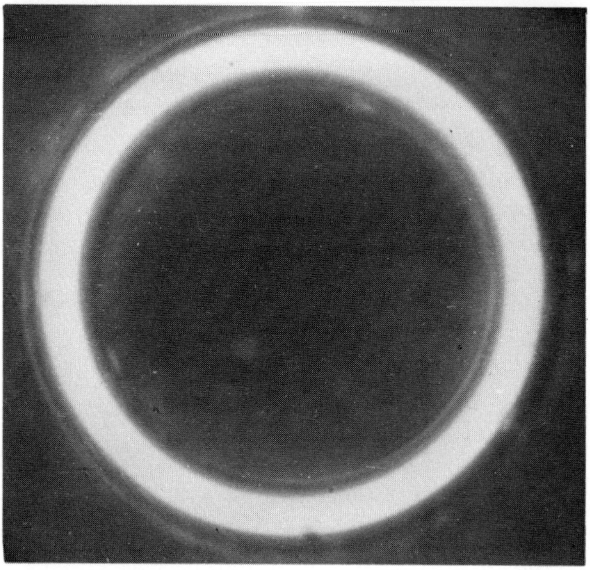

Figure 6.6 Phase rings in microscope out of alignment

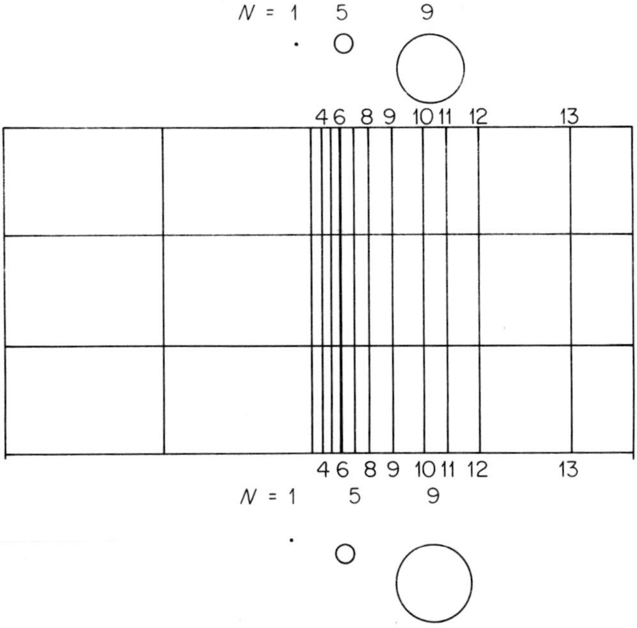

Figure 6.7 Typical graticule designed for the evaluation of spherical particles

described in the Australian and U.S.A. codes of practice, only those fibres on the area of the filter within the graticule grid are evaluated. The standard procedure described by the British Asbestosis Research Council (1971a) and the British Occupational Hygiene Society (British Occupational Hygiene Society Committee on Hygiene Standards, 1968) allows for the full field of view of the microscope to be counted. Recent experiments (Beckett *et al.*, 1976) have shown, however, that for small graticule grids (<20% of the full field area) fibre counts were higher by a factor of about 1.5 for amosite and 2.5 for chrysotile compared with full field counting. The effect was attributed mainly to human errors associated with full field counting, and grid counting is now recommended as producing more reliable and accurate results.

When counting asbestos, fibres frequently cross the edge of the graticule grid, and various methods have been described for determining whether or not these should be included. When a rectangular grid is used the most common procedure is to evaluate only those which cross either one horizontal or one vertical side. Alternatively, the position of a fibre can be designated by its 'lowest point' (Walton and Beckett, 1977). When this lies within the graticule grid area, the fibre is evaluated.

Many fibres are approximately 5 μm in length (Timbrell *et al.*, 1970). It is therefore necessary that their length can be compared easily with a standard length, because only those >5 μm should be counted. Most graticules, however, do not contain markings exactly 5 μm long. In addition, almost all graticules were designed for the evaluation of spherical particles, and are not well suited for counting fibres. A graticule has therefore been designed specifically for this purpose (Figure 6.8) (Walton and Beckett, 1977), and is now in use in laboratories in the U.K. and Scandinavia. In order to obtain markings of the exact sizes required, the graticule must be manufactured for each model of microscope (Graticules Ltd., Sovereign Way, Botany Trading Estate, Tonbridge, Kent, U.K.). Once this has been done, however, the critical size markings of length and aspect ratio are present in the field of view. An additional advantage is that the graticule grid is circular and 5 μm markings on the diameter can therefore be rotated and brought close to any fibre within the circle, without the counting area being moved.

It is, of course, not normally feasible to evaluate the complete sampling area of the filter. A random selection of fields of view is therefore evaluated, and the remainder of the sample is assumed to have an even (Poisson) distribution. Some laboratories initially scan the dust deposit at a low magnification in order to assess the distribution of the fibres. This is not always practical, however, especially when the objective lenses and substage phase rings are individually mounted. The areas selected should be from the whole sample area, and not restricted to the centre or edge of the filter. The choice must be random and, once chosen, the grid area should be counted

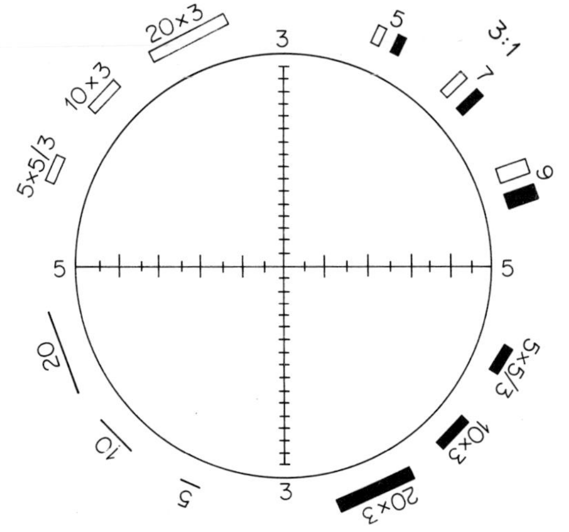

Figure 6.8 'Walton and Beckett' graticule for the evaluation of fibrous dust. (Reproduced with the permission of the British Occupational Hygiene Society from *Ann. Occup. Hyg.* 1977, **20,** 21.)

regardless of the number of fibres observed, unless the field is mainly obscured by a large particle. Some laboratories prefer to scan continuously across the filter and calculate the area evaluated from the vernier on the microscope stage.

When counting, it is very important constantly to adjust the fine focus back and forth, as this brings into focus fibres that might not otherwise be seen. Failure to do so has resulted in counts as low as one fifth of those obtained when adjusting the focus (Beckett and Attfield, 1974). The use of filters having a printed grid facilitates finding the level in which the majority of the fibres occur.

The number of fibres counted, or the area evaluated, depends on the code of practice used. The U.K. recommendation (Asbestosis Research Council, 1971a) is that if the deposit is sparse or unevenly distributed then 100 fields are observed. Otherwise, the number of fields needed for a total count of about 200 fibres is noted. Two U.S.A. methods (Asbestos Textile Institute, 1971; Ewards and Lynch, 1968) recommend that 20 random fields should be evaluated, or the number of fields containing 100 fibres, if this is less. When a wedge-shaped sector of the filter is used, the fields are sometimes selected at random along a radial line (AIHA–ACGIH Aerosol Hazards Evaluation Committee, 1975).

The measured concentration of the original airborne dust is calculated in the following way. If N is the number of fibres counted in n graticule areas, the total number of fibres per unit area estimated to be on the filter would be N/na, where a is the graticule grid area. The total number of fibres on the filter is therefore NA/na, where A is the sampling area of the filter. These were collected out of $V\,cm^3$ of air, so the original airborne concentration was NA/naV fibres/cm^3.

3(viii) Accuracy of the membrane filter method

Errors can occur both in sampling and evaluating the dust samples. Sampling differences are difficult to determine as it is not possible to maintain stable standard dust clouds. The errors are probably similar to those found when sampling spherical dust, and similar precautions should be taken, e.g. the use of isokinetic sampling heads in high-velocity air streams. The effects of electrostatic charges, filter pore size, and sampling flow-rate are not yet fully understood, however. These errors, on the other hand, are probably small compared with those in the microscopical evaluation.

Differences in technique can give rise to large differences in results, for example the full field and graticule grid counting methods already described. A single laboratory, however, can achieve good reproducibility, and standard deviations of counts of approximately 0.2 fibres/cm^3 (Rajhans and Bragg, 1975) or from ± 20 to $\pm 40\%$ in terms of a one standard deviation coefficient of variation (AIHA–ACGIH Aerosol Hazards Evaluation Committee, 1975). Differences were shown between different segments of the same filter unless extreme care was taken when sampling, the use of a normal filter holder producing variations up to ± 1.5 fibres/cm^3 (at a mean 5 fibres/cm^3). This is particularly important when 37-mm filters are used and only one sector is evaluated. When other laboratories with close contact and experienced counters were compared, differences of 3–6 times those within a single laboratory were found (Rajhans and Bragg, 1975). Novice counters produced even wider variations (Beckett and Attfield, 1974) and had difficulty in determining which fibres to evaluate. The Australian guide (National Health and Medical Research Council, 1976) includes diagrams of different types of fibres, e.g. split fibres or fibres with particles attached, together with instructions on how they should be counted. No international standard exists, however, and the most common rule (Walton *et al.*, 1977) for counting clumps of fibres was to move to another field in cases of difficulty! An exchange of mounted and unmounted filter samples carried out internationally between laboratories in nine countries (Walton *et al.*, 1977) gave ratios of highest to lowest of 2.8 for chrysotile and 1.7 for amphibole asbestos. The better agreement for amphibole asbestos was to be expected because chrysotile samples in general contain more finer fibres, and also have a lower contrast because of the smaller difference in refractive index between the type of fibre and mounting medium. The agreement internationally was probably better than that usually found during exchanges within one country, because each laboratory involved was chosen because of its experience in asbestos evaluation, and also because the counts reported were a consensus of the results of several microscopists.

The need for a continuing series of exchanges of slides was emphasized by Beckett and Attfield (1974), when a laboratory which had had no previous contact with other organizations was found to report results approximately one fifth of the level of another group of laboratories. In the U.S.A., a laboratory proficiency testing programme has been initiated by NIOSH to standardize asbestos counting procedures by various agencies and ensure that continuing comparisons take place (Ortiz *et al.*, 1975). Sample concentration is another variable that affects the accuracy of fibre counts. If the fibre or particle concentration is high, overlap problems and difficulty in distinguishing fibres through the particle background will occur. A theoretical study of the effect of overlapping fibres (Knight, 1975) has been published, but when possible sampling times should be reduced in order to prevent this occurring. Several theoretical studies of the reliability of microscopical evaluation of dust samples have been published (Sniegowski, 1966; Reist *et al.*, 1970). These, however, cannot account for the 'human factor', which is usually the chief cause of error in fibre counting.

For the results to be meaningful, the accuracy with which the airborne dust can be measured should be comparable to the variation in dust levels with time. Tests placing two samplers side-by-side gave similar variations to those of a series of samples taken in the same place at different times (Rajhans and Bragg, 1975). It can therefore be concluded that the standard method is sufficiently precise for industrial purposes.

3(ix) Recent developments in fibre evaluation

Microscope counting of fibres is not only prone to error, but is also very tedious and time consuming. Methods have therefore been developed either to try to aid the microscopist, or to automate the procedure. Statistical methods have been used to try to reduce the time required to count a slide. One of these is the most probable number method, as used for bacteria counting (Reist, 1975). In this method, the proportion of grid areas without any fibres is related statistically to the required mean number of fibres per unit area. This means that the observer need only record whether or not a graticule grid contains any fibres. Once one fibre has been detected, there is no necessity to adjust the focus and search for further fibres. The time to count a slide is therefore reduced. Alternative approaches consist in calculating the minimum number of fibres or fields that it is necessary to count in order to obtain the required accuracy (National Health and Medical

Research Council, 1976). This may be very much less for some slides than the 100 graticule areas suggested by the U.K. code of practice (Asbestosis Research Council, 1971a).

Alternatively, the sample can be modified to make the fibres more easily detected. Asbestos fibres of respirable size can be aligned by magnetic fields when in air or liquid suspension (Timbrell, 1972). Although they will not align when in contact with a solid surface, their slow rate of sedimentation when suspended in air or liquid gives ample time for alignment to take place in the field of an ordinary magnet before deposition takes place. Some types of asbestos align parallel to the field, whereas others settle in a plane normal to the direction of the field, or are a mixture of the two. Aligned specimens have advantages when counting asbestos in that they can be more easily sized, and also high-concentration specimens can be prepared with fibre overlapping minimized. This technique also provides a possible approach for automatically evaluating slides (Timbrell, 1975). A thin celloidin film containing aligned fibre, as can be used for light microscopy, can be examined in the system depicted in Figure 6.9. Here a horizontal laser beam illuminates a vertical glass slide carrying the film and the scattered light is collected on a vertical screen. The light scattering patterns obtained depend upon the type of asbestos present through its manner of alignment [see Section 7(iv)]. If the screen is replaced with a photomultiplier, which can rotate about an axis defined by the laser beam (Figure 6.10), light intensity

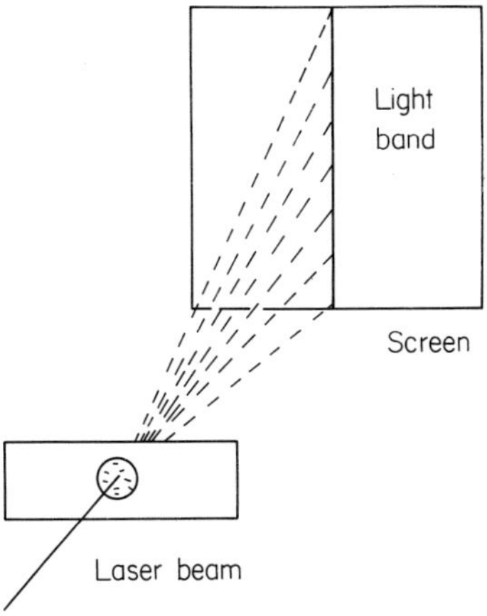

Figure 6.9 System for viewing light-scattering patterns produced by film specimens. (Reproduced with the permission of the British Occupational Hygiene Society from *Ann. Occup. Hyg.*, 1975, **18,** 303)

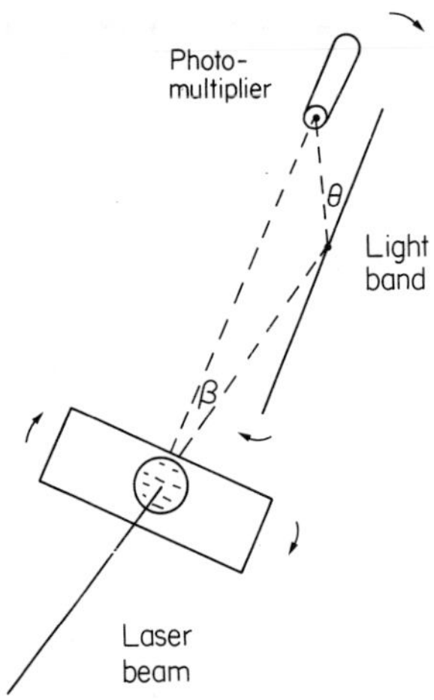

Figure 6.10 System for determination of light intensity distributions produced by film specimens. (Reproduced with the permission of the British Occupational Hygiene Society from *Ann. Occup. Hyg.*, 1975, **18,** 304)

distributions are produced (Figure 6.11). The amplitude of these distributions is proportional to the amount of asbestos present, and so measurement of this provides a possible method of evaluating the sample.

In order to reduce the eyestrain and operator fatigue produced by using microscopes for long periods, systems have been developed in which a camera is used to produce the image of the sample on a television screen. This in turn has lead to semi-automatic counting machines such as the Millipore ΠMC. Here the operator picks out the fibres on the screen using a pointer or light pen. A computer then decides whether the fibres fit the size criteria required, and processes the data. This system is still subject to

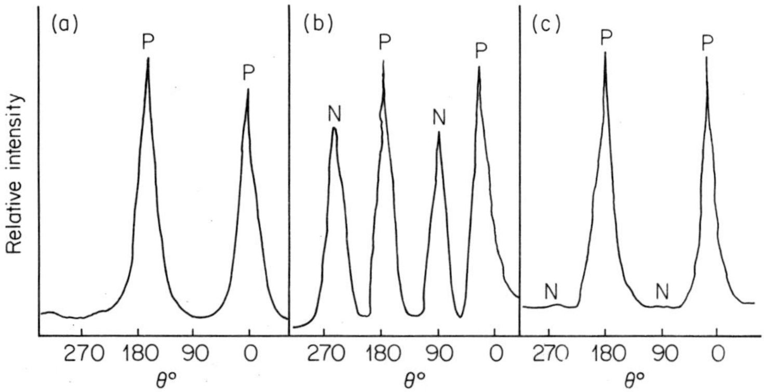

Figure 6.11 Light intensity distributions produced by film specimens: (a) UICC anthophyllite (Finland); (b) UICC amosite (Transvaal); (c) UICC crocidolite (Cape Province). (Reproduced with the permission of the British Occupational Hygiene Society from *Ann. Occup. Hyg.*, 1975, **18,** 304)

human error from failure to adjust fully the level of focus of the microscope or to detect the fibres amongst other particles. Further automation can be introduced by using computer pattern recognition to identify the fibres. One such system is the Quantimet (Harness, 1973). Fibres are identified by their ferré diameters and by calculating the ratio of the area to the square of the perimeter. This is at a minimum for a line and a maximum for spherical particles. Aggregates of asbestos fibres and fibres in contact with particles present difficulties, but this is also true for the manual operator. A major difficulty with these systems is the limited depth of focus of the optical microscope. This cannot at one time include all the levels on which the fibres are present, especially with larger pore size filters. In a fully automated instrument, automatic focusing determines the level of maximum dust concentration. Fibres above or below that level are not recorded and a low count is obtained. This effect can be reduced by using small pore size filters, and mounting methods such as acetone vapour, which tend to produce a sample with fibres more in one plane than most other mounting media. An alternative approach is to sample on a Nuclepore filter and evaluate the sample by using a scanning electron microscope with its greater depth of focus, resolution, and range of contrast. Photomicrographs or direct electronic input signals can then be analysed by the automatic system (Beckett, 1973). When using any automated system, however, care must be taken to obtain its correlation with optical membrane filter counts, on which the present hygiene standards are chiefly based.

(4) DETERMINATION OF VERY LOW ASBESTOS CONCENTRATIONS

Pathological evidence accumulated over the past decade indicates that asbestos is found in the lungs of most city dwellers (Spurny and Stober, 1975) and the need to measure the concentration in the general environment has therefore arisen. The methods described so far are unable to monitor accurately the extremely small amount of asbestos present. In order to obtain sufficient fibres for analysis, the dust must be collected from very large volumes of air. Asbestos, however, constitutes only a very small proportion of the particulate matter present, and any fibres present will be obscured by other material. Ashing can be used to remove organic material, but any remaining fibres must still be analysed to ensure that they are asbestos. Two methods of measuring these low concentrations of asbestos have been developed (Rickards, 1972, 1973). Both involve the collection of airborne solids from approximately 1000 m^3 of air. Sampling has been carried out using an electrostatic device (H. Litton Systems Inc.) in which up to 10 000 l./min are drawn through a 20-kV corona discharge. Particles in the air are electrostatically precipitated on to a plate and concentrated into approximately 100 cm^3 of liquid. The collection efficiency of this type of sampler depends on the size distribution of the particles and the sampling rate. Tests carried out in an asbestos factory (Rickards and Badami, 1971) have shown the efficiency of the Litton sampler to be almost 100% at 2000 l./min, falling to between 25 and 50% at 10 000 l./min. The dust can be filtered out of the liquid and examined using X-ray diffraction or electron microscopical techniques. X-ray diffraction, when used with an external standard, is capable of a detection limit of 10 μg when no gross interferences are present, and can be used with chrysotile and amphibole asbestos. The amount of asbestos fibre found in many environments is below this level, and a more sensitive electron microscopical method can be used to measure chrysotile concentrations down to 10^{-4} $\mu g/m^3$ in air (Rickards, 1973). This is based on the unique morphology and electron diffraction patterns of chrysotile in a transmission electron microscope. The dust collected is ashed and ultrasonically treated to reduce the chrysotile to its ultimate fibrils. These are then analysed and the length of the fibre determined. Because of the narrow diameter distribution of the fibrils, irrespective of their source (Atkinson *et al.*, 1971), it is possible to determine the mass of chrysotile present.

Both methods require extreme care, and are very time consuming. In addition, the method of evaluation enables only a mass measurement of the total airborne asbestos fibre to be made. This may, of course, be very much higher than the mass of the respirable fibre.

(5) DIRECT-READING DUST MONITORING EQUIPMENT

The membrane filter method suffers from the major disadvantage that it is not able to provide an instantaneous measurement of the airborne dust concentration. For engineering and dust suppression purposes, accurate fibre counts are not essential, but the speed with which data become available is of paramount importance. A wide range of devices, mostly based on light-scattering principles, has been developed. These instruments, however, normally record all of the airborne dust present. This may be mainly asbestos, as in an asbestos textile factory, but frequently, e.g. in asbestos–cement production, the non-asbestos particles are the major component of the dust. It is therefore necessary to calibrate this type of instrument against the membrane filter for each type of situation when estimates of asbestos concentrations are required. Two of the many commercially available instruments which are in routine use within the asbestos industry are the Royco Particle Counters and the Rotheroe and Mitchell P3.

The Royco photoelectric particle counter (Royco Instruments Inc., Menlo Park, California) has been developed chiefly for the measurement of very low dust concentrations in so-called 'clean rooms', but Channell and Hanna (1963) gave an account of tests with a number of types of atmospheric dust. The Royco operates on the light-scattering principle, using a very small chamber (approximately 0.002 cm^3 in volume) so that one particle at a time passing through generates a single pulse in the photoelectric detector. This pulse is amplified and fed into a pulse-height discriminator and digital counting system. This enables counts of particles equivalent to spheres from 0.3 μm to over 10 μm to be made in any or all of 15 size ranges on many Royco models. The pulse height depends on the surface area of the particle subtended to the detector and its refractive index, and not its falling speed, and so the meaning of the size selections when used with asbestos is uncertain. Tests in a chrysotile asbestos textile factory (Addingly, 1966) showed that a Royco set to count particles of 5 μm upwards gave a count about 25% lower than the membrane filter. For the 4 μm setting a result was obtained which averaged 15% above that of the membrane filter. These experiments were carried out using a membrane filter head in the exit line from the Royco counting chamber, and thus ensured that the same sample was being tested by each instrument.

In order to ensure that only one particle at a time passes through the chamber, most instruments are fitted with two air-flow controls, one for the flow of the air being sampled and the other to provide clean dilution air. At very high concentrations (>100 particles/cm^3) several particles still pass through the detection volume at the same time, which can result in readings of less than one third of the membrane filter count (Beckett, 1975). It can, however, be used very successfully where the dust is predominantly asbestos at concentrations of the order of 2 fibres/cm^3.

The Rotheroe Mitchell P3 (Greenford, Middlesex) [also known as the Sibata or Weathermeasure Instruments digital dust indicator, Model AP3 (Weathermeasure Corp., Sacremento, California)] operates by drawing dust through an optical chamber, where the light scattered at a 90° angle is monitored by a photomultiplier. A pulse count is produced and displayed on a rate meter, thereby giving a measure of the total fibre concentration and not individual fibres. Over the range 50–500 fibres/cm^3 the instrument gave a good correlation with the membrane filter count in an asbestos dust cloud (Beckett, 1975). The variation in 'dark' count of the instrument, however, makes it unsuitable for monitoring low dust concentrations. Its chief use is to detect high levels due to the failure of dust extraction equipment, or in monitoring mixed dust atmospheres where the total dust count is high, although the asbestos fibre concentration may be relatively low.

(6) MISCELLANEOUS INSTRUMENTS

6(i) Introduction

Most types of dust measuring instruments have been used to monitor asbestos. Although the membrane filter method is now the standard technique in most countries, other instruments are used for specialized purposes or for historical and political reasons. Many data about past exposure to asbestos exist in the form of results obtained by other methods, which have sometimes been retained so that there could be a continuity in the data. Some of the instruments which were commonly used for asbestos monitoring are described in the following sections.

6(ii) The thermal precipitator

This used to be a generally approved method in the U.K. and Europe for the estimation of airborne dusts, and is still used in South Africa. Several types of thermal precipitators were used for asbestos monitoring, but all work on the same principle. When a hot body is placed in a dusty atmosphere and suitably illuminated, it will be seen to be surrounded by a dust-free space, which is wider the greater the temperature difference between the body and the surrounding air. If there is an adjacent cold body, the dust will tend to be deposited on it. The theory of this thermophoretic motion of aerosol particles has been reviewed in detail (Waldmann and Schmitt, 1966). In the normal thermal precipitator, a wire lies horizontally between two surfaces consisting of microscope cover-glasses, and is heated to such a temperature that the width of the dust-free space surrounding the wire is decidedly greater than the distance of the wire from either cover-glass (Figure 6.12). As dusty air is drawn through the instrument, the particles are

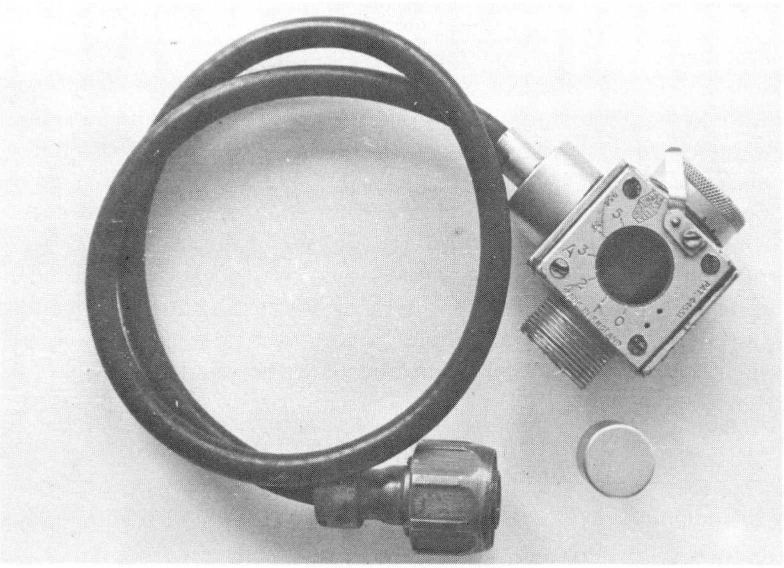

Figure 6.12 Sampling head of thermal precipitator

deposited in a thin line across each cover-glass. The sampling rate is normally very slow, usually less than or equal to 7.0 cm^3/min. The long-running thermal precipitator (C. F. Casella & Co. Ltd., Britannia Walk, London), which has a single dust deposit about 1 cm square and operates at 2 cm^3/min for 8 h, was widely used in U.K. asbestos factories before the development of the membrane filter method.

The dust deposit is normally evaluated by optical microscope, the sample sometimes first being ignited to remove organic material. The thermal precipitator deposit has the advantage over the membrane filter sample that all of the particles lie in the same plane and consequently less adjustment of the microscope is required during counting. In the South African Mining industry, the thermal precipitator has been used for many years, and recent work has been carried out to try to correlate these measurements with membrane filter counts (Du Toit and Gilfillan, 1977).

The thermal precipitator, however, suffers from the disadvantages of having a very slow sampling rate, and being not easily portable, so that it can be used only for long-period static sampling. It has also been suggested that its efficiency falls off for long asbestos fibres (Addingley, 1966).

6(iii) The konimeter

This instrument was designed to take 'snap' samples, and therefore its chief use is in helping to identify working places in which dust concentrations are high and in assessing the effectiveness of dust suppression equipment. Although it has not been widely used in the U.K. or U.S.A. asbestos industries, it was adopted by the South African mining industry, where it is still used.

In the konimeter (Figure 6.13) a small volume of air (of the order of 5 cm^3 or less) is impinged by a sudden expansion through a circular hole on to a movable glass plate, which is coated with a thin film of an adhesive substance such as vaseline or glycerine jelly. The dust spots obtained may be treated to remove carbonaceous material and then examined under a low-power microscope. Sometimes dark-ground illumination is used. The instrument is simple, portable, and robust, and is therefore suitable for underground working conditions. In addition, the short sampling time enables fluctuations in concentrations to be monitored, which are 'smoothed' out by long-period samplers. There are, however, many limitations, one of the main ones being its undersampling of fine particles (Quilliam, 1976). In addition, the high-velocity jet (50–100 m/s) is likely to break up airborne aggregates. A comparison of konimeter and thermal precipitator measurements of asbestos, however, showed a good agreement between the two instruments (Du Toit, 1977).

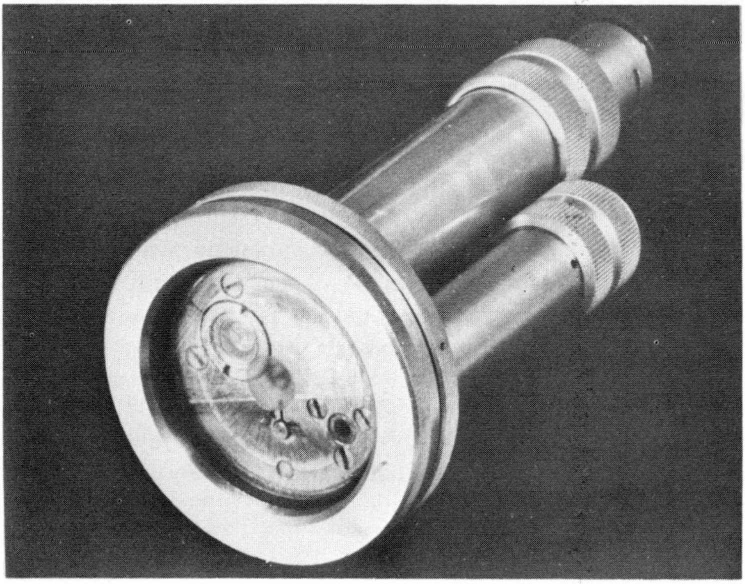

Figure 6.13 Konimeter dust sampler incorporating microscope for on-site evaluation

6(iv) The Owens jet counter

This instrument is similar to the konimeter in both its use and its method of action (Owens, 1922). It does, however, suck the dust through a slit and not a hole, and impinges the dust at a higher speed (200 m/s). Because of a humidifier incorporated in the instrument, condensation takes place on the particles during the expansion of the air. This removes the necessity to coat the impaction plate, which in turn enables a higher power of microscope to be used to evaluate the deposit (LeRoy Balzer, 1972). Efficiency studies have shown that it varies from 20% for particles of 0.1–0.2 μm aerodynamic diameter to 59% for 1–1.25 μm aerodynamic diameter, and then falls again at larger particle sizes. The suction volume was normally 50 cm^3. The instrument, however, was designed to take a series of samples, but the high air velocity on the second and subsequent strokes of the pump tended to disturb the sample deposited during the previous strokes (Addingley, 1966).

6(v) The impinger

This was for many years the standard instrument in the U.S.A., but has never found favour in Europe. Impingers differ from the impactor type of instrument since impingement takes place through a jet submerged in a

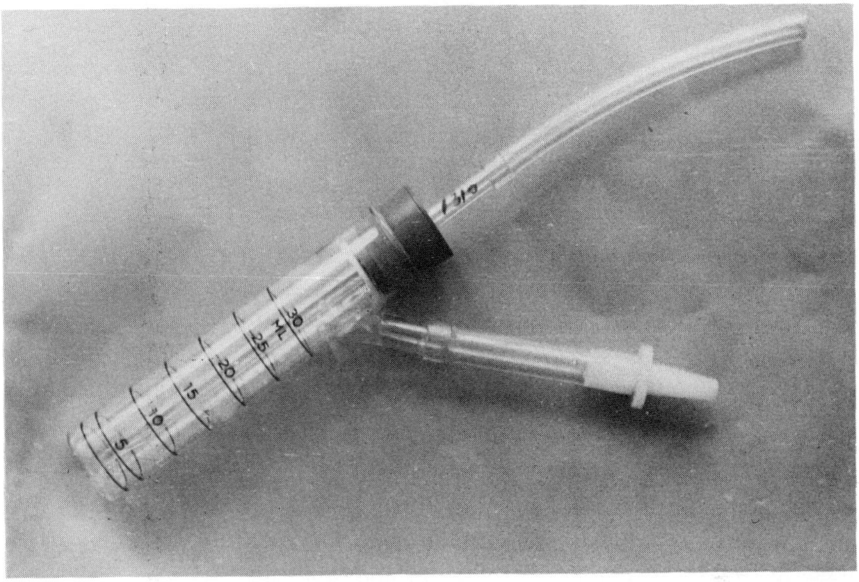

Figure 6.14 Impinger dust sampler

liquid medium (Figure 6.14). They are generally more efficient than konimeter-type impactors, but the efficiency still drops off rapidly below 0.7 μm aerodynamic diameter (LeRoy Balzer, 1972). It was designed, however, to count particles larger than 1 μm. Both the midget and Greenburg-Smith impingers were used to monitor asbestos. These operate by drawing air quickly through a small nozzle and directing it so that it strikes the bottom of a glass flask partially filled with a liquid such as water or alcohol. This high-velocity action causes the dust particles to be 'wetted' and retained in the liquid. The sample is then examined under a microscope and usually a count is made at a magnification of 100× of all the dust particles present. The early epidemiological data on asbestos exposure in the U.S.A. consists almost entirely of impinger counts. These were of total dust and were taken to compare exposure against the then Threshold Limit Value of 5 million particles per cubic foot (Dreesen *et al.*, 1938; American Conference of Governmental Industrial Hygienists, 1969). Attempts to relate these results with membrane filter counts have shown that the relationship depends very strongly on the type of industry being sampled (Lynch *et al.*, 1970). In general, less than 20% of the particles counted were asbestos. In addition, fewer asbestos fibres are seen than with the membrane filter method, because of the greater efficiency of the latter both in collection and in the visibility obtainable during the microscopical evaluation. Tests carried out in an asbestos textile factory gave a count of only 50% of that obtained by the membrane filter (Addingley, 1966), and it is generally thought to provide only a 'poor to fair' index of exposure.

(7) IDENTIFICATION OF AIRBORNE ASBESTOS FIBRES

7(i) Introduction

When determining the dose *versus* disease relationship for asbestos, it is of course necessary to monitor individually the different types of fibre present. Samples taken to check compliance with the Threshold Limit Values may not, however, require the fibres to be identified. In many countries all airborne fibres fitting the definition given in Section 3(ii) are regarded as potentially hazardous, and are therefore evaluated. In the U.K., on the other hand, the standard is for asbestos fibres only (Department of Employment and Productivity, 1970), and so it is necessary to try to distinguish these from other fibres which may be present, e.g. organic fibres, glass, or gypsum (Middleton, 1978). In addition, the separate standard for crocidolite means that this too must be identified separately.

Three different approaches are often used to identify asbestos in airborne dust samples. The first uses optical microscopy and depends upon morphological recognition and optical crystallographical techniques. Their main limitation is that they can only be used with larger fibres. To overcome this, electron microscopical techniques have been developed. These can determine the elemental content of a fibre, but require very sophisticated and expensive equipment. The third method is to use physical or chemical analysis. This is limited to mass measurements, which, as has been explained previously, may vastly overestimate the amount of respirable asbestos present.

Almost all of the above techniques are used to identify asbestos in bulk material, and are described in more detail in the chapter dealing with this problem. The following sections summarize the methods in order to show their relevance to airborne dust sampling.

7(ii) Optical techniques

When counting a membrane filter sample using a phase-contrast microscope, it is possible for the experienced observer to discriminate against some fibres because of their morphology. Fibres with an irregular outline are generally not asbestos, whereas those with large 'bulges' at regular intervals are normally glass-fibre. Crocidolite and amosite are, however, both needle-like in appearance (Figure 6.15) and, although chrysotile is generally curly

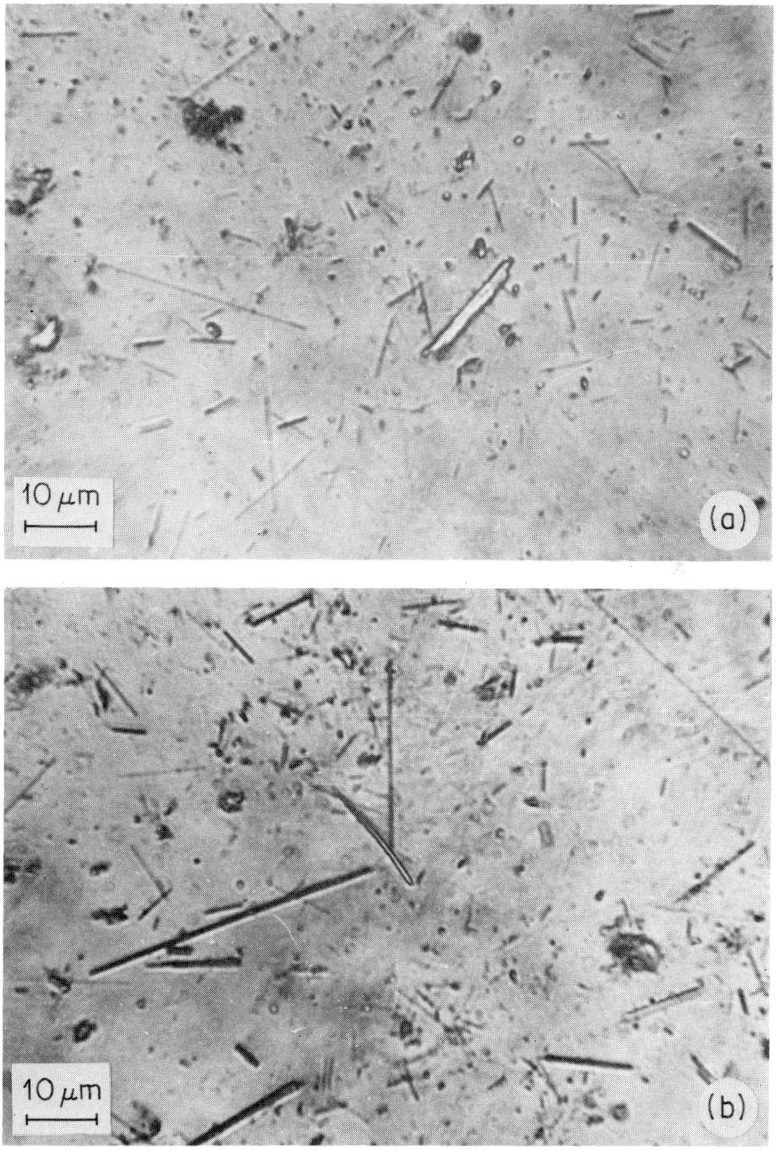

Figure 6.15 Samples of airborne amphibole asbestos under a phase-contrast microscope at 480×: (a) amosite, (b) crocidolite

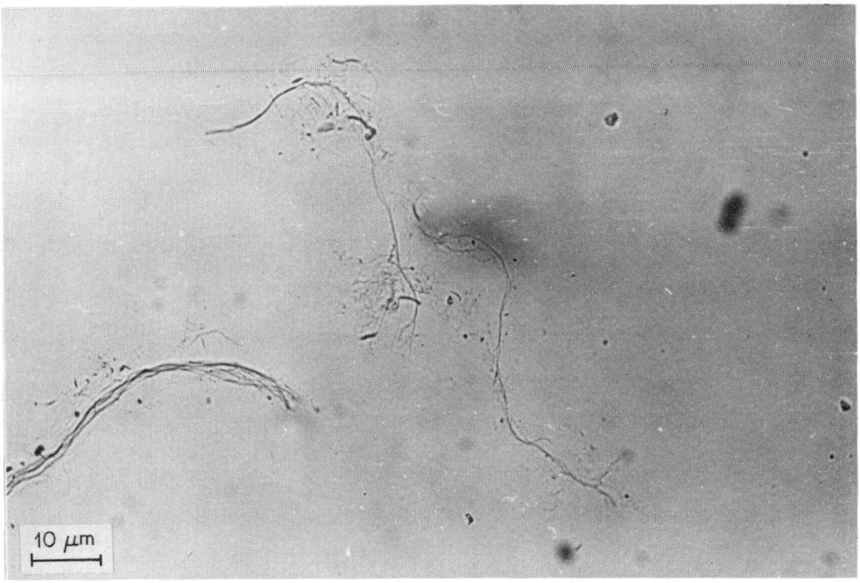

Figure 6.16 Chrysotile asbestos under a phase-contrast microscope at 600×

(Figure 6.16), it cannot be distinguished by morphology from some organic fibres. Each type of asbestos does, however, have a different refractive index and this can be used to help to identify it. Usually liquids of accurately known refractive indices are added to the specimen, which is then studied using a dispersion staining technique (Julian and McCrone, 1970).

A series of liquids with different refractive indices have been used in succession with a single fibre (Vigliani *et al.*, 1976), but this is very tedious and is impractical with large numbers of fibres. Polarizing microscopy can also be used to distinguish between certain types of asbestos (Schmidt, 1960). This uses the property that the refractive index of the fibre measured along its length may be greater or less than that across its diameter according to its type. By rotating the fibre or polarizer, this relationship between the refractive indices can be determined. This process can be aided by the use of accessory plates.

All of the above methods, however, require a fibre diameter of at least 5 μm in order to obtain a satisfactory result (Zielhuis, 1977). Most airborne samples, however, contain few if any asbestos fibres broader than 3 μm, and an analysis of any broader fibres present may not reflect the composition of the respirable portion of the dust. Any results obtained using these methods must therefore be treated with extreme caution, although for broader fibres they can normally be used successfully to distinguish asbestos fibres from non-asbestos fibres.

7(iii) Electron microscopical techniques

Both the scanning and transmission electron microscopes have been widely used to examine asbestos fibres (see, for example, Spurny *et al.*, 1976a; Pooley, 1975). Greater magnification and resolution of the electron microscope compared with the optical microscope enables many more, smaller, fibres to be identified by their morphology.

The scanning electron microscope has the advantage of a wide range of magnification (normally more than three orders of magnitude) and relative ease of sample preparation. Airborne dust can be sampled by normal instruments but using a Nuclepore rather than a membrane filter (Beckett, 1973). The filters may be coated with gold or soaked in hyamine (Mark, 1974) in order to prevent the build-up of electrostatic charge, which could result in fibre loss. After sampling, the filters can be mounted directly on to stubs, and then shadowed with gold or carbon to give the dust deposit a conducting surface.

For the examination of asbestos fibres less than 0.1 μm in diameter, transmission electron microscopes are normally used. The airborne dust samples may be collected on a membrane filter. Small pore size filters are frequently preferred (e.g. 0.05 μm; Rickards, 1973), although pore sizes as large as 0.8 μm have been used successfully. The dust must then be transferred to an electron microscope grid. This can be performed by first acetone clearing the filter, as previously described. The deposit is then coated with chromium and carbon before the cleared filter material is dissolved in acetone liquid (Ortiz and Isom, 1974). Alternative methods involve ashing of the filter, although great care must be taken to avoid sample loss due to explosive combustion of the filter. The transmission electron microscope can be used to provide selected area electron diffraction patterns. These can distinguish chrysotile from other fibre types, but they will not identify individual amphiboles (Pooley, 1975; Zielhuis, 1977). To do this requires chemical analysis of the individual fibres.

The ability to analyse single asbestos dust particles chemically became available with the development of a combination of the electron microscope and X-ray analytical equipment, of both the energy-dispersive and wavelength-dispersive types, to form the electron microscope microprobe analyser. Initially these analytical attachments were combined mainly with scanning electron microscopes, but more recently X-ray spectrometers of the energy-dispersive type have been increasingly used with transmission electron microscopes. These instruments are capable of producing a picture of the fibre together with an X-ray spectrum for elements of atomic number higher than 11 (sodium), enabling different types of asbestos to be identified. Care must be taken, however, to ensure that sufficient fibres are analysed to provide a representative proportion of those present in the airborne dust sample.

7(iv) Physical and chemical analysis

A very wide range of techniques is available which can be used to measure the mass concentrations of airborne dust containing chiefly asbestos. When other dusts are present, however, interference from these may cause misleading results. This is true of X-ray diffraction and infrared spectrophotometry, both of which are used for monitoring airborne asbestos although in rather specialized cases (Rickards, 1972; Beckett *et al.*, 1975). β-ray and ultraviolet absorption have also been proposed, but at present have very limited use.

Chemical analysis is normally used on bulk material, although it can be used to evaluate certain types of airborne dust samples. Usually the analysis is carried out to determine the amount of iron or magnesium present. This assumes, of course, that none of the other dust contains these minerals. Atomic-absorption, X-ray fluorescence, and neutron-activation techniques have all been used, and the measurement of pH has been investigated to determine the presence of chrysotile. Atomic-absorption spectrometry in particular has been used in the U.S.A. to provide a rapid method of routine monitoring for chrysotile workers (Keenan and Lynch, 1970). The sensitivity of the method is sufficient to allow a membrane filter to be divided into two, so that one half can be analysed for its magnesium content, and the other half evaluated using an optical microscope. The results obtained on denser samples were in good agreement with X-ray analytical measurements.

The method of aligning fibres using a magnetic field has previously been described. The diffraction pattern produced by these fibres, when illuminated by a laser, corresponds to the type of asbestos present (Timbrell, 1975) because of the differing modes of alignment. If the specimen is then rotated, light intensity distributions can be obtained (Figure 6.11). When other particles are present, these scatter light in all directions, hence raising the intensity trace with respect to the base-line, but without altering the sizes (areas) of the peaks. This technique is still being developed, however, although it has been used to identify asbestos in lung material. Aligned specimens also improve X-ray diffraction and some other analytical methods.

(8) SUMMARY

The membrane filter method is now being adopted internationally for the monitoring of airborne asbestos dust. For specific applications or historical reasons, other instruments such as the konimeter, impinger, or thermal precipitator are sometimes used. When instantaneous results are required for engineering or dust suppression purposes, light-scattering particle counters are normally preferred, although care has to be taken in interpreting the results from these instruments in mixed dust atmospheres. The identification of the airborne dust fibres using physical or chemical analysis is also very prone to error when other dusts are present. Large fibres can be identified using optical crystallographical techniques, but for airborne dust samples, where most fibres are extremely fine, the combination of an electron microscope and an X-ray analyser offers the most satisfactory method of determining the types of fibre present.

(9) REFERENCES

Addingley, C. G. (1966) Asbestos dust and its measurements, *Ann. Occup. Hyg.*, **9,** 73–82.

Advisory Committee on Asbestos Cancers (1973) *Biological Effects of Asbestos,* Report to the Director of the International Agency for Research on Cancer, *Ann. Occup. Hyg.*, **16,** 9–17.

AIHA–ACGIH Aerosol Hazards Evaluation Committee (1975) Recommended procedures for sampling and counting asbestos fibres, *Am. Ind. Hyg. Ass. J.*, **36,** 83–90.

American Conference of Governmental Industrial Hygienists (1969) *Threshold Limit Values of Airborne Contaminants,* ACGIH Committee on Threshold Limit Values, Cincinnati, Ohio.

Asbestosis Research Council (1971a) *The Measurement of Airborne Asbestos Dust by the Membrane Filter Method (Rev.),* Technical Note No. 1, Asbestosis Research Council, Rochdale, Lancs.

Asbestosis Research Council (1971b) *Dust Sampling Procedures for Use with the Asbestos Regulations 1969,* Technical Note No. 2, Asbestosis Research Council, Rochdale, Lancs.

Asbestos Textile Institute (1971) *Measurement of Airborne Asbestos Fibre by the Membrane Filter Method,* Asbestos Textile Institute, Pompton Lakes, (U.S.A.)

Atkinson, A. W., Gettins, R. B., and Rickards, A. L. (1971) Morphology of chrysotile, Paper presented at the 2nd International Conference on the Physics and Chemistry of Asbestos Minerals, Louvain University, 6th–9th September, 1971 (Unpublished preprints, Paper 2:4).

Bartosiewicz, L. (1973) Improved techniques of identification and determination of airborne asbestos, *Am. Ind. Hyg. Ass. J.*, **34,** 252–259.

Beattie, J. (1961). The asbestosis body, in *Inhaled Particles and Vapours,* (Ed. C. N. Davies), Proceedings of an International Symposium organized by the British Occupational Hygiene Society, Oxford, 29th March–1st April, 1960. Pergamon Press, Oxford, 434–442.

Beattie, J., and Knox, J. F. (1961) Studies of mineral content and particle size distribution in the lungs of asbestos textile workers, in *Inhaled Particles and Vapours,* (Ed. C. N. Davies), Proceedings of an International Symposium organized by the British Occupational Hygiene Society, Oxford, 29th March–1st April 1960, Pergamon Press, Oxford, 419–433.

Beckett, S. T. (1973) The evaluation of airborne asbestos fibres using a scanning electron microscope, *Ann. Occup. Hyg.*, **16,** 405–408.

Beckett, S. T. (1975) The generation and evaluation of UICC asbestos clouds in animal exposure chambers, *Ann. Occup. Hyg.*, **18,** 187–198.

Beckett, S. T., and Attfield, M. D. (1974) Inter-laboratory comparisons of the counting of asbestos fibres sampled on membrane filters, *Ann. Occup. Hyg.*, **17,** 85–96.

Beckett, S. T., Hey, R. K., Hirst, R., Hunt, R. D., Jarvis, J. L., and Rickards, A. L. (1976) A comparison of airborne asbestos fibre counting with and without an eyepiece graticule, *Ann. Occup. Hyg.*, **19,** 69–76.

Beckett, S. T., Middleton, A. P., and Dodgson, J. (1975) The use of infrared spectrophotometry for the estimation of small quantities of single varieties of UICC asbestos, *Ann. Occup. Hyg.*, **18,** 313–320.

Bennett, A. H., Osterberg, H., Jupnik, H., and Richards, O. W. (1951) *Phase Microscopy, Principles and Applications*, Wiley, New York.

Breslin, J. A., and Stein, R. L. (1975). Efficiency of dust sampling inlets in calm air, *Am. Ind. Hyg. Ass. J.*, **36,** 576–583.

Breuer, H. (1964) Erfahrungen mit dem gravimetrischen Feinstaubfiltergerät BAT, *Staub*, **24,** 324–329.

British Occupational Hygiene Society Committee on Hygiene Standards (1968) Hygiene standards for chrysotile asbestos dust, *Ann. Occup. Hyg.*, **11,** 47–69.

British Occupational Hygiene Society Committee on Hygiene Standards (1973) Review of the hygiene standard for chrysotile asbestos dust, *Ann. Occup. Hyg.*, **16,** 7.

British Standards Institution (1971) *Simplified Methods for Measurement of Grit and Dust Emission*, BS 3405, B.S.I., London.

Channell, J. K., and Hanna, R. J. (1963) Experience with light scattering particle counters, *Archs. Envir. Hlth*, **6,** 386–400.

Department of Employment and Productivity, H.M. Factory Inspectorate (1970) *Standards for Asbestos Dust Concentrations for Use with the Asbestos Regulations 1969*, DEP Technical Data Note 13. H.M. Stationery Office, London.

Deutsche Forschungsgemeinschaft: Senatskommission zur Prüfung Gesundheitsschadlicher Arbeitsstoffe (1976) *Maximale Arbeitsplatzkonzentrationen 1976, Mitteilung XII vom 28 Juni 1976*, Deutsche Forschungsgemeinschaft, Bonn-Bad Godesberg.

Dreesen, W. C., Dalla Valle, J. M., Edwards, T. I., Miller, J. M., and Sayers, R. R. (1938) *A Study of Asbestosis in the Asbestos Textile Industry*, U.S. Public Health Service Washington.

Du Toit, R. S. (1977) A review of early results of comparative tests with the konimeter and thermal precipitator in asbestos mines *Ann. Occup. Hyg.*, **20,** 279–281.

Du Toit, R. S., and Gilfillan, T. C. (1977), Simultaneous airborne dust samples with konimeter, thermal precipitator and dosimeter in asbestos mines, *Ann. Occup. Hyg.* **20,** 333–344.

Edwards, G. H., and Lynch, J. R. (1968) The method used by the U.S. Public Health Service for the enumeration of asbestos dust on membrane filters, *Ann. Occup. Hyg.*, **11,** 1–6.

Fuchs, N. A. (1964) *The Mechanics of Aerosols*, Pergamon Press, Oxford, pp. 151–159.

Harness, I. (1973) Airborne asbestos dust evaluation, *Ann. Occup. Hyg.*, **16,** 397–404.

Health and Safety Executive (1976) *Asbestos Hygiene Standards and Measurement of Airborne Dust Concentrations*, HSE Guidance Note, Environmental Hygiene/10, H.M. Stationery Office, London.

Holmes, S. (1965) Developments in dust sampling and counting techniques in the asbestos industry, *Biological Effects of Asbestos* (Ed. H. E. Whipple), New York Academy of Sciences, New York (*Ann. N.Y. Acad. Sci.*, **132,** 288–297).

Jacobsen, M., Rae, S., Walton, W. H., and Rogan, J. M. (1970) New dust standards for British coal mines, *Nature, Lond.*, **227,** 445–447.

Julian, Y., and McCrone, W. C. (1970) Identification of asbestos fibres by microscopical dispersion staining, *Microscope*, **18,** 1–10.

Keenan, R. G., and Lynch, J. R. (1970) Techniques for the detection, identification and analysis of fibres, *Am. Ind. Hyg. Ass. J.*, **31,** 587–597.

Knight, G. (1975) Overlap problems in counting fibres, *Am. Ind. Hyg. Ass. J.* **36,** 113–114.

LeRoy Balzer, J. (1972) Inertial and gravitational collectors, In *American Conference of Governmental Industrial Hygienists. Air sampling Instruments for Evaluation of Atmospheric Contaminants*, 4th Ed., ACGIH, Cincinnati, Ohio, 02–09.

Lynch, J. R., Ayer, H. E., and Johnson, D. L. (1970) The interrelationships of selected asbestos exposure indices, *Am. Ind. Hyg. Ass. J.*, **31,** 598–604.

Mark, D. (1974) Problems associated with the use of membrane filters for dust sampling when compositional analysis is required, *Ann. Occup. Hyg.*, **17,** 35–40.

Middleton, A. P. (1978) On the occurrence of fibres of calcium sulphate resembling amphibole asbestos in samples taken for the evaluation of airborne asbestos, *Ann. Occup. Hyg*, **21,** 91–93.

National Health and Medical Research Council (1976) *Membrane Filter Method for Estimating Airborne Asbestos Dust*, NHMRC, Canberra, Australia.

Ogden, T. L., and Wood, J. D. (1975) Effects of wind on the dust and benzene-soluble matter captured by a small sampler, *Ann. Occup. Hyg.*, **17,** 187–195.

Orenstein, A. J. (Editor) (1960) *Proceedings of the Pneumoconiosis Conference held at the University of Witwatersrand, Johannesburg, 9th–24th February, 1959*, J. & A. Churchill, London, 620.

Ortiz, L. W., and Isom, B. L. (1974) Transfer technique for electron microscopy of membrane filter samples, *Am. Ind. Hyg. Ass. J.*, **35,** 423–425.

Ortiz, L. W., Ettinger, H. J., and Fairchild, C. I. (1975) Calibration standards for counting asbestos, *Am. ind. Hyg. Ass. J.*, **36,** 104–112.

Owens, J. S. (1922) Suspended impurities in the air, *Proc. R. Soc., Ser. A*, **101,** 18–37.

Pooley, F. D. (1975) The identification of asbestos dust with an electron microscope microprobe analyser, *Ann. Occup. Hyg.*, **18,** 181–186.

Quilliam, J. H. (1976) The value of the konimeter for dust control purposes in routine mine sampling, *J. Mine Vent. Soc. S. Afr.*, **9,** 115–116.

Rajhans, G. S., and Bragg, G. M. (1975) A statistical analysis of asbestos fibre counting in the laboratory and industrial environment, *Am. Ind. Hyg. Ass. J.*, **36,** 909–915.

Reist, P. C., VanCamerck, S. B., and Chabot, G. E. (1970) A further note on the reliability of membrane filter dust sample evaluation by microscope counting, *Ann. Occup. Hyg.*, **13,** 201–204.

Reist, P. C. (1975) Counting asbestos fibres by the most probable number method, *Am. Ind. Hyg. Ass. J.*, **36,** 379–384.

Rickards, A. L. (1972) Estimation of trace amounts of chrysotile asbestos by X-ray diffraction *Analyt. Chem.*, **44,** 1872–1873.

Rickards, A. L. (1973) Estimation of submicrogram quantities of chrysotile asbestos by electron microscopy, *Analyt. Chem.*, **45,** 809–811.

Rickards, A. L., and Badami, D. V. (1971) Chrysotile asbestos in urban air, *Nature, Lond.*, **234,** 93–94.

Schmidt, K. G. (1960) Asbestsorten, ihre Untersuchung mit optischen Mitteln und ihre krankmachende Wirkung, *Staub*, **20,** 173–204.

Sniegowski, A. (1966) A note on reliability of membrane filter dust sample evaluation by microscope counting, *Ann. Occup. Hyg.*, **9,** 65–67.

Spurny, K. R., and Stober, W. (1975) Asbestos measurements in ambient air, *Clean Air*, **9,** 38–41.

Spurny, K. R., Stober, W., Opiela, H., and Weiss, G. (1976a) in *Atmospheric Pollution* (Ed. M. M. Benarie) the proceedings of the 12th International Colloquium, Paris, Fance. May 5th–7th, 1976. Organized by the Institut National de Recherche Chimique Applique. Elsevier, Amsterdam, 459–469.

Spurny, K. R., Stober, W., Ackerman, E. R., Lodge, J. P., and Spurny, K. (1976b) The sampling and electron microscopy of asbestos aerosol in ambient air by means of Nuclepore filters *J. Air Pollut. Control Ass.*, **26,** 496–498.

Stanton, M. F. (1973) Some etiological considerations of fibre carcinogenesis, in *Biological Effects of Asbestos*, IARC Scientific Publications No. 8, International Agency for Research on Cancer, Lyon 289–294.

Timbrell, V. (1965) The inhalation of fibrous dusts, in *Biological Effects of Asbestos*, (Ed. H. Whipple), New York Academy of Sciences, New York, (*Ann. N. Y. Acad. Sci.*, **132,** 255–273).

Timbrell, V. (1972) Alignment of amphibole asbestos fibres by magnetic fields *Microscope* **20,** 365–368.

Timbrell, V. (1973) Physical factors as etiological mechanisms, in *Biological Effects of Asbestos*, IARC Scientific Publications No. 8, International Agency for Research on Cancer, Lyon, 295–303.

Timbrell, V. (1975) Alignment of respirable fibres by magnetic fields, *Ann. occup. Hyg.*, **18,** 299–312.

Timbrell, V., Skidmore, J. W., Hyett, A. W., and Wagner, J. C. (1970) Exposure chambers for inhalation experiments with standard reference samples of asbestos of the International Union Against Cancer (UICC), *J. Aerosol Sci.*, **1,** 215–223.

Vickers Instruments Ltd. (1957) *The Baker Interference Microscope*, 3rd edition. York (U.K.), Vickers Instruments Ltd.

Vigliani, E. C., Patroni, M., Occella, E., Rendal, R. E., Skikne, M., and Ellis, P. (1976) Presence and identification of fibres in the atmosphere of Milan, *Medna Lav.*, **67,** 551–567.

Vorwald, A. J., Durkan, T. M., and Pratt, P. C. (1951) Experimental studies of asbestosis, *Archs Ind. Hyg.*, **3,** 1–43.

Waldmann, L., and Schmitt, K. H. (1966) Thermophoresis and diffusiophoresis of aerosols, in *Aerosol Science*, (Ed. C. N. Davies), Academic Press, London and New York, Chapter VI, pp. 137–162.

Walton, W. H. (1954) Theory of size classification of airborne dust clouds by elutriation, in Institute of Physics. *The Physics of Particle Size Analysis*, The Institute, London, (*Br. J. Appl. Phys. Suppl.*, **3,** S29–S40).

Walton, W. H., Attfield, M. D., and Beckett, S. T. (1976) An international comparison of counts of airborne asbestos fibres sampled on membrane filters, *Ann. Occup. Hyg.*, **19,** 215–224.

Walton, W. H., and Beckett, S. T. (1977) A microscope eyepiece graticule for the evaluation of fibrous dust, *Ann. Occup. Hyg.*, **20,** 19–23.

Winters, J. W. (1976) A simple small light-weight personal dust sampling unit for full shift determination of asbestos dust exposure, *Ann. Occup. Hyg.*, **19,** 77–80.

Wood, J. D. (1977) Review of personal sampling pumps, *Ann. Occup. Hyg.*, **20,** 3–17.

Zielhuis, R. L. (*Editor*). (1977). *Public Health Risks of Exposure to Asbestos.* Report of a Working Group of Experts prepared for the Commission of the European Communities, Directorate-General for Social Affairs, Health and Safety Directorate, Pergamon Press, Oxford, (for the Commission of the European Community).

CHAPTER 7

The identification of asbestos in solid materials

A. P. Middleton*

Institute of Occupational Medicine, Edinburgh

(1) SYNOPSIS

The background to the requirement for the provision of suitable methods of analysis for asbestos is considered and in this context the aims of any analysis are discussed. The types of samples which are likely to be encountered are briefly considered, together with some methods of pre-treatment to render them more suitable for analysis. A schematic system for analysis is set out and within this framework various analytical techniques are discussed. The methods considered include optical microscopy, electron-optical methods and X-ray microanalysis, X-ray diffractometry, infrared spectrophotometry, elemental analysis and thermal analysis techniques. The general principles of each method are briefly outlined and the practical applications of the method are reviewed. The advantages and disadvantages of each method, from the point of view of both qualitative and quantitative analysis, are discussed.

* Present address: TBA Industrial Products, Chemical Research Department, P.O. Box 40, Rochdale, Lancashire

(2) INTRODUCTION

The development of techniques for the detection and identification of asbestos in solid materials has been stimulated by the recognition of the potential hazards to personnel involved in the manufacture, application, and removal of such materials. In the U.K. the requirements of the Health and Safety at Work, etc., Act, in conjunction with the provisions of the earlier Asbestos Regulations (Health and Safety Executive, 1975) have provided a special impetus to research into methods of analysis for asbestos.

The need for analysis most commonly arises when materials thought to contain asbestos are to be removed or disturbed in some way. Analysis of new products may also be necessary before they are installed or processed but usually their asbestos content will be available from the manufacturer. In order that appropriate precautions can be taken to protect both those involved directly and others in the vicinity, it is necessary to confirm that the material does in fact contain asbestos and in the context of regulations in the U.K. and some Scandinavian countries to establish whether or not crocidolite (blue asbestos) is present. The identification of the types of asbestos present in bulk materials before they are disturbed and the asbestos becomes airborne is particularly important in view of the difficulties of identifying individual fibres in samples of the airborne dust. The types of products for which analysis is required are extremely varied, ranging from the loosely compacted sprayed asbestos and lagging compositions, through moulded insulations, ceiling tiles, and asbestos–cement, to friction materials and reinforced plastics (Table 7.1). There is therefore a need for flexibility in any analytical procedures.

Many of the samples submitted for analysis will contain mixtures of two or more types of asbestos and, because of the way in which many loosely bonded thermal insulation products were prepared and applied, the distribution of fibres is often markedly inhomogeneous.

There is not an overriding requirement for the analysis to be quantitative since any product containing even small amounts of asbestos, especially crocidolite, will need to be treated with caution. However, a quantitative technique may be required on occasion and the ease with which this can be achieved by different methods will be considered. There seems to be no consensus of opinion on a lower limiting value to the percentage content of asbestos below which the product is not deemed to represent a potential hazard; nor has a lower limit to the proportion of crocidolite which is significant been defined for use in conjunction with the U.K. Asbestos Regulations. Accordingly, the primary requirement of the chosen analytical technique is that it should be capable of detecting and identifying small amounts of asbestos (especially crocidolite) in a wide variety of matrices. In view of the practical requirement for the provision of analyses for large

Table 7.1 Some products containing asbestos*

Product	Asbestos type(s)	Asbestos content (%)	Binding material
Rigid lightweight sectional insulations	Amosite	75–85	Sodium silicate
	Amosite, chrysotile	15–30	Calcium silicate
	Amosite, chrysotile	15	Basic magnesium carbonate
Lagging compositions	Amosite, chrysotile, crocidolite (and non-asbestiform fibres)	10–15	Chalk, diatomaceous earth, clays, cement, etc.
Spray insulations	Amosite, chrysotile, crocidolite	$\geqslant$55%	Cement
Asbestos–cement	Chrysotile (mainly); may include $\leqslant$3% amosite and/or crocidolite	10–15	Cement
Insulation boards	Amosite (mainly); may include $\leqslant$10% chrysotile and/or crocidolite	15–35	Calcium silicate or cement plus diatomaceous earth
Asbestos textiles	Chrysotile, crocidolite	99	
Asbestos rope	Any type within a yarn of chrysotile or crocidolite	99	
Millboards	Chrysotile, crocidolite	80–90	Clays, starch
Friction products	Chrysotile	25–70	Phenolic resins, fillers
Reinforced plastics	Amosite, chrysotile, crocidolite, anthophyllite	20–50	Thermoplastics, thermosetting resins
Gaskets, sealants, bitumen products, etc.	Amosite, chrysotile, crocidolite, anthophyllite	Variable	Variable

* Based on information from Dr. A. A. Hodgson, Cape Asbestos Fibres Ltd. (personal communication).

numbers of samples, the technique should be relatively rapid and, if possible, inexpensive.

Several articles reviewing the application of analytical techniques to the identification of asbestos in both bulk materials and in environmental samples have been published. Keenan and Lynch (1970) considered methods for the preparation and analysis of bulk samples, airborne dusts, and lung dusts, their main aim being the qualitative detection and identification of any asbestos present. More emphasis was laid on the quantitative determination of the asbestos content of samples in review articles by Heidermanns (1973) and Heidermanns *et al.* (1976). A wide variety of techniques were reviewed by these authors, including light microscopy, electron microscopy, X-ray diffraction, infrared spectrophotometry, and several methods of elemental analysis.

The problems associated with the analysis of environmental samples for asbestos are frequently different from those associated with the analysis of bulk samples. In the former case problems of identification often stem from the extremely fine diameters of the fibres, and constraints on the methods of analysis which can be successfully applied often arise because of the small amount of sample which is available for analysis; the use of suitable sample preparation techniques is therefore of great importance. In the analysis of bulk materials, on the other hand, the asbestos is frequently present as coarse fibre bundles and shortage of sample is rarely a problem. Perhaps the major problem is the detection and identification of trace amounts of asbestos in mixtures of highly variable composition. The aim of this chapter is to review some of the methods available for the detection and identification of asbestos, specifically in bulk materials, and to provide an assessment of their particular merits and shortcomings.

Chrysotile is commercially by far the most important type of asbestos, constituting about 93% of total world production (Smither, 1970). Of the other types of asbestos, all of which belong to the amphibole group of minerals, only amosite, crocidolite, and anthophyllite are produced in significant amounts and, accordingly, most attention will be paid to the identification of these amphibole asbestos minerals and chrysotile.

(3) SAMPLING AND PRE-TREATMENT OF SAMPLES

The provision of suitable samples for analysis is of vital importance since clearly a non-representative sample or an insufficient number of samples can lead to erroneous conclusions regarding any potential hazard. Such problems arise most frequently when sampling loosely bonded thermal insulation products and, whenever possible, the history of any installation (e.g. repairs or maintenance which may have necessitated the replacement of particular areas of lagging) should be considered and the sampling programme planned accordingly. It is also important to sample the full depth of the insulation since the inner layers may consist of material of a different composition to that exposed at the surface.

It may be useful to subject samples to some sort of physical or chemical pre-treatment before examination by the chosen analytical techniques. For example, treatment with dilute hydrochloric acid will decompose chalk from lagging compositions and free asbestos fibres from other binders such as calcium silicate. Such treatment will not materially affect the amphibole asbestos minerals but will decompose chrysotile asbestos if it is either prolonged or carried out at elevated temperatures. Organic fibres and binders can be removed by ashing the sample at about 400 °C, although amosite and crocidolite will be oxidized and turn brown. Heat treatment will alter the optical properties of asbestos fibres although they can still be identified by other methods. Low-temperature ashing techniques (Gleit and Holland, 1962) may sometimes be useful as such treatment will have little effect on the asbestos minerals. Some samples may require varying degrees of mechanical disaggregation in order to free fibres from the matrix. Filing or drilling, perhaps allied with ultrasonic dispersion, is often useful in this context.

(4) ANALYSIS OF SAMPLES FOR ASBESTOS

A schematic representation of a possible system of analysis, which will be used as a framework within which to discuss the various techniques, is as follows:

1. Observation by stereo-binocular microscope.
2. Observation of the action of heat on fibres.
3. Analysis:
   - (a) Optical methods.
   - (b) Infrared spectrophotometry.
   - (c) X-ray diffraction.
   - (d) Electron-optical methods (including X-ray microanalysis).
   - (e) Other methods.

Preliminary examination of samples during stages 1 and 2 will frequently provide evidence for the presence of fibrous material and may give an indication of its nature. In order to provide a positive identification of the type(s) of asbestos present, it may be necessary to employ several of the techniques listed under stage 3.

4(i) Observation by stereo-binocular microscope

Using fairly low magnifications of about 10–60×, the sample can be carefully examined, under reflected light, using fine tweezers as an aid to the detection of fibres concealed by matrix materials. The range of magnification available on a simple binocular microscope (see, for instance, McCrone, 1974a) is sufficient to permit the observation of the fibre bundles present in most asbestos products and with experience a preliminary assessment of the type(s) of asbestos present is often possible. Fibres of crocidolite can be recognized by their distinctive lavender-blue colour, but it is important to beware of samples which have been subjected to high temperatures since any crocidolite (or amosite) may have been oxidized to brown oxyamphibole. A further complicating factor may be the presence of black fibres of magnetite, which is often associated with veins of chrysotile ore, or even the rare occurrence of blue-dyed chrysotile asbestos. Chrysotile and amosite asbestos both appear as white to pale green fibres under the binocular microscope, but they can sometimes be distinguished on the basis of their morphology: curly chrysotile fibres *vs.* straight, needle-like amosite. The various types of fibre observed can be picked from the sample using fine tweezers, for subsequent analysis.

If a quantitative assessment of the amount of asbestos present is required, it may be possible to make such an estimate using the binocular microscope by reference to artificially prepared mixtures (e.g. asbestos and chalk) of known composition. Such an estimate will be only approximate and any pre-treatment of the sample which may have produced an 'asbestos concentrate' must be taken into account.

Observation of the sample using the stereo-microscope has a number of advantages as a technique for the detection (rather than for the identification) of asbestos. Apart from the advantages of relative simplicity and low cost, the binocular microscope allows the examination of a relatively large sample, thus reducing the problems of obtaining representative analytical results from inhomogeneous products. In addition, the method has good sensitivity, permitting the detection of trace amounts of asbestos, particularly crocidolite, in a variety of matrices. The technique suffers from the disadvantage that it cannot be readily applied to fine-grained materials (e.g. talc, settled dust).

4(ii) The action of heat on fibres

This test can be easily and quickly applied to fibres picked out during the stereo-microscope examination and may provide an indication of their nature. The fibres are heated in a crucible (in air) to dull red heat and any effects observed. Natural or synthetic organic fibres will perhaps fuse before charring or burning, whereas man-made glass fibres will normally fuse to leave a glassy bead. Some fibres are coated with resins which may char before fusion of the glass occurs. Asbestos fibres and man-made ceramic fibres will neither burn nor fuse but the amphibole asbestos minerals, amosite, and crocidolite will be oxidized to leave brown fibres. Chrysotile normally appears to be unchanged by the heat treatment.

4(iii) Optical microscope methods

Before observation in the optical microscope, the sample must be suitably mounted. Fibres of interest should be carefully picked from the bulk sample, placed on a glass slide, immersed in a fluid of known refractive index and covered with a thin slip of glass. The interpretation of optical properties may be complicated either by the presence of matrix material adhering to fibres or by the examination of very thick fibres. It is therefore important to pick fibres that are as clean as possible for examination and, if necessary, to tease coarse fibre bundles apart with fine tweezers. The fibres should be well dispersed through the fluid since identification may be hindered by the presence of overlapping fibres. Identification of asbestos by optical microscopy depends primarily upon the differences which exist in the refractive indices of the various types of fibre, but other related properties, particularly those which can be observed in a polarizing microscope, are also of use for identification. It should be noted that each type of asbestos exhibits a range of optical properties which may depend on its exact chemical composition and on any degradation which the fibres may have suffered, e.g. thermal decomposition. Zielhuis (1977) noted that the use of optical microscope techniques for the identification of asbestos was restricted to fibres with diameters greater than about 5 μm because of the limitations imposed by the difficulties in observing the optical effects on extremely fine fibres. Other authors (Heidermanns, 1973; Keenan and Lynch, 1970) have, however, quoted a lower limit of about 1 μm for the fibre diameter which can be identified by optical methods. The problems associated with the identification of such fine fibres are more commonly encountered during the identification of respirable fibres in samples of airborne dust, when the fibres of interest are generally <3 μm in diameter by definition, rather than in the analysis of bulk samples.

The phase-contrast method of observation, which is widely used for fibre counting, has been used for the identification of fibrous mineral particles from human lungs (Goni *et al.*, 1975) but the technique is perhaps more flexible when used in conjunction with polarizing microscopy (Schmidt, 1960). The interference microscope (Chao, 1976; Dodgson, 1963) may also have application to the identification of asbestos, but the best documented techniques at present seem to be polarizing microscopy, which is widely used by geologists and mineralogists, and dispersion staining methods, which have been applied to the identification of asbestos by Julian and McCrone (1970).

4(iii).1 *Polarizing microscopy*

Detailed discussion of the construction and use of the polarizing microscope is beyond the scope of this chapter and reference should be made to one of the standard texts, e.g. Bloss (1961) or Hartshorne and Stuart (1970). In order to clarify the discussion below, some features of a polarizing microscope are indicated in Figure 7.1. Fibres of asbestos are biaxial crystals possessing three refractive indices, α, β, and γ. By convention, the symbol α is used to refer to the smallest refractive index, γ to the largest, and β to the index with a particular value intermediate between α and γ. Each refractive index is associated with a particular vibration direction in the crystal. In practice, the index along the fibre length (n_L in Figure 7.2) is normally close to α or γ, while the index measured across the fibres (n_B in Figure 7.2) has some value intermediate between β and γ or between β and α. The values of these indices and the relationship between the crystal morphology provide a basis for identification. The optical properties are well documented for the appropriate mineral species (see, for instance, Deer *et al.*, 1966) and are summarized in Table 7.2.

The observations which may be useful in the identification of asbestos can be divided into two groups:

(a) *Observations under plane-polarized light* (*analyser withdrawn*):
- (i) observation of fibre morphology;
- (ii) observation of colour and pleochroism;
- (iii) determination of refractive indices.

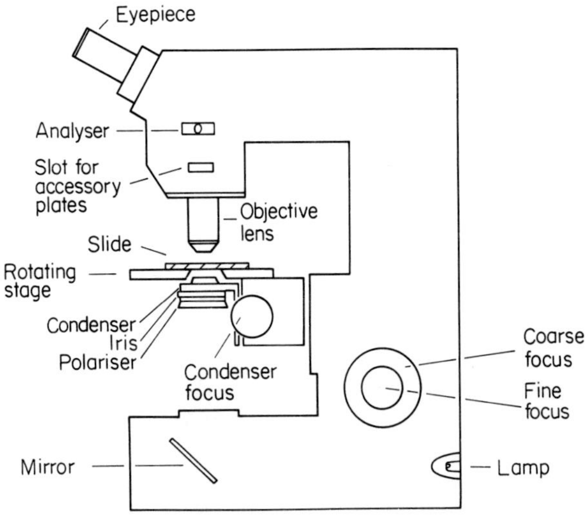

Figure 7.1 Some features of the polarizing microscope

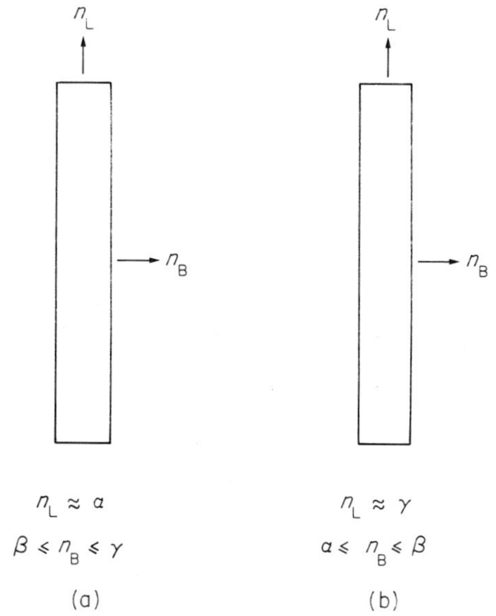

Figure 7.2 Some optical features of a biaxial fibre (a) length fast (negative elongation); (b) length slow (positive elongation)

(b) *Observations under crossed polars* (*analyser inserted*):
  (i) assessment of birefringence;
  (ii) determination of sign of elongation.

(a) *Observations under plane-polarized light.* The morphology of the fibres may in itself provide a strong indication of the type of asbestos present. In particular, the curliness of chrysotile fibres contrasts with the sharp, needle-like shape of the amphibole asbestos minerals. Glass fibres can often be readily distinguished by their monofilamentous character (i.e. lack of cleaved, splayed ends) and the marked parallelism of their sides.

The colours and pleochroic characteristics of the different types of asbestos are given in Table 7.2, but the most useful feature is the distinctive colour and pleochroism of crocidolite. Fibres of this mineral range in colour from deep blue when the fibre length is lying parallel to the vibration direction of the polarizer, to pale blue when the fibre is lying perpendicular to the polarizer. The relationship between the refractive index of the fibre (in the direction parallel to the polarizer vibration direction) and the refractive index of the fluid in which the fibre was immersed on the slide may be determined using the Becke line test (see, for example, Hartshorne and Stuart, 1970, p. 259). If the microscope tube is raised slightly, relative to the sample, from the focused position then a bright halo will be observed round the fibre. This halo will appear to move into or away from the fibre

Table 7.2 Optical properties of some asbestos minerals*

Property	Type of asbestos			
	Chrysotile	Amosite	Crocidolite	Anthophyllite
Colour	Colourless to pale green	Colourless to pale fawn	Blue	Colourless to pale yellow or green
Pleochroism	Nil	Slight	Strong (pale grey–blue to dark blue)	Slight
Refractive indices:				
α	1.493–1.553	1.657–1.688	1.685–1.698	1.578–1.652
γ	1.517–1.557	1.675–1.717	1.689–1.703	1.591–1.676
Birefringence	Moderate	Strong	Weak (often masked)	Moderate
Orientation	Length slow	Length slow	Length fast	Length slow
Dispersion staining colours:				
Fibre parallel to vibration, direction of polarizer	Blue	Red	Red–blue	Green–blue
Fibre perpendicular to vibration direction of polarizer	Red	Yellow	Red–blue	Yellow
Refractive index of immersion fluid	1.560	1.670	1.700	1.610

* Data from Deer *et al.* (1966), Hodgson (1965), and Julian and McCrone (1970).

when the tube is raised, according to whether the refractive index of the fluid is less than or greater than that of the fibre (in the direction of the permitted vibration of the polarizer). Thus, by examination of a series of mounts, the refractive indices may be suitably bracketted. An alternative to the Becke line test is to employ positive phase-contrast illumination in conjunction with plane-polarized light (Schmidt, 1960). In this case, provided that the difference in the refractive indices of the fibre (n_{Fib}) and fluid (n_{Fl}) was small, Schmidt observed the following effects:

$n_{Fl} > n_{Fib}$	fibre white–pale blue with dark halo
$n_{Fl} = n_{Fib}$	fibre delicate blue
$n_{Fl} < n_{Fib}$	fibre distinctly blue with orange halo
$n_{Fl} \ll n_{Fib}$	fibre grey with white halo

When the refractive index difference between fluid and fibre was very large, Schmidt reported that the fibres appeared yellow.

(b) *Observations under crossed polars.* In this configuration of the microscope, only light which has passed through an anisotropic object (e.g. an asbestos fibre) will be transmitted by the analyser, so that the fibres will be seen against a dark background. Depending on its birefringence (the magnitude of the difference between n_L and n_B in Figure 7.2) and thickness, a fibre will exhibit a particular interference colour when it is oriented at about 45° to the vibration directions of the polarizer and analyser. The higher the birefringence and the thicker the fibre, the higher will be the order of the interference colour. In general, for a given thickness of fibre, amosite will give higher order interference colours than chrysotile, which in turn gives higher order colours than crocidolite. Observation of the birefringence of crocidolite is difficult because its blue colour often leads to the appearance of anomalous interference colours. It may be noted that glass fibres are normally isotropic and therefore appear dark in all orientations under crossed polars.

A useful test which can easily be performed under crossed polars is the determination of the sign of elongation. This expresses the relationship between the refractive indices along (n_L) and across (n_B) the fibre (see Figure 7.2) and can be determined with the aid of an accessory plate such as a sensitive tint (first-order red) plate (see, for instance, Hartshorne and Stuart, 1970, p. 292, for details). The crystal is said to be length slow and to possess positive elongation when $n_L > n_B$, or to be length fast and to possess negative elongation when $n_L < n_B$. Crocidolite is the only asbestos mineral which is length fast, so that this test may provide confirmation of its presence and aid in particular its distinction from amosite which has similar morphology and refractive indices.

4(iii).2 *Dispersion staining methods*

Dispersion colours are produced when a particle is immersed in a liquid which has a similar refractive index but which shows significantly greater dispersion of the refractive index with wavelength. The refractive index of the mounting fluid (n_{Fl}) should therefore be close to that of the fibre to be identified, so that it may be necessary to prepare a series of mounts in appropriate liquids: e.g. for chrysotile $n_{Fl} = 1.56$; for anthophyllite $n_{Fl} = 1.61$; for amosite $n_{Fl} = 1.67$; for crocidolite $n_{Fl} = 1.69$. Procedures for the observation of dispersion colours in the microscope and their use in the identification of mineral particles were described by Brown and McCrone (1963) and Brown *et al.* (1963). Normally, the refractive index for both solids and liquids is higher at shorter wavelengths and a solid will show less dispersion of the index than a liquid of similar refractive index. Hypothetical dispersion curves for a fluid and a solid immersed in that fluid are shown in Figure 7.3. It can be seen that the refractive indices of the solid and liquid will be matched for a particular wavelength of light, λ_0. Light of this matching wavelength will pass undeviated through the particle and liquid, while light of other wavelengths such as λ_1 and λ_2 will be deviated (see Figure 7.4).

The focal screening method, attributed by Brown and his co-workers to Cherkasov (1960), permits the observation of either the undeviated or the deviated light*. If a suitable annular screen is placed in the back focal plane

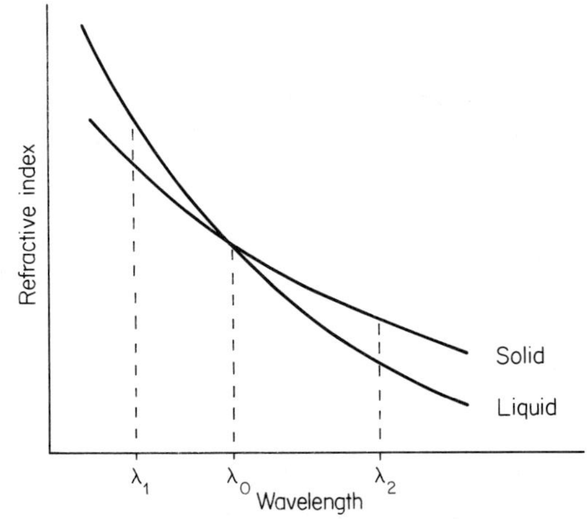

Figure 7.3 Hypothetical dispersion curves for a liquid and solid. (After Brown and McCrone (1963); courtesy of Microscope Publications Ltd.)

* An objective with the appropriate focal screens is supplied by McCrone Research Institute Ltd., 2 McCrone Mews, Belsize Lane, London NW3 5BG.

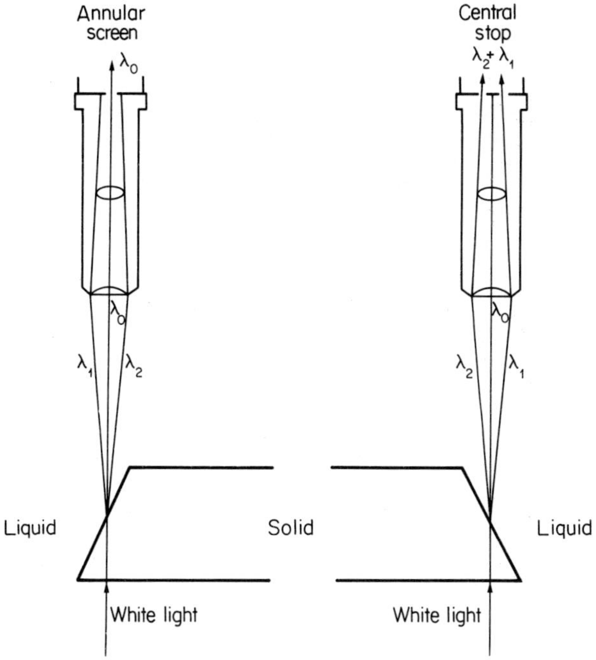

Figure 7.4 Schematic light paths through the liquid and solid of Figure 7.3. (Adapted from Brown and McCrone (1963); courtesy of Microscope Publications Ltd.)

of the objective, then only undeviated light will be observed. If the sample is illuminated by parallel white light, then the edges (i.e. surfaces not normal to the incident light) of the object take on the colour of light of wavelength λ_0. In areas of the solid which present surface normal to the incident light and in areas where no particle is present, light of all wavelengths will pass through the sample undeviated, so that these areas will appear bright by white light. If the annular screen is replaced by a central stop of suitable dimensions, then the image will be formed only by deviated light of wavelengths such as λ_1 and λ_2, and the colour observed will be that formed by the subtraction of light of wavelength λ_0 from white light, i.e. complementary to the colour observed with the annular screen. When the central stop is used the particles will be seen on a dark field.

The specific application of dispersion staining techniques to the identification of asbestos was discussed by Julian and McCrone (1970) and McCrone (1974b). These authors used high-dispersion Cargille* refractive index liquids and the basis of their procedure was to note the matching wavelength,

* R. P. Cargille Laboratories, Inc., 33 Factory Street, Cedar Grove, New Jersey, U.S.A.

λ_0, for each of the different asbestos types, each in a series of liquids. They found that the central stop method of observation gave better resolution and contrast when observing fine fibres.

Brown and McCrone (1963) established dispersion curves for the various types of asbestos and concluded that for identification purposes a series of liquids could be chosen, each of which would impart a characteristic blue magenta colour ($\lambda_0 = 550$–650 nm) to only one variety of asbestos. The use of a microscope with a substage polarizer for the observation of dispersion colours has the advantage that the dispersion characteristics appropriate to the different refractive indices along and across the fibre can be observed by rotation of either the fibre or the polarizer. The lower refractive index (across the fibre, except in the case of crocidolite) will show dispersion colours appropriate to a higher value of λ_0. Thus the sign of elongation [see Section 4(iii).1] can be determined, which may aid in the distinction between amosite and crocidolite.

McCrone (1974b) discussed in more detail the practical application of the method to the identification of asbestos and recommended that standard samples should be prepared for comparison purposes. Each standard should be mounted in a liquid of the same refractive index as that in which the unknown is mounted. McCrone also drew attention to the similarity between the dispersion colours obtained from chrysotile and those from quartz, paper fibres, and talc, further emphasizing the need for suitable standards for direct comparison.

In summary, it may be said that optical methods afford the possibility of detecting and identifying small amounts of asbestos fibre in solid samples. The methods demand a certain degree of expertise if they are to be applied successfully, but they can be used to identify asbestos in mixtures and in fine dusts. A possible disadvantage stems from the fact that the volume of sample examined is very small so that non-representative results may be obtained from inhomogeneous samples. Another consequence of the small amount of sample examined is that it is very difficult to achieve quantitative analysis of bulk materials by optical methods. The need to produce a dispersed fibre sample free of matrix material also contributes to the difficulties in quantitative analysis.

4(iv) Infrared spectrophotometry

The applicability of infrared (IR) spectrophotometry to the identification of minerals was demonstrated by Hunt *et al.* (1950), who used it to study the mineralogy of fine-grained rocks, and more recently by Farmer (1974) in a comprehensive monograph on the subject. Taylor *et al.* (1970) showed that data from IR studies provided a valuable adjunct to information from X-ray diffraction analysis in problems of mineral identification. These authors also considered that the IR method used in isolation was helpful in the analysis of dust samples in the field of occupational health, and they published a series of IR spectra which included those of the commonly used varieties of asbestos. Parks (1971) applied IR spectrophotometry specifically to the identification of asbestos. He noted that the distinction between chrysotile and the amphibole group of asbestos minerals was much sharper than that between the individual amphiboles, but concluded that the method enabled the positive identification of the asbestos minerals.

A commonly used method of presenting the sample to the IR spectrophotometer is in the form of a pressed alkali halide (e.g. potassium bromide) disc. Suitable fibres of interest (*ca.* 250 μg) are picked from the bulk sample after examination by stereo-binocular microscope and mixed thoroughly with the halide carrier. Using a die and press, the mixture is then made into a disc which can be placed in the sample beam of the instrument and its infrared absorption spectrum recorded.

The asbestos minerals are hydroxylated silicates and show two main absorption regions, one associated with stretching vibrations of Si–O groups in the structure, at wavenumbers of *ca.* 850–1150 cm^{-1}, and the other associated with stretching vibrations of OH groups (*ca.* 3500–3700 cm^{-1}). In addition, the amphiboles show absorption bands at wavenumbers $<800\ cm^{-1}$, which have been assigned to vibrations of the silicate chain and to metal–oxygen stretching and Si–O bending vibrations (Patterson and O'Connor, 1966). The exact positions of the absorption bands depends on the influence of cations in the structure, so that fibres of different compositions will have different IR spectra. The spectra shown in Figure 7.5 confirm the similarity noted by Parks between the spectra of the various amphiboles compared with that of chrysotile, but careful examination of the spectra of unknown samples will usually permit a positive identification to be made.

The usefulness of the IR method for identification may be seriously impaired by the presence of interfering substances (Bagioni, 1975). It is therefore important, as with the optical methods already discussed, to select fibres for analysis which are free of matrix material. Difficulties in the interpretation of spectra may also be encountered when mixtures of different types of asbestos (especially two or more amphiboles) are present in the halide disc. This problem can normally be avoided, however, by careful selection of fibres for analysis.

Infrared spectrophotometry has been used for the quantitative analysis of single varieties of asbestos used in animal experiments (Beckett *et al.*, 1975) and for the quantitative analysis of chrysotile asbestos in airborne samples (Gadsen *et al.*, 1970). Use of the method for the quantitative analysis of solid materials such as lagging compositions is not practicable, however, since it is not possible to analyse the bulk material because of the difficulties associated

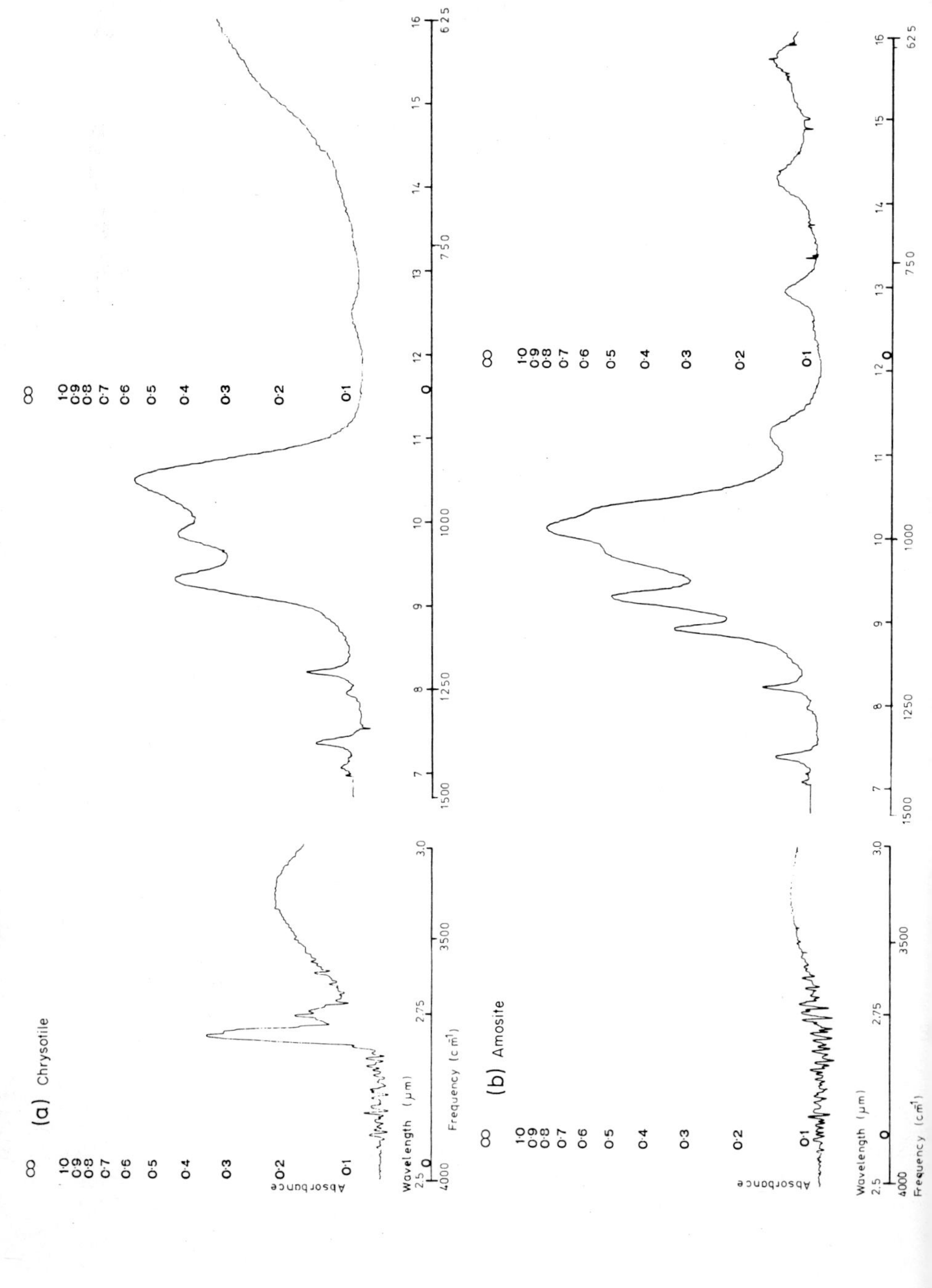
(a) Chrysotile
(b) Amosite
Absorbance
Wavelength (μm)
Frequency (cm⁻¹)
∞
1·0
0·9
0·8
0·7
0·6
0·5
0·4
0·3
0·2
0·1
0
2.5
2.75
3.0
7
8
9
10
11
12
13
14
15
16
4000
3500
1500
1250
1000
750
625

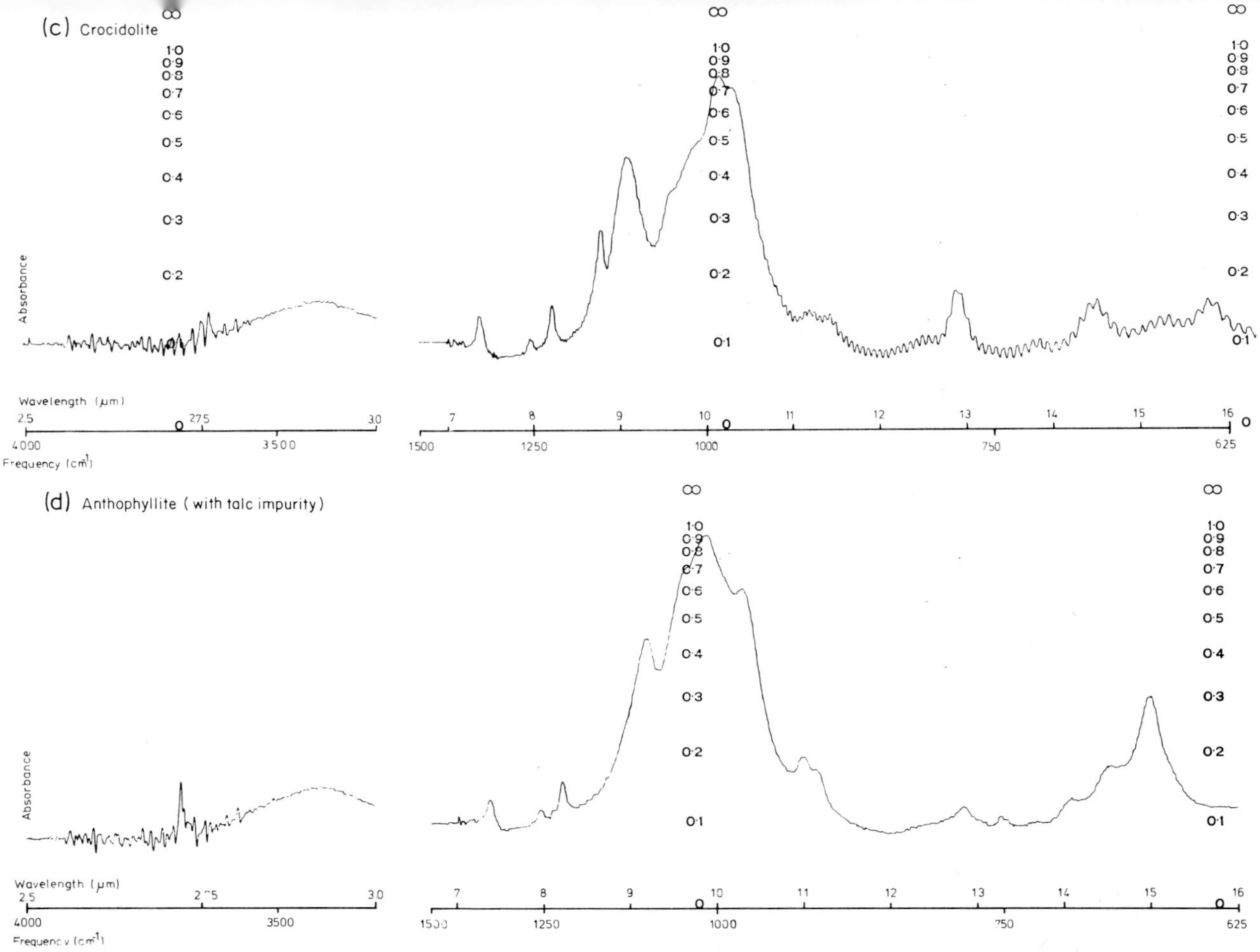

Figure 7.5 Infrared spectra of some asbestos minerals: (a) chrysotile; (b) amosite; (c) crocidolite; (d) anthophyllite (with talc impurity)

with interference and the presence of mixtures of asbestos types mentioned above.

The method has the advantage that it is a relatively rapid form of analysis and requires only a small amount of sample. The apparatus required, although more costly than an optical microscope, is considerably cheaper than the equipment required to carry out X-ray diffraction analysis or electron microscope investigations. In common with the optical methods described above, a certain degree of experience is desirable in the interpretation of the spectra. Perhaps the main drawback to the method, however, is its susceptibility to interference.

4(v) X-ray diffraction analysis

X-ray diffraction (XRD) techniques have been applied extensively to the qualitative and quantitative analysis of minerals and reference should be made to one of the standard texts (e.g. Klug and Alexander, 1954; Nuffield, 1966) for discussion of the theory and practice of X-ray diffraction. The application of XRD to the analysis of industrial dusts was discussed by Talvitie and Brewer (1963b) and Nenadic and Crable (1971). The asbestos minerals each possess distinct crystalline structures, based upon the ordered arrangement of atoms, and are therefore amenable to analysis by XRD. X-ray powder diffraction data for the asbestos minerals have been published (Berry, 1974), although entries for amosite and crocidolite were not included. Instead, reference must be made to the corresponding non-asbestiform mineral species (i.e. grunerite and riebeckite). The observed relative intensities of the diffraction peaks may differ from those quoted in the data book because of the effects of preferred orientation of the asbestos fibres (see below). Representative diffraction patterns are shown in Figure 7.6. Analytical procedures using X-ray diffraction for the identification and quantitative determination of asbestos have been described by several authors (e.g. Champness *et al.*, 1976; Crable, 1966; Goodhead and Martindale, 1969; Rickards, 1972).

The sample for analysis may be in the form of a fine power or a single fibre. The former technique of sample preparation is the most useful in occupational hygiene investigations, particularly in the study of dusts. Single crystal methods, in particular the photographic recording of fibre diffraction patterns, however, would be useful for the identification of individual fibres of asbestos. Such methods, however, require skilled experimental technique and careful interpretation of the resulting diffraction patterns; they are not applicable to quantitative analysis and will not be considered further. There are two well known techniques by which the diffraction pattern from a powdered sample may be recorded: photographically in an X-ray powder camera or graphically on to a strip chart using a diffractometer. A method for the quantitative determination of amosite and chrysotile using photographic recording was described by Goodhead and Martindale (1969). Since their procedure was aimed particularly at the analysis of samples of airborne dust, the amount of material analysed was small (5–30 mg). However, provided that the problems of obtaining a representative sample could be overcome, the method would be equally applicable to the analysis of bulk samples. The authors ground the sample in a mill with silicon as an internal standard (see below), and mounted the resulting powder in a 0.3 mm diameter lithium-glass tube. The intensities of the chosen analytical lines of the standard (I_R) and the asbestos (I_A) were measured on the films using a microdensitometer. The ratio I_A/I_R was then compared with standard graphs prepared from the analysis results of artificial mixtures, and the amount of asbestos in the sample determined.

The use of photographic techniques has the advantage that it provides a record which is easily stored or compared qualitatively with other patterns for identification purposes. The need for photographic processing, which is both time consuming and a possible source of error in quantitative analysis, and the use of a microdensitometer for any quantitative assessments, are, however, significant disadvantages. For these reasons the diffractometer has to a large extent replaced the powder camera as a method of recording diffraction patterns, particularly when quantitative analysis is envisaged. Crable (1966) and Hayashi (1973) described diffractometer techniques for the identification and quantitative determination of chrysotile, amosite, and crocidolite. Although Hayashi was mainly concerned with the analysis of airborne samples taken on glass-fibre filters, the methods of analysis were essentially the same as those of Crable, whose technique included provision for the analysis of settled dusts and bulk samples. In both cases the samples were mounted on filters [either membrane (Crable) or glass-fibre (Hayashi)] and the quantitative determinations were based on the use of external standards. Crable used a technique described by Talvitie and Brewer (1962a) to obtain a representative sample for analysis on a membrane filter. After preliminary grinding in a mill, a known weight of material was dispersed with the aid of a wetting agent and ultrasound in 250 cm^3 of water; a known aliquot was then taken and filtered through a membrane filter. Standard filters with known amounts of chrysotile, amosite, and crocidolite were prepared by the same technique. The areas (in counts) of the chosen analytical peaks were measured and calibration graphs plotted. Normally the strongest diffraction line of each type of asbestos was chosen for analysis, and measurement of the appropriate peak area in the sample pattern gave the weight of asbestos on the filter directly from the calibration graph.

In two later papers (Crable and Knott, 1966a,b) Crable's group described calibration techniques using an internal standard. In this work the sample

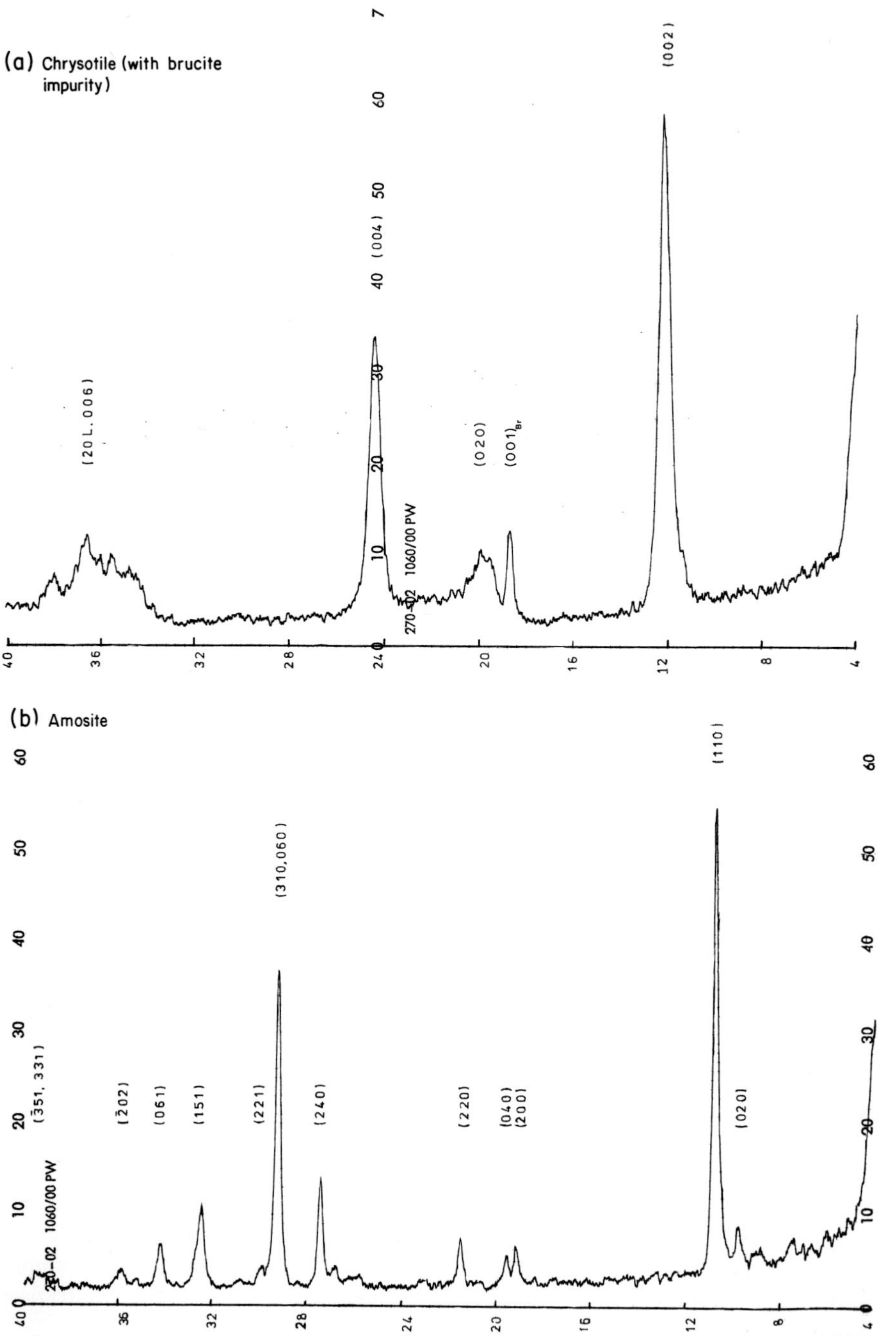

Figure 7.6 X-ray powder diffraction patterns of some asbestos minerals (recorded purity); (b) amosite; (c) crocidolite; (d) anthopyllite (with talc impurity). The chrysotile, 21–543; amosite (Grunerite), 17–725; crocidolite (Riebeckite),

(c) Crocidolite

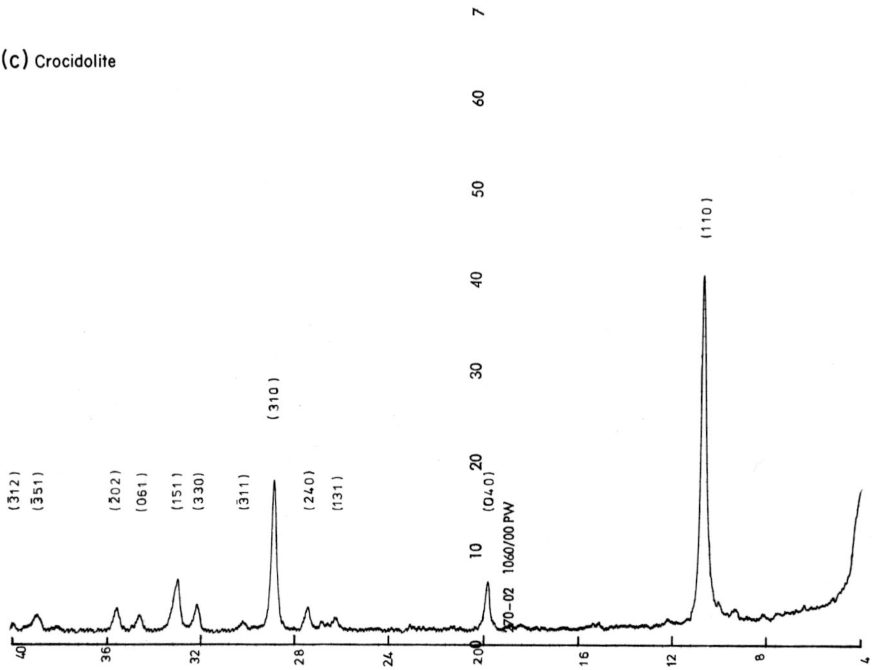

(d) Anthophyllite (with talc impurity)

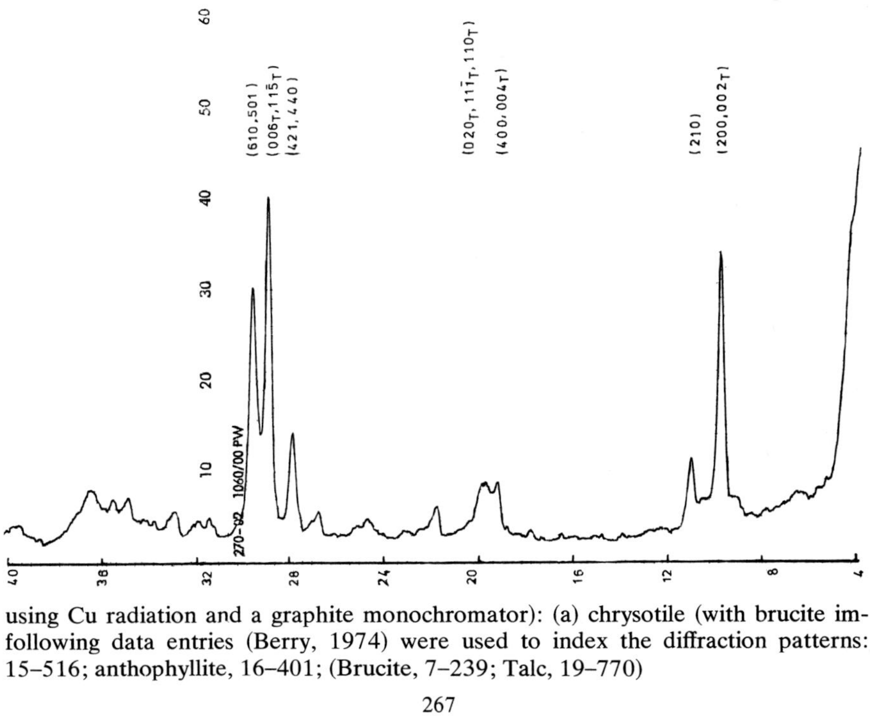

using Cu radiation and a graphite monochromator): (a) chrysotile (with brucite im-
following data entries (Berry, 1974) were used to index the diffraction patterns:
15–516; anthophyllite, 16–401; (Brucite, 7–239; Talc, 19–770)

was dry ground in a mill, mounted in a standard diffractometer cavity holder, and a rapid, qualitative diffraction trace recorded. From this trace the presence and type of asbestos was established and an approximate estimate made of the proportion present. A mixture of the ground sample with either aquamarine (for determination of chrysotile) or quartz (for determination of amosite or crocidolite) was then prepared, the proportion of internal standard being approximately the same as the proportion of asbestos. In the diffractometer this sample was scanned in the region of the asbestos and internal standard peaks and the counts recorded digitally. By reference to standard graphs prepared by the analysis of known mixtures, the proportion of asbestos in the sample was determined.

Although the absolute limit of detection of asbestos is of the order of 10 μg (Hayashi, 1973; Rickards, 1972) using the direct on-filter analysis of dusts and an external standard, the limiting factor for the detection of asbestos in bulk samples is the lowest detectable *proportion* of asbestos in a mixture. Crable and Knott (1966a,b) were unable to detect asbestos in artifically prepared mixtures containing 1–2% of asbestos, and their results indicated that for reliable quantitative determinations the sample should contain $\geqslant$15–20% of asbestos. The application of XRD to the routine identification of asbestos may suffer from problems of interference by other materials in the sample. For example, other members of the serpentine group of minerals (antigorite, lizardite) will interfere with the strong basal reflections of chrysotile. Another possible interference of note is the strong (006) peak of talc, which may be confused with the (310) peak of crocidolite. Examination of the full pattern will normally allow these problems to be resolved.

As was noted earlier, some asbestos products submitted for analysis will have been subject to prolonged high temperatures, which may have oxidized the iron(II) in amosite and crocidolite to produce brown oxy-amphiboles. The oxy-amphiboles have slightly smaller unit cell parameters (and hence slightly different diffraction patterns) than those of the corresponding unaltered species (Hodgson *et al.*, 1965). This problem was discussed by Plowman (1973), who investigated the effect of heat on the *d*-spacings of amosite and crocidolite. In particular he noted that the strong (310) reflection, the position of which is commonly used as a basis for distinguishing amosite and crocidolite, was shifted to such an extent that the (310) reflection of crocidolite was at a spacing appropriate to unheated amosite. In order to avoid confusion of this sort, it is important, once again, to consider the diffraction pattern as a whole, rather than to rely on particular diagnostic peaks.

A further, and potentially more serious, problem particularly if quantitative analysis is to be attempted arises from the effect of grinding on the asbestos minerals. Ocella and Maddalon (1963) showed that dry grinding of asbestos samples in an agate mortar, even for short periods, resulted in a marked decrease in the intensities of the diffraction peaks. For example, after grinding for 4 min the following reductions in intensity (relative to the traces recorded after grinding for 1 min) were noted: chrysotile (004), 45%; amosite (310), 47%; and crocidolite (310), 42%. These observations were confirmed by Heidermanns (1973) for chrysotile and crocidolite.

Preferred orientation of fibres of asbestos may arise during sample preparation and will change the relative intensities of the various diffraction peaks in the trace. Such orientation effects are notoriously difficult to control when preparing mounts of platy of fibrous particles, and differing degrees of orientation between samples and standards may lead to serious inconsistency in quantitative analysis, and in extreme cases even to difficulties in the qualitative interpretation of patterns.

To summarize, X-ray diffraction is a useful technique for the detection and identification of the asbestos minerals in solid materials. The method is capable of distinguishing between chrysotile, amosite, and crocidolite, even in samples which have been subject to high temperatures, which is difficult to achieve by any other method. Although the method is potentially well suited to the quantitative determination of asbestos in mixtures, the results need to be interpreted with caution because of the possible effects of interfering substances, decrease in intensity on grinding and preferred orientation incurred during sample preparation. The method also suffers from the disadvantage of a rather high detection limit of *ca.* 1%, long analysis times, even for purely qualitative assessment of samples, and high capital cost.

4(vi) Electron-optical methods

Resort to the use of electron-optical methods for the identification of asbestos is perhaps more common in problems associated with the examination of airborne or waterborne pollutants than in the evaluation of bulk materials. Nonetheless, such techniques, particularly when used in conjunction with X-ray microanalysis, can be of value in the detection and identification of asbestos in fine powders and settled dusts. Observations may be carried out in either a transmission or a scanning electron microscope. The principles and use of these instruments have been discussed in standard texts such as those by Agar *et al.* (1974) and Hearle *et al.* (1973).

Skikne *et al.* (1971) examined preparations of the UICC standard reference samples of asbestos (Timbrell *et al.*, 1968) by transmission electron microscopy (TEM) and published electron diffraction patterns of the various asbestos types. The distinctive appearance of chrysotile in the TEM, in particular its tubular character, and its characteristic electron diffraction pattern [which includes reflections which are streaked along the layer lines

(Whittaker and Zussman, (1971)] permitted the ready identification of this type of asbestos by Skikne *et al.* The relative ease of identification of chrysotile in the TEM was utilized by Pooley (1972) in a study of inhaled chrysotile asbestos recovered from human lung tissue and by Rickards (1973) in the development of a technique for the determination of submicrogram amounts of chrysotile asbestos collected on membrane filters.

The strong similarity between the diffraction patterns of the amphibole asbestos minerals was noted by Skikne *et al.* (1971), who concluded, however, that in most cases the various asbestos minerals could be distinguished. A similar conclusion was reached by Clark and Ruud (1974) and Ruud *et al.* (1976), who published series of matching electron photomicrographs and electron diffraction patterns of asbestos and other materials. On the other hand, Langer *et al.* (1974), Pooley (1975), and Beaman and File (1976) argued that the diffraction patterns from the amphibole asbestos minerals cannot always be reliably distinguished. These authors advocated the use of the additional information on elemental composition, available from energy-dispersive X-ray microanalysis (see below) in the TEM. In a thorough review of the application of TEM/X-ray microanalysis methods, Champness *et al.* (1976) showed that it may frequently be necessary to base identification on all the information available, viz. fibre morphology, electron diffraction characteristics and chemistry (derived from X-ray microanalysis). The use of a scanning electron microscope (SEM), combined with energy-dispersive X-ray microanalysis, was reported by Rubin and Maggiore (1974). These authors found difficulty in distinguishing chemically between chrysotile and anthophyllite although the different morphologies of these types of asbestos will usually permit their distinction. The use of the SEM for microanalysis may present problems related to the influence of sample X-ray mass absorption and fluorescence on the observed X-ray intensities (see below). On the other hand, the SEM has the distinct advantage that a minimum of sample preparation is required because of the flexibility of sample handling of these instruments. This is in contrast to the requirement to transfer any sample to a small grid for examination by TEM, and to ensure the adequate dispersion of particles across the grid.

X-ray microanalysis in the electron microscope

Electron-beam X-ray microanalysis is widely used by metallurgists and geologists in the form of the electron microprobe (Reed, 1975). Such instruments traditionally used a wavelength-dispersive crystal spectrometer to separate the characteristic X-rays excited by an electron beam from a flat, polished sample. The technique in this form has two significant drawbacks in the application to the problems of the routine identification of asbestos. The first is the requirement to present a polished specimen and the second is the time taken to scan with the crystal spectrometer the X-ray spectrum for the elements of possible interest. The advent of energy-dispersive X-ray detectors (Russ, 1972) which allow the rapid, simultaneous collection of X-ray spectral data for all the major elements of atomic number 11 or higher, and the combination of such detectors with electron microscopes (either TEM or SEM), permit the rapid chemical analysis of single fibres of asbestos. The restrictions on the use of this technique for quantitative analysis were discussed by Champness *et al.* (1976). If the fibre to be analysed is sufficiently thin, then the effects of X-ray absorbance and fluorescence by the sample can be neglected. In this case quantitative analysis can be achieved, with suitable standards, by using a simple ratio technique. If, on the other hand, the fibre is sufficiently thick and its surface flat, then the corrections applied in normal microprobe analysis (ZAF procedures) may be used. In the TEM the selection of suitably thin fibres is relatively straightforward, whereas in the SEM this requirement is difficult to achieve, so that reliable quantitative analytical results will not always be obtained.

Thus, electron-optical methods combined with X-ray microanalysis may be useful on occasion for the identification of asbestos in bulk samples, but for routine application such techniques have significant disadvantages. In addition to the relatively lengthy time taken for sample preparation and analysis, other drawbacks stem from the degree of sophistication of electron microscopes, viz. high cost and the need for skilled operation and interpretation of results. The provision of quantitative estimates of the asbestos content of samples is difficult because of the small amount of material examined, particularly in the TEM.

4(vii) Miscellaneous methods of analysis

4(vii).1 *Elemental analysis*

The elemental analysis of bulk materials can be carried out by using any of several physical analysis techniques (Zussman, 1967), including atomic-absorption spectrophotometry, optical emission spectrography, neutron-activation analysis and X-ray fluorescence, or by classical wet-chemical analysis. Of these methods, atomic-absorption spectrophotometry seems to have been applied most commonly to the analysis of asbestos. Keenan and Lynch (1970) described a technique using quantitative analysis for magnesium by atomic absorption to provide an estimate of chrysotile in airborne samples. The method possessed very good sensitivity and accuracy, and duplicate analysis of samples by atomic absorption and X-ray diffraction were in good agreement.

Table 7.3 Some aspects of possible analytical methods for asbestos in bulk material

Analytical technique	Approximate minimum cost of equipment (£)	Time taken for sample preparation and analysis (min)	Approximate size of sample analysed (mg)	Suitability for approximate quantitative analysis
Stereobinocular microscope	300	20–30	Variable	Good
Optical microscopy (polarizing; dispersion staining)	500–1000	20–30	<0.5	Fair
Infrared spectrophotometry	5000	20–30	<1	Poor
X-ray diffraction	25 000	30–60	<500	Good
Electron microscopy:				
Transmission	30 000	60	<0.001	Poor
Scanning	20 000	30	<0.100	
(for X-ray microanalysis add:	15 000			
Differential thermal analysis	3000	90	<500	Poor–good
Atomic-absorption spectrophotometry	3000	Variable	Variable	Poor–good

Heidermanns (1973) also applied atomic-absorption analysis for magnesium to the determination of chrysotile, in this instance in asbestos–cement. However, Heidermanns noted that the method was only of limited application to the routine analysis of chrysotile in bulk samples, because certain products contain magnesium-rich additives, and thus analysis of the bulk material for magnesium would lead to an overestimate of the amount of asbestos present. Similar problems would apply to any method relying on the elemental analysis of the bulk material, and the detection and differentiation of the amphibole asbestos minerals, especially in mixtures, is likely to be difficult because of the relatively minor chemical difference between these minerals.

4(viii).2 *Thermal analysis*

The asbestos minerals undergo certain changes when they are progressively heated, and these changes may be monitored in two ways. During thermogravimetric analysis (TGA), any changes in weight are recorded as the sample is heated. In a second technique, differential thermal analysis (DTA), a sample cell and a reference cell are heated simultaneously and any difference in temperature between the two cells is recorded. These techniques can give information on the temperature at which volatiles (especially water) are lost from the structure, and on the temperatures of any exothermic or endothermic changes which may occur on heating. Hodgson (1965) published data which indicated that TGA and DTA might enable the asbestos minerals to be differentiated, but there is only limited published information concerning the sensitivity of these techniques and the interfering effects which may arise from contaminating materials or from mixtures of different types of asbestos. However, Schelz (1974) described the use of DTA for the detection of chrysotile asbestos in talc and using this method he was able to detect $\geq 1\%$ by weight of chrysotile. However, at these levels he found that the DTA response was not specific to chrysotile, similar thermograms being obtained from other serpentine minerals, e.g. antigorite. Thus, any sample showing evidence of serpentine in its DTA trace would need to be examined in a microscope in order to confirm the presence of chrysotile fibres.

(5) CONCLUSION

The use of simple analytical techniques such as the stereo-binocular microscope and optical microscope, in conjunction with the observation of the action of heat on any fibres detected, will in most instances be sufficient to provide information on the presence of mineral fibres. In addition, the use of these techniques will usually permit the positive identification of any asbestos fibres present when sufficient material is available to allow several tests (e.g. the preparation of sub-samples mounted in several fluids of known refractive indices) to be carried out on the same bulk sample. As was noted in the Introduction, the amount of material available from samples of airborne dust will normally be insufficient for the full range of optical tests to be applied and problems may be encountered in the distinction of asbestos from other natural and man-made mineral fibres (e.g. ceramic fibre, calcium sulphate, various amphibole and pyroxene minerals, palygorskite, and sepiolite). The introduction of novel man-made mineral fibres (MMMF) as substitutes for asbestos and for new applications may add to these problems, so that the positive identification of mineral fibres may only be possible by the extensive use of sophisticated electron-optical techniques. Judgement will need to be exercized on whether the cost of such determinations, in terms of equipment and personnel, is worthwhile, particularly in the light of increasing evidence that it is the physical character of the fibres, rather than their chemical composition, which is the controlling factor in their ability to cause disease, particularly mesotheliomata. Muir (1976) suggested that any material containing durable fibres of a potentially respirable size, especially those 'having a diameter less than 0.5 μm and a length greater than 10 μm should be regarded as a possible cause of mesothelioma until proved otherwise'. In this case it may not be necessary for the analyst to provide detailed identification of the fibrous material present in any sample, but simply to be able to detect small amounts of mineral fibre (either asbestos or MMMF) which are likely to give rise to these extremely fine fibres when the material is disturbed or processed.

(6) REFERENCES

Agar, A. W., Alderson, R. H., and Chescoe, D. (1974) Principles and practice of electron microscope operation, in *Practical Methods in Electron Microscopy* (Ed. A. M. Glauert), Vol. 2, North-Holland, Amsterdam.

Bagioni, R. P. (1975) Separation of chrysotile asbestos from minerals that interfere with its infra-red analysis, *Envir. Sci. Technol.*, **9,** 262–263.

Beaman, D. R., and File, F. M. (1976) Quantitative determination of asbestos fibre concentrations, *Analyt. Chem.*, **48,** 101–110.

Beckett, S. T., Middleton, A. P., and Dodgson, J. (1975) The use of infra-red spectrophotometry for the estimation of small quantities of single varieties of UICC asbestos, *Ann. Occup. Hyg.*, **18,** 313–320.

Berry, L. G. (Editor) (1974) *Selected Powder Diffraction Data for Minerals, Data Book*, 1st Ed., Joint Committee on Powder Diffraction Standard, Swarthmore, Pennsylvania.

Bloss, D. (1961) An Introduction to the Methods of Optical Crystallography, Holt, Rinehart and Winston, New York.

Brown, K. M., and McCrone, W. C. (1963) Dispersion staining: Part I. Theory, method and Apparatus, *Microscope*, **13,** 311–322.

Brown, K. M., McCrone, W. C., Kuhn, R., and Forlini, L. (1963) Dispersion staining: Part II. The systematic application to the identification of transparent substances, *Microscope*, **14,** 39–54.

Champness, P. E., Cliff, G., and Lorimer, G. W. (1976) The identification of asbestos, *J. Microsc.* **108,** 231–249.

Chao, E. C. T. (1976) The application of quantitative interference microscopy to mineralogic and petrologic investigations, *Am. Miner.*, **61,** 212–228.

Cherkasov, Y. A. (1960) Applications of "focal screening" to measurement of indices of refraction by immersion method, *Int. Geol. Rev.*, **2,** 218–235.

Clark, R. L., and Ruud, C. O. (1974) Transmission electron microscopy standards for asbestos, *Micron*, **5,** 83–88.

Crable, J. V. (1966) Quantitative determination of chrysotile, amosite and crocidolite by X-ray diffraction, *Am. Ind. Hyg. Ass. J.*, **27,** 293–298.

Crable, J. V., and Knott, M. S. (1966a) Application of X-ray diffraction to the determination of chrysotile in bulk or settled dust samples, *Am. Ind. Hyg. Ass. J.*, **27,** 383–387.

Crable, J. V., and Knott, M. J. (1966b) Quantitative X-ray diffraction analysis of crocidolite and amosite in bulk or settled dust samples, *Am. Ind. Hyg. Ass. J.*, **27,** 449–453.

Deer, W. A., Howie, R. A., and Zussman, J. (1966) *An Introduction to the Rock-forming Minerals*, Longmans, London.

Dodgson, J. (1963) Use of interference microscopy for the mineralogical analysis of samples of airborne dust obtained with the thermal precipitator, *Nature, Lond.*, **199,** 245–247.

Farmer, V. C. (Editor) (1974) *The Infra-red Spectra of Minerals*, Mineralogical Society, London.

Gadsen, I. A., Parker, I., and Smith, W. L. (1970) Determination of chrysotile in airborne asbestos by an IR spectrometric technique. *Atmosph. Envir.* **4,** 667–670.

Gleit, C. E., and Holland, W. D. (1962) Use of electrically excited oxygen for the low temperature decomposition of organic substances, *Analyt. Chem.*, **34,** 1454–1457.

Goni, J., Jaurand, M. C., and Caye, R. (1975) Emploi de la microscopie en contraste de phase pour l'identification des particules fibreuses présentes dans la poumon humain, *Bull. Soc. Fr. Minér. Cristallogr.*, **98,** 294–298.

Goodhead, K., and Martindale, K. W. (1969) The determination of amosite and chrysotile in airborne dusts by an X-ray diffraction method, *Analyst, Lond.*, **94,** 985–988.

Hartshorne, N. H., and Stuart, A. (1970) *Crystals and the Polarising Microscope*, Arnold, London.

Hayashi, H. (1973) Quantitative determination of airborne asbestos dust in occupational environment by X-ray diffraction using conventional and rotating anode X-ray tube, *Ind. Hlth.*, **11,** 225–236.

Health and Safety Executive (1975) *Asbestos: Health Precautions in Industry*, 2nd Ed., (Health and Safety at Work, No. 44) H.M. Stationery Office, London.

Hearle, J. W. S., Sparrow, J. T., and Cross, P. M. (1973) *The use of the Scanning Electron Microscope*. Pergamon Press, Oxford.

Heidermanns, G. (1973) Asbestos content determination by optical, chemical, radiographic and IR spectrographic analysis procedures, *Staub* (*Engl.*), **33,** 67–72.

Heidermans, G., Riediger, G., and Schutz, A. (1976) Asbestbestimmung in industriellen Feinstaüben und in Lungenstauben, *Staub*, **36,** 107–111.

Hodgson, A. A. (1965) *Fibrous Silicates*, Lecture series, No. 4, Royal Institute of Chemistry, London.

Hodgson, A. A., Freeman, A. G., and Taylor, H. F. W. (1965) The thermal decomposition of crocidolite from Koegas, S. Africa, *Mineralog. Mag.*, **35,** 5–30.
Hunt, J. M., Wisherd, M. P., and Bonham, L. C. (1950) Infra-red absorption spectra of minerals and other inorganic compounds, *Analyt. Chem.*, **22,** 1478–1497.
Julian, Y., and McCrone, W. C. (1970) Identification of asbestos fibres by microscopical dispersion staining. *Microscope*, **18,** 1–10.
Keenan, R. G., and Lynch, J. R. (1970) Techniques for the detection identification and analysis of fibres, *Am. Ind. Hyg. Ass. J.*, **31,** 587–597.
Klug, H. P., and Alexander, L. E. (1954) *X-ray diffraction Procedures*, Wiley, New York.
Langer, A. M., Mackler, A. D., and Pooley, F. D. (1974) Electron microscopical investigation of asbestos fibres, *Envir. Hlth. Perspect.* **9,** 63–80.
McCrone, W. C. (1974a) Light microscopy, in *Characterisation of Solid Surfaces*, (Ed. P. F. Kane and G. B. Larrabee), Plenum Press, New York.
McCrone, W. C. (1974b) Detection and identification of asbestos by microscopical dispersion staining, *Envir. Hlth. Perspect.* **9,** 57–61.
Muir, D. C. F. (1976) Health hazards of thermal insulation products, *Ann. Occup. Hyg.*, **19,** 139–145.
Nenadic, C. M., and Crable, J. V. (1971) Application of X-ray diffraction to analytical problems of occupational health, *Am. Ind. Hyg. Ass. J.*, **32,** 529–538.
Nuffield, E. W. (1966) *X-ray Diffraction Methods*, Wiley, New York.
Ocella, E., and Maddalon, G. (1963) X-ray diffraction characteristics of some types of asbestos in relation to different techniques of comminution, *Medna. Lav.*, **54,** 628–636.
Parks, F. J. (1971) Infra-red spectroscopy and its use in the identification of asbestos minerals, Paper presented at the 2nd International Conference on the Physics and Chemistry of Asbestos Minerals, Louvain University, 6–9th September 1971 (unpublished preprints, Paper 2.5).
Patterson, J. H., and O'Connor, D. J. (1966) Chemical studies of amphibole asbestos. I. Structural changes of heat-treated crocidolite, amosite and tremolite from IR absorption studies, *Aust. J. Chem.*, **19,** 1155–1164.
Plowman, C. (1973) Effect of heat on the identification of asbestos in X-ray diffraction, *Nature, Lond.*, **244,** 280.
Pooley, F. D. (1972) Electron microscope characteristics of inhaled chrysotile asbestos fibre, *Brit. J. Ind. Med.*, **29,** 146–153.
Pooley, F. D. (1975) The identification of asbestos dust with an electron microscope microprobe analyser, *Ann. Occup. Hyg.*, **18,** 181–186.
Reed, S. J. B. (1975) *Electron Microprobe Analysis*, Cambridge University Press, London.
Rickards, A. L. (1972) Estimation of trace amounts of chrysotile asbestos by X-ray diffraction, *Analyt. Chem.*, **44,** 1872–1873.
Rickards, A. L. (1973) Estimation of submicrogram quantities of chrysotile asbestos by electron microscopy, *Analyt. Chem.*, **45,** 809–811.
Rubin, I. B., and Maggiore, C. J. (1974) Elemental analysis of asbestos fibres by means of electron probe techniques, *Envir. Hlth. Perspect.* **9,** 81–94.
Russ, J. C. (1972) *Elemental X-ray Analysis of Materials EXAM Method*, EDAX Laboratories Division, EDAX International, Inc., Illinois.
Ruud, C. O., Barrett, C. S., Russell, P. A., and Clark, R. L. (1976) Selected area electron diffraction and energy dispersive X-ray analysis for the identification of asbestos fibres, a comparison, *Micron*, **7,** 115–132.
Schelz, J. P. (1974) The detection of chrysotile asbestos at low levels in talc by differential thermal analysis, *Thermochim. Acta*, **8,** 197–204.

Schmidt, K. G. (1960) Types of asbestos, their examination by optical means and their pathogenic action, translated from *Staub*, **20,** 173–204 (1960).

Skikne, M. I., Talbot, J. H., and Rendall, R. E. G. (1971) Electron diffraction patterns of UICC asbestos samples, *Envir. Res.*, **4,** 141–145.

Smither, W. J. (1970) Asbestos and asbestosis, *Ann. Occup. Hyg.*, **13,** 3–5.

Talvitie, N. A., and Brewer, L. W. (1962a) Separation and analysis of dust in lung tissue, *Am. Ind. Hyg. Ass. J.*, **23,** 58–61.

Talvitie, N. A., and Brewer, L. W. (1962b) X-ray diffraction analysis of industrial dust, *Am. Ind. Hyg. Ass. J.*, **23,** 214–221.

Taylor, D. G., Nenadic, C. M., and Crable, J. V. (1970) IR spectra for mineral identification, *Am. Ind. Hyg. Ass. J.*, **31,** 100–108.

Timbrell, V., Gilson, J. C., and Webster, I. (1968) UICC Standard reference samples of asbestos, *Int. J. Cancer*, **3,** 406–408.

Whittaker, E. J. W., and Zussman, J. (1971) The serpentine minerals, in *Electron-optical Investigation of Clays*, (Ed. J. A. Gard), Mineralogical Society, London, pp. 159–191.

Zielhuis, R. L. (Editor) (1977) *Public Health Risks of Exposure to Asbestos*, Report of a Working Group of Experts prepared for the Commission of the European Communities, Directorate-General for Social Affairs, Health and Safety Directorate, Pergamon Press, Oxford (for the Commission of the European Communities).

Zussman, J. (Editor) (1967) *Physical Methods in Determinative Mineralogy*, Academic Press, London.

CHAPTER 8

Dealing with asbestos problems

J. D. Cook and E. T. Smith

Hazardous Materials Service
Harwell

(1) INTRODUCTION

Asbestos deposits were first observed in 1847, but the first Canadian mines did not start production of chrysotile until 1880. Deposits were later found and worked in Rhodesia, South Africa, Cyprus and the U.S.S.R. The term 'asbestos', meaning unquenchable, is applied to a group of fibrous silicate minerals which have excellent thermal and electrical insulation properties. They do not have the same crystallographic form or chemical composition and thus can be identified from each other by chemical analysis and physical techniques such as microscopy. The most common asbestos minerals are shown in Table 8.1.

Of the minerals, actinolite and tremolite are rare in the fibrous habit, non-asbestiform tremolite being a major constituent of talc, which has given rise to difficulties in the formulation of legislative controls (Anon., 1976).

Table 8.1 List of common asbestos minerals

Mineral	Fibrous forms of mineral (asbestos)	Common name	Fibre type
Actinolite	Actinolite	—	—
Anthophyllite	Anthophyllite	—	Brittle
Antigorite	Chrysotile	White asbestos	Fine, silky, flexible
Cummingtonite	Amosite	Brown asbestos	Straight and brittle
Riebeckite	Crocidolite	Blue asbestos	Straight and flexible
Tremolite	Tremolite	—	Straight and flexible

Amosite, anthophyllite, chrysotile, and crocidolite together comprise almost all of the industrial output and of these, chrysotile, which can be easily woven into a textile, forms by far the largest percentage.

Use of asbestos is very widespread but can be broadly classified into four categories:

1. Asbestos–cement products—where cement is used as a matrix and includes corrugated and flat sheet, pipes, and moulded sections.
2. Asbestos-filled products—where a variety of binders are used such as starches, elastomers, and resins, e.g. paper, millboard, gaskets, brake linings floor tiles.
3. Sprayed asbestos—where fibres, usually with a binding agent, are sprayed on to surfaces or into cavities for insulation and fire proofing.
4. Asbestos textiles—where the fibres are woven into cloth or rope and used for lagging, door seals, protective clothing, etc.

Any piece of equipment, from a domestic hair drier to an industrial furnace, that has to withstand high temperatures probably incorporates asbestos. Large amounts are used in the construction industry for building and insulation and in the transport sector brake linings and clutch plates contain asbestos. It pervades both the domestic and industrial environments, finds its way into water supplies (Cunningham and Pontefract, 1971; Legge, 1974; Von Buchstab, 1974), and is a common air pollutant in many areas although at low concentrations (Langer, 1974; Smither, 1974).

As early as 1927 an effect on the lungs similar to silicosis was reported (Cooke, 1927) and termed asbestosis. In 1955 it was established that occupational exposure to the dust could produce lung cancer (Doll, 1955), confirming an observation previously made in 1935 (Lynch and Smith, 1935). During the 1960s, relationships between asbestos and other forms of cancer were demonstrated, including a high risk of a rare tumour called pleural mesothelioma to workers exposed to crocidolite (Wagner *et al.*,

1960). It also became evident that not only asbestos workers were at risk, but also their families and those who lived close to asbestos works or mines. Selikoff *et al.*, (1968) demonstrated that asbestos exposure coupled with cigarette smoking could be synergistic with increased risk of cancer. In 1970 the importation of crocidolite into the U.K. was banned.

It is clear that although asbestos is used extensively, the degree of hazard may be uncertain. Some authorities contend that the material can be used safely, whereas others say that one fibre could be fatal. When using asbestos it is therefore prudent to know what precautions to take and, perhaps more important, to know why they are being taken. It is hoped that the contents of this chapter will enable asbestos problems to be approached with realism and understanding.

(2) REGULATIONS FOR USE OF ASBESTOS

In the U.K. soon after the first observation of the risks associated with asbestos a set of regulations was drawn up and introduced as the Asbestos Regulations 1931, which were implemented in April 1933. These controlled work places only, but the Asbestos Regulations 1969, which came into force in May 1970, were much wider in their cover. Also published was a code of practice for dealing with asbestos and asbestos products by the Asbestosis Research Council, a body set up by the industry. Taken together, these documents give guidelines to the user, which, in some instances are now mandatory. One of the main tasks in the Regulations is to set standards to which the work force adhere. These are generally set out as shown in Table 8.2. The concentrations given in Table 8.2 should be compared with the expected dust emissions from various industries, which are given in Table 8.3.

Such a comparison shows that there would be very few operations where the Regulations do not apply, even unloading sheets from a lorry.

Criticism on the use of the standards specified in the Regulations is often due to the assumption that whenever such standards are put forward they soon become used as magic numbers, below which lies safety and above which lies disaster. Hence such standards must be used with understanding, and any limitations clearly appreciated.

For instance, the standard of 2 fibres/cm^3 (or 0.2 fibres/cm^3 for crocidolite) was promulgated to protect from asbestosis, and it does not necessarily protect from other diseases. It has been suggested that the only safe limit is to have no asbestos dust at all, although this is manifestly unreasonable as asbestos occurs so widely in our environment (Langer, 1974).

Another criticism is that fibres below 5 μm are not counted and there is evidence that these are important (Harwood *et al.*, 1975; Pott *et al.*, 1972).

Table 8.2 Standards for working with asbestos (from Technical Data Note No. 42, *Probable Asbestos Dust Concentrations at Construction Processes*, reproduced by permission of the controller of Her Majesty's Stationary Office)

Average asbestos dust concentration (fibres/cm^3)*	Sampling period	Interpretation
<2	10 min	Regulations are not expected to apply
2–12	4 h	The amount of control will depend how much above 2 fibres/cm^3 the concentration is and how long the exposure will be
12	10 min	Protective respiratory equipment is mandatory

* Where crocidolite is involved these concentrations are reduced 10–fold. The standard of 2 fibres/cm^3 is also accepted by OSHA for fibres longer than 5 μm.

However, sampling, counting, and identification of asbestos fibres together with the interpretation of results remains a specialized matter. No simple asbestos meters have yet been developed to provide continuous control, so industrial firms engaged on work with asbestos tend to instal considerably better protection equipment than would otherwise be necessary. In disposal operations this solution is usually impractical or very expensive and a better solution has to be sought, so that monitoring assumes much greater importance.

For the factory or user who knows the asbestos and asbestos type that is being used, worker protection should be relatively simple. However, the demolition team has a very variable material to deal with. Asbestos substitutes may have been used, or the types of asbestos used could vary from location to location. Some products which give rise to excessive dust on use are no longer manufactured or, if they are, they probably incorporate a dust suppressant. Use in lagging is being discontinued, replacement being by calcium silicates, glass-fibre, slag wool fibre, and rock wool fibre. The use of crocidolite for sprayed asbestos has been discontinued and it has been replaced with mineral wools, but old structures probably still have dangerous forms of asbestos at some location. Therefore, before embarking on a demolition programme, all of these factors have to be evaluated in order to enable a safe demolition scheme to be formulated. Guidelines for using asbestos are available and the bibliography gives a list of relevant U.K. publications. In the U.S.A. the Environmental Protection Agency and the Occupational Safety and Health Administration (O.S.H.A.) issue similar publications. The work in the U.K. is now the responsibility of the Health and Safety Executive, incorporating the Factory Inspectorate. Their recommendations usually cover a particular theme or industry and such an arrangement is advantageous, but it follows that some information (e.g. acceptable dust levels) is repeated in several documents. The private user can also be at risk and an Asbestos Safety Code (Figure 8.1) covers this

Table 8.3 Typical dust concentrations for various processes. These data are taken from Appendix 2 of Health and Safety Executive Technical Data Note No. 42, *Probable Asbestos Dust Concentrations at Construction Processes* (reproduced by permission of the Controller of Her Majesty's Stationary Office). They are illustrative only and should be only used in connection with the points raised in the Note and not taken out of context. Data for various operations are also given in Appendix VI of EEUA (Engineering Equipment Users Association) Handbook No. 33, Addendum No. 1 (1976).

Use	Method employed	Concentration (fibres/cm^3)
Asbestos spraying	Pre-damping	5–10
	No pre-damping	>100
Delagging	Soaking	1–5
	Water sprays	5–40
	Dry	>20
Asbestos–cement sheet and pipes	Machine drilling	<2
	Hand sawing	2–4
	Jig sawing without exhaust ventilation	2–10
	Circular sawing without exhaust ventilation	10–20
	Machine sawing with exhaust ventilation	<2
Asbestos insulation board	Vertical drilling	2–5
	Overhead drilling	4–10
	Sanding, etc.	6–20
	Breaking	1–5
	Hand sawing	5–12
	Jig sawing without exhaust ventilation	5–20
	Circular sawing without exhaust ventilation	>20
	Machine sawing with exhaust ventilation	2–4
	Unloading and handling cut pieces	5–15
	Unloading and handling manufacturers sheets	1–5

The Asbestos Safety Code

It is known that asbestos dust can cause lung diseases and there are strict regulations governing the manufacture and commercial use of asbestos products.

For the home handyman and domestic user of asbestos products, it is very unlikely that harmful quantities of dust will escape in their normal use. As a precaution, however, you are advised to:

Avoid creating and breathing asbestos dust.

The safest way to do this is to follow a few simple rules:

1. **Damp the work:** wet dust does not become airborne and is not inhaled. For instance do not sand wall plugging compounds unless damped. When re-lining your car brakes, remove dust from brake drums with a damp cloth.
2. **Damp any dust that falls to the floor** and pick it up as soon as possible. Place it in a plastic bag and seal the bag.
3. **Work in a well ventilated space,** if possible outdoors, when sawing, drilling, filing or sanding.
4. **Use hand saws and drills,** which produce less dust than power tools.
5. **Renew worn or frayed asbestos insulators** like oven gloves, oven door seals, hot plate cover seals, ironing pads and simmering pads.

Figure 8.1 Asbestos safety code. (Reproduced by permission of the Asbestos Information Committee)

particular need. From October 1st, 1976, products containing asbestos have had to be labelled as shown in Figure 8.1.

Asbestos is now dealt with during the erection or demolition of structures involving asbestos products. In this chapter, we include with demolition the treatment of asbestos products to reduce the hazard as well as its subsequent removal and disposal.

The prime consideration is that no-one, either those working with asbestos, those in contact with asbestos workers, or the general public, should be put at an unacceptable risk. There will be some degree of risk, but in any given situation the maximum possible protection should always be provided and its use ensured.

(3) CONTROL OF ASBESTOS IN A WORKING ENVIRONMENT

Asbestos is most dangerous when it is in dust form. Protection from this danger can be given by suppressing dust and by keeping the dust away from workers by using protective measures or clothing.

Dust suppression can begin at the manufacturing stage by the incorporation of suppressants or by coating the asbestos with dust-proof membranes such as plastics and paints. Keeping sheets wet when handling also reduces the hazard. There is a wide choice of protective equipment available and comprehensive information on approved equipment is given in the Health and Safety Executive's Technical Data Note No. 24 and the Asbestosis Research Council's Control and Safety Guide No. 1 (see Bibliography). The Factory Inspectorate hold lists of approved equipment and asbestos companies will provide advice. Although there is a wide variation in the type of protective equipment available basically four main items are required as described below.

3(i) Respirators

These have to be of a type approved by the Factory Inspectorate and a number of suppliers market 'approved' equipment. A given respirator may be approved up to a certain fibre-count level but some are approved for any level; the latter are invariably high-efficiency, full-face, positive-pressure powered types. The air is supplied from a battery-driven fan forcing air through a high-efficiency filter to the mask.

The powered masks are so designed that the operator can still breathe without difficulty even if the fan fails. The batteries are rechargeable and so sized as to cover a minimum of a 4-h shift.

If the respirator is not of the 'without limitation' type, then it is essential that air sampling is carried out, otherwise the operator may be at risk if the fibre count is in excess of that prescribed for the respirator in use. Anything preventing a complete face seal, such as a beard or spectacles, needs either removal or else a special design of mask.

For some operations, involving difficult access or where it is necessary to lie on one's back, the operator may prefer to use a natural ventilation mask as a fan unit would restrict his mobility or even render access impossible. Approved self-contained breathing apparatus (rescue equipment) can be used for unlimited concentrations of asbestos in air. Compressed air line breathing apparatus is likewise suitable, but in practice one would generally not wish to use such equipment.

3(ii) Overalls

These generally are the one-piece boiler-suit type, fitting closely at the neck, wrists, and ankles. The suit incorporates a hood which is designed to go over the respirator straps and is fitted with a draw-string to ensure a close fit. The aim is to exclude ingress of dust. The overalls are of a fine weave, usually nylon, and often with a plastic impervious finish. Man-made fibres are preferable as there is less tendency for asbestos particles to adhere to the material.

Overalls should only be laundered by firms who have experience in handling asbestos-contaminated clothing.

Some operatives wear their 'street' clothes under their overalls but in at least one company the sole undergarment consists of a swimming costume. There is a strong case for not wearing any clothing which will subsequently be worn outside the work area and the wearing of minimal underclothing greatly facilitates contamination control on leaving the work area.

Some firms use the same overalls inside and outside the work area but this is not to be recommended. The preferred method would be to have separate overalls, preferably distinctively marked or of a different colour to those used elsewhere.

3(iii) Footwear

The ideal choice is gumboots (the calf-length variety) worn with the overall trousers outside the gumboots. Shoes or boots, even of the safety type, are undesirable. As with the overalls, operators should not wear the same footwear outside the work zone, since it is likely to be heavily contaminated. Gumboots, because of their smooth surface, are easily decontaminated by vacuuming or by washing. Shoes or boots are much more difficult to decontaminate and the lack of an effective seal between boots and trousers can lead to contaminated socks.

3(iv) Gloves

Fingernails are a notorious trap for contaminants and gloves of some form should be used. Gloves made of cotton, rubber, or plastic are suitable. Disposable gloves can be used if desired.

3(v) Machinery controls

Control of machinery and working areas is of prime importance and it is a requirement that in any premises or plant handling asbestos there should be strict control of dust formation and clean conditions should be maintained to rigorous standards to ensure the safety of the operatives. These requirements are covered in detail in the Asbestosis Research Council's Control and Safety Guides Nos. 7 and 9, and the Health and Safety Executive Technical Data Notes Nos. 1 and 35 (see Bibliography).

The first approach is to prevent dust being created at all, either by modifications to the process or by complete enclosure of the working area. If this is impracticable, then a dust control system is required, and traditionally high-volume, low-velocity control systems have been used. If the dust is localized then a suitable hood can extract the dust as it is formed. If the dust generation is over an area rather than a precise location, then booths should be constructed to prevent the dust escaping. All openings should be kept as small as possible consistent with ease of operation, as wide openings need higher extraction rates.

In general, fabric filters should be used in the dust collection system. The filters are made of felt or woven cloth, depending on the design of the system. The system can be either continuous or intermittent, depending on the work pattern of the plant. The intermittent design is simpler but needs plant shut-down (which could be at meal breaks) to change a filter.

Fans and motors should be installed after the filter to ensure that the filtration system is under negative pressure as a safeguard against leaks and to simplify maintenance.

More recently, low-volume, high-velocity systems have been introduced particularly for hand tools and machining operations. Use is then made of close-fitting nozzles and hoods connected to small-bore pipework. Capture velocities vary between 10 000 and 12 000 ft/min, but the volumetric flow-rate is only 10–250 ft^3/min.

On any filtration system, the extract should be started before operations are carried out and it should be left running for several minutes after completion of the work.

The filtered air can be recirculated back into the work area so as to minimize heat loss from the buildings, but it must meet statutory requirements to avoid recirculation of dust.

It is a requirement that all machinery, equipment, plant, internal surfaces of buildings, and external surfaces of exhaust ventilation systems are kept clean. Cleaning should be carried out as far as possible by methods which do not create dust (that is, using vacuum cleaners or wet methods). Should this not be possible, then protective clothing and respirators should be worn. The external surfaces of the vacuum cleaners must be kept clean.

Buildings used for the first time after May 1970 for a scheduled process must have smooth, impervious interior surfaces. They must also have a centralized fixed vacuum-cleaning system with pipes to which portable cleaning heads can be attached.

Smooth, crevice-free surfaces facilitate and improve the efficiency of cleaning operations, whether by dry (vacuum) or wet methods. If wet methods are used, then proper disposal of the waste water must be arranged so that a hazard is not present after the surfaces/materials dry out. If the water is to drain to a sewer, then the asbestos particles it contains must be removed by filtration.

Where older buildings are in use and fixed vacuum systems are not compulsory, then portable vacuum cleaners can be considered. Portable vacuum cleaners exhaust back into the working area and require a two-stage filter to prevent escape of asbestos dust. The use of portable machines has disadvantages in that they have limitations in power and dust capacity plus the need for a trailing cable.

A fixed vacuum system is tailored to meet the needs of the plant lay-out. Operators can undertake and be responsible for the cleanliness of their own machines. The dust is entirely removed from the work area and the filters and collecting units can be cleaned under controlled conditions. There can be many other advantages. The principal disadvantages are the high capital cost and lack of flexibility in meeting changes in plant and layout.

As with any plant, it is imperative that it is regularly checked and kept in safe and efficient working order.

Exhaust equipment has to be inspected every 7 days and examined, tested, and reported upon by a competent person once every 14 months. Requirements are also laid down on the use and maintenance of respirators and protective clothing.

General aspects of the handling of asbestos fibre are covered in Asbestosis Research Council Control and Safety Guide No. 3 (see Bibliography). Common sense and good housekeeping are the basis for most of the operations.

Asbestos fibre, as far as is possible, should be transported only in closed receptacles. Bags should be handled as little as possible and protected from moving vehicles to eliminate as far as possible any chance of puncturing or tearing. Respirators and protective clothing should be available and worn as necessary. Likewise, a supply of impermeable bags (for slipping over damaged bags) and a vacuum cleaner should be kept close at hand to deal with broken bags and spillages.

Plant and equipment should be so designed as to minimize and contain any dust generated. The empty bags are classified as asbestos waste and have to be placed in closed receptacles for disposal.

Whenever specially designed premises and equipment are used for asbestos, then it is relatively easy, although expensive, to carry out manufacturing operations safely. However, most of the difficulties occur with demolition or remedial work and the remainder of this chapter will be confined to this aspect.

(4) GENERAL ASPECTS OF STRIPPING AND FITTING ASBESTOS INSULATION MATERIALS

Operations and the necessary protective clothing and equipment are described in Asbestosis Research Council Control and Safety Guides Nos. 3 and 1 (see Bibliography), respectively. Removal aspects will be amplified later.

The operators must wear approved overalls, head covering and footwear. A respirator approved for the expected dust level must be chosen and carefully fitted. If crocidolite (blue asbestos) is known or suspected to be present, then samples must be taken for analysis. If found, then 28 days' notice of intention to do the work must be given to the Factory Inspector. In this case the respirators used must be ten times more efficient than for 'white asbestos', that is, a mask approved for 800 fibres/cm^3 would be limited to 80 fibres/cm^3 for crocidolite.

Before commencing operations the asbestos should be soaked by injecting water or, if this is impossible, fine water sprays should be installed and kept in continuous operation. Sawyer (1977) has shown that the addition of a wetting agent to the water is highly advantageous. Spraying with such water was most effective just preceding the removal of asbestos. The wetting time was brief, penetration was quick and thorough, yet little run-off of water took place and visible dusting was rare. A 1 : 1 mixture of polyoxyethylene ester and polyoxyethylene ether was used at a concentration of 0.15% v/v (1 oz per 5 U.S. gallons).

If water cannot be used because of electrical or other problems (for instance the presence of sodium, or pipes carrying steam or other high-temperature fluids), then there is a need for local dust extraction near the work position. It is best if the asbestos can be cut at strategic points and removed in blocks. This is feasible with pre-formed insulation but sprayed asbestos presents problems. During the work good housekeeping is of paramount importance in order to reduce dust levels and minimize the spread of contamination.

The material when removed should not be allowed to fall but should be placed directly into plastic bags. If this is not possible, a plastic sheet should be slung or spread beneath the work. Droppings on this sheet should be continually removed and placed in bags. Any asbestos slurry should not be left to dry out but placed in a container.

When pre-formed blocks need to be fitted, the operation is capable of generating significant clouds of dust. This should be collected together with off-cuts, and on no account should the debris be trodden under foot. If finishing cement or other bonder needs to be used, water should be added gently to obtain a wet mix and to avoid dust being ejected by a strong water jet. All of the containers, bags, and excess of asbestos on tools must be classified as waste and treated as such, and must be sent for approved disposal.

(5) DISPOSAL OF ASBESTOS

Recommended methods of disposal are published by the Asbestosis Research Council (see Bibliography).

Since asbestos is a rock mineral it is not toxic in the same sense as potassium cyanide and there is no basic reason why the minerals should not be disposed to landfill. However, there must be safeguards, particularly to prevent dust formation. This can be achieved by burial at a sufficient depth to ensure that the surface does not dry out.

The present practice at a landfill is to have a face over which the waste is tipped. On well run sites this is immediately covered with soil or other fill such as general refuse as the face advances. However, there are two problems. In tipping down a face, dust can be formed from loose sheets, or bags may puncture, and rain can expose asbestos from a shallow fill. For these reasons, in our opinion it is better to dig trenches about 3 m deep and bury the waste in these. In this way the asbestos is safe and non-toxic provided that the site remains undisturbed. There is some concern that asbestos fibres may get into water supplies from landfill. If the material is buried as recommended then any fibres should be filtered by the rock and soil beneath them. However, there is a danger if they come to the surface that surface waters may pick up the fibres.

The waste itself can be in many forms. Dust, swarf, sweepings, and loose wool must be put into heavy-gauge polyethylene or PVC bags sealed and labelled ASBESTOS. It is helpful if the bags are a distinctive colour, e.g. red for white asbestos and blue for blue asbestos. It may be dangerous to break large pieces so it is probably better to soak them and transport them as one piece. Asbestos–cement articles are less dangerous if care is taken, but damping is still a sensible precaution. No special transport is necessary if bags are used but if sheets are transported the vehicle storage surface should be designed without pockets and be easily cleaned. Drivers should be provided with overalls and masks if they are to go on to a landfill when asbestos is being buried, as should other operators on the site involved in the burial.

At the disposal site the asbestos waste should be placed with care either at the foot of a tip face or in a deep trench and immediately covered with suitable fill. A minimum depth of 25 cm is recommended and more would be advantageous. This process is repeated until no more waste can be added. Enough final cover should then be provided to prevent the asbestos being disturbed by vehicles (Keen and Mumford, 1975). It is unlikely that less than 1 m would be sufficient for this purpose. All asbestos wastes need to be notified in the U.K. under the Control of Pollution Act 1974, but if a company is operating continuous disposal they may be able to arrange for a season ticket. Slurry wastes may be transported in tankers and pumped into a suitable landfill.

To sum up, asbestos must be buried as deep as possible and the site chosen must then stay undisturbed. Planners should be informed so that no building or major earth-moving operations are carried out in the area in the future.

(6) DEALING WITH AN ASBESTOS PROBLEM

In this section, it is intended to amplify some aspects of the preceding sections by using as an illustration an operation involving the stripping of lagging or the removal of fire-control insulation from girders and other steel work. Although many jobs are similar, no two are alike and the success of an operation is greatly dependent upon good planning.

The prime aim in any operation involving asbestos is to ensure that the risk of exposure is kept as low as possible, but this will depend on the site and the costs. In a densely populated area, precautions to contain asbestos would have to be much stricter than at an isolated site. Protection is achieved by taking suitable measures for those directly involved. Effective containment control to keep asbestos dust within a defined area is the means used to protect those outside the working area.

Asbestos dust is extremely fine and some particles are too small to be seen by the naked eye. The fine dust is noted for its mobility and there have been instances where persons external to the work area have been put at risk due to inadequate precautions. Because there is no suitable direct-reading instrument for monitoring asbestos concentrations as with normal work places, it must be ensured that effective measures are taken from the start and that subsequent monitoring is carried out.

Specialist firms usually undertake stripping as part of the overall operation. The contractor has the responsibility of ensuring that areas external to the operation zone do not become contaminated and that the asbestos is disposed of in a safe manner. It is often impossible or impracticable to remove all of the asbestos but the removal aspects and any post-removal treatments should be such that the area is safe and will remain safe when re-occupied by unprotected personnel. The work needs to be closely specified for tender action is problems are not to be experienced later. There are advantages in engaging the services of consultants in drawing up the job specification if a firm does not posses such expertise within its own organization.

6(i) Preliminary aspects

Premises routinely handling asbestos are designed to facilitate the removal of any contamination. Walls and floors are constructed of materials (or coated with special finishes) to provide a smooth, impervious surface from which contamination is relatively easily removed. The ventilation systems, containment, entry and exit routes, and change-room and washing facilities are designed for safe working and to contain any contamination.

Asbestos lagging and/or stripping operations are essentially a one-off situation and the buildings and plant are generally not specifically designed to expedite the operation. Conditions vary widely, for example a roof-void is one of the worst areas especially as regards stripping operations. Here the asbestos may have been applied to the girders before the installation of the ceiling. The ceiling is often of the suspended variety, and is not intended to carry any load. The upper surface of the ceiling may be insulated with fibre-glass mats which lie below catwalks made of metal grills. Invariably there are various service ducts and other systems running through the roof-void. Service ducts and underground areas also present problems. Lighting is usually poor and the access is often limited (e.g. vertical ladders, restricted headroom), and in some locations it is easy to become disorientated owing to a network of girders and pipes.

Faced with such a situation (and especially if blue asbestos is involved), one needs to consider the problems very carefully before starting any de-lagging operations. A suspended ceiling, for instance, is not a leak-tight structure and fibre-glass mats are almost impossible to decontaminate.

On boilers or process plant, it may be impossible or uneconomical to shut off the plant. The use of water is almost impossible in these situations and the work will have to be carried out dry. As the asbestos is removed, there will be a marked rise in ambient temperature, making working conditions most unpleasant, especially in areas of poor access and ventilation.

The presence of asbestos (even crocidolite) is not in itself a valid reason for removing it. The safest long-term solution may well be to encapsulate the asbestos rather than try to remove it and replace it with other materials. There are a number of high-build coating materials which can be applied using a 'full' brush or sprayed on to the asbestos. These films are thick, tough, have a very high coefficient of elasticity, and meet exacting ageing and heat-resistance tests. Their application results in a system which is very resistant to normal wear and tear while still maintaining the insulation requirements of the asbestos coating. The coatings can be reinforced with glass-cloth if desired. If used, it is desirable to label the coating at the end of the work to show the presence of asbestos. The encapsulated, asbestos-clad steelwork (or piping) can then be removed at such time when the building is demolished; the steel can then be cut into lengths for disposal and negligible stripping would be involved.

Assuming a situation where the removal of the asbestos appears to be the best solution, then the general procedure recommended is as follows.

The job is carefully surveyed and an operational plan drawn up. Problem areas should be anticipated as far as possible and solutions devised.

The following are some aspects that need consideration; the list is not exhaustive.

(a) The means of sealing off and isolating the work zone and carrying out the work;

(b) is there any fire risk necessitating the use of fire-retardent plastic sheeting?;

(c) will the wind and weather affect 'tents' if these are outside a building?;

(d) the provision and location of change-rooms, toilet and washing facilities, and rest areas;

(e) contamination control procedures and choice of protective equipment;

(f) method of transporting asbestos waste from the work zone, through the barriers to the transport vehicle;

(g) identification of a disposal site for wastes;

(h) access for contractors equipment, scaffolding, etc.;

(i) extra lighting and the provision of low-voltage supplies for power tools; conversely, the isolation of some electrical cources;

(j) water supplies for damping the asbestos if the use of water is permissible;

(k) ventilation and air filtration;

(l) filters to be placed in drainage systems;

(m) laundry facilities;

(n) monitoring facilities and frequency of samples;

(o) alterations to access to work area and, where necessary, marking out exit routes;

(p) facilities for re-charging power-packs on the respirators and maintenance of equipment;

(q) security passes for contractors' employees;

(r) do smoke detectors need to be isolated?;

(s) are all the non-asbestos safety aspects of work covered, e.g. scaffolds, ladders, toe-boards?

6(ii) Operational aspects

Once all of the preliminaries have been cleared and the contractor is ready to start, then the standard practice is to sheet off the area by using

Figure 8.2 Asbestos stripping team removing asbestos from Battersea 'A' power turbines. The work is carried out inside dust-proof sheeting. (Photography by courtesy of George Cohen, Sons & Company Ltd.)

suitable gauge PVC or polythene sheeting (Figure 8.2). The sheeting should be strong enough to withstand any stresses applied to it, such as winds in an open situation. The sheeting-off has two roles: one is to contain the asbestos in a given location and the other to prevent its ingress into any items such as electrical motors which cannot be conveniently removed from the work zone. The sheets should overlap and have the joints sealed either by heat or with adhesive tape. On some jobs it may be acceptable to sheet-off openings such as windows and doors, provided that the walls, ceiling, and floor are smooth and impermeable, but in many cases it will be advantageous to sheet all surfaces.

Ideally the sheeting should be smooth and without folds, as this facilitates decontamination during the work and at the final stage. If some piece of equipment has to be protected from ingress of dust, it is often advantageous to build a simple box structure around it which is then easily sheeted and sealed, rather than try to wrap it up in the manner of a painter's dust sheet.

When the area in which work is being carried out is large, it is common practice to sub-divide it into smaller units. Any mishaps in operations will then contaminate only a small, defined area.

Access into the work area will be through an 'air lock' system which also incorporates changing and washing facilities.

The sheet area should be kept below atmospheric pressure by an extract fan fitted with an 'absolute' filter so that the direction of the air flow is from the change room into the work zone. It may be necessary to fit ventilation flaps into the tent system at selected locations to improve ventilation in a particular area or where high extraction rates are required.

Since the chief hazard from asbestos work arises from the generation and spread of dust, dust-control measures are paramount. The principal method is to dampen the asbestos by the use of water (preferably containing a wetting agent, as described earlier), which is injected into the material using special high-pressure tools; this also facilitates its removal. Some formulations and painted asbestos cannot easily be wetted and water sprays are used to reduce dust formation.

If the use of water is impossible or very restricted, the conditions will be more dusty and potentially more hazardous. Local air circulation and filtration can be used to ameliorate the problem in such cases, but in general the fibre count will be much higher than that under wet conditions.

Air samples should be taken inside and outside the work area (especially the latter) to ensure that all is well and the risk is acceptably low.

The removal of the asbestos is achieved by the use of scrapers and wire brushes, used manually and as power tools. Powered tools must be low-voltage versions in the presence of water and be fitted with a local vacuum extraction system. High-pressure water jets are also used when possible.

The loosened asbestos must not be left lying around but should be loaded into bags of the prescribed type, which are sealed effectively as soon as they are full, otherwise the asbestos will dry out and be prone to cause a dust hazard. No asbestos should be left unbagged at the end of a working day.

Asbestos waste should be securely bagged in approved labelled bags, which are sealed with tape. The bags are probably best kept inside the work area until a sufficient number have been collected for a transfer operation. The bags, as with personnel, need a contamination control procedure. As the bags are made of plastic, the exterior surface can readily be cleaned by wiping it with a damp cloth or by washing it with a water spray. The bags during 'posting out' will be handled at one end by fully protected operators and at the other end by operators wearing conventional protective clothing. For blue asbestos, the bags are commonly placed inside another bag.

It should be noted that the managers of many disposal sites prefer asbestos waste to be delivered by prior arrangement early in the day so that it can be promptly covered with other fill materials. The waste requires notification under the Control of Pollution Act 1974, where there is a statutory obligation to allow 3 days' notice before removal of the material.

One company employs a system whereby all of the asbestos waste is transferred under vacuum from the work area via a flexible pipeline to two vacuum skips, which eliminates the use of bags and simplifies contamination control.

It is generally impossible to remove all asbestos, and residues will remain on all surfaces despite vigorous wire brushing. There will also be problem areas of restricted access where girders cross and in crevices. These presented no great problem when the asbestos was applied but effective removal can be impossible. Such areas must be covered with a suitable sealant.

It is common practice to wash-down or vacuum (or a combination of both) all of the surfaces at the end of the stripping operation. Drains should have been sealed off or fitted with effective filters so that no asbestos is lost to surface drains or to sewers. The clean-up needs to be carried out in a planned manner to ensure that all areas are covered and no clean areas are re-contaminated. This usually involves working from the top to the bottom and from the farthest point away from the entrance.

After a final inspection, all surfaces, including sheeting, should be mist-sprayed with an emulsion-type paint to 'fix' any fine residual asbestos, particularly if blue asbestos is involved.

Before the removal of the isolation tent, the area should be subjected to air checking and very low results should be achieved if the job has been carried out properly. The tent is carefully rolled up and sent for disposal by landfill.

The area may now give acceptably low fibre counts but that in itself may be inadequate. For example, in a car-park (especially an enclosed one) some asbestos may be ingrained into the concrete floor and it would be prudent to carry out tests to see if conditions become unacceptable once cars start to move over the floor and generate dust.

6(iii) Barrier-control procedure

The control procedure between the working area and the exterior environment is of prime importance. It is here that the operatives change into and out of their protective clothing and also where the bagged waste is passed out for subsequent disposal. Ineffective procedures can be detrimental to the safety of the operators and also result in the spread of asbestos dust to clean areas.

The removal of his contaminated clothing and respirator by an operator is potentially a very hazardous operation. If an operator comes out of the working area with clothing contaminated with blue asbestos and pushes through a flap in the 'tent', removes the hood, pulls off the respirator and walks into the clean area, then not only has risk been incurred by the operator but can also be a spread of contamination to other people. This emphasizes the need for careful and standarized barrier procedures.

It is not always possible to fit in a changing area of an ideal pattern, but it should be incorporated whenever possible even if in a modified form. The change-room separates the clean, uncontrolled exterior from the working area. Ideally, the change-room should be divided into a minimum of three rooms, each of which can be isolated from the others (Figure 8.3).

There should be a standard documented procedure for a passage into and out of the working area through the change-room and the work-force should be thoroughly trained in the procedure. There should be a standard

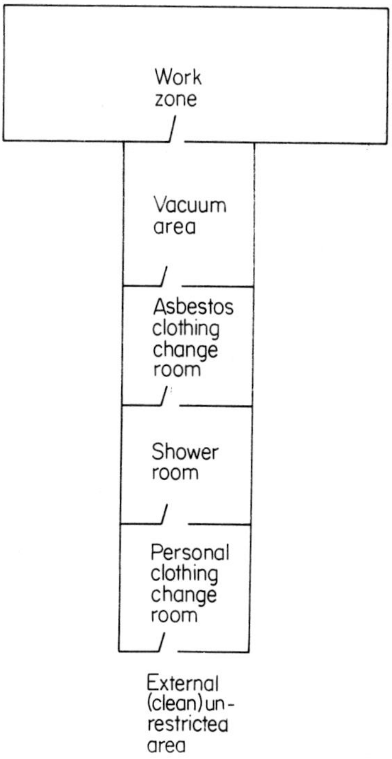

Figure 8.3 Change-room facilities (diagrammatic)

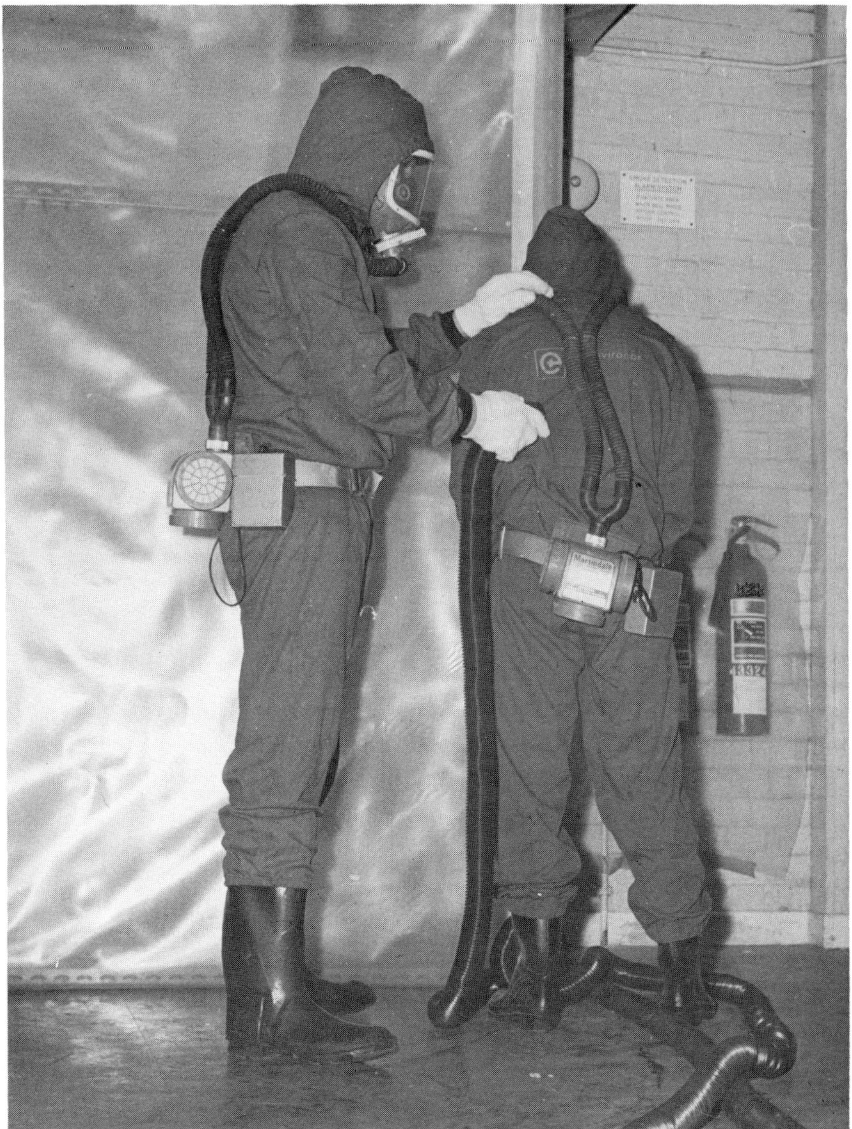

Figure 8.4 Before leaving work area, strippers vacuum off excess asbestos fibres from red overalls, masks, etc., then pass into an airlock to change into blue 'transition' overalls to walk to the de-contamination cabin, still wearing respirators. Dirty red overalls are left in the airlock. (Photography by courtesy of Envirocor Ltd.)

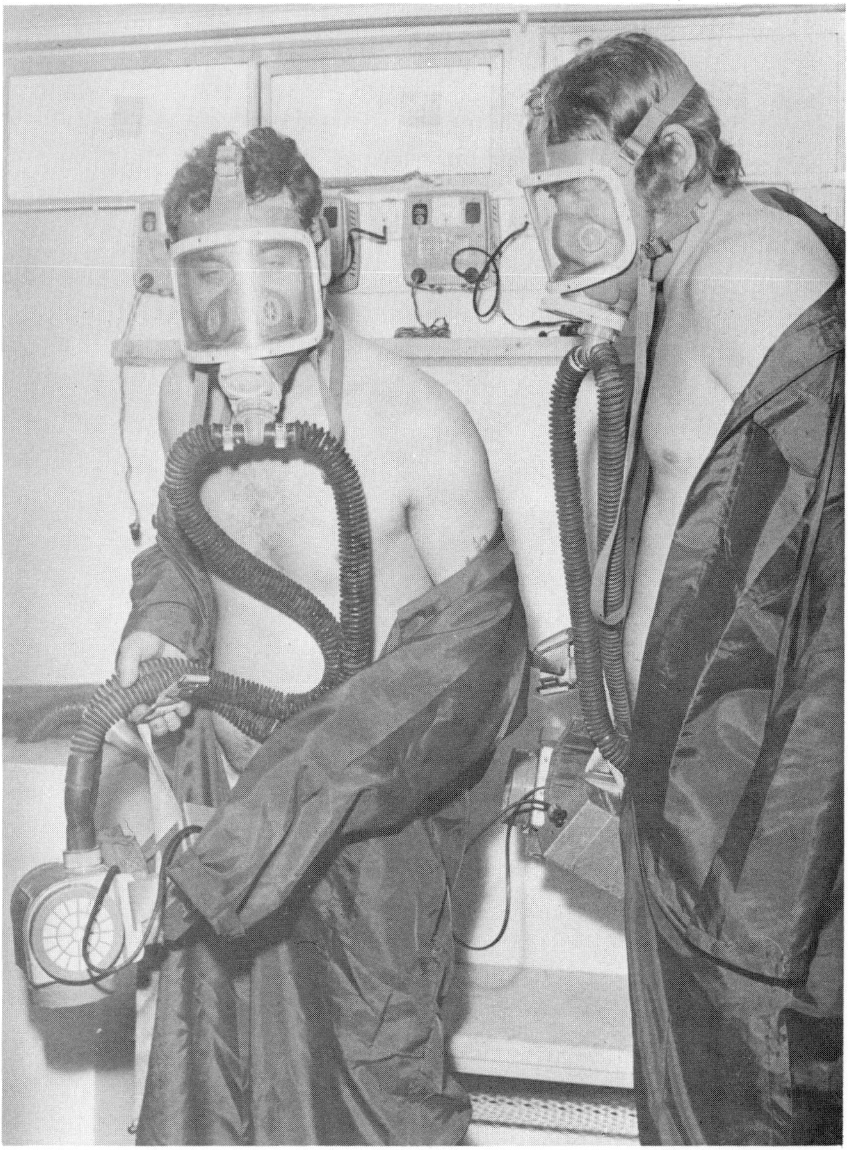

Figure 8.5 Upon entering 'dirty' end of the decontamination unit, operators shed 'transition' overalls, take a deep breath, remove respirators and pass through air-tight door to shower room. Supervisor wearing mask then washes respirator face pieces and seals them in plastic bags ready for re-use. (Photography by courtesy of Envirocor Ltd.)

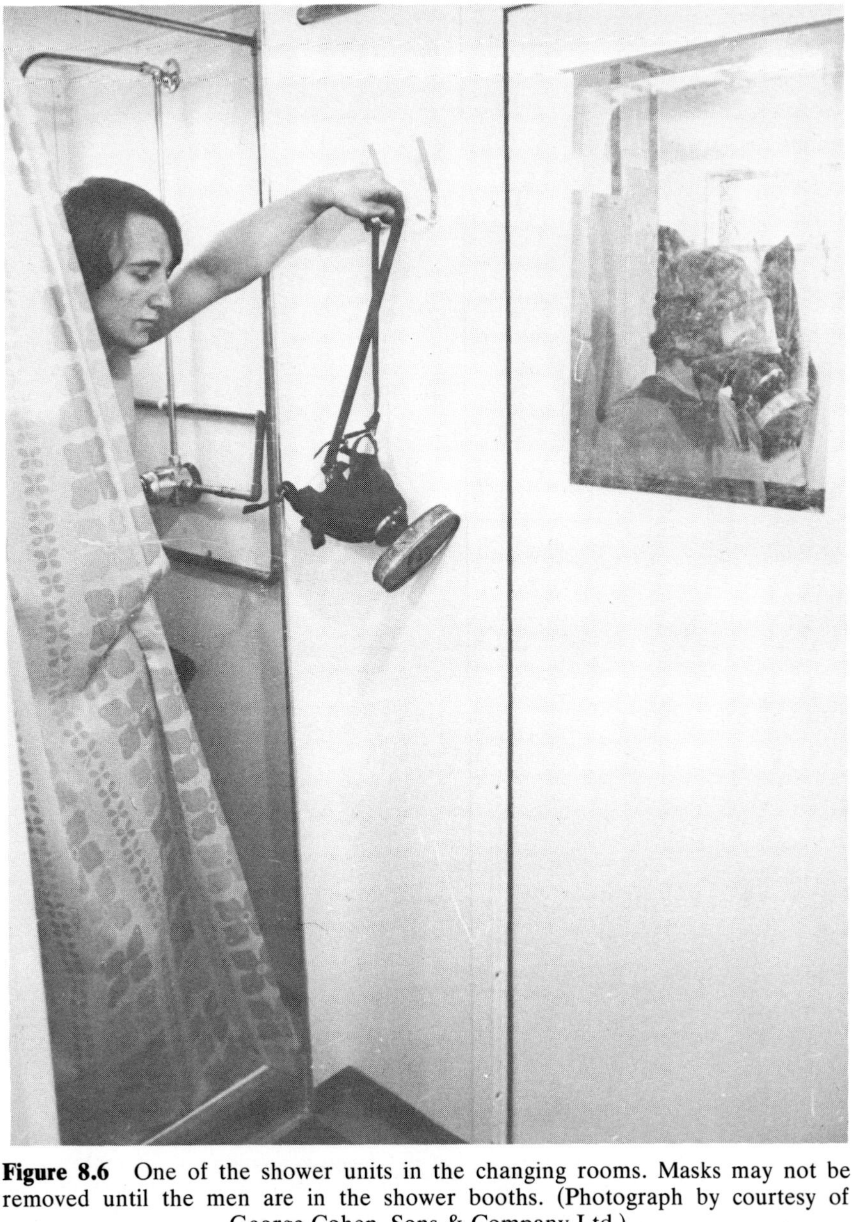

Figure 8.6 One of the shower units in the changing rooms. Masks may not be removed until the men are in the shower booths. (Photograph by courtesy of George Cohen, Sons & Company Ltd.)

documented procedure for a passage into and out of the working area through the change-room and the work-force should be throughly trained in the procedure. There should be separate sets of clothing for the 'clean' and 'dirty' sides of the change-room, preferably colour differentiated. Underwear is minimal and if anything other than a swimming costume is worn as an undergarment, e.g. in cold conditions, then this should be regarded as contaminated and should stay on the dirty side of the change room. Each operator will have his own fitted respirator and sets of clothing.

It is preferable if operators leave the work zone in pairs. Each worker carefully vacuum-cleans the other's clothing (Figure 8.4) and external surfaces of the respirator face piece before entering the first part of the change-room. Here they continue to wear their respirators while removing their overalls, which they place in individual lockers or bags.

From here there are two major variations in practice, which are illustrated by Figures 8.5 and 8.6. Either the operator takes a deep breath, removes his mask, which he leaves behind, and then steps through into the shower room, or he retains his respirator until he is in the shower, at which point he then removes it and quickly turns on the shower. It is essential to shower thoroughly, paying particular attention to the hair and fingernails before passing into the clean side, where a change into 'clean-side' overalls or street clothes can be made before passing out to the uncontrolled exterior.

Respirators are cleaned before re-use and suitably protected.

When going from the clean side to the dirty side, one procedure (in outline) is as follows. The operator removes all clothing at the clean side and dons a swimming costume. A clean respirator is picked up and put on, while walking through the shower room to the contaminated area where overalls used for stripping are put on. The hood is carefully fitted over the respirator and the elastic legs of the overall over the wellington boots to give complete protection.

Other procedures could be equally effective provided they achieve the dual aims of restricting asbestos to the 'dirty' side of the system and preventing exposure of the operator and other staff to asbestos dust. All control systems need good supervision to ensure that the correct procedure is being followed and that the equipment is in good working order. It will also be necessary to carry out checks for cross-contamination and carry out regular clean-ups.

(7) CONCLUSION

Asbestos is so unique in its properties that it probably cannot be replaced completely with substitutes. Its continued use will depend on its particular value in the application proposed and the degree of safety needed. The balance can only be ascertained by continued research. Provided that the technology we have is used properly, that common sense prevails, and that operatives are properly informed and trained, then asbestos can be dealt with successfully.

(8) REFERENCES

Anon. (1976) Is talc wrongly classed with asbestos as a hazard?, *Ceram. Ind.* **106,** 24–26.

Cooke, W. E. (1927) Pulmonary asbestosis, *Brit. Med. J.*, **2,** 1024.

Cunningham, H. M., and Pontefract, R. (1971) Asbestos fibres in beverages and drinking water, *Nature*, Lond., **232,** 332–333.

Doll, R. (1955) Mortality from lung cancer in asbestos workers, *Brit. J. Ind. Med.*, **12,** 81.

Harwood, C. F., Oesteich, D. K., Sierbert, P. and Stockham, J. D. (1975).

Asbestos emissions from baghouse controlled sources, *Am. Ind. Hyg. Ass. J.*, **36,** 595–603.

Keen, R. C., and Mumford, C. J. (1975) A preliminary study of hazards to toxic waste disposal operators on ten landfill sites in Britain, *Ann. Occup. Hyg.* **18,** 213–228.

Langer, A. M. (1974) The subject of continuous vigilance, *Envir. Hlth. Perspect.* **9,** 53–56.

Legge, R. F. (1974) The Great Lakes Advisory Board report on asbestos—a preview, *Can. Res. Dev.*, **7,** 19–20, 54.

Lynch, K. M., and Smith, W. A. (1935) Pulmonary asbestosis. III. Carcinoma of lung in asbestos workers, *Am. J. Cancer*, **24,** 56.

Pott, F., Huth, F., and Friedrichs, S. (1972) Tumours of rats after i.p. injection of powdered crysotile and benzo(*a*]pyrene, *Zbl. Bakt, Parasitenkd. Infektionskr. Hyg., Abt. 1, Orig., Reihe B*, **155,** 463.

Sawyer, R. H. (1977) Asbestos exposure in a Yale building, *Envir. Res.* **13,** 146–149.

Selikoff, I. J., Hammond, E. C., and Churg, J. (1968) Asbestos exposure, smoking and neoplasia *J. Am. Med. Ass.*, **204,** 106.

Smither, W. J. (1974) Asbestos in the workplace and the community, *Envir. Hlth. Perspect.*, **9,** 327–329.

Von Buchstab, V. (1974) Asbestos, fibres, dollars and sense, *Can. Res. Dev.*, **7,** 33–36.

Wagner, J. C., Sleggs, C. A., and Marchand, P. (1960) Diffuse pleural mesothelioma and asbestos exposure in North Western Cape Province, *Brit. J. Ind. Med.*, **13,** 250.

(9) BIBLIOGRAPHY

9(i) Publications by Asbestosis Research Council

(a) *Codes of Practice*

For Handling Consignments of Asbestos Fibre (Rev. Aug. 1975).
For Handling and Disposal of Asbestos Waste Materials (Rev. Mar. 1973).

(b) *Technical Notes*

No. 1: *The Measurement of Airborne Asbestos Dust by the Membrane Filter Method* (Rev. Sep. 1971)

No. 2: *Dust Sampling Procedures for Use with the Asbestos Regulations* (Issued Jan. 1971)

(c) *Control and Safety Guides*

No. 1: *Protective Equipment in the Asbestos Industry (Respiratory Equipment and Protective Clothing)* (Rev. Jan. 1977)
No. 2: *The Application of Sprayed Asbestos Coatings* (Rev. Mar. 1972)
No. 3: *Stripping and Fitting of Asbestos-containing Thermal Insulation* (Rev. Mar. 1973)
No. 4: *Asbestos Textile Products, CAF/Asbestos Beater Jointings and Asbestos Millboard* (Rev. Dec. 1971)
No. 5: *Asbestos-based Materials for the Building and Shipbuilding Industries and Electrical and Engineering Insulation* (Rev. Sep. 1975)
No. 6: *Handling, Storage, Transportation and Discharging of Asbestos Fibre into Manufacturing Processes* (Rev. Dec. 1971)
No. 7: *Control of Dust by Exhaust Ventilation* (Rev. April 1973)
No. 8: *Asbestos Based Friction Materials and Asbestos Reinforced Resinous Moulded Materials* (Rev. Mar. 1977)
No. 9: *The Cleaning of Premises and Plant in Accordance with the Asbestos Regulations* (Rev. Jan. 1977)

(d) *Miscellaneous*

Safety Posters
Leaflets *Safe Working with Asbestos* series:
(*a*) *Stripping Asbestos Thermal Insulation.*
(*b*) *Care and Accommodation of Clothing and Masks.*
(*c*) *Handling and Disposal of Asbestos Waste.*
(*d*) *Using Asbestos Cloth.*
(*e*) *Asbestos Cement and Insulating Board.*
(*f*) *Friction and Anti-friction Materials.*

9(ii) Health and Safety Executive (Formerly Department of Employment and Productivity)

(a) *Problems Arising from the Use of Asbestos*, Memorandum of the Senior Medical Inspector's Advisory Panel (1968).
(b) *Health and Safety at Work Booklet*, No. 44: *Asbestos Health Precautions in Industry.*
(c) *Technical Data Notes*
No. 1: *Dust Control (Revised) Low Volume, High Velocity System* (being revised).
No. 13: *Standards for Asbestos Dust Concentration for Use with Asbestos Regulations 1969* (Revised). [Now replaced: see under (f).]
No. 24: *Asbestos Regulations 1969: Respiratory Protective Equipment.*
No. 35: *Control of Asbestos Dust* (being revised).
No. 42: *Probable Asbestos Dust Concentrations at Construction Processes.*
No. 52: *Health Hazards from Sprayed Asbestos Coatings in Buildings.*
(d) *Precautions in the use of Asbestos in the Construction Industry* (1974).
(e) *Asbestos Health Hazards and Precautions* (an interim statement by the Advisory Committee on Asbestos) (1977).

(f) *Guidance Notes*

Environmental Hygiene/10 (December 1976).
Asbestos-hygiene Standards and Measurement of Airborne dust Concentrations.

9(iii) Legislation

Asbestos Regulations 1931.
Asbestos Regulations 1969.
Deposit of Poisonous Wast Act 1972.
Control of Pollution Act 1974.
Health and Safety at Work Act 1974, particularly Sections 2, 3, and 4.

9(iv) Asbestos Information Committee

(a) *Asbestos—Its Special Attributes and Uses.*
(b) *Asbestos Dust—Safety and Control* (June 1976) (questions and answers on the subject).
(c) *Asbestos and Health* (a general survey of the medical situation).
(d) *Asbestos and Your Health* (for the guidance of workers).
(e) *Asbestos—Public Not at Risk.*
(f) *Why Asbestos?*
(g) *Is Brake Lining Dust Harmful?*
(h) *Asbestos and the Docker.*
(i) *Safety of Buildings Incorporating Asbestos.*
(j) Report of the Advisory Committee on Asbestos Cancers to the Director of the International Agency for Research on Cancer, Lyon, October 1972.
(k) Factory posters.

9(v) British Standards Institution

(a) BS 4275 (1974), *Recommendations for the Selection, Use and Maintenance of Respiratory Protective Equipment.*
(b) BS 3958, *Thermal Insulating Materials.*

9(vi) Miscellaneous

(a) *Recommendations for Handling Asbestos,* EEUA (Engineering Equipment Users Association) Handbook No. 33 (Revised 1971), plus Addendum No. 1 (1976).
(b) *Q+A; Safe Living and Working with T.A.C.'s Asbestos Products* (T.A.C. Construction Materials Ltd.).

CHAPTER 9

The use of asbestos and asbestos-free substitutes in buildings

Robert Derricott*

Greater London Council

* Robert Derricott, B.Arch., ARIBA, AI. Arb., is Deputy Technical Policy Architect in the Greater London Council's Department of Architecture and Civic Design, and is a member of the Building Control Panel of the Royal Institute of British Architects. The unattributed views expressed in this review are those of the author and do not necessarily reflect those of either the Greater London Council or the Royal Institute of British Architects. Although note is made where appropriate of the situation prevailing elsewhere, the review relates to considerations in and legislation pertaining to the United Kingdom up to March 31st, 1977.

(1) SYNOPSIS

The use of asbestos in buildings is a subject which is highly topical; there is, however, generally a lack of knowledge of the risks, legislation and necessary precautions associated with its use or removal. The object of this review is to collate information on its use in buildings so as to reflect an up-to-date summary of current medical and technical knowledge relevant to that industry, and to the work of those who may have to deal with enquiries from tenants or building owners, developers, site labour, health and safety inspectors, public authorities, councillors, building and demolition contractors, specifiers, etc. The review therefore considers such questions as how and why asbestos-based building materials have been used or changed over the years; what hazards and labour problems are peculiar to the use or presence of asbestos in the building industry and its corollary, demolition; the extent to which asbestos-free substitutes may be considered either available or safe; and the extent to which risks and needs may be safely balanced in the continued use of some forms of asbestos. Finally, the current situation is assessed in terms of likely trends with regard to the use of asbestos and asbestos-free substitute materials in buildings.

(2) INTRODUCTION

Asbestos in its various forms, and because of its unique properties, has long been, and still is, widely used in buildings. Extensive publicity, particularly over the last 2 years, has led to a greater public awareness of associated risks, and to an active involvement of a wide cross-section of the community—manufacturers, architects, building users, building operatives, trades unions, general and specialist contractors, building control officers—in the question of its use or replacement. Not all of this publicity has been well informed, but the hazards associated with the use of asbestos in buildings have certainly led to changes in legislation and in production and machining methods, to limitations of the use of asbestos, and to a demand for asbestos-free substitutes.

(3) ASBESTOS IN BUILDINGS

Asbestos is the only known naturally occuring fibrous mineral. It is a generic term which is collectively applied to a number of such minerals, characterized principally by their resistance to decay, vermin, and heat, leading to their use in thermal insulation and fire-resisting applications, and by the length, strength, and flexibility of their fibres. Asbestos fibres generally do not degrade until a temperature of about 500 °C, when dehydration and major strength loss occurs. 'Asbestos' sheet is really a misnomer; asbestos fibres used in buildings in sheet form must be, unless they are woven, used in conjunction with binders. Further, not being indigenous to the U.K., all asbestos fibre and some asbestos-based products used in the U.K. building industry are imported, the significance of which will be referred to below.

The types of asbestos fibre which are of greatest importance in building are chrysotile (commonly called white asbestos), amosite (brown asbestos) and crocidolite (blue asbestos), their common names broadly corresponding with their natural colour, although this may be obscured to some extent when used in small proportions with some binders. Other forms are anthophyllite and the weaker tremolite and actinolite fibres; some of these asbestiforms may be present in small amounts in buildings from their use in other industries, for example in containers of cheap talc containing industrial-grade talc.

Of these so-called asbestiforms, chrysotile asbestos has the longest and strongest fibres, which enable them to be spun and woven or to be used in the reinforcement of other materials such as cement, since chrysotile is also alkali-resistant, and polymers to form a composite. The remaining types are all comparatively inflexible or generally too short in fibre length to be spun or woven, but tend to be chemically more stable and more resistant to acids than is chrysotile asbestos. Fibre characteristics have thus helped to determine the choice of asbestiform for different building applications, just as they have also affected the degree of risk which the various asbestos-based building materials present.

The precise extent to which the three asbestiforms of most importance to the manufacture of building products have been used is open to some conjecture, although all sources agree that chrysotile is by far the most extensively used, comprising about 92% of total annual production, taking the mean of figures published by Green and Pye (1976), the Asbestos Information Committee (1976a), and TBA Industrial Products Limited. It is understood that the remaining 8% is divided roughly equally between amosite and crocidolite asbestos.

For reasons which have been well publicized, and which are noted below, crocidolite generally has not been imported into the U.K. as a raw material or used in the U.K. manufacture of asbestos-based building products since 1970 or soon after. This does not necessarily follow, however, in respect of products manufactured outside and then imported into the U.K. as finished goods, nor does it necessarily follow in respect of the possible use of then existing stocks of crocidolite used, for example, in spraying applications. For a period during the 1960s, crocidolite was for some applications a cheaper alternative to chrysotile asbestos. The Asbestos Information Committee (1976b), however, has published the amount of asbestos by type which is contained in U.K. manufactured building products, and these figures are given in Table 9.1, data from which are used in the following text.

Table 9.1 Amounts of asbestos by type which are contained in U.K. manufactured building products and products used indirectly in the building industry [based on information published by the Asbestos Information Committee (1976b), by permission]

Building product	Approx. content (%)	Type
Asbestos–cement building products (except pipes)	12–15	Chrysotile
Fire-resistant insulating boards	25–40	Amosite (with a small percentage of chrysotile in some cases)
Asbestos paper	70–95	Chrysotile
Asbestos millboard	45–98	Chrysotile
Asbestos insulation blocks and pipe sections	55	Amosite
Asbestos-reinforced thermosetting plastics	55	Chrysotile
Asbestos jointings and packings	25–85	Chrysotile
Asbestos textiles	85–100	Chrysotile
Asbestos–cement pipes	12–14	Chrysotile with a small proportion of amosite
Vinyl/asbestos floor tiles	$5–7\frac{1}{2}$	Chrysotile

The extent to which asbestos-based building materials may or may not be acceptable to individual users or specifiers will, of course, depend on all of the relevant circumstances. The Greater London Council (1976a, 1977c) has published its own policy on the subject. In order to assess the use of asbestos, however, both in the construction and in the use of buildings, the range of relevant asbestos-based components is set out below in four broad categories for the sake of convenience.

3(i) Higher density hard-surfaced materials

For asbestos–cement building products, about 12–15% of chrysotile fibre is normally added to hydrated Portland cement; with pipes a small proportion of amosite fibre is introduced also. The end product is then compressed into flat or corrugated sheets and other moulded goods in which the asbestos fibre is cementitiously bound in the resultant strong but brittle, durable, economic, non-combustible composite.

Semi-compressed asbestos–cement flat sheets are available in a naturally light grey colour and with a single smooth face. They are often specified for use as wall and ceiling linings, particularly in agricultural and sectional buildings. Fully compressed flat sheets are available with a range of factory-applied coloured finishes and are specified for use in the facing of composite walling panels, in weatherboarding, etc. Thermal conductivities and densities should be checked with individual manufacturers but the thermal conductivity will be of the order of 0.65 W/m °C and the density about 1200 and 1600 kg/m^3 for semi-compressed and compressed asbestos–cement flat sheets, respectively. U.K. examples are the Poilite and Glasal ranges of TAC Construction Materials Limited and Eternit Building Products Limited, respectively.

'Asbestos wood', so named because of its workability and not its content–it contains a higher proportion, believed to be about 25%, of chrysotile asbestos fibre in hydrated Portland cement—is used in general thermal insulation work and in the fire protection of structural steelwork.

Semi- and fully compressed asbestos–cement flat sheets are also available with the addition of 'pulp fibre', giving the boards a light tan appearance. Uses include internal partitions and roof and wall linings in industrialized buildings. The Turnall range of TAC Construction Materials Limited includes these types of sheets and also asbestos wood. The density of fully compressed sheets is about 1500 kg/m^3 and of semi-compressed and asbestos wood sheets about 1200 kg/m^3. The thermal conductivity of all three types of board is about 0.3 W/m °C.

Fully compressed asbestos–cement is also available in integrally coloured, extruded, hollow-section form. The variety of uses include cills and copings in an unfinished state for external application and internal skirting boards, shelves, supported working surfaces, etc., which if desired can be wax polished. The Massal range of Eternit Building Products Limited includes these products. This company's Pierrite Sheet is similarly a fully compressed asbestos–cement sheet. With a density of the order of 2000 kg/m^3 it has been specially developed for use in laboratory engineering and foundry buildings and in electrical installations. Here a limited choice of colour is available and uses include bench surfaces and fume-cupboard linings.

UK Marinite, a rigid insulating composite of autoclaved amosite asbestos-reinforced hydrated lime and silica, has been developed for the manufacture of glazing channels by the Timber Research and Development Association (TRADA) under licence from Cape Boards and Panels Limited as Marinite TRADA Firecheck Channel. This material assists the retention of areas of glazing in fire conditions and may be finished with intumescent paint in the sealing of gaps in panels, screens, and door constructions. It is also available in board form for internal insulation and fire protection applications where boards thicker than standard asbestos insulating boards are required or when veneering or edge screwing is specified.

3(ii) Lower density soft-surfaced materials

For lower density asbestos-based thermal insulating boards, about 25–40% of amosite asbestos fibre is normally added to hydrated Portland cement. The composite is then semi-compressed into boards which, on account of their characteristic softness and higher asbestos fibre content in comparison with asbestos–cement products, are more susceptible to fibre release on being handled or worked. This lower density (around 700 kg/m^3), softness, and higher asbestos fibre content give the boards an ability to offer not just fire protection but also insulation against heat (the thermal conductivity is about 0.1 W/m °C) and sound transmission. Being light in weight, off-white in colour, acid-resistant, smooth-faced on one side, easy to cut to size or shape, the boards are often used for internal wall and ceiling linings, duct and door linings, and in the fire protection of structural steelwork. Timber-framed partitions faced with asbestos insulating boards will offer about 35 decibel sound reduction, depending on the board thickness and cavity dimensions and/or content. U.K. examples are Asbestolux, produced by Cape Boards and Panels Limited, and Turnabestos, produced by TAC Construction Materials Limited.

With a chrysotile asbestos fibre content of about 45–98% and with other non-combustible agents and fillers, asbestos millboard can also be used for thermal insulation applications. These include linings for low-pressure steam pipework, the thermal insulation of sheet metal ductwork and industrial ovens, and, in the home, rests for electric clothes irons and the insulation of the irons themselves. U.K. manufacturers include Bestobell Engineering Products Limited.

3(iii) Sprayed asbestos

'Sprayed asbestos' generally refers to a gun-applied hydrated asbestos–cement containing about 60% of asbestos fibre. Chrysotile, amosite, and crocidolite fibres have all been used in this form of providing insulation particularly to structural steelwork protected prior to 1970 in the U.K. against fire and heat, in providing protection against cold bridges and condensation in external wall construction, and insulation against sound. Sprayed asbestos is not now normally specified in the U.K.

3(iv) Other Asbestos-based building materials and components

In the consideration of the use of asbestos in buildings, those concerned by its presence must account not only for the materials with which a building is constructed and serviced but also for the materials introduced whenever a building is used.

It has been said above that chrysotile asbestos fibres have a flexibility, length, and strength which enable them to reinforce other binding materials or to be spun and woven. Asbestos textiles, for example, normally have a chrysotile asbestos fibre content of the order of 85–100% and are used in the manufacture of fire-protective clothing—overalls, gloves, gaiters, aprons, etc.—and fire blankets, curtains, and drapes. Asbestos textiles have long offered significant protection against heat. One manufacturer claims a degradation rate better than steel plate at over 2000 °C. Asbestos yarns can be reinforced with other yarns or filaments to improve or alter the properties of the textiles. Plaited asbestos tubings are also in common use as a flexible fire-insulative sleeving for electricity wires and cables. Asbestos ropes are frequently used to seal gaps around pipework in order to maintain the integrity of compartment walls and floors.

Asbestos-based jointings and packings normally have a chrysotile asbestos fibre content in the range 25–85%, and are used to provide integrity for joints between, for example, composite wall panels in industrialized system-built structures which have a fire-resisting requirement, and to seal joints in pipework and doors in kilns, ovens, and autoclaves.

The relevance in buildings of asbestos papers, which normally have a chrysotile asbestos fibre content of 70–95%, is generally restricted to its manufacture and supply as a raw material for other production processes, except for its common use in electrical installations.

Vinyl asbestos floor tiles contain 5–7½% of chrysotile asbestos fibre and consist of a polymeric binder encapsulating asbestos fibres in order to limit surface spread of flame and to use in conjunction with under-floor heating systems.

Asbestos-based roofing felts consist of a bitumen binder encapsulating asbestos fibres in order to resist flaming brands and the transmission of radiant heat, and are referred to in Schedule 9 of the Building Regulations 1976 and in Part VI of the London Building (Constructional) By-laws 1972. Some damp-proof courses also contain asbestos fibres.

A very different material, which was a dry powder containing 4% of asbestos fibre and needing the admixture of water for hydration, was Artex Compound, manufactured by Artex Products (Manufacturing) Limited. The product offers a ceiling finish with protection against flame spread and also offers a means of caulking plasterboard joints. The same U.K. manufacturer offered a product for similar applications known as W14, with a further reduced asbestos content of about 1.8% and with an emulsion binder. Since June 1976 and November 1976 respectively, however, the company has offered both products in asbestos-free form, the latter now being named Artex washable W14.

Asbestos fibres are also present in some high-temperature-resisting mastic and caulking compounds, and foamed asbestos may be used to give acoustic insulation combined with low spread of flame characteristic in mechanical sound attenuators in properties sited close to high noise sources.

The foregoing is intended to describe the classes of use of asbestos-based building materials and, as such, it constitutes a representative rather than a complete list of product types. Manufacturers' product data, and where necessary corroborating copy test certificates, should be consulted in respect of the satisfaction of fire test standards; in the U.K. these will be, as appropriate, Regulation E15(1)(e) and (f) of the Building Regulations 1976 and British Standard 476, Parts 3(1975), 4(1970), 5(1968), 6(1968), 7(1971), and 8(1972). Any of the above categories of asbestos-based materials may be used—worked, handled or stored—in a wide range of building types. Asbestos clothing and drapes in laboratories, foundries, workshops, and other industrial premises, fire blankets in all of these buildings and schools, domestic kitchens, and others, some asbestos rope, jointings, and packings, and applications in electrical engineering may be specified in the widest range of building types. Those products within which the asbestos fibre is not totally encapsulated may not be present in their raw state; they may have been aluminized to improve radiant heat characteristics—as with drapes in workshops and boundries—given a polymeric coating to render them waterproof, painted, plastered, laminated, or otherwise boxed-in and concealed from view.

(4) THE RISKS TO HEALTH IN THE USE OF ASBESTOS IN BUILDINGS

Until recently, it had been generally accepted that the risks of building workers and users contracting asbestos-related diseases, principally asbestosis, mesothelioma, or lung cancer, were related to the degree of exposure to the inhalation of asbestos fibres, to the type of asbestos fibre present, and to the relative health or resistance of the exposed person. The consideration that the person may also be a smoker is also thought to increase the risk significantly; and fibre size is also accepted as being of great relevance. TBA Industrial Products Limited state:

> 'Fibres can be successfully split into finer fibres with a diameter of less than 0.000 05 mm (0.000 002 in), many hundreds of times finer than a human hair'

and the Greater London Council (1976a) that fibres 5–100 μm* in length and less than 2 μm in diameter constitute a hazard to health. It is generally accepted that such fibres having a length to diameter ratio of at least 3:1 present the greatest danger. In building processes, drilling, scribing, sawing,

* μm = 0.000 04 in.

cutting, fitting, and breaking out serve to expose and vandalize the fibres and a knowledge of and respect for the hazards facing the site operative and even the do-it-yourself building user are essential.

Much work has already been done by the manufacturers of building materials, not only in seeking alternatives but also in dust supression of asbestos fibres. Fortex yarns are such a development. Although they do not discuss the life expectancy of the process, TBA Industrial Products Limited state of Fortex:

> 'The raw asbestos is "opened" by a chemical process to form a colloidal dispersion. This eliminates the need for the traditional mechanical opening and carding; the dispersion is extruded and coagulated, and the strand so formed is then twisted or spun into purer, finer, stronger, and more uniform asbestos yarns than have formerly been avialable...
> Extensive tests on dust supressed asbestos textiles have shown at least a 5-fold reduction in dust emission in comparison with non-supressed textiles'.

In recent months, however, the relationship of the degree of exposure to the various asbestiforms to the relative resistance of those exposed—often referred to as the dose–response—has been challenged. A single, unfortunate, chance fibre may be all that is required; just as a single, equally unfortunate, chance burst tyre at speed may be all that it takes to cause disability or death. Experiments, albeit on animals, have even been widely reported to suggest that the risks may not necessarily be restricted to inhalation alone. The building industry affects many people: asbestos factory workers; merchants; contractors and site labour; architects and site supervisory staff; those who clean and maintain, for example, rainwater gutters to asbestos-based roofing materials; those who inhabit, work in, and use buildings; and those who enjoy the environment around all of these factories, shops, homes, building sites, cars with asbestos brake linings, and so on. These factors, therefore, will be examined where they have particular relevance to the construction, demolition, or use of buildings.

Two things are reasonably certain. Firstly, given that all of the conditions are right for the contracting of an asbestos-related disease, a dose of asbestos fibres, freed from their binding material or woven state, by handling, mishandling, working, or the action of vandals, will not produce an immediate disease response. A period of several decades may pass before symptoms of disease may reveal the effect of exposure. Secondly, crocidolite asbestos fibres, which have been subject to a voluntary ban by U.K. building material manufacturers since the late 1960s or early 1970s, seem to be capable of penetrating into the lung more easily than other asbestiforms and are capable of causing mesothelioma, a rare and, as yet, incurable form of cancer. Because of this delayed effect, however, and because building and building material manufacturing processes are cleaner and companies are more health conscious than in earlicr decades, the results of earlier health tests may be used to illustrate the health hazard today with a certain factor of safety.

Tests in the asbestos industry carried out by the British Occupational Hygiene Society in 1966 resulted in a standard of 2 fibres/cm^3 being applied to working conditions, affecting, for example, asbestos factory and building workers, as giving less than a 1% risk of contracting asbestosis, in respect of chrysotile and amosite asbestos fibres. A factor of 10 was applied in the U.K. to this level where crocidolite asbestos was involved, again for a 1% risk, to give a maximum concentration of 0.2 fibres/cm^3.

While these standards related to a 1% risk, the British Occupational Hygiene Society (1968) applied a standard of 0.4 fibres/cm^3 for chrysotile asbestos dust as giving a *negligible* risk applied to a lifetime's working exposure. A factor of 10 was again applied to this level to give a maximum concentration of 0.04 fibres/cm^3 for negligible risk in a non-working situation affecting, for example, the homeowner or the pedestrian.

Recent argument on the hazards to which those involved with buildings are or may be exposed has been very wide ranging. It has been argued, for example, that the application of a factor of 10 to the standard for crocidolite asbestos is arbitrary and that, since a single fibre may be all that is required to induce mesothelioma, there can be no acceptable standard for crocidolite asbestos fibres released in the air. Nevertheless, Byrom *et al.*'s (1969) dust survey in buildings in which various asbestos-based building materials were present or were being used, is claimed by the Asbestos Information Committee (1976c) to show that even the level of *negligible* risk of contracting asbestosis was not being exceeded in 93% of the buildings sampled.

Rickards and Badami (1971), however, under the sponsorship of the Asbestosis Research Council, determined that much more sensitive analytical techniques were required, particularly when examining the common environment–the air which all building users, not just asbestos factory or building workers, breathe.

Gillie *et al.* (1977) reviewed a report by J. Peto (in press) of the Oxford University Cancer Epidemiology and Clinical Trials Unit:

> 'The new report suggests that if the standard remains unchanged as many as one asbestos worker in 14 spending a lifetime in the industry may die prematurely...
>
> Peto's most worrying suggestion is that one man in 100 exposed to a dust level half the official standard—one fibre per centimetre—will die of asbestosis after "a lifetime's" exposure'.

These sentiments have been expressed also in the U.K. technical press. Stevens (1976), in commenting upon an earlier American study by Johns-Manville, says:

> 'The current British standard of 2 fibres/cm^3 is virtually worthless

until an accurate and reliable sampling technique can be perfected. To raise the standards... to say 0.5 fibres/cm^3 would require even more care to ensure the standards could be enforced'.

The Commission of the European Communities (1976a) has also expressed a concern for the reliable analysis of asbestos fibres.

In constructing and living with buildings in which asbestos is used, therefore, the relative ease with which fibres of the three main asbestiforms—chrysotile, amosite, and crocidolite asbestos—may be released is of great relevance; so too is the relative composition of the asbestos-based material, whether in high-density hard-surfaced materials, lower density soft-surfaced materials, sprayed asbestos, or asbestos textiles, or in a particular building material; so too are the intermittency, degree, and period of exposure of the operative, whether or not asbestos is involved in a building or insulation process, or in a demolition or de-lagging activity, whether or not the asbestos-based material has been wetted prior to being worked, and whether or not the tools used to work the asbestos composite are powered. The Department of Employment (1973) has considered all of these factors and has described the respective fibre emissions which may be associated or expected from each material type or process. Similarly, the Asbestosis Research Council indicates safe durations for various building tasks. The adequate wetting of sprayed asbestos or the use of approved exhaust ventilation when power sawing, for example, reduces fibre emission in some cases by more than 10-fold.

Perhaps the greatest hazard in building and demolition contracting arises during the demolition of buildings and during work requiring part or total removal of old asbestos-based thermal insulation of structural steelwork from chemical or heating plant or pipework, particularly where crocidolite asbestos is present. Indeed, in the U.K. any person engaged in a process involving crocidolite asbestos must give 28 days' prior written notice to the Inspectorate under the Asbestos Regulations 1969. The utmost respect is therefore recommended when dealing with crocidolite asbestos, and asbestos of any type in loose or sprayed form. Working on asbestos-based building products by mechanical means is likely to necessitate dust-extraction equipment, and it is always preferable, even when not using mechanical tools, to work, where practicable, in the open air rather than in a confined space and to cut rather than tear asbestos textiles. Again, where practicable, wetting of the material to be worked should always be considered, and friable material such as asbestos rope should be transported in cut lengths in dust-proof containers such as polythene. Waste material and off-cuts containing asbestos should also be collected into and transported, in sealed, double-skinned, dust-proof bags suitably labelled to identify the material easily, to the local refuse disposal facility by arrangement*. Here,

* The disposal of high-density material such as asbestos–cement does not necessitate such pre-sealing measures.

of course, there is a dichotomy of interests: asbestos-based building materials cannot be disposed of by burning; nevertheless, the requirements in the U.K. of the Asbestos Regulations 1969, the Control of Pollution Act 1974, and the Health and Safety Inspectorate must be satisfied.

Frequently the asbestos hazard may be due as much to the asbestos itself as to publicity (or lack of it), supervision (or lack of it), or intervention by the building user or by a local combination of workers. There have been cases of children being kept from school, workmen being told to vacate premises before the work on asbestos is complete so that the affected room can be used, and so on. There have been instances where the use of asbestos-based building materials has necessitated remedial action in locations where damage, by balls for example in gymnasia or by window-opening poles, cannot be avoided. Some have suspended the checking, by shaking out, of the condition of asbestos fire blankets provided within buildings in high fire risk areas such as commercial kitchens and laboratories. With the benefit of hindsight, the Greater London Council (1975a) made a choice understatement; in advising on precautions in the use of asbestos in building it suggested that the subject might, 5 years after the Asbestos Regulations 1969 became law in 1970, lack topicality! Kinnersley (1975) described some other reaction following the Asbestos Regulations 1969 in the building field in the U.K.

> 'The P.O. (Post Office) move is the latest in a series in the public sector. First off the mark was the Central Electricity Generating Board. In 1969 it banned all asbestos materials in its buildings, not just high risk areas...
> In 1971 the Property Services Agency of the Department of the Environment (DoE) issued a general instruction to discontinue one of the most hazardous practices spraying asbestos insulation and fire proofing... In 1973 the DoE issued a further instruction to stop using loose asbestos in all buildings provided for the Post Office...
> British Rail also claims to be cutting down on asbestos in its building....'

The Asbestos Information Committee (1976d) reported that in 1972 at Lyon the World Health Organization considered that there was, as yet, no evidence of risk to building users of contracting mesothelioma. More recently, however, the Times Staff Reporter (1975), commenting on the 1975 International Congress on Occupational Health at Brighton, referred to research at the Hebrew University of Jerusalem into mesothelioma being contracted without a working exposure. In Sweden the manufacture of asbestos-based building materials is understood to have been banned during 1976, but that this decision is being contested by union members potentially out of work. Similar industrial reaction in the U.K. was reported by Gillie *et al.* (1977) and by Hodkinson (1977). In Europe generally, the Commission of the European Communities (1976a) wished to reduce the level of asbestos fibres in the environment to the lowest practical value, and to establish European codes of practice. Building Design (1977) reported that the California Department of Health had banned asbestos-based building materials and referred to eight legal cases involving asbestosis in American building material factory workers. In the U.K. the House of Lords (1977) noted that while importation of some raw asbestos material may be subject to customs identification in respect of fibre type, this is not normally the case with foreign manufactured products. Indeed, legally these could include, even now, crocidolite asbestos fibre, although this is not necessarily the case. Nevertheless, in the building and demolition industries neither laboratory nor factory conditions prevail. Even if just crocidolite asbestos was banned, there would still be the question of the existing building stock. Also, while it is possible to wet and bag, there are questions of degree, practicability, bonus, profit, progress of work, and so on. What might be acceptable in a house or a church may not be acceptable in a school or factory with a high damage rating, or in a hospital where the health of the occupants is already of concern. Further, what might be unacceptable in one specification may be acceptable in another. For example, an architect using a standard specification for the majority of his contracts may also involve himself in package deals, where a contractor offers speed and low-cost building to which his own economic specification applies. Similarly, for what might be acceptable on one building site, labour may not be procurable on another, or at a moment's notice may be no longer available. A specialist contractor may consider it perfectly reasonable to wear breathing apparatus; a general contractor may not. In these cases, an architect may be faced with a demand for re-specification which he will have to consider in terms of original intent (relative to means of escape, flame spread, thermal insulation, protection against fire, and the like), possible conformity with both statutory considerations, and current advice pertaining to the use of asbestos in the original specification. He may wish to respond to such a demand with a call for the contractor's own proposal for approval at no extra cost in cases where the specification conforms with both the law and current advice; alternatively, he may wish to invite the manufacturer to attend on site to reply to doubts and objections.

(5) SAMPLING OF INSTALLED BUILDING MATERIALS

It is a fact of life perhaps that most organizations, on account of limited resources, do not store information in a readily retrievable form except on less common materials. Unfortunately, it is some of these everyday materials which, after a considerable period of use, in some applications have been seen to give hitherto unexpected and unacceptable performance: calcium chloride as an additive to concrete and polyurethane are two examples, in addition to asbestos-based building materials. In this situation, as with asbestos, the only alternative is to sample in order to identify the material present and, in turn, to assess properly the degree of associated risk. Indeed, the 'asbestos problem' is not simplified by the fact that the material is frequently painted or laminated or that the natural distinctive colour of the various asbestiforms may be lost owing to its use in small proportions with binders and fillers or simply owing to the effects of dirt, grime, and age. Excepting these difficulties for the moment, it is essential that those directly concerned with building should familiarize themselves with the wide range of asbestos-based building materials by studying samples. Where identification is not readily possible or where there may be abnormal problems, due perhaps to vandalism, the suspected presence of crocidolite asbestos, the effect of an unrelated building defect, or a change of use of the building, sampling by a specialist contractor and an experienced analyst, to ensure safe removal and transportation, representative material, and acceptability of results, should be instigated.

Much has been written elsewhere about the use of X-ray diffraction and microscopy by the various independent laboratories who undertake these analytical contracts, and of the need for a new method of recording trace fibres, particularly whenever crocidolite asbestos is present. Certainly, according to the U.K. technical press, there is a great demand for authoratitive advice; in *Building* (1976), for example, it was stated:

> 'Some measure of the concern felt by local authorities over the asbestos problem is shown in the overwhelming response to the offer announced recently by Harwell Atomic Research Establishment that it was prepared to advise on tests identifying blue asbestos in buildings (*Building*, 30 April)'.

In the U.K. at least, the implications of the Health and Safety at Work etc. Act 1974 are that when, for example, remedial or maintenance contracts are undertaken and the construction or engineering installation suggests the presence of asbestos when this possibility is visually confirmed by a responsible examination, and particularly when the colour of material or date of erection suggests crocidolite fibres, the main contractor should be required to call in an approved specialist to identify the materials in the areas to be worked. This possibility would be properly allowed for by the inclusion of a contingency sum in the contract documents drawn up for the work.

(6) GENERAL CONSIDERATIONS OF HEALTH AND SAFETY

The Asbestos Regulations 1969 became law in the U.K. in May 1970 and applied to all occupational use of asbestos. The earlier Asbestos Regulations

1931, which applied only to asbestos factories following a 2-year inquiry into the health of asbestos workers, were thereby revoked. Under the 1969 Regulations anyone engaged in any process involving crocidolite asbestos is required to inform the Inspectorate at least 28 days before the work commences. Provisions also relate to exhaust ventilation, protective equipment, cleanliness, storage and distribution, and special premises.

Already in 1957, however, three leading British manufacturers of asbestos-based building materials had agreed jointly to sponsor research and development into dust control and had formed the Asbestosis Reasearch Council. Among other things, this body has published a series of Control and Safety Guides intended to protect the health of workers and that of those who they may affect. Further, in 1967, they formed the Asbestos Information Committee in order to disseminate data on the safe use and handling of their products. The number of participating companies has greatly increased to the present day.

Still in the U.K., under the Control of Pollution Act 1974, 3 clear working days' notice must be given to the responsible authorities prior to the disposal of asbestos. The Department of Employment (1974) issued advice on health and safety in the building industry and this was reported by the Greater London Council (1975a). The Health and Safety at Work etc. Act 1974 was promulgated to ensure proper safety standards and awareness applicable to all people 'at work' and, it follows, the general public who might be affected by the working processes. Relative to Section 6 of this Act, the Asbestos Information Committee (1976b) advises associated companies to warn their customers to consider the requirements of the Asbestos Regulations 1969 and the precautionary recommendations of the Asbestosis Research Council. The Greater London Council (1977a) has made similar recommendations to contractors since 1976. Other liabilities may exist—in the U.K., for example, certain provisions of the Occupiers Liability Act 1957.

Against this background, in the House of Commons (1976) the British Secretary of State for Employment announced an advisory committee on asbestos with members and working groups drawn from the fields of industry, unions, medicine, science, law, and user, and invited evidence on matters related to the use of asbestos to be submitted through the Health and Safety Commission. In making their interim statement, the Advisory Committee on Asbestos (1977) did not recommend any change in the standards applicable to dust levels in the air.

(7) LABELLING SCHEMES FOR ASBESTOS-BASED BUILDING MATERIALS

In 1976, the Asbestosis Research Council and the Asbestos Information Committee jointly devized a scheme to advise customers that their products were asbestos-based and that the observance of safety rules could protect the health of users in the building or do-it-yourself trades. The scheme, described by the Asbestos Information Committee (1976e, f), involved a similarly jointly devised warning label which would be voluntarily applied by the manufacturers of asbestos-based building materials to their products, and these have been applied from about October 1976. Figure 9.1 illustrates the industry's symbol, and it is available in tie-on or adhesive form.

This, therefore, was a voluntary arrangement and was not retrospective; products already installed or having been despatched by the manufacturer were not labelled as part of the scheme. Furthermore, products were not identified in the wording of the label according to their fibre content or their readiness to release asbestos dust, and even if they were there would be occasions, which will be considered below, when the potentially more hazardous asbestos-based building materials might be obscured from inspection and from the immediate attention of building users or maintenance or other non-allied building tradesmen.

For this reason, the Greater London Council (1977b) drew attention to the industry's symbol and at the same time announced that where these potentially more hazardous materials were to be retained in this way two labels would be displayed. The first would be the industry's label, and the second would be based on the accepted warning symbol for toxic substances published by the Commission of the European Communities (1976b). Figure 9.2 illustrates the Greater London Council's label.

(8) OTHER SAFETY PRECAUTIONS FOR BUILDING OPERATIVES AND USERS

Although it is extremely doubtful that the pinning of a notice on a wall of a drawing office building site, operatives' messroom, or public building will exonerate an employer from his rights or duties under such legislation as the Health and Safety at Work etc. Act 1974, the dissemination of clear, concise warnings or informative statements to operatives, for example, will do no harm—and may do much good—on a material about which there has been so much written and, in the form in which they may well have received it, in an incomplete, dramatized fashion.

All organizations concerned with building, such as a manufacturer, a building contractor, a large architects' office, or a public direct-labour organization, normally have a single person—even if he heads a small team—nominated to act in the role of safety officer; he will liaise with the Health and Safety Inspectorate and advise management, supervisors and operatives on safety matters; with asbestos he will be the source or disseminator of information on topics such as the types of asbestos currently in use or likely to be found in the processes or contracts with which his

Figure 9.1. The warning label forming the basis of the voluntary labelling scheme devised by the Asbestosis Research Council and the Asbestos Information Committee. (Reproduced by permission of the Asbestosis Research Council and the Asbestos Information Committee)

Figure 9.2. The warning label used, in addition to the industry's label (Figure 9.1), by the Greater London Council to draw attention to potentially dangerous forms of asbestos-based building materials. (Reproduced by permission)

organization is involved; he will advise on what is to be done when asbestos or an asbestos-based material *appears* to be present; which materials are permitted or not permitted from time to time, in terms of acceptable safety standards; how asbestos or asbestos-based building materials may be safely—and legally—disposed of; whether and by whom scientific analysis is necessary in order to identify the asbestiform present; when protective clothing should be worn; or when sealed protective screening is necessary; what should be done while test results are awaited; and with what statutory provision must one comply.

The Greater London Council (1975a, 1976a, 1977b, c) illustrated some of the ways in which one major public authority has acted in these directions.

(9) REMEDIAL CONSTRUCTION AND MAINTENANCE WORK

From the foregoing, it may be deduced that an investigation of the use of asbestos-based materials in any building stock is capable of division into two broad categories:

(1) materials which are free from damage and where future damage on account of building use, occupancy, or material protection is unlikely;

(2) materials which have been damaged or are likely to sustain future damage, and which in consequence will place building users or operatives at risk on account of uncontrolled fibre release.

Those materials which do not give cause for concern when considering a real risk of damage, wanton or otherwise, may be surface sealed to reduce further the risk to health, for example from the simple act of brush cleaning. On the other hand, where damage is likely, one of two possibilities exist: either the material is so badly damaged that its original usefulness is already lost and replacement is necessary, or a more substantial protection of the surface is called for in order to both seal what is there already and also to provide the mechanical barrier against further damage.

Various allied considerations must be taken into account at an early stage. For example, for what period will the work interfere with the normal use of the building? To what action, including maintenance work, is the material likely to be exposed? Will a surface primer be sufficient? Will steel plate be over-providing for even the remotest possibility? Is the type of damage to which the protection will be exposed equivalent to the action of a mis-directed window-opening pole or of a fully loaded fork-lift truck? To what extent can complex engineering installations be adequately cleaned? Did the original requirement which led to the initial selection of the asbestos-based building material involve any or all of the requirements: fire protection, thermal insulation, and accoustic insulation? Who will clean the building for re-occupation? Is the exposed asbestos-based material soft or hard surfaced, of high or low fibre content? Will the building have to be evacuated while the work is in hand or can the affected areas be effectively and safely sealed off? What facilities will be required for the operatives to equip themselves, change clothes, cleanse themselves, wet the materials, and monitor air pollution? Has this type of work been adequately costed to include for B factor payments*.

The Greater London Council (1977c) has indicated for its own work that the type of primer acceptable for surface sealing a number of asbestos-based building materials not subject to damage would be alkali-resistant, oleo-resinous, and compatible with both the asbestos-based substrate and the imposed decorative system. It should be understood, however, that materials used for acoustic insulation will suffer in performance from any surface treatment.

When the material has been or is likely to be so badly damaged that removal is necessary, the question of substitution arises, and this is considered in depth below. Nevertheless, this consideration raises the possibility of the removal of asbestos-based building materials; the problem of creating a health hazard where none existed previously or of increasing the magnitude of the problem where fibres were already being released through damage. In general, removal should always be carried out by specialist contractors who will be properly equipped to contain and limit airborne dust.

It may be, of course, that it is impracticable to carry out full and proper removal whatever the asbestiform present. There will be cases where removal will create a greater hazard than sealing-in and where sealing-in is not a practical solution in terms of maintenance work, for example that associated with some engineering installations. A greater depth of coating may, of course, be necessary in order to contain fibres in an abraded surface than would be achieved with a primer such as that to which reference has already been made. Here, a hard-setting cementitious compound such as the LD range manufactured by TAC Construction Materials Limited offers a low-flexibility solution; a polyvinyl acetate emulsion overspray such as Decadex Firecheck manufactured by Liquid Plastics Ltd. offers a more flexible alternative. In respect of the fire performance of both these and other materials, the manufacturers make significant claims, but obviously these should be checked with the relevant local and fire authority against the original requirements of the job in hand. Also, a metal mesh reinforcement may be first applied to the surface to be oversprayed, particularly where this is of low-key characteristic. In situations where further protection against mechnical damage is required, hardboard, plywood, or steel sheet may be necessary, according to the specific type of damage in question. In order to improve fixing conditions for the operative, prior use of a surface sealer or

* For those not directly associated with the colloquialisms of the building site, the term B factor payment may alternatively be described as inconvenience money.

overspray will normally be recommended; this will limit fibre release while battens or other anchors are being fixed for flame-retarded hardboard and plywood sheathing, or steel sheet. Cover strips are advisable with these materials at joints and edge details.

Reference has already been made to the requirement to give 28 days' prior notice, to the Inspectorate, of work involving crocidolite asbestos. The identification of competent experienced specialist contractors willing to start a contract immediately can be agreed, but the preparation of precise contract documentation and the obtaining of acceptable tenders for the work can add significant additional delay to that which may have already occurred as a result of a detailed visual inspection, subsequent analysis, and consideration of the implications. Temporary protection in agreement with the Inspectorate may be considered necessary in order to contain areas of exposed loose fibrous material such as might have resulted from mechanical damage. A simple example which might prove successful as a temporary expedient is the papering or patching of the damaged material with a lining paper, but the effectiveness of this or any other temporary measure will depend upon the type and extent of the damage, the relative accessibility of the damaged area, the extent to which the temporary expedient may create problems in the final solution, and so on.

In the context of remedial construction and maintenance work to existing buildings of unknown age, two specific problems should be stressed. Firstly, in buildings with a high engineering content, for example hospitals, chemical plants, and power stations, text-book detailing of what to do and how to do it is not always relevant and much will depend upon a thorough and competent examination of the installation by appropriately equipped persons, experienced in the work in hand. Secondly, in voids, for example those above suspended ceilings, used as plenum chambers, ventilation air may be mechanically drawn over asbestos-based building materials and then blown into occupied areas, thus presenting a hazard perhaps abnormal to that otherwise associated with the particular material or its present condition.

Both remedial construction and maintenance work could involve limited or extensive demolition. Here, while the degree of hazard has already been referred to above, it is worth re-emphasizing that the large amounts of water which are recommended by the Asbestosis Research Council (1973), in practice to contain dust emission to an acceptable level, may give rise to extensive implications in the context of damage to the fabric of some buildings and add to the problems and costs of re-occupying others after adequate drying-out delays.

(10) THE SUBSTITUTION OF ASBESTOS IN BUILDINGS

In simple terms, the substitution of asbestos in buildings concerns the search for another material which will offer the same properties which are available in asbestos, principally that it should be:

non-combustible;

resistant to decay, vermin, and many acids;

constituted of long, flexible fibres which can be spun or woven;

have sufficient strength to permit the reinforcement of other binding materials;

unaffected by temperatures up to 500°C;

resistant up to 2000 °C for short periods;

economic;

—in short, a hazard-free 'wonder material'! With the possible exception of the last-mentioned property, not all of these qualities may be required all of the time and there are alternative building materials which may be considered as asbestos substitutes when a requirement is not necessarily all-encompassing. These alternative materials will now be reviewed, for the sake of clarity using the same four broad categories as previously. The question as to whether the substitution of asbestos is necessary, however, will be left until the following section.

10(i) Higher density hard-surfaced materials

Depending on which of the properties of asbestos–cement, asbestos wood, U.K. Marinite, and so on may be considered superfluous to the requirements of the job in hand, wall and ceiling linings, partitions, fire-door constructions, cills, skirtings, and the like may alternatively be constructed with such hard-surfaced materials as cementitious composites reinforced with, in order of ascending extra cost, glass, steel, or carbon, or constructed with glass-reinforced gypsum, metal sheeting, glass-reinforced polyester, or thermoplastics.

In glass-fibre-reinforced cement (GRC), for example, about 5% of relatively expensive alkali-resistant glass-fibres are normally added to hydrated cement and other fillers. The resulting composite can be manufactured by different processes including extrusion, hand, or mechanical spray-up and moulding and can be vacuum de-watered. The composite was introduced by Majumdar and Nurse (1974), and the Building Research Station (1976) reported, on a more recent assessment, that although there is a tendency to embrittle on wet storage and long-term strength forecasts are difficult to make, the material has a high impact strength and is non-combustible and has consistently satisfied a 1-h fire integrity in indicative tests. Thermal insulation materials can be encapsulated and claims of up to 4-h integrity are not unknown. The commercial development of this composite has been taken up by Pilkingtons as Cem-FIL under licence from the National Research Development Corporation (NRDC). Various external finishes are available, and the density is of the order of 2000 kg/m^3.

Glass-reinforced cement or reinforced concrete is also offered as a substitute for large-diameter pipes which may otherwise be specified in asbestos–cement; smaller diameter pipes are available in unplasticized PVC and various metals. In each case the viability of the alternative material may depend upon its ability or failure to cover the properties of asbestos; with pipes embedded in the ground, soil contamination may restrict the choice available.

During the investigations which led to the introduction of Cem-FIL by Pilkingtons, glass-reinforced gypsum (GRG) was made using commercially available cheaper E glass-fibre and a non-alkaline (gypsum) binding material. The NRDC is responsible for encouraging its manufacture under licence. Again, the material offers a high impact strength, a density of about 1000 kg/m^3 and excellent fire resistance, although on account of the gypsum base it is not available for external applications. Manufacturing processes and applications are otherwise similar, and both GRC and GRG have been reviewed by the Greater London Council (1975b, 1976b,c). Both materials have a thermal conductivity of about 0.2 W/m °C, and are offered for use as wall and ceiling linings, in fire door constructions, cills, skirtings–etc., GRG being restricted to internal use, and, as with asbestos–cement, these materials are available in composite construction encapsulating thermal insulation.

Sheet metals are also available for all of these applications, aluminium or steel sheet, for example, in flat or moulded sections, being generally strong but susceptible to impact damage and twisting and buckling on heating, are offered for roof and wall cladding materials, cills, and skirtings. Various finishes, including colour coating, are possible. Being a good conductor of heat, however, metals necessitate protection against cold bridges and the formation of condensation, and again the end result is higher cost.

Glass-reinforced polyester (GRP) may be considered where non-combustibility is not a requirement, for example in flat or moulded wall and roof cladding panels, cills, and the like, where its properties can be accommodated. Generally GRP has a poor scratch resistance and a low impact strength; it is a thermoset and so chars on heating and produces large amounts of smoke and noxious fumes on burning, although its fire properties and weathering characteristics vary greatly according to the formulation. Allowance must be made for thermal expansion in properly detailing fire stops and against the ingress of water. Apart from thermosets, thermoplastics such as acrylics also scratch relatively easily, have a higher impact strength than GRP, and melt on heating. Depending on the thermoplastic, smoke production and the release of molten droplets in fire conditions may be a problem.

An important use of asbestos-based building materials is in the upgrading of internal doors to give fire resistance and hence improve the conditions of escape. Instead of replacing the door with a fire door or facing with asbestos insulating board, medium-grade and oil-tempered hardboard may be used to flush up and face, and hardwood to lip, panelled doors; in this case the local building control officer should be consulted at an early date in respect of fixings to stiles and rails, overall width, frame requirements, and so on.

10(ii) Lower density soft-surfaced materials

Two leading U.K. building material manufacturers have recently marketed alternatives to their asbestos insulating boards. Asbestolux and Turnabestos, manufactured by Cape Building Products Limited and TAC Construction Materials Limited, respectively, are now offered at extra cost in asbestos-free form: Supalux and TAC Limpet Insulation Board, respectively. the Greater London Council (1977d,e) has prepared a report on each new product and in each case relates properties and characteristics to the asbestos-based building material it is intended to substitute.

Cape's Supalux is said to be organic fibre-reinforced calcium silicate with other fillers. In comparison with Asbestolux, Supalux has a higher density, different moisture movement characteristics, lower impact and tensile strengths, and a slightly greater thermal conductivity. The extent to which these differences will critically affect a design prepared for Asbestolux will depend on the individual circumstances of use and handling. The manufacturer offers independent test certificates to demonstrate fire performance. TAC Limpet Insulation Board, on the other hand, is said to be non-asbestos fibre-reinforced Portland cement with density modifiers. In comparison with Turnabestos, Limpet Insulation Board also has a higher density and different moisture movement characteristics, but this time a higher impact strength, a lower tensile strength, and a slightly greater thermal conductivity. Again, independent test certificates are available from the manufacturer in respect of performance in fire conditions and it is, as with Supalux, impossible to generalize on the effect of the different properties of the asbestos-free direct replacement board.

Ceramic fibre board is also available. Based on natural aluminosilicate china clay, the fibres which, it is claimed, will resist continuous temperatures of over 1200 °C, have an *average* diameter of 2.8 μm, can be attacked by certain acids, and can be processed to form asbestos-free insulation board. An example is Triton Kaowool ceramic fibre board available in the U.K. from Morganite Ceramic Fibres Limited, and applications include refractory linings and high-temperature thermal insulation.

Cape Boards and Panels Limited offer Cape Monolux as an asbestos-free alternative to U.K. Marinite for applications requiring a rigid, non-combustible, monolithic, asbestos-free lining material for use in ventilation ducts and fire-door construction, which can be veneered and is resistant to attack by vermin and mould.

Resin-bonded mineral wool fibres are available in various densities from 30 to 200 kg/m^3 and thicknesses and are used for the non-combustible sound and fire resistance of plant pipework and buildings. An example is Rockwool in slab form, designed for use at temperatures up to about 700 °C, with a thermal conductivity of the order of 0.03–0.15 W/m °C depending on the temperature; it is formed from various stones into an inert rock wool fibre and is manufactured by the Rockwool Company (UK) Limited.

Vermiculite is also used with inorganic binders to provide asbestos-free non-combustible insulating boards, which are used, for example, for the protection of structural steelwork. They are relatively light and have a density of about 400 kg/m^3. The boards are rigid and resistant to chemical attack and rot. An example is Vicuclad*, the U.K. manufacturer of which, William Kenyon & Sons (Vicuclad) Limited, claims up to 4 h fire resistance, depending upon thickness.

Glass is also used to form rigid, non-combustible insulating products which are impervious and rot and vermin proof, but may be expected to soften at about 730 °C. Pittsburgh Corning United Kingdom Limited, for instance, market the Foamglas cellular glass insulation range. With a density of about 130 kg/m^3 and a thermal conductivity (dependent on temperature) of about 0.05 W/m °C, the material is offered for use in such applications as in roofing, curtain walling, and pipe insulation.

Perlite (volcanic rock expanded by heat) is also used together with mineral fibres and binders to form a rigid, strong, non-combustible insulation board. Celo-therm, manufactured in the U.K. by Celotex Limited, is typical of this type; with a density of about 175 kg/m^3 and a thermal conductivity of 0.05 W/m °C, the board is offered as a roofing insulant.

10(iii) Sprayed or floated materials

Asbestos-free sprayed insulating materials include products based on mineral fibres. A spray of mineral fibres, inorganic binders, and water is directed on to most surfaces to provide a simple and economic means of providing thermal and acoustic insulation and protection, for up to 4 h depending on thickness, against fire and against condensation in, for example, structural steelwork. The density and thermal conductivity are about 200 kg/m^3 and 0.04 W/m °C depending on temperature, respectively. Examples of mineral fibre-based spray materials include TAC Sprayed Limpet Mineral Fibre GP Grade and Ceramospray, manufactured in the U.K. by TAC Construction Materials Limited and R. B. Hilton Ltd., respectively.

* Recently, Cape Boards and Panels Limited have announced a new product, Vermiculux, in this catagory.

The manufacturer should be consulted in respect of adhesion, and the fixing of a light mesh reinforcement which may be required.

Vermiculite is the basis of a number of hard-set materials for the thermal insulation and fire protection of structural steelwork such as Mandolite P20 and Mandostal, produced by Mandoval Limited, and Pyrok, produced by Pyrok Industrial Marine Coatings Limited, and in the manufacture of acoustic plasters, such as Audex A and Audex G, produced by Mandoval Limited.

Tilling Construction Services Limited offer ready-mixed Perlited Plastering Mortar midway between dense sanded and lightweight plasters. The manufacturer will advise on the thickness required for fire protection or acoustic insulation.

10(iv) Other building materials and components

Glass, wool, and ceramic fibres may be used to form flameproof fabrics as a substitute for those using asbestos fibres. Again, careful attention must be paid to the relative characteristics, and manufacturer's data sheets and, where appropriate, supporting evidence should be consulted. Fire blankets, aprons, gaiters, gloves, protective curtains and welding drapes, and so on, can be made by using, as appropriate, glass or ceramic textiles, flame-proof wool, even heat-resistant leather or nylon, or flame-resistant hessian. The requirements of the job in hand must always be carefully considered, of course. For example, fire blankets made from glass-fibre may not be regarded by some fire authorities as a satisfactory substitute for asbestos in intense fires such as might be encountered when using oxy-acetylene equipment; those made from ceramic fibre may be acceptable at significant extra cost. Generally, the substitution of building materials traditionally manufactured with a high asbestos content depends on the questions of whether heat is of the essence and, if so, the temperature to which the material is to maintain its integrity. Considering the extent of the cost penalty, possible substitute hazards in use can be queried.

Whereas glass-fibre fire blankets may not be acceptable at high temperatures, blankets made of ceramic fibre such as Durablanket, produced by the Carborundum Company, are offered for use at service temperatures of over 1200 °C.

Rope made with ceramic fibre is offered as an alternative to asbestos rope, depending on the required temperature resistance, for some fire-stopping applications. In the U.K., Morganite Ceramic Fibres Limited, for instance, offer ceramic rope with diameters from 12.5 mm.

Safety gloves are offered in the U.K. by Safety Equipment Centres and are made from Du Pont Nomex fibre, a heat-resistant nylon. The material is claimed not to support combustion and to offer sustained exposure at a temperature of 250 °C and short exposure to thermal shock at 350 °C. Comfort, since the product is lighter, abrasion, and chemical resistance, except against concentrated acids, are said to be better than with asbestos gloves.

Fire blankets are available, manufactured in glass fabric. Tutor Safety Products Limited offer Sentilock 42, which is towel-like to touch, and claim non-combustibility and retention of tensile strength when exposed to a temperature of 200 °C for 1 h. Tutor also offer a neoprene rubber and glass fabric composite, Neoglass 180, with claimed resistance to chemical, oil, and water attack, and to the support of combustion, for application in screening spark-producing industrial processes and for protective clothing. Heat-resistant leather for similar applications is claimed by the same company to be self-extinguishing and to maintain flexibility when exposed to a temperature of 200 °C for 2 h. For combustion-resistant screens and clothing and for low-cost screens, Tutor also offer flame-proof wool and flame-resistant hessian, respectively.

Asbestos loose-fill insulation has long been abandoned in the U.K. on health grounds. Alternatives include vermiculite, perlite such as Tilcon Insulation Perlite from Tilling Construction Services Limited, blown mineral wool, and, depending on fire performance requirements, foamed plastics beads.

(11) THE ASBESTOS HAZARD IN PERSPECTIVE

Hodgkinson (1976), in assessing the size of the asbestos hazard, concluded:

> '. . . according to some of the latest laboratory work, artificial mineral fibres produced as substitutes for asbestos can have the same physical structure that is thought to make crocidolite so dangerous. If their use became widespread because of panic action on asbestos, the whole sad cycle of death and disease could reappear after the 20 or 30 years that these illnesses usually take to develop.'

Certainly asbestosis patients, healthy asbestos workers, and the general public have been caused a lot of concern. Is that concern justifiable when considered in perspective? Is asbestos the danger that sensational reporting makes it out to be? Would the risk be removed if asbestos were banned? Are substitutes adequate and safe? The consideration of the asbestos risk in perspective must relate to different asbestiforms: crocidolite is surely a greater hazard than chrysotile asbestos; asbestos rope surely requires more health care than asbestos–cement; machining asbestos in a confined space requires more stringent precautions than hand tooling outdoors; the necessity to sweep damp asbestos dust into sealed bags must have some implication for the continued existence of the brush. Nevertheless, 'ordinary dust' when sweeping a garage, traffic exhaust fumes, gases generated in a burning building constructed and furnished with the materials of a technological age, the siting of buildings using or manufacturing highly explosive materials, the wearing of shoes designed to follow trend and not function, the use of weedkillers and pesticides, the sale of firearms, road death statistics, and many other factors represent a hazard to one degree or another, and all are accepted to one degree or another on account of the principle of probability. One thing is certain: asbestos is a natural resource. Like coal or North Sea oil it has a limitation. Just as new sources of energy will have to be sought, so substitutes for asbestos sooner or later will have to be found. Until then, even if it is banned in one country or even on one housing estate, the air we breath knows no such boundaries; maintenance and demolition work and adequate protection against fire will still be required. The ideal situation in theory would be one in which all danger to human life was completely eliminated. In practice this is not possible: demolition contractors, steelwork erectors, steeplejacks, and others take risks each day in the course of building operations which many would not accept; all people, building workers and the unemployed alike, take risks each time they cross a road; the stability of world economy is based on the rational use of currency; and there are issues of risk assessment and cost benefit and this is by no means limited to the use or substitution of asbestos-based building materials. It has to be possible to calculate and determine a limit of acceptable risk to be assured that persons using a building from which asbestos is removed are not being subjected to an unacceptable hazard as a result. The difference, furthermore, between the hazard associated with the decision to step out on to a partly fixed steel girder 200 ft above ground and the decision to release, say, crocidolite asbestos fibres into the atmosphere is that in the first only the operative's health is at stake. In February 1977, in a near parallel philosophical exercise, the British Government declined to pre-empt a Royal Commission then examining the alleged danger in administering hooping cough vaccine to children; the reason for declining was that, in the light of current knowledge, the gains from using the vaccine far outweighted the risks. A similar situation is held by many to apply with some applications of some asbestiforms in building.

Certainly, to ban asbestos is easy. To ignore the benefits which the material offers, the difficulty in replacing all of those benefits with a single substitute, and the relationship of today's asbestos-related diseases with a much less safety-conscious industry and people of several decades previously, is also easy. To be constructive in criticism is always slightly more difficult; an 'asbestos widow' will always put a different value on the material than the wife whose husband's life has been saved by asbestos in a burning building. Yet to call for the wholesale removal of asbestos-based building materials would be to ignore the substantial hazard which that act would create, if only to the vast number of workmen engaged on the removal. Nevertheless, the asbestos dilemma is inevitable. The emotive, irresolute considerations of contemporary ever-escalating standards of legislation, health, and safety make it so; unfortunately, these measures, as far as the use of asbestos in building is concerned, will be neither proved nor disproved for some time to come—perhaps longer than it will take to consume the remaining natural reserves of asbestos.

(12) THE FUTURE FOR ASBESTOS IN BUILDING

In the U.K., the asbestos-based building material manufacturer is at present free to make his own risk evaluation; he may use as much asbestos as he considers expedient, provided that his employees are adequately protected and provided that his products meet the ambient dust standard; his warning labels advise care in avoiding inhaling dust; he has sought and developed asbestos-free substitutes. The architect is charged with providing a safe, functional, weatherproof, aesthetically pleasing building within set cost limits and within set planning and building control requirements pertaining at the time; he must use his skill, knowledge, and experience in this regard with a whole host of manufacturers, some more reputable than others, advertizing their products, for which often, only short-term test results and no British Standard apply, as better than others. The architect, in addition to practising in an area of commerce which concerns profit and the survival of the fittest, may also be called upon from time to time to act on specific technical directions from his client; in this regard, apropos of the asbestos dilemma, he will have to use his professional integrity in order to satisfy himself and his client that his design meets the stringent safety requirements of the brief and the law, which may require the facility to evacuate totally a building in a certain time, or to evacuate and save the building in the event of fire.

Certainly, for an enforced substitution of asbestos-based building materials to occur tomorrow, a national acceptance of more costly alternative material, which will not necessarily fulfil all of the requirements with which asbestos has been associated in buildings, will be necessary today; so too it must be accepted that the danger to health through specifying asbestos, at least in some of its building applications, will be replaced by a danger to health and life owing to the provision of a lower performance capability in fire conditions and the use without a secure knowledge that new hazards will not be identified later, of relatively new materials; so too, the acceptance by nations and trades unions of the closure of the asbestos industry, except in so far as it has already taken the initiative to develop and market substitute materials will be necessary—always remembering that the asbestos dilemma cannot be considered in isloation, as the air we breath knows no boundaries.

Nevertheless, the 'asbestos' problem is one of fibrotic diseases. Current international opinion seems to agree that any fibre having a diameter less than 2 or 3 μm may constitute a great danger; this, therefore, is not necessarily limited to diseases induced by asbestos. Of this two-facet asbestos dilemma, Green and Pye (1976) reported:

> '(a) certain specialized applications cannot be met with existing alternative fibres;
> (b) it is not considered that the health hazards which may be encountered with fibres other than asbestos have been evaluated with any certainty'.

Indeed, one manufacturer of ceramic fibre goods confirms an average fibre size of 2.8 μm in his sales literature, and Gosney (1977) referred more recently to a working party announced in March 1977 in the U.K. and set up by the Health and Safety Commission to respond to concern that glass fibres constitute as great a danger to health as asbestos. He suggested:

> 'When worked, glass fibre is said to break transversely, unlike asbestos which breaks longitudinally into particles of smaller diameter . . . Glass fibres are produced by spinning and drawing. Dr. P. Elmes of the (Medical Research Council) Pneumoconiosis Unit states that this process produces strands with elongated ends . . . "within the respirable range, having a diameter below two microns . . . the first to break off when the material is handled"'.

It is, of course, only reasonable to say that eminent scientists on both sides of the Atlantic have both supported and rejected this conclusion in respect of non-asbestos fibres. The Association of British Manufacturers of Mineral Insulating Fibres (1976), however, was emphatic:

> 'The inhalation of man-made mineral fibres, unlike asbestos, does not produce fibrotic lung disease. There is no evidence of any increased incidence of lung cancer and no case of mesothelioma has been recorded in relation to exposure to man-made mineral fibres . . . (Nevertheless) the two European Trade Associations, the European Insulation Manufacturers Association (ERIMA) of which Eurisol-UK is a member, and the Comité International de la Rayonne et des Fibres Synthetiques (CIRFS) have joined forces to form a Joint European Medical Research Board (JEMRB) to ensure that the necessary research is carried out'.

Against this background, the Trades Unions Congress (1976) in Britain has determined to pursue a progressive programme of statutory substitutes of asbestos applications in the U.K. for which safe alternative materials exist:

> 'In the shorter term the TUC proposes the outright banning of the importation of all products containing crocidolite into the U.K.... recommends the (interim) establishment of a Maximum Allowable Concentration of asbestos fibres of 0.2 fibres/cm^3 for those asbestos applications for which there is no safe and practical substitute available.
>
> (However) harmful skin effects including dermititis often result from the direct handling and application of glass fibre products. For these reasons the TUC recommends that glass fibre products should not be used in large scale insulation activities and that other safe materials should be used'.

Hence the nature of the asbestos substitution problem is that alternative materials must be both safe to use and give safety in use. In practice, they will cost more and perhaps not do the job so well. The inevitability of the exhausting of asbestos reserves confirms the necessity of substituting asbestos-based building materials; the call for an early enforced substitution, however, does not invalidate a need to assess fully the adequacy of the alternative materials in terms of quality control and workmanship in manufacture and on site, and the kind of misuse which all but the most expensive precautionary measures would fail to eliminate in, for example, some rented property, and in terms of the relevance of indicative or accelerated testing.

Current knowledge relative to asbestos substitutes seems to preclude, at this time, definitive conclusions on replacement. Much more development work will be necessary before that stage is reached. The Asbestos Information Committee (1975) asserts:

> 'More is known about the potential hazards of asbestos and the proper means for controlling them that is the case with materials that might be used to replace them'.

Even if a definitive statement on substitution could be offered, installed asbestos-based materials will, like most other materials, continue to degrade through wear, misuse, vandalism, and modernization and maintenance work, and the demolition of buildings containing crocidolite asbestos, sprayed asbestos, and asbestos pipe lagging and the disposal of waste will present a hazard for generations to come.

Certainly it would seem prudent to adopt the following:

(i) prohibit the use or importation of products containing crocidolite asbestos fibres and value-analyse a reduction of the 2 fibre/cm^3 standard for non-crocidolite asbestos, to 0.2 fibres/cm^3, the present limit for crocidolite asbestos, until such time as the situation relative to substitute materials permits a responsible phased programme of substitution;

(ii) pursue the development of air measurement techniques and define an atmospheric or environmental concentration level, as distinct from maximum allowable working levels of hazardous dust; perhaps this might involve the arbitrary safety factor of 10 to give a value of 0.02 fibres/cm^3, although it is important to ensure that regulations can be policed;

(iii) demonstrate an intelligent risk evaluation associated with the intended asbestos-based or substitute building material; here the trades unions, the public, and specifiers alike will have a vital role to play and, since derogatory conclusions are normally those remembered, much will depend also on the national and technical press;

(iv) recommend the restriction of the friable high-asbestos content materials for which there exists a satisfactory substitute and for the use of which there is a requirement to protect against fire or severe thermal shock;

(v) pursue the identification, recording, and even labelling of the more hazardous materials so that subsequent maintenance or modernization work will not generate a hazard where, with proper precautions, skill, and equipment, none need exist.

(13) ACKNOWLEDGEMENTS

The co-operation of the following for permission to use published texts is greatly appreciated:

Asbestos Information Committee (W. P. Howard, Deputy Chairman);
Builder Group (Cornelius Murphy, Editor, *Building*);
Eurisol-UK (Ian Munro, Secretary General);
Fulmer Research Institute Limited (A. M. Pye);
Greater London Council (M. Gordon);
IPC Magazines Limited (J. Watt, Rights and Permissions Department, New Science Publications, *New Scientist*);
IPC Press Limited (Jerry Gosney, Technical Editor, *Contract Journal*);
Morgan Grampian Limited (Peter Murray, Editor, *Building Design*);
TBA Industrial Products Limited (D. Tomlinson);
Times Newspapers Limited (Miss B. Orawski, Syndication Department, *The Times* and The *Sunday Times*);
Trades Union Congress (Lionel Murray, General Secretary);

and to the following for access to background data:

TAC Construction Materials Ltd (J. D. Simpson);
Tutor Safety Products Limited (K. W. Smith).

(14) REFERENCES

Advisory Committee on Asbestos (1977) *Asbestos: Health Hazards and Precautions, Interim Statement*, HMSO, London.

Asbestos Information Committee (1975) *Asbestos and Health*, Asbestos Information Committee, London.

Asbestos Information Committee (1976a) *Asbestos Dust and Your Health*, Asbestos Information Committee, London (republication of a 1975 paper: *Asbestos and Your Health*).

Asbestos Information Committee (1976b) *Asbestos Dust—Safety and Control*, Asbestos Information Committee, London.

Asbestos Information Committee (1976c) *Safety of Buildings Incorporating Asbestos*, Asbestos Information Committee, London (republished *Asbestos in Buildings*).

Asbestos Information Committee (1976d) *Asbestos Kills? A Commentary on Mrs. Nancy Tait's Thesis*, Asbestos Information Committee, London.

Asbestos Information Committee (1976e) *Take Care with Asbestos: Manual Governing the Design and Use of the Asbestos Safety Symbol*, Asbestos Information Committee, London.

Asbestos Information Committee (1976f) *Asbestos Products Labelling Scheme*, Asbestos Information Committee, London.

Asbestosis Research Council *Working with Asbestos*, Asbestosis Research Council, London.

Asbestosis Research Council (1973) *Stripping and Fitting of Asbestos Containing Thermal Insulation*, Control and Safety Guide No. 3. Asbestosis Research Council, London.

Association of British Manufacturers of Mineral Insulation Fibres (1976) *Health Aspects of Man-Made Mineral Fibres*, Eurisol-UK.

British Occupational Hygiene Society (1968) *Hygiene Standards for Chrysotile Asbestos Dust*, Pergamon Press, Oxford.

Building (1976). 'Sensible' new asbestos regulations, *Building*, 14 May 1976.

Building Design (1977). Asbestos ban in California, *Building Des.*, 4 February 1977.

Building Research Station (1976) *A Study of the Properties of Cem-Fil/OPC Composites*, Building Research Establishment Current Paper CP 38/76, Dept. of the Environment, Building Research Station, Garston.

Byron, J. C., Hodgson, A. A., and Holmes, S. (1969) A dust survey carried out in buildings incorporating asbestos-based materials in their construction, *Occup. Hyg.*, **12,** 144–145.

Commission of the European Communities (1976a) *Off. J. Eur. Commun.*, C294/26, Written Question 440/76 (Lord Bethel).

Commission of the European Communities (1976b) *Off. J. Eur. Commun.*, Information and Notices, **19,** C96.

Department of Employment (1973) *Probable Asbestos Dust Concentrations at Construction Processes*, Technical Data Note 42, HMSO, London.

Department of Employment (1974) *Precautions in the Use of Asbestos in The Construction Industry*, Sub-Committee of the Joint Advisory Committee on Safety and Health in the Construction Industries, HMSO, London.

Department of Employment (Rev. 1974) *Standards for Airborne Dust Concentrations for Uses with the Asbestos Regulations* 1969, Technical Date Note 13, HMSO, London.

Gillie, O., Gillman, P., and May, D. (1977) Official safety limit for asbestos may put one in 14 at risk, special report, *Sunday Times*, 30 January 1977.

Gosney, J. (1977) Is the dust deadly?, *Contract J.*, 10 March 1977.

Greater London Council (1975a) Precautions in the use of asbestos in the construction Industry, *Dev. Mater. Bull.*, No. 83, 7.

Greater London Council (1975b) Departmental Steering Group for the development of products of glass reinforced cement and gypsum, *Dev. Mater. Bull.*, No. 86, 2.

Greater London Council (1976a) Asbestos-based materials, their use in GLC buildings, *Dev. Mater. Bull.*, No. 93, 5 [updated as Greater London Council (1977c)].

Greater London Council (1976b) Departmental Steering Group for development of products of glass reinforced cement and gypsum—2nd Interim Report, *Dev. Mater. Bull.*, No. 91, 1.

Greater London Council (1976c) Departmental Steering Group for development of products of glass reinforced materials—3rd Interim Report, *Dev. Mater. Bull.*, No. 100, 5.

Greater London Council (1977a) *Preambles for Bills of Quantities*, 8th Ed., Clause B41.3, GLC, London.

Greater London Council (1977b) Warning symbols for asbestos, *Dev. Mater. Bull.* No. 101, 8.

Greater London Council (1977c) Asbestos-based materials—GLC use in new buildings and for alteration or maintenance work, *Dev. Mater. Bull.*, No. 103, 4.

Greater London Council (1977d) An asbestos-free alternative to asbestos insulating board, *Dev. Mater. Bull.*, No. 102, 8.

Greater London Council (1977e) Another asbestos-free alternative to asbestos insulating board, *Dev. Mater. Bull.*, No. 105, 5.

Green, A.K., and Pye, A.M. (1976) *Asbestos—Characteristics, Applications and Alternatives*, Fulmer Special Report 5, Fulmer Research Institute, Stoke Pages, Slough.

Hodgkinson, N. (1976) Asbestos: how big is the risk? *The Times*, 12 March 1976.

Hodgkinson, N. (1977) Asbestos dangers threaten a £200 m industry, *The Times*, 28 March 1977.

House of Commons (1976) *Hansard*, **911,** c1218–1225.

House of Lords (1977) *Hansard*, **380,** 34, c611–614.

Kinnersley, P. (1975) What future for asbestos after Post Office ban?, *New Scientist*, 19 June 1975 (Technology Review).

Majumdar, A.J., and Nurse, R.W. (1974) *Glass Fibre Reinforced Cement*, Building Research Establishment Current Paper CP 79/74, Department of the Environment, Building Research Station, Garston.

Newhouse, M.L., and Berry, G. (1976) Precautions of mortality from mesothelioma tumours in asbestos factory workers, *Brit. J. Ind. Med.*, **33,** 147–151.

Peto, J. (in press) in *Environmental Health: Quantitative Methods*, Society for Industrial and Applied Mathematics of Pennsylvania, Heyden and Son, London.

Rickards, A.L., and Badami, D. V. (1971) Chrysotile asbestos in urban air, *Nature, Lond.*, **234,** 93–94.

Stevens, E. (1976) TUC to urge tighter control on asbestos, *Build. Des.*, 12 November 1976.

TBA Industrial Products Limited, Product Information, Asbestos Textiles Division, Rochdale, Lancs.

Times Staff Reporter (1975) Health: dangers of asbestos (Science Report), *The Times*, 17 September 1975.

Trades Union Congress (1976) TUC Evidence to the Advisory Committee on Asbestos. in *Selected Written Evidence to the Advisory Committee on Asbestos*, HMSO, London (1977).

Other background material

Asbestos Cement Manufacturers Association (1976) Submission to the Advisory Committee on Asbestos, Asbestos Cement Manufacturers Association, in *Selected written Evidence to the Advisory Committee on Asbestos.* HMSO, London (1977).

Health and Safety Executive (1976) Health Hazards from *Sprayed Asbestos Coatings in Buildings*, Technical Data Note 52, HMSO, London.

CHAPTER 10

Alternatives to asbestos in industrial applications

A. M. Pye

Fulmer Research Institute, Slough, Berks.

(1) INTRODUCTION

The numerous applications of asbestos are a consequence of its desirable physical and chemical properties, combined with a low material cost. It is this unique combination that makes the replacement of asbestos very difficult in many applications.

Some of the properties of asbestos are summarized in Table 10.2, together with the comparable properties of some of the synthetic fibre materials that have been suggested as replacements for asbestos in some applications. Some of these properties require further comment.

1(i) Thermal properties

The most widely known property of asbestos is its heat and fire resistance, although this resistance is not as great as is popularly believed. Asbestos cannot be classed as refractory, although normally its properties are sufficient to withstand super-heated steam and other high-temperature industrial environments. Degradation of the crystal structure of asbestos and major loss of strength occur at temperatures in the range 300–500 °C. However, a useful performance can be obtained at higher temperatures than this; specified working temperatures for some asbestos products may be as high as 600 °C. The reasons for this are unclear, but some points of significance are apparent.

Chrysotile contains 14% by weight of hydroxyl groups, which are lost from its structure as water vapour ($2OH^- \rightarrow H_2O + O^{2-}$) at temperatures greater than 450 °C. The latent heat of vaporization of this water content is thought to be a potent heat sink, protecting the remaining undegraded fibre. Further, the solid decomposition products are inert and of low thermal conductivity, providing additional protection to the remaining fibres, and maintaining structural integrity. It has been shown that, in some cases, asbestos can maintain its integrity at temperatures up to 1700 °C.

1(ii) Mechanical properties

It can be seen from Table 10.1 that the values quoted for the strength of asbestos fibres are very high. However, even the average values quoted in the table may not tell the whole story. The measurements of strength are inevitably derived from testing in controlled laboratory conditions, and the values obtained may not be representative. Discussion with suppliers of asbestos has suggested that the reliable strength of chrysotile fibres, as produced and used commercially, is no higher than approximately 700 MN/m^2.

1(iii) Other properties

Various other properties make asbestos a valuable material. For instance, its resistance to chemical and biological attack is valuable in applications involving hostile environments, and in achieving a useful service life.

The friction and wear characteristics of chrysotile and its thermal decomposition product forsterite, a non-fibrous silicate, make chrysotile a widely used material in such applications as friction clutches, brake linings, and bearings.

The high aspect ratio of asbestos fibres makes them useful as a mechanical reinforcement in both polymer- and cement-based products.

1(iv) Price and availability

This favourable combination of properties in one material, which is obtainable at a price significantly lower than that of its competitors in

Table 10.1 United Kingdom breakdown of asbestos fibre usage—1973

Use	Chrysotile		Amosite		Anthophyllite		Crocidolite		Total	
	Metric tons	%	Metric tons	%	Metric tons	%	Metric tons	%	Metric tons	%
Asbestos–cement for building products	55 600	37.5	—	—	—	—	—	—	55 600	32.2
Asbestos–cement pressure sewage and drainage pipes	7800	5.3	1200	5.0	—	—	—	—	9000	5.2
Fire-resistant insulation boards	3000	2.0	19 500	81.3	—	—	—	—	22 500	13.1
Insulation products including spray	1300	0.9	2700	11.2	—	—	—	—	4000	2.3
Jointings and packings	11 400	7.7	—	—	—	—	—	—	11 400	6.6
Friction materials	17 000	11.5	—	—	—	—	—	—	17 000	9.9
Textiles products not included in friction materials	8300	5.6	—	—	—	—	—	—	8300	4.8
Floor tiles	16 200	10.9	—	—	—	—	—	—	16 200	9.4
Moulded plastics and battery boxes	2200	1.5	600	2.5	—	—	—	—	2800	1.6
Fillers and reinforcements (belts, millboard, paper, underseals, mastics, adhesives, etc.)	25 300	17.1	—	—	300	100	—	—	25 700	14.9
Totals	148 100	100	24 000	100	300	100	—	—	172 500	100

specific applications, makes asbestos an extremely attractive material. However, as noted in Table 10.1, prices of asbestos are likely to rise. Further, since asbestos is a limited natural resource, with an estimated resource life of about 25 years, prices are likely to continue to increase, and alternative materials will have to be found on availability and cost as well as on health grounds.

1(v) Applications

The breakdown of U.K. usage of asbestos (1973) is given in Table 10.1. In Section 2 the function of asbestos in the different industrial applications and the alternatives that are available are discussed.

(2) INDUSTRIAL APPLICATIONS OF ASBESTOS PRODUCTS

2(i) Asbestos textiles

Chrysotile fibre forms the basic raw material for almost all of the activities of the asbestos textile industry. The length and flexibility of the longer grades of chrysotile are such that spinning into yarn and cloth weaving are possible. Two basic types of yarn are produced: plain, possibly braced with an organic fibre; and reinforced, which incorporate either wire or another yarn such as nylon, cotton, or polyester. The wire-reinforced yarns and textiles can retain their mechanical properties at temperatures up to 600 °C. Recently developed textiles combined with resins and ceramic binders have successfully withstood short-term exposure to temperatures up to 2200 °C. The main applications of asbestos textiles are represented in Figure 10.1. Some of the applications and their alternatives are considered in other sections as indicated.

2(i).1 *Fire and heat protection clothing*

These garments are manufactured from asbestos cloth which is aluminized to give a heat-reflecting surface. The metallic layer is bonded to the cloth by a thermosetting resin.

As an alternative, clothing made from temperature-resistant nylon fibre has found application in fire-fighting and foundry work and as protective underclothing for racing-car drivers. The materials in suitable form can provide short-term protection from exposure to temperatures up to 1370 °C and also for protection against molten metal impingement. Gloves made from this material are suitable for use with contact temperatures up to 300 °C. The product is marketed and manufactured by Du Pont under the name Nomex.

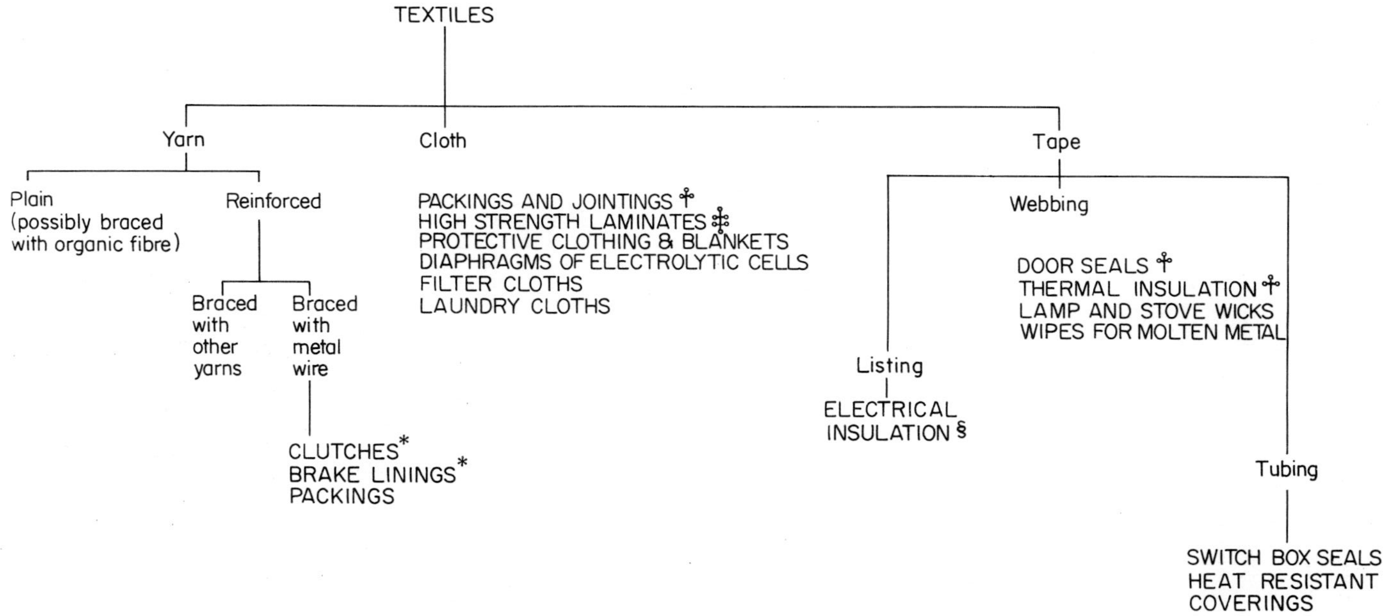

Figure 10.1 Applications of asbestos textiles

* Clutches and brake linings are discussed in Section 2(v).
‡ High strength laminates are discussed in Sections 2(vii) and 2(viii).
† Packings, seals and insulation are discussed in Section 2(ii).
§ Electrical Insulation is discussed in Section 2(vii).

Nomex is suitable for protection against most chemical hazards, with the exception of some strong acids, and can be laundered with little deterioration in properties. Nomex is about 3 times more expensive than asbestos. However, the ability to launder Nomex clothing means that its useful life is much longer than asbestos clothing.

Clothing for heat protection is also made from special wool blends, such as Multitect, manufactured by Multifabs Ltd., Derby. These materials are suitable for direct heat and metal splash protection and have been tested for protection against splashes of molten steel at 1500 °C. Aluminized grades are available for greater protection from radiant heat. The materials are resistant to chemical attack and can be laundered with no deterioration in properties. The wool fibres are surprisingly resistant to ignition and flame-spread and the clothing is competitive in price with asbestos-based products.

Bleached Teflon fluorocarbon fibres, also manufactured by Du Pont, are advertized as being 'the most fire-resistant organic fibre in oxygen-rich and high-pressure atmospheres'. Garments woven from this material have been worn by Apollo astronauts and for missile fuel handling. These suits protect the wearer from extremely high flash temperatures and corrosive missile fuels.

2(i).2 *Fire blankets, curtains, and aprons*

In general, the materials detailed above are also suitable for these applications. Additionally, a recently developed product, a sandwich of a layer of ceramic fibres [Section 2(ii)] between two woven glass cloths has become available from Marglass Ltd. This has been shown to be effective against flame and molten metal hazards. In the latter case, the surface glass layer melts and the ceramic fibres provide the protection. In certain circumstances blankets or rolls of mineral wool or ceramic fibres may be used, although these may tend to disintegrate more readily than the woven products.

2(i).3 *Electrical insulation*

Most cases in which asbestos textiles are employed for electrical insulation also demand a degree of thermal and/or chemical protection. Electrical insulation is discussed in Section 2(vii).

2(i).4 *Filters*

Asbestos cloths are widely used for filtration of bulk liquids such as beer, which has latterly been the cause of some concern in relation to possible cancers of the stomach and gut. Expanded Perlite has been used successfully as a substitute for asbestos in some applications. Vermiculite products are similarly employed.

Filter bags of woven fabrics and needled felts of 100% Teflon fluorocarbon fibres are used in many filtration applications where temperatures up to 300 °C coupled with corrosive chemical environments are encountered.

2(i).5 *Ropes, yarns, tapes, etc.*

In general, satisfactory substitution of asbestos may be made with glass for many of the applications of these materials, provided that the softening of the glass at about 300 °C is not significant. For higher temperature applications, textile forms of the continuous ceramic and silica fibres as described in Section 2(ii) may be suitable replacements if price permits.

2(i).6 *Other applications*

Other textile applications for which Nomex is a suitable substitute include press covers and pads for laundry dry-cleaning and the textile industry, racing-car seats and iron rests on ironing boards. Teflon is employed for wicking felts and fuel cell membranes. The small number of minor textile applications for which no satisfactory alternative exists at present include lamp and stove wicks, wipes for molten metal, diaphragms for some of the electrolytic cells currently employing asbestos, and some filter cloths.

2(ii) Thermal insulation and high-temperature applications

The use of asbestos for insulation purposes encompasses three main areas: asbestos insulation board, asbestos spray, and asbestos for lagging in high-temperature applications. Asbestos insulation boards and asbestos sprays are primarily building products and are dealt with in Chapter 9. For most high-temperature insulation applications chrysotile fibre is the basic constituent because it combines the properties of resilience, strength as a reinforcement, flexibility, and heat resistance. In some cases amosite fibres are used, such as in the shaped block-type lagging that can be applied to high-temperature pipes, in which a lime/silica binder is used.

However, it should not be believed that the high-temperature properties of asbestos are in any way outstanding; major loss of strength occurs in the region 300–500 °C and it is rare that a continuous working temperature in excess of 600 °C, even in unstressed situations, can be specified. This section will show that adequate substitutes are commercially available for thermal insulation purposes, most of which already find extensive application in higher temperature regions where asbestos products cannot be used.

Figure 10.2 illustrates the relative temperature capability of insulating materials.

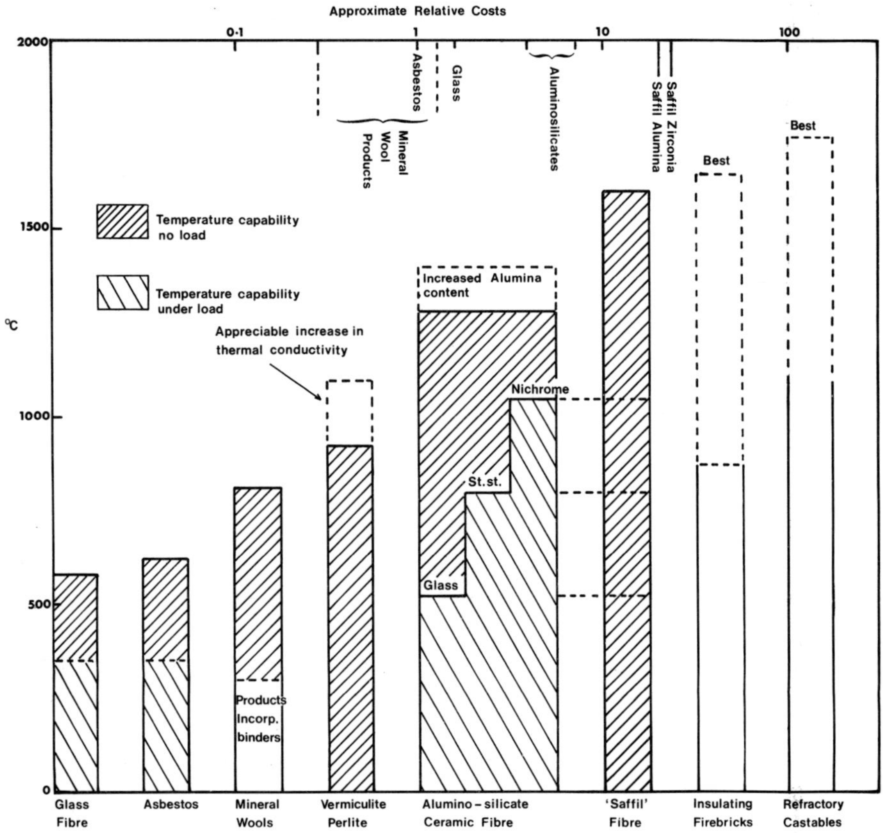

Figure 10.2 High-temperature insulants

2(ii).1 *Dry asbestos packings*

These are used mostly in conditions of dry heat, and are useful as flame spread barriers, e.g. in the construction of prefabricated buildings. Other typical applications are as furnace and kiln door seals and as high-temperature caulking. Both unwoven and woven blanket materials are used as packings and laggings for thermal insulation, and also as heat and flame spread barriers, such as in heat shields in the engine compartment of vehicles, for which applications the asbestos may be aluminized for added protection.

2(ii).2 *Asbestos jointings*

Asbestos in bulk fibre, woven, or plaited form can be bonded with various elastomers such as rubbers and polychloroprene, which can be selected for resistance to specific oils and solvents. A high proportion of fibre is used and creates a high- (or low-) temperature resistant material in sheet form which has sufficient tensile strength to withstand pressure when used as gaskets. The ability to seal imperfections in metal flange faces by flow is also required, whilst maintaining clamping pressures to achieve leak-tight joints. Typical uses are in flange joints and boiler connections, pump, valve, and pipe-joint seals in oil refinery and chemical plant, seals in internal combustion engines, ovens, and autoclave doors, tank seals, and hydraulic systems.

2(ii).3 *Gaskets*

These are a specialized form of jointing for use in internal combustion engines. The polymer/fibre mix is used because of its resilience, low cost, and resistance to the effects of hot gases, hot oils, and hot water/antifreeze mixtures. The familiar type of cylinder head gasket uses a sandwich construction of copper around asbestos paper or millboard, to cope with the severe working conditions of sealing high intermittent pressures combined with the previously mentioned requirements.

2(ii).4 *Linings and insulating blocks*

Asbestos is not very refractory but it is used in some refractory insulating bricks and blocks. One such material consists of a mixture of powdered diatomaceous earth and asbestos. The maximum temperature capability of this product is 870 °C, and is therefore easily substituted by almost all of the conventional refractory insulating materials (e.g. insulating firebrick or calcined diatomaceous earth). It should be noted that many of the ceramic fibre materials highlighted in Section 2(ii).5 are used as refractory insulators. Calculations show that thinner walls are needed than with conventional insulators with corresponding reductions in weight, cost, and heat storage.

2(ii).5 *Ceramic and mineral fibres*

Packed or sprayed asbestos is replaceable by many of the synthetic ceramic fibre product forms currently available, such as bulk fibre, strips, rope, spray, pre-formed shapes, felts, mats, and blankets. The thermal stability is generally superior to that of asbestos and the chemical stability is usually good. Asbestos has a superior chemical stability in some situations, such as in hot, reducing atmospheres, and under hot, stressed conditions, but careful prediction of the service conditions usually allows a ceramic fibre to be selected that is adequate for the job. Although the cost of these materials is higher than that of asbestos, this is frequently offset by an increased service life, less maintenance, and increased process efficiency due to the ability to operate satisfactorily at higher process temperatures.

Cheaper mineral fibre products (the price of bulk wool is about £160/ton) are available in many product forms. Products incorporating mineral wool are usually cost competitive with similar asbestos products but their mechanical and insulation properties may be inferior (although the loose wool fibres can withstand a temperature of 830 °C in some applications, some products incorporating binders are not suitable for service above 300 °C).

In some furnace lining applications where thick sections of insulation are required, the cost of an all-ceramic fibre installation could be prohibitive. An approach which has been employed successfully in situations such as this is to use a thick section of insulation based on the cheaper mineral wool products in the cooler areas, and to face it with a thinner section of the relatively expensive ceramic fibre insulation in the critical higher temperature regions.

Various lime/silica and magnesia bonded products containing unspecified non-asbestos fibres (e.g. Paratemp) are also widely available.

Ceramic fibre packings, mastics, and die-cut and pre-formed shapes are often suitable for sealing and gasket applications.

A considerable variety of ceramic fibres is available, the precise application deciding which will be necessary. In general, they all possess good thermal and chemical stability, low heat storage, resistance to thermal shock, low thermal conductivity, and incombustibility. Ceramic fibre products generally have a thermal conductivity comparable to that of the corresponding asbestos products and, especially in blanket form, are being used increasingly for their acoustic properties, particularly at elevated temperatures where vibration is a problem. They are in widespread use as refractory insulation, seals, gaskets, and fire-protective components in industry.

(a) *Aluminosilicate fibres.* Some of these are briefly described in Table 10.2. They are available in a wide range of product forms and in two grades with continuous use temperatures of 1260 and 1400 °C (20% higher alumina content). They are affected by hydrofluoric and phosphoric acids and concentrated alkalis, and also by hot, reducing atmospheres owing to loss of iron from their structure.

(b) *Alumina and zirconia fibres* (Saffil). These are also described briefly in Table 10.2. The maximum use temperature rating for both is 1600 °C. They are available as bulk fibre, blanket, board, pre-formed shapes, and paper- and resin-bonded expansion board, are suitable for use in hot, reducing atmospheres, and resist general chemical attack. The thermal shrinkage of Saffil alumina fibres, often taken as a measure of refractoriness, at 1400 °C is four times lower than that of the best aluminosilicate fibre. Zirconia has greater resistance and a lower thermal conductivity and additionally is opaque to infrared radiation.

(c) *Continuous ceramic fibres.* These formulations are available from the 3M Corporation. They are based on the alumina–boria–silica, alumina–chromia–silica, and zirconia–silica systems, and are specified for continuous use up to 1300, 1400, and 1000 °C, respectively. They are available in a variety of product forms, such as yarn, roving, fabric, braid, and chopped fibres, but are expensive. Fibre geometry considerations indicate a minimal health hazard (Section 3).

(d) *Silica fibres.* These are available in continuous fibre form as Refrasil and in various yarn and textile products made from the fibres. The mean fibre diameter is 10 mm, indicating minimal health problems (Section 3). The material is specified for a maximum continuous use temperature of 1000 °C and costs £40 000–50 000 per ton (1977) in cord form. Bulk fibre is also available at a price of about £12 000 per ton (1977) and in two diameter grades, 5–10 and 0.75–1.6 mm, the latter of which may possibly constitute a health hazard. Refrasil is attacked by hydrofluoric and phosphoric acids and by mild alkalis. Typical applications are for flexible electrical insulation in textile form at up to 1000 °C and as flexible packings in bulk and felt form between refractory bricks in coke ovens. Another grade containing about 3% chromia (Refrasil C–1400) is available in fabric form for continuous use applications at up to 1400 °C.

Ceramic fibre textiles may contain insert materials to increase the fabric tensile strength. Alloy wire inserts are available for obtaining maximum tensile strength at elevated temperatures. Glass filament inserts are used in applications where metal is undesirable (e.g. dielectrics). The following insert materials are available:

Glass: service to 520 °C;
Stainless steel: service to 800 °C;
Nichrome wire: service to 1050 °C.

Where tensile strength is not important the materials may be used to their maximum service limit of 1260 °C (or higher).

2(ii).6 *Vermiculite and Perlite*

Vermiculite for high-temperature insulation is normally bonded with a high-alumina cement or fireclay and is therefore competitive with ceramic fibres for hot-face lining applications. Many vermiculite products can withstand temperatures of 1100 °C. Expanded Perlite is used as an insulating cover on the surface of molten metal to prevent excessive heat loss during pouring, to produce refractory blocks and bricks, and to top off ingots. The material is used occasionally to 1100 °C although above 925 °C the thermal conductivity increases appreciably.

2(ii).7 *Solid ceramics*

Additionally, in certain specialized applications solid sintered ceramics can be used, e.g. alumina and silicon nitride. These can be fabricated into components such as insulation washers and rotary pump seals, which would

Table 10.2 Characteristics

Fibre property	Chrysotile asbestos	Crocidolite asbestos*	E-glass	A-S carbon (Grafil)
Specific gravity	2.5	3.4	2.54	1.76
Degradation temp. (°C)	450†	300†	Softens at ~300‡	>3500
Max. service temp. (°C)	†	†	~600‡	~1000 in non-oxidizing environments
Fibre strength (MN/m^2)	3100§	3500§	1700	2610
Fibre modulus (GN/m^2)	160	190	70	200
Fibre diameter (μm)	Fibril 0.02 Fibre 0.03–100	~0.5	10–12	9
Fibre length (mm)	Fibril 0.25–5 Fibre 1–80	<5	Chopped or continuous	Chopped, 1 3 and 6 continuous
Price‖ (£/ton)	40–1700¶**	300–400**	600	Chopped 26 000
Chemical constitution	Hydrated magnesium silicate	Hydrated sodium-iron silicate	Calcium alumino-silicate	Graphite

* The properties of amosite are similar to crocidolite, with a higher temperature capability, and lower strength.

† Degradation temps. are those at which water of constitution begins to be lost. Service temps. can be higher. See text.

‡ At 340 °C E-glass retains 75% of its strength, and up to 20% strength may be retained at 600 °C.

§ These values often quoted as average properties. Exceptionally 5800 MN/m^2 has been reported for chrysotile and 8600 MN/m^2 for crocidolite. These quoted strength values may not be attained in practice. See text.

commonly be machined out of a high-density asbestos board of the Sindanyo type. Bulk raw materials costs can be low [Al_2O_3+0.5% NaO–£150/tonne, Al_2O_3+0.1% NaO–£450/tonne, Nominally pure Al_2O_3–£1000/tonne. January, 1978], but the unconventional processing route and specialized manufacturing techniques necessary for solid ceramics would necessitate the purchase of finished components from a manufacturer, so that cost comparisons are not directly possible. However, it is probable that the cost of a solid ceramic replacement would be much higher than that of the asbestos component. This may be offset to some extent by superior properties, e.g. better wear resistance in rotary and sliding seals.

2(ii).8 *Further information*

Data sheets and design information are available from the following manufacturers:

McKechnie Refractory Fibres Ltd.	aluminosilicate fibres
Morganite Ceramic Fibres Ltd.	aluminosilicate fibres and Saffil fibres

of ceramic and mineral fibres

Saffil alumina (ICI)	Saffil zirconia (ICI)	Triton (Morganite)	Mackechnie T.I.	Fiberfrax (carborundum)
3.4	5.55	2.56	2.56	2.63
>2000	>2700	1760	1780	1790
1600	1600	1260	1260	1260
1000	700	1400	2500	2760
100	100	120	103	110
Mean	Mean	Mean	1–10	Mean
3	3	2.8	Mean 2.5	2–3
		Mean 100	25–200	40–250
		up to 250		
7900	Development product, price on application	1000–1500	1300	1000–5000
Alumina	Zirconia	Alumino-silicate	Alumino-silicate	Alumino-silicate

‖ Prices quoted for bulk fibre. Prices for products, e.g. textiles, felts, blankets, will be higher.
¶ Chrysotile price is very fibre length oriented. £40/ton is cost of 'dust' grade, and £1700/ton is the price for the best textile grade of fibre lengths of a few centimetres. 'Cement' grades cost £300–400/ton (1977).
** Prices likely to rise due to increased investment in mining industry, coupled with resource depletion.

Carborundum Company	aluminosilicate fibres
ICI Mond Division	Saffil fibres and products
Darchem Ltd.	Refrasil silica fibres
Tilling Construction Services Ltd.	Perlite
Mandoval Ltd.	Vermiculite
Advanced Materials Engineering	silicon nitride

2(iii) Asbestos millboard

This is one of the most versatile asbestos materials used in industry. Features of millboard that contribute to this versatility are ease of cutting or punching to shape, useful thermal insulation properties, impregnability with bonding agents or cement, ability to be wet-moulded, and compressability.

Typical uses are for the fabrication of rollers for transport of hot materials in the steel and glass industries, as formers for wire-wound electrical resistances, flange gaskets for joints in ducting and trunking used for high-volume/low-pressure gas transport, cylinder head gaskets, insulating linings to minimize heat losses from ovens and moulds, plugs and stoppers for molten metal containers, and in resin-impregnated form for clutch facings.

2(iii).1 *Substitutes*

For many applications, millboards made from aluminosilicate fibres can provide a direct replacement if the extra cost is acceptable. They are made by a suction method from an aqueous slurry of fibres, and are available in thicknesses up to 50 mm. They are also available with a high-temperature silica binder. These aluminosilicate fibre millboards have been successfully used for the process rollers in plate glass manufacture. For insulation applications, if the thermal conditions are not so severe as to necessitate the use of ceramic fibres, one of the several types of mineral fibre block and slab products should be an effective substitution. Cobalt rollers for the steel heat-treatment application have been used as a millboard substitute, but are considerably more expensive than the asbestos product.

Further information on aluminosilicate ceramic fibres and mineral fibres is given in Section 2(ii). Electrical insulation is discussed in Section 2(vii) and friction materials in Section 2(v).

2(iv) Industrial applications of asbestos–cement

Building applications of asbestos–cement are reviewed in Chapter 9. In an industrial context, the major application of asbestos–cement is for pipes used in the transport of corrosive fluids. Asbestos–cement pipes tend to be favoured economically in the middle range of diameters, extruded plastics and metals being prepared for small bores and conventional concrete for very large sizes of pipe.

The pipes have a laminar structure, such that any internal or external corrosive attack has a very gradual effect, and this structure enables holes to be cut for branch pipes with very little risk of cracking beyond the cut area.

The primary function of the asbestos in asbestos–cement products is to act as a reinforcing fibre. The properties that make it useful in this role are a high fibre strength and modulus and a large aspect ratio (ratio of length to diameter). The last parameter is one of the factors that determines the reinforcement efficiency of fibres in composite systems. The asbestos fibres produce useful tensile and flexural properties in the asbestos–cement, enabling it to be used in pressure pipes; the high tensile strength is particularly important. For this reason, most of the crocidolite (blue) asbestos consumed is used in asbestos–cement pipes owing to its superior strength compared with chrysotile asbestos, which has replaced it in the U.K. following the stringent restrictions imposed upon the import of crocidolite; more than 90% of the crocidolite used in the U.S.A. goes into pipework applications. The initial stages of asbestos–cement manufacture involve the mixing of water, cement, and fibres into a slurry. The fibres have a great affinity for water and additionally contribute favourable and controllable drainage properties to the mix, enabling the asbestos–cement pipes and sheets to be produced by a relatively simple and flexible process similar to that used in paper-making.

The main attractions of asbestos–cement products to the user are durability, resistance to weathering attack, and cost-effectiveness; the non-combustibility is significant but is often of secondary importance.

2(iv).1 *Substitutes*

Glass-reinforced plastics (GRP) pipes are often suitable for use in corrosive environments. GRP-wound concrete is often satisfactory as a sewage or pressure pipe and the traditional cast-iron pipe performs satisfactorily. Most of these products are more expensive than asbestos–cement products, although certain thermoplastics, such as rigid PVC and high-density polyethylene, may be cost-competitive in specific applications.

Glass-reinforced cement (*GRC*). The most promising route for achieving the objective of a universal replacement for asbestos–cement is the development of glass-reinforced cement technology. Conventional E-glass fibre (which is commonly used in glass-reinforced plastics) cannot be used in conjunction with the cement owing to degradation by the severely alkaline conditions. However, within the last 10 years, a high-zirconia, alkali-resistant glass-fibre has been developed at the Building Research Establishment, Garston, England. This fibre is marketed by Pilkington Brothers Ltd. of St. Helens under the name of Cem-FIL.

Glass-reinforced cement usually contains between 3% and 7% of fibres and typically a material containing 5% of fibres would have a modulus of rupture similar to that of asbestos–cement containing 15% of fibres. Additionally, GRC is considerably more impact resistant than asbestos–cement of comparable strength. However, several drawbacks exist in the use of this material:

(1) The fibre is approximately four times as expensive as the asbestos used in cement-based products and this, coupled with the less favourable economics of the production process, results in a price for GRC about 70% greater than that for asbestos–cement, even allowing for the smaller fibre content.

(2) The drainage characteristics of a cement–water–glass-fibre mix are poor and do not permit the manufacturing process for asbestos–cement to be used in an unmodified form. Glass-reinforced cement is usually made by a process referred to as 'spray suction'. In this process, a special spray head sprays a mixture of chopped glass-fibre bundles a few centimetres long with a fine cement slurry on to a porous bed. When spraying is complete, a vacuum is applied to the underside of the bed to remove excess of water. A

similar procedure is used to make pipes, the spray head being directed on to the inside of a rotating tubular form with the excess of water being similarly removed. Consequently, the manufacture of glass-reinforced cement is relatively costly and requires a high capital investment.

However, the manufacturers of glass-reinforced cement are expending considerable effort on attempts to make the material on minimally modified asbestos–cement plants. The impetus for this effort was the result of Japanese legislation about 2 years ago whereby asbestos was placed in the 'dangerous chemicals' category of materials. Consequently, severe restrictions apply to any product sold in Japan that contains more than 5% of asbestos, which includes standard asbestos–cement products. It was found that the standard asbestos–cement plant could produce a good quality board from a cement slurry containing both glass and asbestos fibres in suitable proportions, so as to keep the asbestos content below 5%, this producing glass-reinforced asbestos–cement (known as GRAC). The research effort is presently concentrated on eliminating the asbestos content completely and this has been achieved by suitable plant modifications. It is likely that this equipment will shortly become available in the U.K., eliminating one of the principal drawbacks to the use of glass-reinforced cement.

(3) Some attack of the glass fibre by the setting cement is necessary to achieve good fibre bonding and this requirement has raised doubts concerning the long-term durability of glass-reinforced cement. Most existing uses of glass-reinforced cement have been restricted to non-load-bearing components (e.g. cladding), but more demanding structural applications are gradually being attempted as a knowledge of, and confidence in, the material grows. It is likely that the uses for glass-reinforced cement will increase since it allows interesting concepts that are impractical with asbestos–cement. A good example is the use of continuous glass-fibre roving to filament wind large-diameter pipes that are much lighter, owing to the thinner wall required, than a steel-reinforced concrete pressure pipe of comparable strength and price.

Various other fibres have been studied as reinforcements for cement. However, only steel wire (chopped or continuous) can fairly be considered as a replacement for asbestos–cement for load-bearing situations. The fibres are two to three times more expensive than asbestos but are usually incorporated only to the extent of 1–2%. However, the system suffers from the same drawback as the Cem-FIL process in that the drainage characteristics of the mix are such as to necessitate the use of specially developed production techniques. Thus a high capital outlay is involved.

2(iv).2 *Further reading*

See Green and Pye (1976, 1977).

2(v) Friction materials

Asbestos-based brake and clutch linings and pads are in widespread use, particularly, but not exclusively, in the automotive industry. Drum brake linings are usually moulded from a mixture of short chrysotile fibres, phenolic resin, and various mixtures of fillers. Brake pads may be similarly made, or can use a woven asbestos cloth which may be reinforced with brass wire, impregnated with phenolic resin, as is commonly used in clutches.

The chrysotile fibre performs a complex function in these applications which is not completely understood. The fibres stiffen and strengthen the filled phenolic resin matrix and can maintain these properties at the high temperatures generated during, for example, braking and clutch slipping. The friction and wear characteristics of the asbestos fibres and their decomposition product, forsterite, are thought to affect the braking efficiency and service life of the products. Forsterite is a non-fibrous silicate and is not thought to constitute a health hazard. Investigations have established that the dust produced from the wear of asbestos friction materials contains only 1–2% of asbestos fibres. Repeated exposure to this dust is a potential hazard to persons working in close proximity to these products (for example, garage mechanics) and it is strongly recommended that components should be cleaned with suitable vacuum extraction equipment, or a damp rag, rather than the more accustomed practice of blowing dust away with a compressed-air line.

While this debris is not believed to constitute a health hazard to the general public, much effort has been concentrated on searching for substitutes for asbestos in this application. The various types of linings, pads, and facings in commercial application are presented in Table 10.3.

2(v).1 *Substitutes for asbestos in friction materials*

(a) *Sintered products.* Sintered metal, ceramics, and metallized bonded products have been available for many years. However, the high thermal conductivity of these materials creates a risk of heating and boiling of brake fluid, which can cause erratic performance and inconsistent behaviour between cold and hot conditions.

(b) *Vermiculite.* Delaminated vermiculite is used in friction materials which are available commercially throughout Europe. The materials maintain their strength at high temperatures, and, since vermiculite is compatible with the commonly used phenolic resins for this application, little alteration in manufacturing methods is required. Other fibres are sometimes used with vermiculite and some developments are mixtures of asbestos and vermiculite, aimed at reducing the asbestos content.

(c) *Silicon nitride.* This material was used for the brake pads in the Concorde prototypes 001 and 002. It was found to have a higher service life

Table 10.3 Properties of friction materials

Form	Type	Manufacture	Typical dimensions	Coefficient of friction	Wear rate at 100 °C (mm^3/J)*	Temperatures: Max. (°C)	Temperatures: Max. operating (°C)	Working pressure (kN/m^2)†	Applications
Linings‡	Woven cotton	Closely woven belt of fabric impregnated with resins which are then polymerized.	As rolls	0.50	12.2×10^{-6}	150	100	70–700	Industrial drum brakes, mine-winding equipment, cranes, lifts.
	Woven asbestos	Open woven belt of fabric is impregnated with resins which are then polymerized. May contain wire to scour the surface.	Radiused linings	0.45	9.2×10^{-6}	250	125	70–700	Industrial band and drum brakes, cranes, lifts, excavators, winches, concrete mixers, mine equipment.
	Moulded: Flexible	Asbestos fibre and friction modifiers mixed with thermosetting polymer and mixture heated under pressure.	Linings	0.40	6.1×10^{-6}	350	175	70–700	Industrial drum brakes.
	Moulded: Semi-flexible			0.35	3.0×10^{-6}	400	200	70–700	
	Moulded: Rigid			0.35	1.8×10^{-6}	500	225	70–700	Heavy-duty drum brakes, excavators, tractors, presses.
Pads	Resin/asbestos	Similar to linings; greater choice of resin as flexibility [illegible]	Pads up to 1 in. thickness or on backplate to fit proprietary [illegible]	0.32	1.2×10^{-6}	650	300	350–1750	Heavy-duty brakes and clutches, press brakes, earth-moving equipment.

	Sintered metal	Iron and/or copper powders mixed with friction modifiers and sintered.	As above.	0.36	Used at higher temperatures.	650	300	350–3500	As above.
	Cermets	As above, but large proportion of sintered ceramic present.	Supplied in buttons, cups.	0.32	Used at higher temperatures.	800	400	350–1050	As above.
					Power rating (W/mm^2)				
Clutch facings	Woven	As linings	—	0.35–0.40	0.3–0.6	250	150	175–520	Industrial band, plate and cone clutches, cranes, lifts, excavators, winches and general engineering applications.
	Mill-board	Resin impregnated	—	0.40	0.3–0.6	250	150	175–520	Mainly automotive and light commercial vehicles.
	Wound tape/yarn	—	—	0.38	0.3–0.6	350	200	175–700	Mainly automotive and light commercial vehicles.
	Moulded	As linings	—	0.35	0.6–1.2	350	200	175–200	Automotive, commercial vehicles, agriculture and ind. tractors.
	Sintered	As pads	Facings are sintered on to core plates or backing plates.	0.36 static, 0.30 dynamic	1.7	500	300	350–2800	Tractors, heavy vehicles, road rollers, winches machine tool applications
	Cermets	As pads	Supplied as buttons in steel cups.	0.4	4.0	500	300	700–1400	Heavy earth moving equipment, crawler tractors, sweepers, trenchers, graders.

Table 10.3 (*Continued*)

Form	Type	Manufacture	Typical dimensions	Coefficient of friction	Wear rate at 100 °C (mm^3/J) *	Temperatures Max. (°C)	Temperatures Max. operating (°C)	Working pressure (kN/m^2)†	Applications
Oil immersed	Paper	—	—	0.11	2.3	—	—	700–1750	Automotive and agricultural automatic transmission.
Oil immersed	Woven	—	—	0.08	1.8	—	—	700–1750	Band linings and segments for automatic transmissions.
Oil immersed	Moulded	—	—	0.04–0.06	0.6	—	—	700–1750	Industrial transmissions and agricultural equipment.
Oil immersed	Sintered	—	—	0.11 static, 0.06 dynamic	2.3	—	—	700–4200	Power-shift transmission, presses, heavy-duty general engineering applications.
	Resin/graphite	—	—	0.10	3.5	—	—	700–4200	Heavy-duty automatic transmissions.

* $1\ mm^3/J = 165\ m^3/h.p.\ h$.
† $1\ kN/m^2 = 0.14\ lbf/in^2$.
‡ Linings: many lining materials are also supplied as large pads which can be bolted or otherwise attached to band or shoe; the pads can be moved along the band or shoe as wear occurs, and so maximum life can be obtained from the friction material, despite the uneven wear along the band or shoe.
§ The friction materials manufacturer should be consulted at an early stage in the design of brake or clutch. He should also be consulted concerning stock sizes. Standard sizes are much cheaper than non-standard.
|| The information in this table is intended as a guide for comparison purposes only.

than asbestos, and a higher thermal conductivity which in this application was desirable. The material is expensive, the powder costing about £5000 per ton, and the components were heavier than the carbon-reinforced carbon composites eventually adopted.

(d) *Carbon/carbon composites.* These materials are used for the brakes on the production Concordes, and have also been used on racing cars where they are more efficient than asbestos materials under the high service temperature conditions.

The carbon-fibre reinforcement used is a cheaper type than the conventional R.A.E. type fibres, and is formed by charring wool, rayon, nylon, or any organic material in the form of yarn, felt, or cloth. This charred fibre is infiltrated with resin or pitch which is subsequently charred. Alternatively, an organic gas may be passed through the charred fibres while they are heated. These processes are repeated several times, and the final product is denser than most other composites of this type and oxidizes only very slowly at about 800 °C.

The anisotropy of these materials produces uneven heat flow and thermal expansion coefficients and, additionally, they have relatively low tensile and impact strengths. Further, their high cost (the development price quoted is greater than £10 000 per ton) justifies their application only in specialized circumstances.

(e) *Other fibres.* Various other fibres have been used in phenolic binders, such as steel, glass, mineral wool, and aluminosilicates. They all have drawbacks and none is yet as good as asbestos, especially for higher temperature applications such as in disc-brake pads.

However, a considerable volume of work is currently in progress on the use of steel and glass fibres. Vehicle trials are underway, and it is estimated that asbestos substitute materials are available for about 30% of the friction products market, the exception being the heavier duty applications, such as disc-brake pads where high-temperature surface conditions are encountered.

The first use of steel fibres in friction materials was by Germany, prior to and during World War II, due to the difficulty in obtaining imports of asbestos. It was found that the brakes worked well, with little loss of efficiency when wet, but that the wear debris was abrasive and severely damaged the counterface. It has been found easier to use steel than glass in the manufacturing process, which uses the same volume fraction of steel as asbestos. The steel fibre is usually of the fine steel wool type and is about four times as expensive as the asbestos used in friction applications.

Glass-fibre based friction materials also produce a very abrasive wear debris, with corresponding damage problems. At high temperatures, the glass can melt, which can produce a sudden undesirable loss of friction, since the sliding surfaces can be effectively lubricated by the molten glass. Conventional E-glass fibre is approximately half the price of steel wool.

Both glass and steel fibres behave satisfactorily in the wet-mix manufacturing process, but in the dry-mix process steel fibres tend to segregate and glass fibres orientate themselves in preferred directions, leading to loss of optimum properties.

Brakes using steel and glass fibres tend to be considerably noiser than their asbestos counterparts.

Non-asbestos-based products use proportions of fibre comparable to the conventional product and are therefore more expensive. Thus, although asbestos-based friction materials are technically and economically preferable at present for the majority of applications, it seems probable that these will eventually be replaced as a result of concern over the health hazards involved with asbestos.

2(v).2 *Further information*

See Neale (1973).

2(vi) Dry-rubbing bearings

A significant application of asbestos composite materials is in plain rolling bearings. In this application a matrix of thermosetting phenolic resin is used to impregnate asbestos cloth or yarn. The principal advantage of these materials is that, although they can be lubricated with oil or grease, and in some cases are supplied impregnated with up to 7% of mineral oil, they are also able to function effectively without lubrication, or lubricated by *in situ* process fluids or seawater.

The life of dry rolling bearings is comparable with that of lubricated bearings only at low sliding speeds, in practice, principally in oscillating applications. Their use is generally advantageous for applications with sliding speeds below 1 m/s and where the use of a lubricated bearing is impossible or unattractive owing to:

(a) inability of lubricants to survive in the operating environment due to factors such as high or low temperatures, or high levels of ionizing radiation;

(b) penalties of cost, complexity, size, or weight associated with the provision of adequate lubricant feed, for example where bearings are inaccessible and cause relubrication and maintenance difficulties.

Typical applications of dry rubbing bearings include automotive steering column bushes, brake and clutch pedal bushes, bearings for textile spinning frame spindles, food slicer bearings, copying machine rollers, bearings for conveyor rollers, compressor thrust washers, aircraft control and undercarriage bearings, large railway bearings, dock equipment, lock gates, and as seals and liners for metallic ball and roller bearings in, for example, railway axle boxes.

Asbestos bearings in particular find extensive application in large marine bearings, where high loads and low sliding speeds are encountered and seawater may be used as a lubricant. Important examples are rudder and steering gear bearings and stern shaft bearings. Historically, lignum vitae in the form of axial staves with a gunmetal bush was commonly used. This arrangement was also suitable for lubrication by seawater but the hard, oily wood suffered from a tendency to swell in water and the life expectancy was low when high loads were encountered. The next generation of marine bearings employed bronze or white-metal bearings, which are still in extensive use. Metallic bearings are suitable for a wide range of applications when cost is not of prime importance and constitute excellent bearings when loads are moderate and speeds sufficiently high to create a continuous lubricant film. However, under the high-load, low-speed regimes encountered in marine bearings, the consequent marginal (or 'boundary') lubrication can lead to excessive galling and seizure. The reinforced plastics bearing materials also have other major advantages in this application: white metal is inherently weak under shock loading conditions; in marine environments galvanic corrosion is likely to occur with metal bearings; and plastics matrices are less stiff (lower Young's modulus in compression) than bearings metals, which means that misalignment is a less serious problem.

These advantages lead to a reduced maintenance time and longer life expectancy, which, when combined with good machinability and ease of fitting, contribute to significant cost reductions in this application.

Asbestos bearings additionally are tolerant of line and spot loading. This good performance as a bearing material is not well understood, but it is thought that the tendency of the flexible chrysotile fibres to 'brush-out' on the bearing surface is significant.

2(vi).1 *Substitutes for asbestos-reinforced thermosets in bearing applications*

When materials are to be evaluated in connection with a specific problem the test conditions required are unambiguous: they should simulate those of the application as closely as possible. Preliminary evaluation of materials on a wider basis, however, is much more difficult. For a reasonably complete evaluation the following tests are required:

1. determination of the coefficient of friction and wear rate at a moderate load and a relatively low speed of sliding for which frictional heating is negligible (Figure 10.4);

2. determination of the maximum temperature limit at low speeds of sliding (Figure 10.5);

3. determination of the critical load and speed limits at which either friction or wear become excessive as a result of frictional heating (Figure 10.6).

The following factors are also important to a specific application and design: environment, thermal conductivity, expansion coefficient, resistance to shock loading, and cost.

Intelligent use of design options, for example, in the provision of seals against the ingress of corrosive fluids, can increase the range of suitable materials for specific applications. However, the remainder of this section is concerned briefly with the material properties of potential substitutes for asbestos-composite plain bearings, operating under high-load, low-sliding speed conditions.

(a) *Reinforced thermoset.* Asbestos-composite bearings belong to the reinforced thermoset family of plain bearing materials. Among the other members of this family which do not contain asbestos are:

(a) polyester-bonded textile laminates with molybdenum disulphide or graphite;

(b) cellulose-fabric based phenolic laminates with uniformly distributed polytetrafluoroethylene (PTFE) or graphite.

The members of the family have broadly similar physical and chemical properties. However, the maximum operating temperature of asbestos-reinforced composites is generally higher—up to 175 °C compared with 100–130 °C for the other composites (Figure 10.3)—and the coefficient of linear expansion is lower.

The substitution of asbestos-reinforced composites by either of these two classes of alternatives is therefore heavily dependent upon the exact nature of the application and whether the marginal loss in properties (particularly the maximum operating temperature) is acceptable. For applications in which such substitutes are not acceptable alternatives it is necessary to consider other families of plain rubbing bearing materials.

(b) *Other plain bearing materials.* The relative properties of other plain rubbing bearing materials are illustrated in Table 10.4 and Figures 10.4–10.6. These data allow the selection of a suitable class of bearing materials for any combination of bearing pressure, sliding speed, and operating temperature.

It is beyond the scope of this book to discuss in detail a large variety of specific circumstances but considering the high load, low speed, low wear rate, and modest temperature category into which the majority of asbestos bearing applications fall, the potential substitutes are:

(a) polyimides;
(b) woven and resin bonded PTFE fibre;
(c) graphite-impregnated metals;
(d) PTFE-impregnated metals.

In marine applications, the last two, which are metal-based, would not be preferred for much the same reasons as white metal, and one is therefore left with a choice between polyimides, or woven and resin bonded PTFE

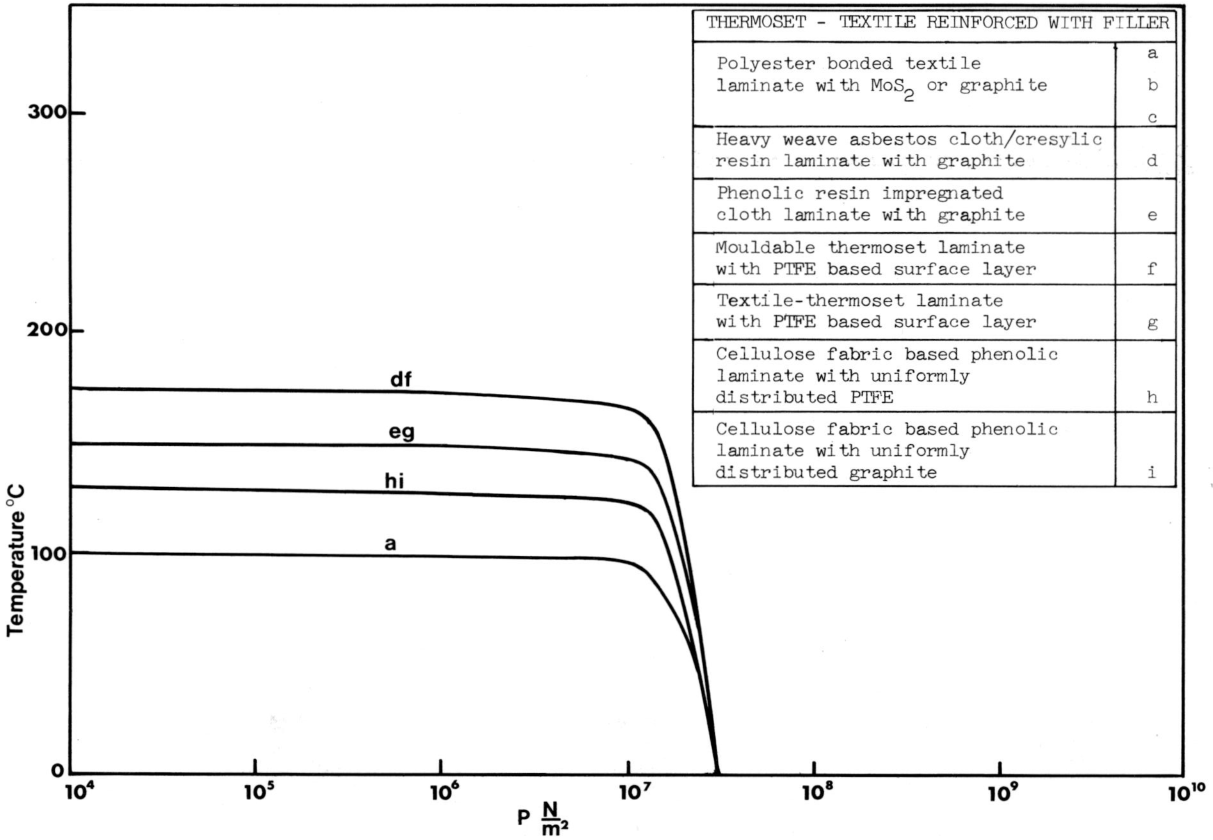

Figure 10.3 Reinforced composites

Table 10.4 Properties of plain rubbing bearing materials

Type	Examples	Maximum static load (MN/m^2)	Maximum service temperature (°C)	Thermal conductivity (W/m K)	Special features	Relative cost in bearing form*
Thermoplastics	Nylon/polyamide- monocast extruded: plain reinforced solid lubricant filled (graphite/MoS_2; PTFE)	10	100	0.24	Inexpensive Solid lubricants reduce friction	1
	Polyacetal plain solid lubricant filled (graphite/ MoS_2; PTFE)					
	Metal-backed nylon on steel					2–2.5
	Polyacetal and additive—impregnated in porous bronze on steel					
Reinforced thermosetting plastics	Phenolics, epoxies + asbestos, textiles, PTFE	30–50	130–175	0.4	Reinforcing fibres improve strength	2–3.5
Filled PTFE	Glass, mica, graphite, bronze–graphite, bronze-lead oxide, ceramics	2–7	250	0.25–0.5	Very low friction	3–8
Polyimides	Glass filled Solid lubricant filled (graphite/ MoS_2)	30–50	320	—		6–12
Carbon–graphite	Varying graphite content: may contain resin	1–3	500	10–50	Chemically inert	10–15
Carbon–metal	With Cu, Ag, Sb, Sn, Pb	3–5	350	15–30	Strength increased	
Metal–solid lubricant	Graphite-filled irons + MoS_2 Graphite-filled bronzes + MoS_2 Ag–PTFE	30–70	250–300	50–100	High temperature capability	12
Special non-machinable products	PTFE-impregnated metal—steel/ porous bronze/PTFE	300	275	42	Need to be considered at the design stage	3–5
	Woven PTFE fibre: thermoset reinforced with cotton/glass/wire weave; may contain graphite	500	250	0.24		6–20
	Thermoset + PTFE surface	50	150	0.3		3–8
	Metal + filled PTFE liner	7	275	0.3		

* Based on 1000 components (1976).

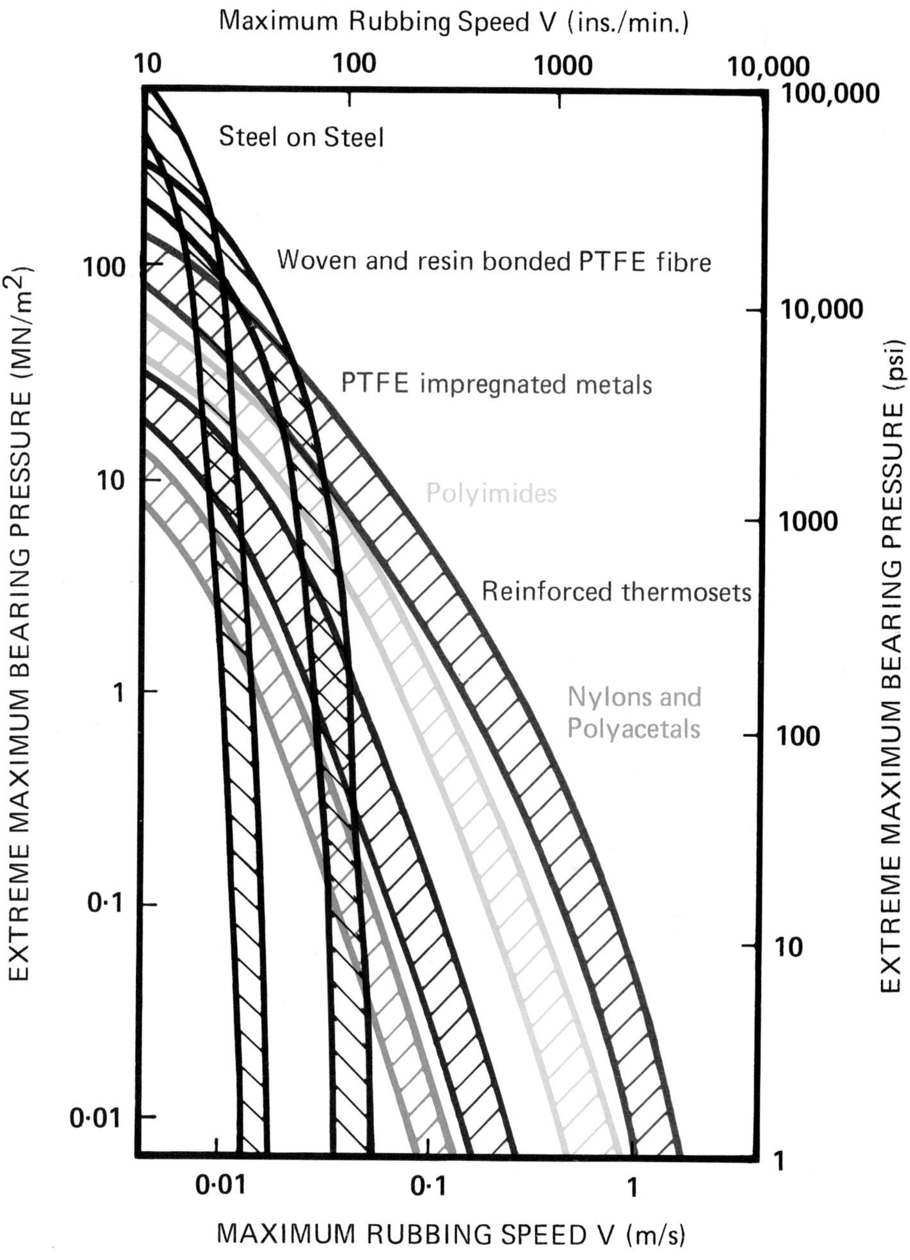

Figure 10.6. A guide to extreme maximum bearing pressures and maximum rubbing speeds for typical rubbing bearing materials against steel

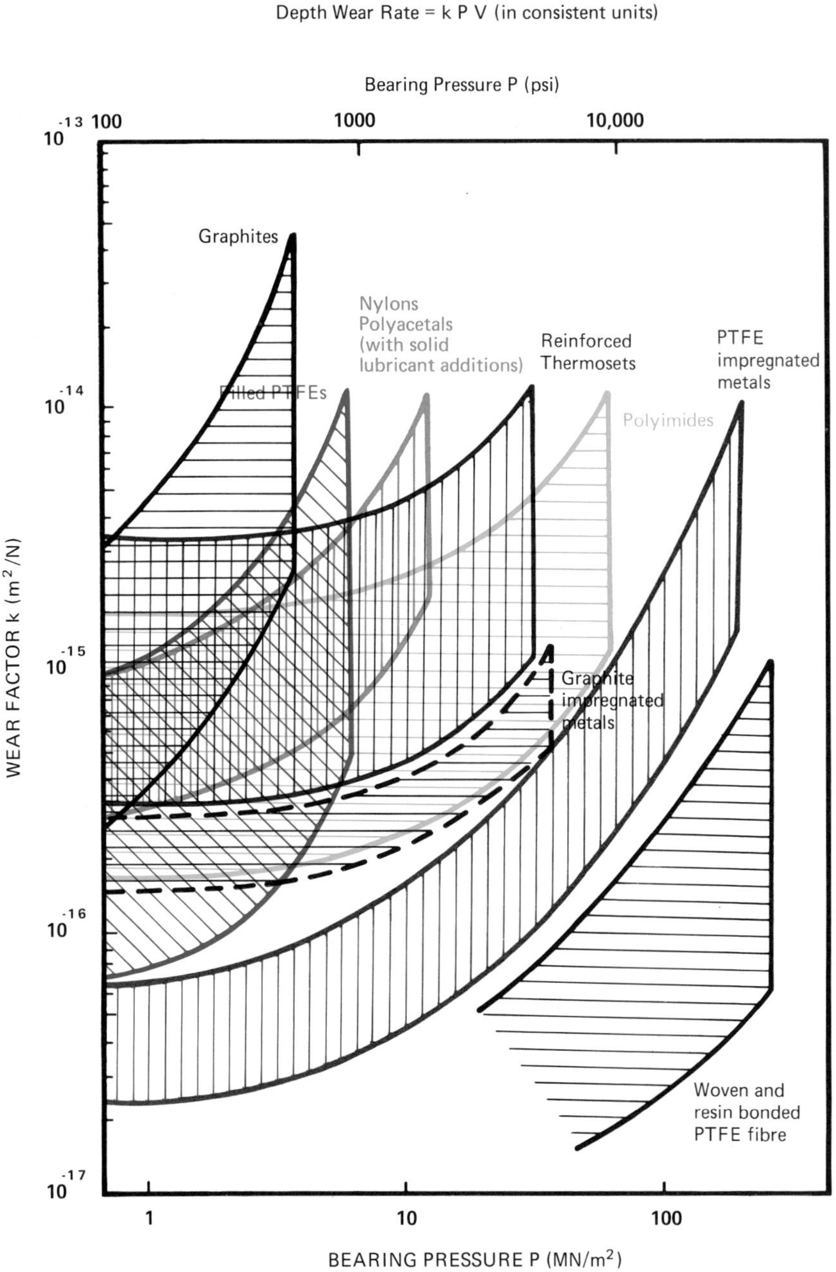

Figure 10.4. A guide to the wear rates at room temperature of typical rubbing bearing materials at various bearing pressures against steel. Depth wear rate = kPV (in consistent units)

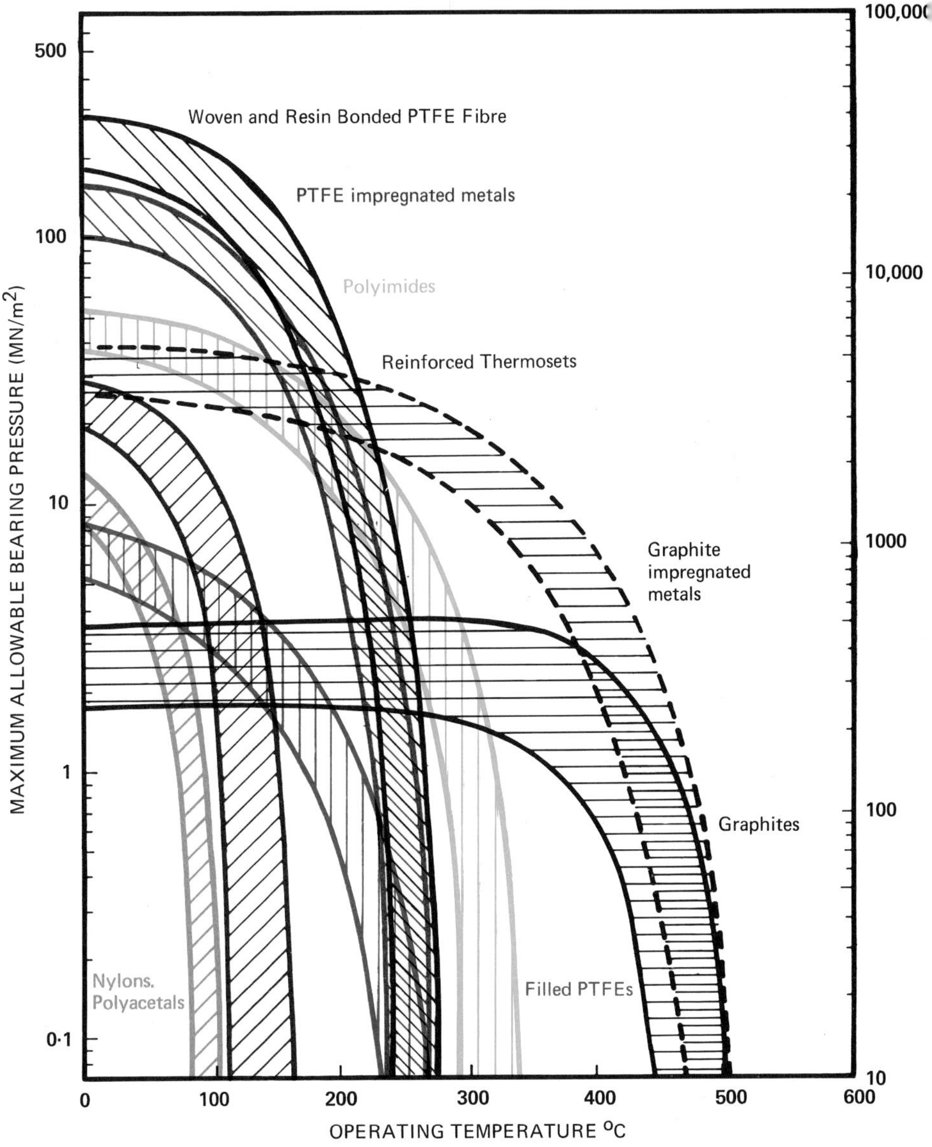

Figure 10.5. A guide to pressure and temperature limits for rubbing bearing materials against steel

fibre, both of which carry a heavy cost penalty compared with asbestos composites. (Table 10.4). In applications where the environment and/or engineering conditions are not so demanding, the impregnated metals would be an economic and technically satisfactory proposition, although in general many such applications could be adequately covered by the family of oil-impregnated porous metal bearings.

2(vi).2 *Conclusion*

There is a range of bearing applications in which the combination of mechanical, thermal, and environmental conditions make asbestos-reinforced thermosets the preferred material; substrates which are currently employed in service (white metal, lignum vitae) are technically inferior. Some applications can be satisfied with little cost penalty by the use of other reinforced thermoset materials, but where these are ruled out on grounds of environmental resistance or operating temperature, one is forced to invoke substitutes, the extra cost of which has militated against their commercial development and production. For this reason alone, a period of time will elapse before such materials could be adopted on a commercial scale.

2(vi).3 *Acknowledgements*

The assistance of the following personnel and organizations in the preparation of this section is appreciated. Dr. N. A. Waterman (Fulmer Research Institute), Mr. M. J. Neale (Michael Neale and Associates, Tribological Consultants), and Engineering Sciences Data Unit.

Figures 10.4, 10.5, and 10.6 are reproduced with permission from Fulmer Materials Optimizer.

2(vi).4 *Further reading*

See Neale (1973); Engineering Sciences Data (1976); *Proc. Inst. Mech. Engrs.* (1967–68).

2(vii) Electrical insulation

Asbestos is widely used in the electrical industry in the form of paper, tape, cloth, and board. It is frequently applied as a felted material and as a filler for natural and synthetic insulating resins. In dielectric applications the most common impregnant is a solid resin. The impregnation of the asbestos with resin increases its dielectric strength, improves its mechanical properties, and results in a moisture-proofing material in order to offset the hygroscopic properties of the asbestos itself. Laminates suitable for use up to 180–200 °C can also be made from felts or woven cloths with appropriate high-temperature resins. They are used for low-voltage transformers, armature slot wedges, furnace parts, domestic heating equipment, and similar applications. Resin-bonded papers and boards use, for example, phenol–formaldehyde, polyvinyl acetal, epoxy, or silicone resins according to the temperature of service and may also contain glass or other fibres. Asbestos papers and mat are made by methods similar to cellulose paper.

Most cases in which asbestos textiles and papers are employed for electrical insulation also demand a degree of thermal and/or chemical protection. Chrysotile asbestos and smaller amounts of amosite and tremolite are used by the electrical industry for insulation purposes as textiles. In this form asbestos is employed for the insulating of wires and cables, especially those which are designed for low-voltage, high-current use under severe temperature conditions, arcing barriers in switches and circuit breakers, and for braided sleevings for electrical appliances and insulated conductors where fire protection and resistance against mechanical abrasion are sought. Asbestos in the form of cloth tapes and sleeving may be reinforced with glass or natural fibres.

Asbestos–cement products are used in the electrical industry for the construction of mechanically strong and heat-resistant rods, tubes, cylinders, and plates. Applications include panel boards, arcing barriers, and insulating tubes and cylinders used in the construction of air-cooled transformers.

Cheap asbestos boards bonded with starch and calendarized are not very strong but will withstand fairly high temperatures. Asbestos millboard is used for applications such as formers for wire-wound electrical resistances.

For electrical use, the choice of asbestos free from iron and the removal of any iron oxide or metallic particles is important. Asbestos materials are not suitable for high-voltage insulation, or for high-frequency insulation because of their high dielectric loss even when dry. Their main electrical use is in low-voltage, high-temperature situations and for confinement of arcs.

2(vii).1 *Substitutes for asbestos products in electrical insulation*

A brief summary of the classes of electrical insulating materials as detailed in BS 2757:1956 is given below. The subsequent discussion deals with the most probable substitute materials for the asbestos insulation products employed at temperatures exceeding those covered by the Standard.

Class Y (formerly 0), up to 90 °C:	Unimpregnated paper, cotton or silk, urea–formaldehyde, aniline–formaldehyde, vulcanized natural rubber, various thermoplastics limited by softening points.

Class A, up to 105 °C:	Paper, cotton, or silk impregnated with oil or varnish, or laminated with natural drying oils, resins, or phenol–formaldehyde; esterified cellulose fibres; polyamides and a variety of organic varnishes and enamels used for wire coating and bonding.
Class E, up to 120 °C:	Phenol–formaldehyde and melamine–formaldehyde mouldings and laminates with cellulosic materials; poly(vinyl formal), polyurethane, epoxy resins and varnishes; cellulose triacetate, poly(ethylene terephthalate), oil-modified alkyd.
Class B, up to 130 °C:	Inorganic fibrous and flexible materials (mica, glass or asbestos fibres and fabrics, etc.), bonded and impregnated with suitable organic resins; shellac bitumen, alkyd, epoxy, phenol– or melamine–formaldehyde.
Class F, up to 155 °C:	As Class B, but with resins approved for this class: alkyd, epoxy, silicone–alkyd, and certain others.
Class H, up to 180 °C:	As Class B, but with silicone resins, silicone rubber (note that a number of high-temperature resins have appeared since this classification was drawn up).
Class C:	Mica, asbestos, ceramics, glass, alone or with certain inorganic binders or silicone resins; polytetrafluoroethylene.

N.B., The original four-class system (0, A, B, C) was drawn up when it was widely believed that no organic resin could withstand temperatures higher than 130 °C even in combination with inorganic materials. The addition of the extra classes E, F, and H has not simplified the problem, while it is now more important than ever to optimize materials for compactness and economy. The user who is not willing to accept a machine manufacturer's judgement must undergo detailed discussion and evaluation, and in this respect, the existence of a classification system may be irrelevant.

2(vii).2 *Substitute electrical insulants at high temperatures*

Chrysotile fibres deteriorate noticeably at 450 °C but suffer some loss of strength at 300 °C; prolonged heating in air at temperatures even below 200 °C will ultimately produce loss of strength. Although these fibres are more resistant than cellulose to temperatures above 100 °C, glass fibres are more resistant above 200 °C. Consequently, in applications such as insulation sleeving for cables and battery separators, direct replacement with glass fabrics is usually satisfactory.

Glass fabrics of all kinds (cloth, tape, tubes, cords) are woven from yarn and used in resin-bonded laminates. Glass chopped strand mat is also used for moulded or hand-load composites. The fibres do not alter in mechanical properties up to 200 °C and their strength is roughly halved at 350 °C; they are not suitable for applications where severe flexing is involved.

Glass fibres served on to wire and treated with suitable resins are used in high-temperature windings. This type of insulation is more vulnerable to abrasion than most other wire coverings.

Polyimide and polyethersulphone are examples of high-temperature thermosetting materials which are used as wire and cable coatings suitable up to 200 °C and 260 °C, respectively, where ability to withstand high temperatures combined with toughness and flexibility justify the cost. The latter has the additional advantage of withstanding soldering temperatures for short periods.

For cable and wire insulation up to 530 °C, ceramic fibre cloth, tape, or sleeving may be used with glass filament inserts to maintain high-temperature strength.

2(vii).3 *Further reading*

See Sillars (1973); Clark (1962).

2(viii) Asbestos composites (not including frictional, bearing, and electrical applications)

If we exclude composite products that are dealt with in other sections, we are still left with a wide range of asbestos/polymer composites in use across the entire engineering spectrum. The matrix is usually of the thermosetting type, although composites using thermoplastic matrices such as nylon and PVC are in fairly widespread use. The high aspect ratio of asbestos fibres is significant in producing an effective reinforcement of the polymer matrix, enhancing strength, stiffness, and toughness. Typical applications in which the fibres are randomly aligned in these directions include small machine parts, usually made by an injection moulding process such as vehicle distributor caps, fans, fan shrouds, small casings, and other similar products. In some applications, combinations of glass and asbestos fibres are used.

The most common matrices (or resins) used with asbestos fibres are the phenolics (high strength, heat resistance to 325 °C) and the silicones (electrical properties, heat resistance to 400 °C). Polyester resins are also widely used owing to their generally good properties, ease of handling, and low cost, and epoxies for applications where chemical resistance is important.

A widespread application for an asbestos/thermoplastic composite is for junction, destination, and mileage signs used on roads and motorways. These are manufactured from asbestos in a PVC matrix and consist essentially of a random array of asbestos fibres in two dimensions.

Composites using asbestos in various degrees of alignment have been made, such as those based on Durestos and Pytotex materials, and using woven cloths of asbestos which utilize the asbestos in an efficient manner and enable a significant proportion of the ultimate properties of the fibres to be achieved in practice. However, these materials use the more expensive, longer fibre grades of chrysotile and an expensive alignment process, so that it is frequently simpler and cheaper to use continuous glass or other fibre for this type of high-performance composite application.

Mention should also be made of the use of asbestos paper in low-cost, non-structural thermal and electrical insulation.

2(viii).1 *Alternatives to asbestos composites*

In general, glass fibre provides an adequate replacement for asbestos in many of the injection-moulded component applications. Longer fibre lengths are necessary with glass than with asbestos to achieve comparable aspect ratios, and this necessitates careful manufacturing control to avoid fibre damage during component fabrication. Glass fibres have a lower modulus than asbestos; consequently, if component stiffness is important, the volume loading necessary with glass fibres may be higher than with asbestos fibres. A wide range of injection-, dough-, and sheet-moulding compounds is available and a suitable asbestos alternative may often be found immediately. Generally the maximum continuous operating temperature for a glass-reinforced composite is comparable to that of the corresponding asbestos-reinforced material.

For two-dimensional random sheet applications, glass chopped strand mat may be effective. The stiffness of glass-fibre sheets may be increased, if necessary, by ‘hybridization’, i.e. the addition of a higher modulus (usually carbon) fibre to the composite in small amounts. Carbon fibres are expensive, but often in this type of application the weight saving possible for a particular design stiffness, and the consequent economy of glass-fibre, make hybrid structures cost effective with all glass composites.

Fillers are often used with polyester resins when mats, preforms, and chopped fibres are the reinforcement. The types frequently used include calcium carbonate, silicate clays, and Perlite. Some of these are coated to improve wetting by the resin.

Synthetic fibres such as nylon and rayon, and natural fibres such as cotton and paper, impart less strength to the plastic than does glass or asbestos. Maximum service temperatures are also significantly reduced. They are generally employed when strengths intermediate between those of the base matrix and the reinforced composite are satisfactory. Cotton and paper reinforcements have been widely used for many years with phenolic and melamine resins for electrical and decorative purposes.

Such materials as high-silica glass, quartz, graphite, and other unusual reinforcements are used to meet extreme high-temperature demands. Typical applications include high-silica glass as insulation in rockets, quartz for structures exposed to long-term elevated temperatures, and graphite for abrasion-resistant mouldings in rockets.

2(viii).2 *Further reading*

See Dubois and John (1967); *Plastics Engineering Handbook* (1960).

2(ix) Miscellaneous applications

Asbestos papers are used extensively in the electrical industry [see Section 2(vii)]. Other special uses include electrolytic cell diaphragms and in some copper sandwich type cylinder head gaskets.

Asbestos felts are used for roofing, impregnated with bitumen, and in damp-proof courses. For these applications the rot resistance and dimensional stability of the asbestos are the important qualities.

Mastics are used for caulking in high-temperature applications, using asbestos fibres in a high-temperature cement binder.

Floor tiles contain up to 10% of short fibres, together with other fillers, in various polymer matrices. Apart from reinforcing the final product, the fibres give the polymer sheets 'wet strength' during the manufacturing process.

2(ix).1 Substitutes

Adequate replacements are available for these applications, which often use asbestos only on cost grounds. Conventional cellulose-fibre papers and felts are frequently used in impregnated form as a replacement for asbestos papers. Glass fibres can often be substituted for asbestos in many other applications, and ceramic fibres where elevated temperatures are involved—mastics based on aluminosilicate fibres in a colloidal silica binder have recently become available for refractory caulking.

(3) HEALTH HAZARDS OF SUBSTITUTE MATERIALS

If it is required to replace asbestos on health grounds, then it is sensible to ascertain whether or not the suggested replacement itself constitutes a health hazard. Some of the suggested synthetic fibre replacements for asbestos in specific applications are listed in Table 10.2 with details of fibre lengths and diameters.

For all practical purposes, the risk from fibrous materials is limited to inhalation of the fibres. Recent research on asbestosis has suggested that the fibrosis is caused principally by fibres between about 5 and 100 μm long. Longer fibres than this settle out in the environment and are not inhaled, or are deposited by interception in the narrow upper airways and seldom reach the finest bronchioles in the lungs. The falling speed of fibres is governed by the fibre diameter rather than length, and this produces a further size separation in the respiratory passages. Fibres greater than 2 μm in diameter (which also tend to be the longest) are mostly deposited in the upper respiratory tract, and are eliminated in the sputum. The definitions of fibre size in the Asbestos Regulations 1969, Appendix, Technical Data Note 13, is based on these observations of causative fibre sizes for asbestosis, and probably also for bronchial cancers.

For the mesotheliomas, less information is available. What evidence exists suggests that the biologically active fibres are straight, up to about 1 μm in diameter and perhaps 10 μm long. Some of the recent data suggest that fibre size and shape are more important than chemical composition in producing carcinogenic effects. A consequence of this is that exposure to airborne fibre dust of any type with diameters less than 0.5 μm and several micrometers long is potentially hazardous.

With these observations in mind, examination of the fibre diameters in Table 10.2 suggests that the various synthetic refractory type fibrous materials that can be used to substitute for asbestos in elevated temperature applications may themselves constitute a health hazard. A mean fibre diameter in the range 2–3 μm is quoted for all of these fibres, suggesting that a significant proportion of them have diameters less than the 2 μm level thought to be significant in the production of asbestosis. It is also possible that a small proportion of the fibres have diameters less than 0.5 μm, the suggested causative size for mesothelioma to be a potential hazard. The as-manufactured length of these fibres is much longer than the suggested hazardous lengths, but of course, as with asbestos, manufacturing and handling processes can create fibre fragments that may disperse as airborne dust. All of these fibres are comparatively new products, having been available in the U.K. for less than 10 years, and since the induction period for asbestosis may be anywhere in the range 15–40 years, one cannot state with certainty that refractory ceramic fibres do not constitute a health hazard. However, a 30-year study in the U.S.A. has shown no health problem with workers in aluminosilicate fibre manufacture, and an investigation into Triton has indicated no mesothelioma potential, and a minimal tendency towards lung fibrosis.

It would appear that the fibre diameters of glass and carbon, which are reproducible with a very small spread of larger and smaller diameter proportions, constitute no health hazard of the pneumoconiosis or bronchial cancer type. Fibres in glass wool materials as supplied for insulation purposes average 5–6 μm in diameter (as also do some more recent types of glass fibre intended for reinforcement purposes) and so far industrial surveys have shown no pattern of a related disease except for a higher incidence of bronchitis among retired workers.

Research in the U.S.A. has shown a carcinogenic effect by surgically implanting glass fibres of less than 1.5 μm diameter and longer than 8 μm into the chest walls of rats. However, the greater than 30-year history of industrial use of glass fibre, with no carcinogenic relation apparent, suggests that this artificial situation does not represent the situation with the more commonly used sizes of glass fibre.

Glass fibres have been found in some cases to be more irritant to the skin than asbestos, occasionally making their handling during manufacturing processes unpleasant.

A limited amount of data is also available on some of the minerals suggested as asbestos replacements in various applications. Mineral wool fibres, with a diameter of about 5 μm, are unlikely, on the basis of the discussion of fibre sizes presented previously, to constitute a health hazard.

It has been shown by observation of miners that a pneumoconiosis disease is associated with the extraction of mica.

Studies have also been performed on vermiculite that suggest that it has no tendency towards carcinogenicity or the production of fibrosis in the lungs.

Expanded Perlite can generate airborne dust. This has been examined as a possible health hazard, and is thought to cause no problem, except for nuisance. Experience in the U.S.A. going back nearly 30 years has indicated no related disease pattern with the use of Perlite.

(4) CONCLUSIONS

For most applications of asbestos, it is possible to find an alternative, often in glass or ceramic fibre. It should be realized that substitution for asbestos on health grounds often involves a penalty in performance and almost always a penalty in cost; in some cases health hazards may be associated with substitutes.

Although the consumption of asbestos could be reduced significantly, it is unlikely that its use could be eliminated, and a problem would still exist with handling existing installations. There is consequently a need for stringent enforcement of safety regulations. It is perhaps paradoxical that those asbestos products which are used in 'life-saving' applications (e.g. asbestos–cement, friction materials) tend to be those which present a health hazard in service, while in many of the products in which asbestos is used purely for economic reasons the fibre is imparted into the product in such a way as to constitute little or no health hazard to the general public.

The most difficult industrial asbestos products to substitute are:

1. insulation applications subject to loading, flexure, or corrosive environments;
2. specialized bearings;
3. a few textile applications.

(5) REFERENCES

Clark, F. M. (1962) *Insulating Materials for Design and Engineering Practice*, Wiley, New York, London.

Dubois, J. H., and John, F. W. (1967) *Plastics*, Reinhold, New York.

Engineering Sciences Data, Item No. 76029 (1976).

Green, A. K., and Pye, A. M. (1976) *Asbestos, Characteristics, Applications and Alternatives*, Fulmer Special Report No. 5, Reedprint.

Green, A. K., and Pye, A. M. (1977) The search for asbestos substitutes, *Asbestos*, **58** (11), 1977.

Neale, M. H. (Ed.) (1973) *Tribology Handbook*, Butterworths, London.

Plastics Engineering Handbook of the Society of the Plastics Industry Inc. (1960) 3rd Ed., Reinhold, New York.

Proc. Inst. Mech. Engrs., (1967–68), **182,** Part 1, No. 2, 33–51.

Sillars, R. W. *Electrical Insulating Materials and their Application*, I.E.E. Monograph Series, No. 14, Peter Peregrinus Ltd., London.

CHAPTER 11

The literature relating to asbestos

Peter Warren,

The London Hospital Medical College

(1) INTRODUCTION

The range and extensive use of asbestos in all branches of science and technology has meant that the literature relating to these applications is vast and growing. In order to make the reading and understanding of this literature as simple as possible, it is considered here under four separate headings. Section 2 deals with the identification, physical properties and measurement of the asbestos dust containing fibres. The importance of the parameters of physical size and type of airborne fibres will be seen in the medical literature on asbestos-related diseases discussed in Section 3. Section 4 deals with those industrial processes which use asbestos in the manufacture of fire-resistant materials and heat and acoustic insulation in the building industry. The range of applications of asbestos, from ironing boards to aircraft carriers, shows that it is used or is present as frequently in the home as it is in the manufacturing and construction industry. The implications of this could be considerable if it was shown that a single exposure to a low fibre density in air could significantly increase the risk of mesothelioma of the pleura. The literature produced by the medical profession indicating the degree of risk is considered in Section 3 in some detail.

Section 5 considers information available to the general public via the press, television, and special pamphlets.

(2) IDENTIFICATION, MEASUREMENT, AND PHYSICAL PROPERTIES OF ASBESTIFORM MINERALS

2(i) Measurement of dust and fibre samples

Minerals of the asbestos type are all fibrous silicates and, although they can produce considerable amounts of fine dust under manufacturing conditions, they also produce fibres of varying sizes. Fibres are defined as material having a length greater than 5 μm and a length to breadth ratio of at least 3:1. Fibre dimensions are considered to be highly significant, especially where the risk of fibres entering the human lung is concerned. It will be seen later that a fibre length of 5 μm proves to be metastable in air and enters the lung mainly in that form. Large, heavy fibres cannot float in air and fall to the ground. In dust sampling techniques used to assess the concentration of asbestos fibres per cubic centimetre of air these criteria are important (Asbestosis Research Council, 1971a). In air measurements the sampling orifice is placed at approximately head height and a filter suitable to trap fibres of the metastable size is mounted at right-angles to the air flow. This prevents excessively large particles not of biological interest being deposited on the filter and blocking it. Cyclone precipitators are often supplied by manufacturers to prevent this (C. F. Casalla and Co. Ltd.). Sampling using a white gridded membrane filter of a standard 25 mm diameter with a pore size of 0.8 mm requires fixing and mounting the sample so that a count of significant fibres can be made. This is carried out by using a microscope fitted with a micrometer eyepiece. Details of all the technical procedures required to collect, fix, mount, and count these fibres were given by the Asbestosis Research Council (1971b), together with the names and addresses of filter and equipment suppliers. H. M. Factory Inspectorate (1971) has given practical guidance on the sampling of airborne asbestos dust. The Asbestosis Research Council (1971a,b) have adopted these in their Technical Notes 1 and 2. Two types of sampling are proposed, the static type and the personal monitor sample. In the static type, the general work area is measured whereas with the personal sampler, a portable mobile type of instrument is attached to the worker so that an air sample containing asbestos can be taken in the breathing zone. The duration of the sampling period can be (a) a snap sample taken at random using a hand-operated pump of the Draeger type (Draeger Safety Ltd., Kitty Brewster, Blyth, Northumberland; (b) a 10-minute sample, i.e. a sample taken with a continous flow of at least 250 cm^3 of air through the apparatus for 10 minutes; this gives a good indication of the concentration of airborne asbestos dust; or (c) the more accurate 4-hour sample in which a continous flow of at least 200 cm^3 of air is passed through the sampling apparatus for 4 h. Technical Note 2 (Asbestosis Research Council, 1971b) also explains the general considerations to be applied to the evaluation of dust levels, the methods of calculation of time-weighted average dust concentrations and a useful list of firms in the U.K. who produce dust sampling equipment.

2(ii) Physical properties and behaviour of asbestos fibres

Fibrous particles with an axial ratio of greater than 3:1 behave in different ways, depending on their diameter (Harris *et al.*, 1975). The influence of fibre shape on lung deposition is very important; large particles with extremely long fibre lengths when released into the air tend to fall rapidly to the ground. Asbestos dust and short fibres, i.e. $<10\ \mu m$ long, can easily enter the lung alveoli (Timbrell, 1970), where they may be taken up by phagosomes and ferrutin deposition on them producing the so-called ferruginous bodies (Gross, *et al.* 1967). Similar bodies can be produced experimentally with aluminium silicate fibres or glass fibres (Davis *et al.*, 1970).

These bodies can be seen very easily by examining lung sections using the electron microscope. The body appears to be an asbestos fibre on to which an envelope is attached, possibly composed of a mucopolysaccharide. Many bodies appear like a knitting needle on to which balls of mucopolysaccharide 'wool' have been attached. There appears to be some speculation as to whether the fibre coating is a protective mechanism against the presence of fibres in the lung. Pooley *et al.* (1970) showed that in advanced asbestosis many fibres are not coated and formed into asbestos bodies. This may be due to the excessive numbers present, viz several million fibres per gram of lung tissue. These asbestos bodies are of considerable interest since they occur in 20–60% of urban dwellers with no known industrial exposure to asbestos (Thomson *et al.*, 1963; Cauna *et al.*, 1965; Anjivell and Thurlbeck, 1966; Selikoff and Hammond, 1970). The aerodynamic properties of the fibres were carefully considered by Parkes (1973) in his excellent review of asbestos-related disorders. Fibres deposited in the upper bronchioles are cleared by ciliary action whereas those which penetrate beyond this region are not and may be retained by the lungs. The pliable, curly nature of chrysotile fibres makes them liable to be trapped in the upper respiratory tract whereas the rigid amphibole fibres can travel easily in the inspired air and can reach the peripheral parts of the lung. A considerable amount of work has been done on determination of fibre types and concentrations in lung tissue. It is important to realize that all fibres in lung are not necessarily of an asbestos type. The use of techniques such as light microscopy

(Richards *et al.*, 1976), electron microscope examination (Churg *et al.*, 1977), micro-incineration techniques and the use of electron diffraction (Seshan, 1977), electron probe analysis (Rubin *et al.*, 1974), infrared spectrometry (Beckett *et al.*, 1975), and X-ray microanalysis (Omoto *et al.*, 1975) have all played their part in the identification of the type of fibrous structures.

Many research workers have carefully considered the types and amount of asbestos dust and fibres found in the air and the lung. Lung dust problems have been studied by Pooley (1974, 1976), Davis *et al.* (1976), Heidermans *et al.* (1975a,b), Pott (1975), and Rüttner (1975). The problems of airborne dust measurement and evaluation have been considered by Beckett *et al.* (1974), Dement *et al.* (1976), Sawyer (1977), and Rajhous *et al.* (1975).

(3) LITERATURE RELATING TO THE TOXICITY OF ASBESTOS TO MAN AS SHOWN BY HUMAN AND ANIMAL STUDIES, AND STUDIES ON ASBESTOSIS AND MESOTHELIOMA

As one might expect, the literature on the clinical features of asbestos-related diseases is immense and is well reviewed elsewhere in this book. In recent years, two disorders associated with the inhalation of asbestos fibres have been extensively studied. They are asbestosis and malignant mesothelioma of pleura or peritoneum. Asbestosis is a fibrosis of the lungs caused by inhaled asbestos dust and is commonly found in asbestos workers (Weill *et al.*, 1975; Hughes, 1974; Bellotte *et al.*, 1976; Rodriguez-Roisin, *et al.* (1976). Asbestosis is diagnosed by a number of clinical features, but principally radiographic changes seen in the lungs (Horai, *et al.*, 1975); Liddell *et al.*, 1975; Weyer, 1975), the presence of asbestos in the sputum (Attia, 1975; Greenberg *et al.*, 1976; Shishido *et al.*, 1976), the appearance of pleural plaques (Mattson *et al.*, 1975; Mattson, 1976), by transbronchial biopsy (Kane *et al.*, 1977; Fukui *et al.*, 1976), or by detection of asbestos bodies at autopsy (Matsuda, 1975). Radiological evidence is used internationally to assess the degree of severity of the asbestosis using the criteria laid down by the International Labour Office (1970). The earliest provisional diagnosis between those who have asbestosis and others is best made by measuring the subject's lung vital capacity (Becklake *et al.*, 1970; Britton *et al.*, 1977). This value can be reduced up to 15 years before moderate radiological changes are seen (Bader *et al.*, 1970; Seaton, 1977).

In the last few years, surveys have been made of patients who had been diagnosed as asbestosis cases (Britton *et al.*, 1976), either from exposed workers (Webster *et al.*, 1975; von Chossy, 1976 or as a result of non-occupational risks (Truneviski *et al.*, 1974). Current information on the problems and health risk with asbestos are not confined to the major mining and manufacturing countries (Elmes, 1976; Hany, 1975; Kelly, 1976). The epidemiology and implications for clinical practice have been studied by Jones (1974), Gilson (1974), Hain (1975), Selikoff (1975), Gillam et al. (1976), Becklake (1976), and Saracci (1977). Enterline (1976) examined 38 papers and discussed the problems concerned with epidemiological research. Benign asbestosic pleurosies and pleural calcification have been considered by Boutin *et al.* (1975), Yazicioglu (1976), and Moigneteau *et al.* (1977). The relationship between lung carcinoma and asbestos was first recognized in the 1930s, become a significant cause of death in British male asbestos workers by 1947 and in recent years has accounted for about 50% of all cases of asbestosis (Buchanan, 1965). Papers by Hajdukiewicz *et al.* (1975), Fears (1976), Martischnig *et al.* (1977), and Woitowitz *et al.* (1977) discuss this problem. Malignant mesothelioma of pleura and peritoneum was first demonstrated in people living in North West Cape Province, an asbestos mining area of South Africa (Wagner *et al.*, 1960). The condition is lethal and, unlike asbestosis, can be produced by a low-dose short exposure to crocidolite. There is some suggestion that crocidolite fibres if swallowed are able to penetrate the intestinal wall and enter the peritoneum (Enticknap and Smither, 1964).

The latent period of development of the tumour is very long. Periods of 20–40 years have been reported, and because of this there have been many reports of the occurrence of malignant mesotheliomas following asbestos exposure (Pylev, 1974; Otto, 1975; Milne, 1976; Haider *et al.*, 1975; Bianchi *et al.*, 1976; Jones *et al.*, 1976; Platedydt, 1975; Hasan *et al.*, 1977). All types of asbestos if placed experimentally in the pleura of animals produce mesothelial tumours Pylev *et al.* (1976) Pylev, 1976; Engelbrecht *et al.* (1975), but crocidolite appears to produce more tumours than amosite (Harington *et al.*, 1971). It is important to remember that a small proportion (<10%) of these tumours do occur in cases in the U.K. without apparent exposure to asbestos.

The effect of asbestos fibres in both the workplace and the environment has stimulated considerable study during the last few years (Selikoff, 1974, 1977; Rohl *et al.*, (1977). Publications such as those by Aharonson *et al.* (1976), Bignon *et al.* (1974), Gibbs (1975), and Morley (1976) illustrate the work being carried out to study pollution and fibre behaviour. Britton *et al.* (1976) and Kogan (1977) have discussed the exposure to asbestos dust and the maximum permissible exposure level of asbestos powders. Safety measures when handling asbestos in industry were considered by Smither (1974), McDonald (1975), Cross (1976), and Wagg (1976); in brake maintenance and repair shops by Newhouse (1977); naval dockyards by Harries (1976); and in asbestos mines by Kogan (1977). Hazards relating to asbestos spraying and disposal of asbestos waste were covered by Barnes (1976) and Kinsey *et al.* (1977). In the last 3 years, research publications have tended to show an interest in examination of animal models to indicate carcinogenicity of various inhaled fibrous substances (Bolton *et al.*, 1976; Tetley *et al.*, 1976; Rahman *et al.*, 1976; McDermott *et al.*, 1975; Pott *et al.*, 1976). No literature survey on the medical aspects is complete without reference to the possible immunological responses to asbestos. Pulmonary asbestosis and autoimmunity were discussed by Toivanen (1976), frequency of HLA antigen in asbestos workers by Evans *et al.* (1977), and asbestosis immunology by Kagan *et al.* (1977a,b). Burrell (1974), Wilson *et al.* (1977), and Millerk *et al.* (1977) have also contributed work on this topic.

(4) INFORMATION ON THE HAZARDS CAUSED BY THE USE OF ASBESTOS-CONTAINING MATERIALS IN INDUSTRY

There are six varities of asbestos which are used in industry Hodgson (1966). Their use in manufacturing industry depends to a large extent on their physical and chemical properties (International Labour Office, 1971). Chrysotile ($3MgO.2SiO_2.2H_2O$), a white serpentile form of asbestos, is the softest form and is extremely suitable for spinning and weaving into asbestos cloth or tape. Many products containing this form of asbestos are made; figures for 1966 show that about 2 800 000 tonnes of this substance were used. Of the five asbestiform amphiboles, tremolite ($2CaO.5MgO.8SiO_2.H_2O$) and actinolite ($2CaO.4MgO.FeO.8SiO_2.H_2O$) are used as additives in the production of amorphous magnesium silicate as commercial talc. Amosite ($5.5FeO.1.5MgO.8SiO_2.H_2O$ is a brown-coloured iron magnesium silicate whose long fibres and harshness make it unsuitable for spining but excellent for use as a heat insulation material. Considerable amounts of this material are used in the construction industry, about 86 500 tonnes being used in 1966.

The white magnesium silicate known as anthophyllite ($7MgO.8SiO_2.H_2O$) is used in relatively small amounts in industry, about 12 400 tonnes annually being processed. It consists of short fibres that are not suitable for spinning but because, like crocidolite, it has more resistance to acid than the common chrysotile, it is used in the manufacture of electric battery cases.

Crocidolite ($Na_2O.Fe_2O_3.3FeO.8SiO_2H_2O$) is a silicate containing sodium and iron but no magnesium. It has a characteristic blue colour with moderately long fibres ideal for spinning. Production in 1966 of 142 000 tonnes shows that it is used in considerable amounts. This material is often used in boiler lagging, particularly in ships. The particular dangers to health in the handling and use of crocidolite are now well known (Harries, 1971a,b) and have been dealt with in the literature mentioned in Section 3. These forms of asbestos have different properties that can be usefully employed in the manufacturing industry. In general, asbestos provides a cheap, easily handled fire- and heat-resistant material which stands up to abrasion. It can be manufactured into cloth forms for fire-resistant clothing and lamp wicks. The cheapest forms can be used as fillers or coatings in the building industry.

The universal application of the material means that all branches of the construction industry will face potential hazards with asbestos. An excellent survey has been carried out by Berkovitch (1976). He deals with much information of considerable use generally in the building industry based on the requirements of the U.K. Asbestos Regulations 1969. The U.K. Health and Safety Executive (1970, 1975, 1976) has published technical data covering health and safety precautions relating to these regulations in Booklet 44 and Technical Data Notes 24 and 52. Any organization or person who is intending to use or handle asbestos, particularly in the fabrication of asbestos-containing structures, should seek advice from the appropriate Government Inspectorate. In the U.K., the local Health and Safety Inspectorate office would be the appropriate place. If crocidolite (blue asbestos) is being used, the Inspectorate *must be informed* because of the serious health hazards associated with the use of this material. A good indication of the probable level of dust concentrations for various construction processes was given by the U.K. Department of Employment (1973) in Technical Data Note 42. Management should ensure that chargehands and foremen have read and understood the Employment Medical Advisory Service leaflet MS(A) 3 relating to practical aspects of handling asbestos materials at work. In their turn, they should read and follow the guidelines of health precautions given by the Health and Safety Executive (1970) in Health and Safety Booklet 44.

Information and guidance in the use and handling of asbestos materials is also given by the asbestos industry's sponsored Asbestosis Research Council. Its Environmental Control Committee have published a number of very useful technical notes and control and safety guides, as listed in the References, copies of which are obtainable from the Council. Particularly useful for industry are the nine control and safety guides covering all of the major processes and operations involving asbestos. These booklets have been produced with advice and criticism from the Health and Safety Inspectorate, and in consequence do overlap with some U.K. Government publications. The Asbestos Information Committee is also financed by the asbestos industry and has produced a number of semi-scientific publications relating to the risk to health of the general public by asbestos. The titles of these pamphlets suggest that they were produced to counter the public disquiet in the U.K. about the mishandling and misuses of asbestos in the manufacturing and construction industry.

(5) INFORMATION SUPPLIED TO THE GENERAL PUBLIC BY THE PRESS AND TELEVISION CONCERNING ASBESTOS IN THE UNITED KINGDOM

The use and abuse of asbestos materials in industry, the dumping of waste containing asbestos and its use in the home have become important social and political matters in the U.K. It is extremely difficult, when reviewing the literature, articles, and television programmes on asbestos, to be impartial since both the industry and their critics have put forward very strong cases. The Asbestos Information Committee has produced a number of useful booklets which give the general public and lay persons an indication of the problem, together with practical advice on how to solve difficulties, (Asbestos Information Committee, 1975, 1976a,b,c). If management and workers behaved reasonably and took precautions as simple as those outlined in the documents produced by the asbestos industry, a substantial improvement in the present position would be seen. Unfortunately, because of lack of enforcement and deterrents, the laws that exist to protect the workforce and the public are not effective. In the U.K. in the first year of the Health and Safety at Work etc. Act 1974, there were 1666 prosecutions for all infringements following 7599 prohibition and improvement notices. These notices are issued by Inspectors of the Health and Safety Inspectorate (HSI) if they find unsuitable or dangerous practices or equipment present in workplaces.

In spite of the work of the Inspectorate, during 1975, 61 persons died of mesothelioma and 50 died of asbestosis (*Transport Review*, 1977). It has been estimated that about 30 000 persons die each year as a result of all occupational diseases. The same sort of problems of lack of effective control are seen in the U.S.A. The Office of Safety and Health Administration (OSHA) formed in 1971 has the duty to inspect factories and plants. However, since there are only 1400 inspectors to protect more than 62 million workers in 5 million plants, offices, and shops, the chances of receiving an inspection are rare (*Economist*, 1977). To be fair, both the HSI or OSHA will spend considerably more time over dangerous chemical processes than, for example, ladies shoe shops, but unless the control is self-imposed by management and workers there cannot be effective control of safety standards by the Government. Prosecutions, when they do occur, usually result in very small fines which cannot deter the offender. British Leyland were fined £300 for failure to maintain a clean work area in one of their plants. The clean-up required to rectify the problem produced 132 sacks filled with dust containing 12% of asbestos (*Birmingham Evening Mail*, 1978; *Daily Telegraph*, 1978). The press and public are also concerned about asbestos removal by demolition firms sometimes referred to as 'asbestos cowboys', who do not appear to take proper care (*Yorkshire Post*, 1977). The problem also extends to the tipping and disposal of asbestos-containing waste. The public concern is as much about the volume as the type of material dumped. The villages of Charlbury near Oxford opposed an application to fill a quarry with approximately 1 million cubic yards of waste containing some asbestos (*Oxford Mail*, 1977). At another site near Warrington, 180 000 tonnes of waste were tipped in 1 year, of which 57 000 tonnes were 'hazardous or difficult waste' (*Warrington Guardian*, 1977).

Five journals and newspapers reported the fine of £400 imposed on a waste disposal company who left bags containing blue asbestos waste in Folly Quarry at Greenside (*Dundee Courier and Advertiser* 1977; *Sheffield Morning Telegraph*, 1977c; *Times*, 1977; *Newalls Publicity Department Journal* 1977; *Surveyor*, 1977).

Since the publication of articles in the scientific literature concerned with the hazards of using asbestos, particularly blue asbestos (crocidolite), the press have given prominence to any reports of the presence of this material in buildings (*Beckenham and Penge Advertiser*, 1977; *Sheffield Morning Telegraph*, 1977b; *Scottish Sunday Express*, 1977; *Luton Evening Post*, 1977), in the cloth used to make the head codes situated in the driving cabs of railway diesel and electric locomotives (*Evening Standard*, 1977a,b), and even its absence (*Sutton and Cheam Advertiser*, 1977). Eyewitness accounts and personal and individual stories of tragedy always interested the newspapers. 'Widow lives in fear of asbestos' was a headline in the *Sheffield Morning Telegraph* (1977a) relating to blue asbestos lining found in the loft of a Mrs Brewer's house. Because of lack of action by the Local Authority, she was reported as being prepared to pay the cost of its removal. The Cape Asbestos factory at Barking, East London, prompted an article by the *Barking and Dagenham Post* (1978) and a story relating to the same factory by *The Guardian* (1978c) and a television visit and interview (BBC Nationwide, 1978) to Henry Steggles, who dug up lumps of blue asbestos in his garden near to the former Cape Asbestos factory. Mrs Wendrop is reported as working in the factory in 1930 when she was 17 years old. Her husband and uncle died of asbestosis, her sister and sister-in-law both had asbestosis (*The Guardian*, 1978c). Members of the House of Commons have been reported by the press in sharing considerable concern as to the reporting and incidence of asbestosis among the workers in the industry, and the compensation as a result of disablement or death. One report suggests that death certificates had been deliberately falsified to hide cancer caused by asbestos, i.e. mesothelioma described as asbestosis. Mr. Roland Moyle (Minister, Department of Health), in answer to Mrs Margaret Bain (Scottish Nationalist MP) said, 'there are some cases where death has actually been due to cancer, due to asbestos exposure, but for humanitarian reasons the true nature of the condition has been withheld' (*Daily Mirror*, 1977). The MP for Barking was reported as wishing to find the incidence of asbestosis among workers at the Cape Asbestos factory in his constituency (*Barking and Dagenham Advertiser*, 1978). The *Sheffield Star* (1977) reported 'City MPs urged to back fight against asbestos disease'. They wished asbestos to be phased out or banned completely, and considered that the Government regulations were not tough enough. Concern was also expressed by Mr. Robert Bean, MP for Rochester and Chatham, a constituency which has a naval dockyard, that the dockyard workers were at risk from asbestos diseases (*Chatham News*, 1977; *Kent Evening Post*, 1977). One of the saddest stories reported by the newspapers concerns the way in which claims for compensation by asbestos disease sufferers are treated. The Acre Mill asbestos factory in Hebden Bridge, Yorkshire, was closed in 1970 because of problems with asbestosis. Claims for compensation for affected workers in this factory have been so delayed that Mr. Max Madden, MP for Sowerby, told the Hebden Bridge Asbestos Action Group that steps would be taken to speed up claims for compensation (*Halifax Evening Courier*, 1978). He also stated that the European Commission, following the adoption of a nineteen-point resolution on asbestosis by the European Assembly, was expected to propose action during the year. This would include temporary limits to the degree of exposure to asbestos to which workers were subjected. *The Guardian* (1978b), commenting on this problem stated that at present the negligence of the employer(s) has to be proved before compensation can be given. It was also considered that examination by medical consultants of workers who might be affected by asbestos should be more readily available. Mr. Madden was also given an assurance by Mr. Stanley Orme, Minister of Social Security, that relaxation of the existing appeals procedure would be considered when a report was received of the first year's operation of the new system (*Halifax Evening Courier*, 1978). The whole matter might be simplified if the Royal Commission on Civil Liberty and Compensation for Personal Injury, in its report expected in 1978, proposed the setting up of a fund paid for by a surcharge on all asbestos products produced by the industry. A worker who on medical evidence had contracted asbestosis would then be entitled to immediate compensation from this fund. A similar proposal will be made by the Royal Commission regarding a petrol surcharge and compensation for injuries sustained as a result of road accidents.

The Times (1978a) and *The Guardian* (1978a) reported that the Employment Medical Advisory Service had produced figures which showed that mesothelioma cases has risen 68% between 1970 and 1976, and suggested that the rate was still rising. In 1975 there were 260 cases reported. Recently Dr. Mann at a meeting of the Royal College of Physicians showed that when the Hebden Bridge asbestos factory was closed in 1970, 2200 workers were employed there; 12% of the labour force had developed asbestosis compared with 1.05% of workers in the same company's factories in the U.K. (*The Times*, 1978b).

Television has given coverage to the problems of the health risks from asbestos. One of the most significant and well presented programmes in the U.K. was produced by Granada Television TV Network Ltd. (1974) in the 'World in Action' series entitled 'Killer dust—a standard mistake'. The programme gave a simple, clear history of the establishment of the 2 fibres/cm^3 standard for asbestos in air. In 1951, the Turner and Newell

Rochdale factory pioneered the study of dust levels in factories. This survey was carried out by two employees: Dr. John Knox, who carried out the assessment of medical evidence, and Dr. Stephen Holmes, who monitored the factory environment and made dust measurements. In 1966, the British Occupational Hygiene Society (BOHS) set up a Sub-Committee to consider the problem and were given the results of the Rochdale Survey. Of the 290 persons studied, 8 showed clear signs of asbestosis and the fibre count in air was $3\frac{1}{2}$–6 fibres/cm^3. In 1969 The Asbestos Regulations adopted 2 fibres/cm^3 as the acceptable standard for factories and workplaces; 2 fibres/cm^3 represents 16 million fibres per 8-h working shift. At that stage no check was made on Dr. Knox's X-ray evidence on which the asbestosis assessment was made. The Rochdale figures were apparently accepted without question by the BOHS, i.e. the evidence of one survey in one company set the standard even though it was known that dust counts by available methods gave inaccurate values, with variations of as much as ±30%. In 1970, Dr. Lewinsohn of Turner Brothers Asbestos (TBA) re-examined the men studied by Knox and produced very different figures:

	Cases studied	Signs of asbestosis
Knox assessment	290	3%
Lewinsohn assessment	290	50%*

Dr. Lewinsohn felt that the safety standard of 2 fibres/cm^3 was wrong and published his views (Lewinsohn, 1972).

Professor Irvine Selikoff, an American authority on asbestosis and lung diseases, also thinks that the 2-fibres 1 cm^3 standard is wrong. Selikoff wrote to Lewinsohn in 1972 for all the TBA data from the Rochdale study, and questioned the safe level publically. Neither Dr. Holmes nor Dr. Lewinsohn was pleased about this. Dr. Lewinsohn suggested that Professor Selikoff was making a political campaign and a series of heated exchanges occurred. The interpretation of the X-ray pictures is at the heart of the dispute. Dr. Lewinsohn has suggested that all the changes seen by X-ray investigation of the lungs could not be attributed to asbestos exposure, e.g. some effects would be due to ageing or smoking or atmospheric pollution. Selikoff believes that the appearance of lung scarring in X-ray pictures must be related to asbestosis. In 1972, TBA produced new figures for the 290 persons surveyed:

	Original 290 persons	Number with asbestos
Still employed	162	12
Left TBA	95	11
Died	28	3

* This value represents signs of lung changes or scarring that could lead to asbestosis.

Of the 28 persons that had died, seven had died of cancer and three of mesothelioma. The BOHS could not quantify the cancer risk when they proposed the 2 fibre/cm^3 standard. Even today the risk of lung cancer cannot be assessed accurately in terms of fibre in air concentration. Dr. Lewinsohn considers that approximately 50% of those workers with asbestosis (26 cases) will probably die of lung cancer but he feels that the lung cancer deaths as a proportion of the 290 original workers investigated is the same as for the general population. One of the difficulties brought home to the viewers in this television programme was the difficulty TBA had in ensuring that workers used the correct protective clothing while at work.

Very recently Peto (1978) has published a paper based on the report of the Advisory Committee on Asbestos (1977). He made three important points in this paper:

(1) previous studies on the basis of the 2 fibre/cm^3 standard have underestimated the risk of morbidity following exposure to low levels of asbestos dust;

(2) if excess mortality from asbestos-related disease is proportional to the dust level, approximately 100% of all male asbestos workers might die after a 50-year exposure at the 2 fibre/cm^3 level;

(3) there is now evidence that chrysotile alone can cause pleural-type mesothelioma.

Owing to the enthusiasm of individuals and the concern of the section of the general public referred to as the 'environmental lobby', groups have been formed to try to ban or to restrict severely the use of asbestos. Foremost in this campaign is Mrs. Nancy Tait, who has produced two booklets (Tait, 1976, 1977), in which abstracts from research papers and other documents covering various aspects of the asbestos problem are set out. The booklets express her concern regarding the dangers of using asbestos, and that asbestos-containing products are used in the home. The Asbestos Information Committee (1976d, 1977) have produced two documents in reply to Mrs. Tait's comments. They represent a confrontation at the charge and counter-charge level. Mrs. Tait's booklets might be criticized as being emotionally biased against the asbestos industry, and selecting or stressing certain aspects of research work which aids the presentation of her case. The latter criticism can equally be levelled at the Asbestos Information Committee. Probably the most significant quotation with which to conclude this chapter on the literature relating to asbestos was produced by the International Agency for Research on Cancer (1977): 'By the time cancer potential of asbestos had been recognized and defined (1935–1960), asbestos had penetrated much of modern industry, and indeed modern society, with thousands of products being manufactured and utilized throughout the world, in circumstances that we now understand were inadequate for the control of occupational diseases, including cancer. As a result, we are now faced with a double dilemma, of how to deal with the consequences of previous inattention and error, in terms of human disease, and of how to avoid further exposure which could produce disease in the future'.

(6) ACKNOWLEDGEMENTS

I am grateful to Mrs. Hubbard of the Asbestos Information Centre for her help with the press cuttings and to Mrs E. M. Hart for assistance in preparing the manuscript.

REFERENCES

Aharonson, E. F., *et al.* (Ed.) (1976) Air Pollution and the Lung, Wiley, New York.

Anjivel, L., and Thurlbeck, W. M. (1966) The incidence of asbestos bodies in the lungs of random necropsies in Montreal, *Can. Med. Ass. J.*, **95,** 1179.

Asbestos Information Committee (1975) *Asbestos and the Docker*, Asbestos Information Committee, London. Asbestos Information Committee (1976a) *Asbestos—Public not a Risk*, Asbestos Information Committee, London. Asbestos Information Committe (1976b) *Asbestos Dust and Your Health*, Asbestos Information Comittee, London. Asbestos Information Committee (1976c) *Asbestos Dust Safety and Control*, Asbestos Information Committee, London.

Asbestos Information Committee (1976d) *Asbestos Kills? A commentary on Mrs. Nancy Tait's Thesis*, Asbestos Information Committee, London.

Asbestos Information Committee (1977) *Asbestos Kills—New Facts. Commentary on Mrs. Nancy Tait's Pamphlet*, Asbestos Information Committee, London.

Asbestosis Research Council (1971a) *The Measurement of Airborne Asbestos Dust by the Membrane Filter Method*, Technical Note No. 1, Asbestosis Research Council, Rochdale, Lancs.

Asbestosis Research Council (1971b) *Dust Sampling Procedures for Use with the Asbestos Regulations 1969*, Technical Note No. 2, Asbestosis Research Council, Rochdale, Lancs.

Asbestosis Research Council (1977) *Protective Equipment in the Asbestos Industry* (*Respiratory Equipment and Protective Clothing*), Control and Safety Guide No. 1. Revised January 1977, Asbestosis Research Council, Rochdale Lancs.

Asbestosis Research Council (1972) *The Application of Sprayed Asbestos Coatings*, Control and Safety Guide No. 2. Revised March 1972. Asbestos Research Council, Rochdale, Lancs.

Asbestosis Research Council (1973) *Stripping and Fitting of Asbestos-containing Thermal Insulation Materials*, Control and Safety Guide No. 3. Revised March 1973. Asbestos Research Council, Rochdale, Lancs.

Asbestosis Research Council (1977) *Asbestos Textile Products, CAF/Asbestos Beater Jointings and Asbestos Millboard*, Control and Safety Guide No. 4. Revised April 1977, Asbestosis Research Council, Rochdale, Lancs.

Asbestosis Research Council (1977) *Asbestos Based Materials for the Building and Shipbuilding Industries and Electrical and Engineering Insulation*, Control and Safety Guide No. 5, Asbestosis Research Council, Rochdale, Lancs.

Asbestosis Research Council (1971) *Handling, Storage, Transportation and Discharging of Asbestos Fibre into Manufacturing Process*, Control and Safety Guide No. 6. Revised December 1971, Asbestosis Research Council, Rochdale, Lancs.

Asbestosis Research Council (1973) *Control of Dust by Exhaust Ventilation*, Control and Safety Guide No. 7. Revised April 1973, Asbestosis Research Council, Rochdale, Lancs.

Asbestosis Research Council (1970) *Asbestos Based Friction Materials and Asbestos Reinforced Resinous Moulded Materials*, Control and Safety Guide No. 8. Revised December 1970, Asbestosis Research Council, Rochdale, Lancs.

Asbestosis Research Council (1977) *The Cleaning of Premises and Plant in Accordance with the Asbestos Regulations*, Control and Safety Guide No. 9. Revised January 1977, Asbestosis Research Council, Rochdale Lancs.

Attia, O. M., *et al.* (1975) Sputum picture in workers at an Egyptian asbestos cement pipe factory, *J. Egypt. Med. Ass.*, **58,** 227–233.

Bader, M. E., Bader R. A., Tierstein, A. S., Miller, A., and Selikoff, I. J. (1970) Pulmonary function and radiographic changes in 598 workers with varying duration of exposure to asbestos, *Mt. Sinai J. Med.*, **37,** 492.

Barking and Dagenham Post (1978) Clothes were covered in dust, (15th February 1978).

Barking and Dagenham Advertiser, (1978) Death from asbestosis. Why I need help in this enquiry. MP, (13th January 1978).

Barnes, R. (1976) Asbestos spraying and occupational and environmental hazard, *Med. J. Aust.*, **2,** 599–602.

BBC Nationwide (1978) Visit to Whiting Avenue, TV probes the deadly garden, (7th February 1978).

Beckenham and Penge Advertiser (1977) Deadly asbestos sealed, (22nd September 1977).

Beckett, S. T., *et al.* (1974) Inter-laboratory comparisons of the counting of asbestos fibres sampled on membrane filters, *Ann. Occup. Hyg.*, **17,** 85–96.

Beckett, S. T., *et al.* (1975) The use of infrared spectrophotometry for the estimation of small quantities of single varities of U.I.C.C. asbestos, *Ann. Occup. Hyg.*, **18,** 313–320.

Becklake, M. R. (1976) Asbestos related diseases of the lung and other organs: their epidemiology and implication for clinical practice, *Am. Rev. Resp. Dis.*, **114,** 187–227.

Becklake, M. R., Fournier-Massey, G., McDonald, J. C., Seimiatycki, J., and Rossiter, C. E. (1970) Lung function in relation to chest radiographic changes in Quebec asbestos workers, in *Pneumoconoisis: Proceedings of the International Conference, Johannesburg*, 1969, (Ed. H. A. Shapiro), Oxford University Press, Cape Town, 233.

Bellotte, J. A., *et al.* (1976) Asbestosis in West Virginia and Eastern Ohio, *W. Va. Med. J.* **72,** 341–344.

Berkovitch, I. (1976) *Hazards of Asbestos in Construction Practice: a Review of U.K. Sources of Information and Advice*, Construction Industries Research and Information Association, London.

Bianchi, C., *et al.* (1976) Diffuse mesothelioma of the peritonium and exposure to asbestos (reflections on 2 cases), *Pathologica*, **68,** 975–976.

Bignon, J., *et al.* (1974) Long term pulmonary clearance of fibrous particles in man, in *Réactions Bronchopulmonaires aux Pollutants Atmosphériques*, (Ed. C. Voisin), INSERM, Paris.

Birmingham Evening Mail (1978) Plant inches deep in asbestos dust—expert, (16th January 1978).

Bolton, R. E., *et al.* (1976) The short term effects of chronic asbestos injestion in rats, *Ann. Occup. Hyg.*, **19,** 121–128.

Boutin, C., *et al.* (1975) Benign asbestosic pleutisies (apropos of 3 cases), *Poumon Coeur,* **31,** 111–118.
Britton, D. C. (1976) Exposure to asbestos dust, *Lancet,* **ii** 175.
Britton, M. G., *et al.* (1976) A survey of patients diagnosed as having asbestosis, *Med. Sci. Law,* **16,** 279–284.
Britton, M. G., *et al.* (1977) Seral pulmonary function tests in patients with asbestos, *Thorax,* **32,** 45–52.
Buchanan, W. D. (1965) Asbestosis and primary intrathoracic neoplasms, *Ann. N. Y. Acad. Sci.,* **132,** 507.
Burrell, R. (1974) Immunological reflections on asbestos, *Envir. Hlth. Perspect.* **9,** 297–298.
C. F. Casalla and Co. Ltd., General Purpose Personal Sampler, *Instruction Leaflet,* 3110/TA/1, C. F. Casalla and Co. Ltd., p. 2.
Cauna, D., Totten, R. S., and Gross, P. (1965) Asbestos bodies in human lungs at autopsy, *J. Am. Med. Ass.* **192,** 371.
Chatham News (1977) Dockyard workers in disease danger says MP, (23rd December 1977).
Churg, A., *et al.* (1977) A simple method for preparing ferriginous bodies for electron microscope examination, *Am. J. Clin. Pathol.,* **68,** 513–517.
Daily Mirror, (1977) Lies that hide cancer peril, (8th November 1977).
Daily Telegraph (1978) Plant inches deep in asbestos dust—expert, (17th January 1978).
Davis, J. M. G., Gross, P., and De Treville, R. T. P. (1970) 'Ferriginous bodies' in guinea pigs, *Arch. Pathol.,* **89,** 364.
Davis, J. M., *et al.* (1976) *The mineral dust load of human lungs, Bronchopneumologie,* **26,** 103–113.
Dement, J. M., *et al.* (1976) Discussion paper: asbestos fiber exposures in a hard rock gold mine, *Ann. N.Y. Acad. Sci.,* **271,** 345–352.
Department of Employment (1973) Probable Asbestos Dust Concentrations at Construction Processes, Technical Data Note No. 42, HMSO London.
Dundee Courier and Advertiser (1977) Fine of £400, (13th October 1977).
Economist (1977) Lawsuits—A new field, (31st December 1977).
Elmes, P. C. (1976) Current information on the health risk of asbestos, *R. Soc. Hlth. J.,* **96,** 248–252.
Engelbrecht, F. M., *et al.* (1975) Mesothelial reaction of asbestos and other irritants after intraperitoneal injection, *S. Afr. Med. J.,* **49,** 87–90.
Enterline, P. E. (1976) Pitfalls in epidemiological research. An examination of the asbestos literature, *J. Occup. Med.,* **18,** 150–156.
Enticknap, J. B., and Smither, W. J. (1964) Peritoneal tumours in asbestosis, *Brit. J. ind. Med.,* **21,** 20.
Evans, C. C., *et al.* (1977) Frequency of HLA antigen in asbestos workers with and without pulmonary fibrosis, *Brit. Med. J.,* **i,** 603–605.
Evening Standard (1977a) Deadly asbestos found in trains, (22nd September 1977).
Evening Standard (1977b) Crew refuses to move blue asbestos train, (7th October 1977).
Fears, T. R. (1976) Cancer mortality and asbestos deposits, *Am. J. Epidemiol.,* **104,** 523–526.
Fukui, T., *et al.* (1976) Biopsy study of asbestosis found 18 years after 1 year exposure, *Jap. J. Thorac. Dis.,* **14,** 17–20.
Gibbs, G. W. (1975) Physical parameters of airborne asbestos fibres in various work environments—prelimary findings, *Am. Ind. Hyg. Ass. J.,* **36,** 459–466.

Gillam, J. D., *et al.* (1976) Mortality patterns among hard rock gold miners exposed to an asbestiform mineral, *Ann. N.Y. Acad. Sci.*, **271,** 336–344.

Gilson, J. C. (1971) Asbestos, in *Encyclopaedia of Occupational Health and Safety*, Vol. 1, A to K, International Labour Office, Geneva.

Gilson, J. C. (1974) Proceedings: Biological effects of asbestos, unanswered questions posed by epidemiological studies, *Clin. Sci. Mol. Med.*, **47** (3) 11P.

Granada TV Network Ltd. (1974) *World in Action*, Killer dust—a standard mistake, (14th October 1974).

Greenberg, S. D., *et al.* (1976) Sputum cytopathological findings in former asbestos workers, *Tex. Med.*, **72,** 39–43.

Gross, P., Cralley, L. J., and De Treville, R. T. P. (1967) Asbestos bodies, their non-specificity, *Am. Ind. Hyg. Ass. J.*, **28,** 541.

The Guardian (1978a) Sharp rise in cancers linked with asbestos, (9th February 1978).

The Guardian (1978b) Asbestos victims to see Minister on claims, (13th January 1978).

The Guardian (1978c) Blue asbestos found in garden near former factory site, (20th January 1978).

Haider, M., *et al.* (1975) Mesothelioma cases and asbestos exposure in Austria, *Hefte Unfallheilkd.*, (126), 547–549.

Hain, E. (1975) Current results of epidemiological studies on the asbestos problem in Northern Germany, *Hefte Unfallheilkd.*, (126), 536–538.

Hajdukiewicz, Z., *et al.* (1975) Pulmonary asbestosis and lung neoplasms, *Patol. Pol.*, **26,** 551–556.

Halifax Evening Courier, (1978) Madden given asbestos pledges, (1st February 1978).

Halifax Evening Courier, (1978) Asbestosis: 1978 is the big year, (6th January 1978).

Hany, A. (1975) Asbestos problems in Switzerland, *Hefte Unfallheilkd.*, (126), 542–545.

Harington, J. S., Gilson, J. C., and Wagner, J. C. (1971) Asbestos and mesothelioma in man *Nature, Lond.*, **232,** 54.

Harries, P. G. (1971a) Asbestos dust concentrations in ship repairing: a practical approach to improving asbestos hygiene in naval dockyards, *Ann. Occup. Hyg.* **14,** 241.

Harries, P. G. (1971b) *The Effects and Control of Diseases Associated with Exposure to Asbestos in Devonport Dockyard*, Institute of Naval Medicine, R. N. Clinical Research Working Party, Alverstoke, Hants.

Harries, P. G. (1976) Experience with asbestos disease and its control in Great Britain's naval dockyards, *Envir. Res.*, **11,** 261–267.

Harris, R. L., Jr., *et al.* (1975) The influence on fibre shape in lung deposition—mathematical estimates, *Inhaled Particles*, **IV,** Pt. 1, 75–89.

Hasan, F. M., *et al.* (1977) The significance of asbestos exposure in the diagnosis of mesothelioma, a 28 year experience from a major urban hospital, *Am. Rev. Resp. Dis.*, **115,** 761–768.

Health and Safety Executive (1970) *Asbestos: Health Precautions in Industry*, HSE Health and Safety at Work Booklet No. 44, HMSO, London.

Health and Safety Executive (1975) *Asbestos Regulations 1969: Respiratory Protective Equipment*, HSE Technical Data Note No. 24, 2nd revision, HMSO London.

Health and Safety Executive (1976) *Health Hazards from Sprayed Asbestos Coatings in Buildings*, HSE Technical Data Note No. 52, HMSO, London.

Heidermanns, G. *et al.* (1975a) Asbestos determination in industrial micro dusts and in lung dusts, *Hefte Unfallheilkd.*, (126) 574–584.

Heidermanns, G., *et al.* (1975b) Asbestos determination in industrial micro dusts and in pulmonary dusts, *Hefte Unfallheilkd.*, (126) 617–623.

H.M. Factory Inspectorate (1971) *Standards for Asbestos Dust Concentration for use with the Asbestos Regulations 1969*, Technical Data Note 13, HMSO, London.

Hodgson, A. A. (1966) *Fibrous Silicates*, Lecture Series, 1965, No. 4, Royal Institute of Chemistry, London.

Horai, Z., *et al.* (1975) Correlation between asbestos dust concentration and frequency of radiographic detection of pulmonary asbestosis, *Jap. J. Thorac. Dis.*, **13,** 33–39.

Hughes, D. T. (1974) Lung disease related to exposure to asbestos, *Med. Sci. Law*, **14,** 147–151.

International Agency for Research on Cancer, (1977) *Evaluation of the Carcinogenic Risk of Chemicals to Man, Vol. 14, Asbestos*, International Agency for Research on Cancer, Lyon.

International Labour Office (1970), *International Classification of Radiographs of Pneumoconioses, Revised (1968)*, Occupational Safety and Health Series, No. 22, International Labour Office, Geneva.

International Labour Office (1971) *Encyclopaedia of Occupational Health and Safety*, Vol. 1, Atok, 120–124. International Labour Office, Geneva.

Jones, J. S. (1974) Pathological and environmental aspects of asbestos associated diseases, *Med. Sci. Law*, **14,** 152–158.

Jones, J. S., *et al.* (1976) Factory populations exposed to crocidolite asbestos a continuing survey, *IARC Sci. Publ.*, No. 13, 117–120.

Kagan, E., *et al.* (1977a) Immunological studies of patients with asbestosis. 1. Studies of cell-mediated immunity, *Clin. Exp. Immunol.*, **28,** 261–267.

Kagan, E., *et al.* (1977b) Immunological studies of patients with asbestosis. II. Studies of circulating lymphoid cell numbers and humoral immunity, *Clin. Exp. Immunol.*, **28,** 268–275.

Kane, P. B., *et al.* (1977) Diagnosis of asbestosis by transbronchial biopsy. A method to facilitate demonstration of ferruginous bodies, *Am. Rev. Resp. Dis.*, **115,** 689–694.

Kelly, R. T. (1976) Asbestos health hazards in perspective. Constructional uses, *R. Soc. Hlth. J.*, **96,** 246–248.

Kent Evening Post, (1977) Union says beat disease, (22nd December 1977).

Kinsey, J. S., *et al.* (1977) A preliminary survey of the hazards to operators engaged in the disposal of asbestos waste, *Ann. Occup. Hyg.*, **20,** 85–89.

Kogan, F. M., *et al.* (1977) Working conditions of women track maintenence workers in asbestos mines, *Gig. Sanit.*, (9), 19–23.

Kogan, F. M. (1977) Maximum permissible exposure level (MPEL) of asbestos powders, *Med. Lav.*, **68,** 142–148.

Lewinsohn, H. C., (1972) The medical surveillance of asbestos workers, *Jl. R. Soc. Hlth.*, **92,** 69–77.

Liddell, D., *et al.* (1975) Radiological changes over 20 years in relation to chrysotile exposure in Quebec, *Inhaled Particles*, **IV,** Pt. 2, 799–813.

Luton Evening Post (1977) £40,000 to clear blue asbestos, (8th October 1977).

Martischnig, K. M., *et al.* (1977) Unsuspected exposure to asbestos and bronchogenic carcinoma, *Brit. Med. J.*, **1,** 746–749.

Matsuda, M. (1975) Asbestos bodies 1. Detection of asbestos bodies in the lung at autopsy, *Jap. J. Thorac. Dis.*, **13,** 40–44.

Mattson, S. B., *et al.* (1975) Pleural asbestoses, *Lakartidningen*, **72,** 3802–3804.

Mattson, S. B. (1975) Editorial. Pleural plaques and asbestos *Lakartidningen*, **73,** 496–497.

McDermott, M., *et al.* (1975) The effects of inhaled silica and chrysotile on the elastic properties of rat lungs: physiological, physical and biochemical studies of lung surfactants, *Inhaled Particles*, **IV,** Pt. 2, 415–427.
McDonald, J. C. (1975) Problems in the determination of safety standards for asbestos exposed workers, *Hefte Unfallheilkd.*, (126), 603–607.
Millerk, K. *et al.* (1977) Immune adherence reactivity of rat alveolar macrophages following inhalation of crocidolite asbestos, *Clin. Exp. Immunol.*, **29,** 152–158.
Milne, J. E. (1976) 32 cases of mesothelioma in Victoria, Australia: a retrospective survey related to occupational asbestos exposure, *Brit J. Med.*, **33,** 115–122.
Moigneteau, C., *et al.* (1977) Asbestosic pleural calcifications and the associated pathology (32 cases), *Poumon Coeur*, **32,** 101–106.
Morley, G. E. (1976) Asbestos in the air, *Lancet*, **i,** 1075.
Newalls Publicity Department Journal (1977) Fine of £400, (13th October 1977).
Newhouse, M. L. (1977) Asbestos content of dust encountered in brake maintenance and repair, *Proc. R. Soc. Med.*, **70,** 291.
Omoto, M. *et al.* (1975) Application of X-ray microanalyser to analysis of asbestosis *Jap. J. Hyg.*, **30,** 1.
Otto, H. (1975) Mesothelioma and asbestos exposure, *Hefte Unfallheilkd.*, (126) 555–559.
Oxford Mail (1977) Villagers prepare to repel rubbish, (14th October 1977).
Parkes, W. R. (1973) Asbestos related disorders, *Brit. J. Dis. Chest*, **67,** 261–300.
Peto, J. (1978) The hygiene standard for chrysotile asbestos, *Lancet*, **i,** 484–489.
Plantedydt, H. T. (1975) Mesothelioma and asbestos in the Netherlands, *Hefte Unfallheilkd.*, (126), 549–555.
Pooley, F. D., Oldham, P. D., Chang-Hyun Um, and Wagner, J. C. (1970) The detection of asbestos in tissues, in *Pneumoconiosis: Proceedings of the International Conference, Johannesburg, 1969*, (Ed. H. A. Shapiro), Oxford University Press, Cape Town, 108.
Pooley, F. D. (1974) Proceedings: The recognition of various types of asbestos as minerals, and in tissues, *Clin. Sci. Mol. Med.*, **47** (3), 11P–12P.
Pooley, F. D. (1976) An examination of the fibrous mineral content of asbestos lung tissue from the Canadian chrysotile mining industry, *Envir. Res.*, **12,** 281–298.
Pott, F. (1975) Significance and size of inhalable fibres and their carcinogenic effect, *Hefte Unfallheilkd.*, (126) 593–594.
Pott, F., *et al.* (1976) Results of animal experiments concerning the carcinogenic effect of fibrous dusts and their interpretation with regard to the carcinogenesis in humans, *Zentralbl. Bakteriol. Parasitentd. Infektionskr. Hyg., Abt. 1, Orig., Reihe B*, **162,** 467–505.
Pylev, L. N., *et al.* (1976) Mechanism of the induction of asbestos mesotheliomas in the pleura of rats, *Vopr. Onkol.*, **22,** 63–68.
Pylev, L. N. (1976) A histochemical study of mesothelioma in the pleura induced by asbestos in rats, *Vopr. Onkol.*, **22**(5), 53–57.
Pylev, L. N. (1974) Carcinogenic action of commercial serpentine asbestos, *Vopr. Onkol.*, 20(10), 87–88.
Rahman, Q., *et al.* (1976) Biochemical changes caused by asbestos dust in the lungs of rats, *Scand. J. Work Envir. Hlth.* **1,** 50–53.
Rajhous, G. S., *et al.* (1975) A statistical analysis of asbestos fiber counting in the laboratory and industrial environment, *Am. Ind. Hyg. Ass. J.*, **36,** 909–915.
Richards, R. J., *et al.* (1976) Light microscope studies on the effect of chrysotile asbestos and fiber glass on the morphology and reticulin formation of cultured lung fibroblasts, *Envir. Res.*, **11,** 112–121.
Rodriguez-Roisin, R., *et al.* (1976) Asbestos exposure and small airways disease, *Scand. J. Respir. Dis.*, **57,** 318.

Rohl, A. N., *et al.* (1977) Environmental asbestos pollution related to use of quarried serpentine rock, *Science, N.Y.* **196,** 1319–1322.

Rubin, I. A., *et al.* (1974) Elemental analysis of asbestos fibres by means of electron probe techniques, *Envir. Hlth. Perspect.* **9,** 81–94.

Rüttner, J. R. (1975) Asbestos problems from the morphological and dust-analytical viewpoint, *Hefte Unfallheilkd.*, (126) 559–564.

Saracci, R. (1977) Asbestos and lung cancer: an analysis of the epidemiological evidence on the asbestos smoking reaction, *Int. J. Cancer,* **20,** 323–331.

Sawyer, R. N. (1977) Asbestos exposure in a Yale building: analysis and resolution, *Envir. Res.*, **13,** 146–169.

Scottish Sunday Express (1977) 'Killer dust' secret in school's locked room, (2nd October 1977).

Seaton, D. (1977) Regional lung function in asbestos workers, *Thorax*, **32,** 40–44.

Selikoff, I. J. (1974) Environmental cancer associated with inorganic microparticulate air pollution, in *Clinical Inplications of Air Pollution Research*, (Ed. A. J. Finkel and W. C. Duel), Publishing Sciences Groups, Acton, Mass., 49–66.

Selikoff, I. J. (1975) Epidemiologic investigations of asbestos exposed workers in the United States, *Hefte Unfallheilkd.*, (126), 512–520.

Selikoff, I. J. (1977) Air pollution and asbestos carcinogenesis: investigation of possible synergism, *IARC Sci. Publ.*, No. 16, 247–253.

Selikoff, I. J., and Hammond, E. C. (1970) Asbestos bodies in the New York City population in two periods of time, in *Pneumoconoisis: Proceedings of the International Conference, Johannesburg, 1969*, (Ed. H. A. Shapiro), Oxford University Press, Cape Town, 180.

Seshan, K. (1977) Explanation for the insensitivity to tilts of the electron diffraction patterns of amphibole asbestos fibers, *Envir. Res.*, **14,** 46–58.

Sheffield Morning Telegraph (1977a) Widow lives with fear of asbestos, (23rd September 1977).

Sheffield Morning Telegraph (1977b) Blue asbestos taken out at four schools, (24th September 1977).

Sheffield Morning Telegraph (1977c) Fine of £400, (13th October 1977).

Sheffield Star (1977) City MPs urged to back fight against asbestos disease, (7th October 1977).

Shishido, S., *et al.* (1976) Asbestos pollution of the lung. 1. Incidence of asbestos bodies found in the lung and sputum, *Jap. J. Thorac. Dis.*, **14,** 728–735.

Smither, W. J. (1974) Asbestos in the work place and the community, *Envir. Hlth. Perspect.* **9,** 327–329.

The Surveyor (1977) Fine of £400, (20th October 1977).

Sutton and Cheam Advertiser (1977) Buildings are free of the deadly blue, (20th September 1977).

Tait, N. (1976) *Asbestos Kills*, The Silbury Fund, London.

Tait, N. (1977) *Asbestos Kills: New Facts 1977*, The Silbury Fund, London.

Tetley, T. D. *et al.* (1976) Chrysotile induced asbestosis: changes in the free cell population, pulmonary surfactant and whole lung tissue of rats, *Brit. J. Exp. Pathol.*, **5,** 505–514.

Thomson, J. G., Kaschula, R. O. C., and MacDonald, R. R. (1963) Asbestos as a modern urban hazard, *S. Afr. Med.*, **37,** 77.

Timbrell, V. (1970) The inhalation of fibres, in *Pneumoconoisis: Proceedings of the International Conference, Johannesburg, 1969*, (Ed. H. A. Shapiro), Oxford University Press, Cape Town, 3.

The Times (1977) Plant inches deep in asbestos dust—expert (13th October 1977).

The Times (1978a) (9th February 1978).

The Times (1978b) (17th March 1978).

Toivanen, A. (1976) Pulmonary asbestosis and autoimmunity, *Brit. Med. J.*, **1,** 691–692.

Transport Review (1977) Not so healthy nor so safe, (7th October 1977).

Trunevski, M., *et al.* (1974) Non-occupational asbestosis, *God. Zb. Med. Fak. Skopje*, **20,** 451–459.

von Chossy, R. (1976) Pneumoconoisis due to silica and asbestos dust, *ZFA (Stuttgart)*, **52,** 1667–1670.

Wagg, R. M. (1976) Safety measures when handling asbestos, *R. Soc. Hlth. J.*, **96,** 252–255.

Wagner, J. C., Slegg, C. A., and Marchand, P. (1960) Diffuse pleural mesothelioma, *Brit. J. Ind. Med.*, **17,** 260.

Warrington Guardian (1977) Grateworth? A tip top tip, says county councillor, (9th September 1977).

Webster, I., *et al.* (1975) Control of asbestos-exposed workers in South Africa, *Hefte Unfallheilkd.*, (126), 614–617.

Weill, H., *et al.* (1975) Lung function consequences of dust exposure in asbestos cement manufacturing plants, *Arch. Envir. Hlth.*, **30,** 88–97.

Weill, H. *et al.* (1977) Differences in lung effects resulting from chrysotile and crocidolite exposure, *Inhaled Particles*, **IV,** Pt. 2, 799–813.

Weyer, R. V. (1975) Radiologic aspects of asbestosis, *J. Belge Radiol.*, **58,** 347–361.

Woitowitz, H. J., *et al.* (1977) Asbestosis and asbestos related tumours, assessment of disablement, *Prax. Pneumol.* **31,** 153–159.

Wilson, M. R., *et al.* (1977) Activation of the alternative complement pathway and generation of chemotactic factors by asbestos, *J. Allergy Clin. Immunol.*, **60,** 218–222.

Yazicioglu, S. (1976) Pleural calcification associated with exposure to chrysotile asbestos in South East Turkey, *Chest*, **70,** 43–47.

Yorkshire Post, (1977) Safety war on the asbestos 'cowboys', (31st October 1977).

CHAPTER 12

Asbestos-related diseases Part 1: Introduction

William D. Buchanan
Rickmansworth, Herts.

There can be little doubt that no substance in use today, whether in industry of elsewhere, has attracted the attention, concern, and even alarm that are currently given to asbestos. This chapter seeks to set out, as objectively as this emotional subject permits, the facts about asbestos and about asbestos-linked disease, the outcome of human exposure.

Although there is evidence that asbestos in a crude form was used in Stone Age pottery, asbestos is very much a mineral of the 20th Century. World-wide consumption of asbestos during the past 80 years has been explosive, increasing by at least a 1000-fold, which may be contrasted with the increase in consumption of petroleum products during the same period of around 80-fold. Currently, in excess of 3 million tons of freshly mined asbestos are used each year world-wide, of which about 200 000 tons are used in the U.K.

(1) THE NATURE OF ASBESTOS

Asbestos has as its derivation a Greek word meaning 'unquenchable' and its fire-resisting property is one of its best known. A more recently appreciated problem, also stemming from its stability, is that of disposal of asbestos waste. In one sense, the name 'asbestos' is rather unfortunate as it covers several quite distinct minerals which share the common property of being capable of breaking down into fibres when crushed and processed. However, all of these naturally occurring minerals fall into one or other of two main classes:

1. Serpentine minerals—white asbestos (chrysotile) and the talcs
2. Amphibole minerals—blue asbestos (crocidolite) and brown asbestos (amosite), together with other much less used fibrous minerals of which anthophyllite is the most important.

Chrysotile, or white asbestos, accounts at present for more than 90% of the total world production. Its main sources are Canada and the U.S.S.R., with secondary sources in many places elsewhere, including Italy, Cyprus, Rhodesia, and the Republic of South Africa. Chemically, it is a hydrated magnesium silicate and it consists of fine silky fibres which spin well. Most woven asbestos articles thus consist of chrysotile and it is the only variety of asbestos used in brake linings and clutch plates. However, its main use lies in the production of asbestos–cement products. Unfortunately, it is readily attacked by mineral acids.

Crocidolite, or blue asbestos, contains iron and chemically is a sodium ferrosoferric silicate. It produces fine, resilient fibres of a typical blue colour. Blue asbestos comes almost exclusively from the Republic of South Africa although small deposits are found in Australia and Bolivia. It does not spin, has rather poor heat-resistant properties, but resists acid attack. Much blue asbestos was formerly used in insulation work, including spray applications. Amosite, or brown asbestos, comes entirely from South Africa. It produces long, rather harsh fibres. Like crocidolite, it contains iron but this time as a ferrous magnesium silicate. Its uses are mainly in asbestos–cement and in heat-resistant products. Anthophyllite is used to a limited extent in cheap 'fillers' or where chemical resistance is required.

Asbestos fibres can divide along their length many times, ultimately ending as fibrils. When these are examined under very high magnification, white asbestos fibrils are seen to be hollow tubes, whereas those of blue and brown asbestos are solid. The finest tubes of white asbestos are about 0.01 μm in diameter but those of blue and brown asbestos are coarser and up to 0.1 μm in diameter.

(2) HISTORY OF ASBESTOS-LINKED DISEASE

Before 1930, asbestos had been used without much thought of possible harmful effects, although we now realize that there had been earlier indications of ill effects. Before condemning this inaction too strongly, we must remember that this had been the traditional attitude in those days—new materials were assumed to be safe until guilt was proved. It is essentially in recent years that the reverse approach of assuming guilt until innocence is established has been increasingly adopted.

What is generally accepted as the first case in the U.K. of lung disease to be associated with asbestos exposure was noted in 1900, but was not reported until 6 years later. A long gap followed before a second case was reported in 1926. The next few years saw much activity, culminating in the Report by Dr. E. R. A. Merewether (Merewether and Price, 1930), a Medical Inspector of Factories, published in 1930, which established beyond doubt that asbestosis was a serious and wide-spread disease affecting workers in the asbestos textile industry. The Asbestos Industry Regulations quickly followed and asbestosis was made a compensatable industrial disease within the Workmen's Compensation Act. Merewether believed that with the dangerous properties of asbestos firmly established, control measures could be instituted which would speedily bring to an end the development of fresh cases. To quote his own words: 'From consideration of the nature of the processes in the asbestos industry, and other relevant matters, it is felt that the outlook for preventive measures is good. That is to say that in the space of a decade, or thereabouts, the effects of energetic application of preventive measures should be apparent in a great reduction in the incidence of fibrosis'. What has happened subsequently has shown only too clearly and tragically how far out Merewether's forecast has proved to be. Yet he was undoubtedly correct in his statement of principle. What went wrong was that everybody—the industry, the Inspectorate, and others, failed to realize just how serious is the effect of inhalation of asbestos dust. This was not helped by the essentially long-term character of such effects. The temptation has always existed to assume that the continuing cases of asbestosis could all be related to the conditions ruling before control measures, however imperfect they may now seem, with hindsight, to be, had been introduced. It was the development of a hitherto rare tumour, first associated with asbestos around 1960, which finally shook people out of this complacency.

However, before that date, the development of lung cancer was coming to be linked with asbestosis and perhaps even with asbestos exposure. This was first suggested as a possibility as far back as the mid-1930s although it was not until 1947 when a report was published (Merewether, 1947) based on a study of 235 death certificates recording asbestosis, of which no less than 13.2% also had a record of lung cancer, that this associated risk was really established. By comparison, a list of death certificates over a comparable period recording another dust disease, silicosis, also recorded only 1.32% lung cancers. No one has yet suggested that silicosis protects one from getting lung cancer!

In the succeeding years, this association between asbestosis and lung cancer has been amply confirmed in many countries. Currently, over half of the British death certificates recording asbestosis also record a lung cancer.

A note of caution is necessary, however, before we take death certificates at their face value. They are not necessarily an accurate indicator of the severity of the asbestosis or the extent to which its presence directly contributed to the death or was merely an incidental finding. For that reason, it is better to regard death certificates as a measure of people who have died with asbestosis present rather than of people who have died directly as a result of asbestosis. For somewhat similar reasons, there may be an unconscious in-built exaggeration in the high proportion of people dying from asbestosis (or, more accurately, whose death certificates record asbestosis) who also have a lung cancer. Lung cancer in an asbestosis victim is itself presumed, for compensation purposes, to have been caused by the asbestosis even although the latter is only minimally present. Now, by means of modern X-ray examination techniques, it is possible to identify effects due to asbestos in a considerable proportion of all asbestos workers, even although many will have no clinical evidence of the disease. Should a worker in this stage of minimal X-ray asbestosis develop a lung cancer, it is possible that it will be presumed to be due to asbestos and he, in due course, will appear in the vital statistics as a death from asbestosis complicated by lung cancer. We do not really know at present whether lung cancer occurs more frequently in these minor cases of asbestos-induced lung changes, but by selecting those who do have a cancer and including them in the total of deaths featuring asbestosis, we may unwittingly, as suggested, be exaggerating the importance of lung cancer as a complicating feature of asbestosis.

One final point about statistics relating to people dying with asbestosis ought to be cleared up. Depending on where one obtains the data, very conflicting and, unless their origin is understood, confusing figures are likely to be produced. There are three principal sources of such data:

1. Those published from time to time in Annual Reports of the Chief Inspector of Factories or the Employment Medical Advisory Service. As we have seen already, such statistics, based on examination of death certificates, are best regarded as referring to people who have died with asbestosis present.

2. Deaths recorded by the Registrars General. These are deaths in which asbestosis is regarded as the proximate casuse. For example, a death from lung cancer with asbestosis present would normally be recorded as being due to lung cancer and not asbestosis. On the other hand, a Coroner's unqualified verdict following an Inquest of death caused by asbestosis would thus appear in these statistics even if the asbestosis did not play a major causative role. It will thus be obvious that the Registrars General totals will be less than those of death certificates recording somewhere the word 'asbestosis', and this fact is well brought out in the comparison in Table 12.1.

3. Yet a third set of data on 'asbestosis' deaths are those for which the Department of Health and Social Security has made a death benefit award to an eligible dependant. Not everyone dying from asbestosis necessarily leaves a dependant, and this total tends to fall between the other two.

Table 12.1 Comparison of data recording deaths from asbestosis

Year	Death certificates recording asbestosis	Deaths attributed to asbestosis by the Registrars General
1967	61	22
1968	80	13
1969	77	21
1970	85	23
1971	85	27
1972	105	32
1973	107	28
1974	147	25
1975	169	46

Table 12.2 indicates the disturbing fashion in which new cases of asbestosis have increased in recent years. These totals are of first assessments by the Pneumoconiosis Medical Panels, i.e. they exclude those appearing for check-up and re-assessments of previously accepted disease.

It will be noted how these annual totals have shown two main spurts—in the mid-1950s and again in the mid-1960s. This in each case suggests that some new factor may have been introduced and, as is so often the case with medical statistics, we are entitled to question the validity of these figures—could they correspond to a greater awareness of the problem or an easing of the criteria for acceptance of a case for injuries benefit? This latter factor almost certainly plays a part, at least from the mid-1960s when more formal methods of assessing X-rays were developed and taken into use. It will be noted that in the most recent years the totals appear to have levelled off, again characteristic of a change in diagnostic standard.

Table 12.2 First assessments for asbestosis

Year	Cases	Year	Cases
1954	31	1965	82
1955	48	1966	114
1956	31	1967	168
1957	56	1968	128
1958	27	1969	134
1959	37	1970	153
1960	29	1971	145
1961	43	1972	125
1962	52	1973	143
1963	67	1974	139
1964	83	1975	161

(3) THE MESOTHELIOMA STORY

A mesothelioma is a malignant growth or cancer affecting the lining membrane of the lung surface or inner chest wall (pleura), or a similar membrane lining some abdominal organs and the inner abdominal wall (peritoneum). Until the late 1950s, cancers of these membranes were thought to be excessively rare. The position changed dramatically in 1960 when Wagner in South Africa recorded a series of such cases, most of which initially had been thought to be tuberculosis of the lungs, which he was able to demonstrate convincingly had all in some way or another been exposed to blue asbestos (crocidolite). In some cases the exposure had been a residential one, i.e. the subjects lived in the vicinity of the Cape Province asbestos mines, and others were limited to subjects passing through the area in which these mines were situated. Their findings were not immediately accepted by everyone but did stimulate much interest and research, including work in the U.K. The reader should be reminded here that South Africa is unique in being the only country in which all three main types of asbestos are mined in distinct districts, and these mesothelioma cases were found only in the vicinity of mines recovering crocidolite.

The interest caused by this report quickly brought to light similar tumours in other countries and, although these were uncommon, they turned out to be by no means as rare as had hitherto been believed. A re-examination of old records traced 50 cases of mesothelioma in this country over the first 50 years of the century, but 200 cases were identified between 1950 and 1966 with, in a large proportion, a direct or indirect link with asbestos exposure. There were real fears at that time that we might have been on the threshold of an 'epidemic' of asbestos-linked cases of mesothelioma, the cases occurring up till then merely representing the much smaller population exposed to asbestos 30–40 years ago. Fortunately, more recent evidence indicates that the risk of an epidemic of mesotheliomata can be discounted, although the annual incidence of new cases is now about 300 a year and apparently still rising, which is still a very serious matter when account is taken of the considerable concentration of these cases in the relatively small group of occupations with either direct or indirect exposure to asbestos. Undoubtedly there has also been an underestimate of cases in the earlier years, for even among expert pathologists there is sometimes disagreement on whether a tumour is a mesothelioma or a lung cancer.

A study of all identified deaths from mesothelioma and newly diagnosed cases during 1967 and 1968 brought out a number of interesting facts, although leaving other issues still unresolved. The findings showed that there was considerable inaccuracy in the standard of certification in that some cases diagnosed clinically were not recorded as such on death certificates, while conversely, cases certified as deaths from mesothelioma were not always confirmed when post-mortem material was re-assessed by a panel of expert pathologists. However, the study confirmed the very significant association of asbestos exposure at some time and the development of this disease, 85% of the agreed cases having had such an exposure. In some cases, the period of exposure had been very brief and, as most people, at some time or another, have brief non-occupational exposures to asbestos products, it is difficult to be certain as to the significance of such minor exposures without a comparison with properly matched control populations not suffering from mesothelioma. The resources available, unfortunately, were insufficient to make the study of such control groups possible. In 14 of the cases, the exposure to asbestos was non-occupational and this aspect is considered in the next section. As might be expected in conditions where in most instances the type of asbestos which had been used was unknown or known to have been mixed in type, the mesotheliomas could not be attributed to any particular type of asbestos. Nearly half of the total of confirmed mesotheliomas with a definite occupational history of exposure to asbestos were in previous or currently employed shipyard workers, an industry where asbestos is or was much used for insulation and fire-proofing purposes and where control is often difficult. Next in frequency were asbestos-using factories, these two groups together accounting for almost 70% of all the definite cases. Table 12.3 shows the industry or job responsible for such definite cases.

Table 12.3 Mesothelioma cases associated with definite asbestos exposure (adapted from Greenberg and Lloyd Davies, 1974)

Industry or job	Number of mesothelioma cases
Shipyard worker	75
Asbestos factory worker	39
Insulation worker	13
Boiler house worker	5
Chemical worker	5
Docker	4
Welding rod manufacturer	4
Building worker	3
Electrician	3
Sack cleaner/repairer	3
Welder/plater	3
Battery box manufacturer	2
Electricity generating industry	2
Gas worker	2
Railway coach/loco builder	2
Motor mechanic	1
Refuse worker	1
TOTAL	167

(4) NON-OCCUPATIONAL AND NEIGHBOURHOOD ASBESTOS DISEASE

People in general can usually accept that there may be a greater risk of a specific disease or accident associated with their own specific job than occurs naturally. Society itself frequently imposes such a situation and it is up to it to pay the price to make the job as safe as possible and, directly or indirectly, to compensate its unavoidable victims. Soldiers in peace-time are more likely to be killed as a result of terrorist activity than people in general, aircrews more likely to die as a result of air crashes, and, in a similar context, asbestos workers, certainly at least in the past, from asbestosis and its various complications. However, it seems wrong and socially unacceptable that other people should, without choice or consultation, be put at risk, however slight, from activities with which they are unconnected and over which they have no control. A great deal of emotion has therefore been generated around non-occupational and neighbourhood factors in asbestos-related disease.

This problem of non-occupational exposure to asbestos can be considered from the points of view of people living in urban areas generally and of those families with a member working in an asbestos factory.

Let us begin with a dogmatic statement that the possibility of members of the public developing clinical asbestosis is, for practical purposes, non-existent. Assuming on the one hand that lung cancer is a complicating feature of asbestosis (and studies in hand will finally provide the answer to this debatable question), we can be equally confident that environmental asbestos plays no role in the present high incidence of lung cancer in this country. If, on the other hand, asbestos comes to be shown as an indisputable primary carcinogen, the position will be less certain. This uncertainty would stem from the evidence that asbestos in exceedingly low concentrations is widely dispersed as an environmental contaminant.

The most compelling reason for expressing some measure of concern about asbestos as an environmental contaminant, however, is the finding, in a high proportion of human lungs examined routinely at autopsy, of 'asbestos bodies'. These are asbestos fibres of microscopic dimensions, around which a coating of brownish yellow staining and iron-containing material has been deposited, probably as a protective reaction, by the tissues. These bodies are present in considerable numbers in the lungs and even in the sputum of persons who have been occupationally exposed to asbestos, although they are regarded as an indicator of exposure rather than of disease. In the public generally, however, about 40% of all city-dwelling males have similar bodies in their lungs, but these are few in number and often found only after a prolonged search. The more prolonged such a search, the greater is the proportion of lungs with positive findings which have been reported as high as 60% in one series. The proportion of females found to have asbestos bodies is usually lower.

Other fibres besides those of asbestos can, however, produce very similar 'bodies' in the lungs, and these can only be distinguished by a procedure based on X-ray diffraction analysis. Since the occurrence of mesotheliomas is not well related to the recorded past level of exposure to asbestos, concern has been aroused lest the presence of asbestos bodies in the lungs of people generally is an indication of some risk from asbestos to the general public. Bearing in mind, however, that the proportion of mesotheliomas of around 15% not attributed to occupational exposure is low and seems to be dispersed over the whole population, present indications are that the chances of someone who does not work with asbestos, or does not come into family contact with someone who does, of developing a mesothelioma are very small indeed.

A special category of the general public is made up of those living in the vicinity (which could mean up to 1 mile away) of an asbestos factory or in whose household someone works in such a factory. There is convincing evidence that there has, in the conditions ruling in the past, been an increased chance of such people developing a mesothelioma as, for example, the 14 cases referred to in the previous section. It is a reasonable assumption, however, that the present practice of discharging only filtered air from such factories and providing protective clothing for asbestos workers for which there are special laundering arrangements and which may not be taken home will have effectively ended any such risk. The better control of dumping grounds for asbestos waste is a further safeguard.

Asbestos happens to be an excellent filtration medium, for which purpose it is used to clarify some beers and spirits. Asbestos fibres have been detected in the filtered fluids and this has caused some concern lest it entail a slight risk of peritoneal mesothelioma. This possibility seems very unlikely when the issue is seen in perspective. Assuming an asbestos worker is exposed throughout his working day to the current U.K. threshold level of 2 fibres per cubic centimetre of air, he will inspire around 1000–2000 fibres with each breath. He breathes in this amount perhaps 15 times per minute for 480 min (8 h) during the course of the working day, i.e. a total of at least $1000 \times 15 \times 480$ fibres daily. By no means all remain in the lungs, but this total of over 7 million fibres can be contrasted with estimates of, at most, a few thousand fibres in a pint of beer!

(5) THE EFFECTS OF ASBESTOS IN THE LUNGS

With particulate dusts, only those particles within a certain size range can normally gain access to the lung alveoli, where they produce the initial damage. The great majority of such particles do not exceed 5–6 μm in diameter. Since the falling speed in free air is the critical parameter, the position with fibrous dusts is different. The falling speed of fibres is dependent on their diameter rather than on their length, and asbestos fibres with lengths up to 50 μm or more can be inhaled provided that their cross-sectional diameter is not much more than 3 μm. Once inhaled, unfortunately, the normal mechanisms for getting rid of alveolar deposited foreign material are not very efficient and the fibres tend to remain in the lung air spaces. A reaction is set up, leading in many instances to the formation of asbestos bodies. Over the years, these tend to break down, releasing pieces of fibre and the lung tissue is gradually destroyed and replaced with scar tissue. In some cases, the changes are particularly marked in the pleura, which becomes thickened. It is not known whether such cases are those subjects who in later life are more liable to develop a mesothelioma or not.

Cases of asbestosis are diagnosed from a combination of a history of exposure to asbestos, a clinical examination, and a chest X-ray. This is not as simple as it might appear and, especially with early cases, there is frequently disagreement even between experts. Although most cases have been exposed for up to 20 years or even more, in a minority, exposure may have been for a much shorter period and have ended many years before the first evidence of disease appears. The main complaint is one of increasing breathlessness as lung tissue is destroyed. Cough may be troublesome and there is commonly a feeling of perpetual tiredness. A peculiar thickening of the ends of the fingers (finger 'clubbing') is frequently found, but this also occurs in certain other chronic chest or heart diseases. Developed asbestosis is always a serious disease and, unlike most other types of pneumoconiosis, tends to progress even although exposure may have been ended.

As noted earlier, a large proportion of asbestosis cases ultimately succumb from lung cancer. Mesothelioma may result from even low exposure levels, particularly with crocidolite dust, although almost certainly the other types of asbestos can also cause this variety of cancer. The exposure may not only have been brief but may also have occurred many years previous to the appearance of the mesothelioma.

As the lung tissue is progressively destroyed, an increasing strain is put on the heart to pump blood through the lungs, and heart failure is the other common cause of death in victims of advanced asbestosis.

(6) THE ASBESTOS REGULATIONS 1969

It was against the above background that the Asbestos Regulations 1969 were formulated. It was realized that the earlier Regulations of 1931 were ineffective, partly because of their main emphasis on the textile industry and failure to control the main uses which in the interim had sprung up, and also because of their absolute requirement that there should be no dust liberated—a condition impossible to meet and hence impracticable to enforce. The Asbestos Regulations 1969 apply to all factories and any other premises to which the Factories Act itself applies and where asbestos or articles composed wholly or partly of asbestos are used, unless the process is one in which 'asbestos dust' cannot be given off. Similar precautions are now necessary elsewhere even where the Factories Act does not apply, although they are not enforced through the Asbestos Regulations but rather through application of Section 2 of the Health and Safety at Work etc. Act. The mistake of attaching an absolute interpretation to emission of asbestos dust is no longer made. Instead, 'asbestos dust' is defined in the Regulations as meaning 'dust consisting of or containing asbestos to such an extent as to be liable to cause danger to the health of employed persons'. Clearly, this implies some level and, although this in the last resort would be for the Courts to decide, the Factory Inspectorate has issued a Guidance Note (Health and Safety Executive, 1977) indicating that it currently interprets such a level as being 2 fibres per cubic centimetre of air other than for crocidolite, where the more severe standard of 0.2 fibres/cm^3 is used. The first of these standards is a recommendation of the British Occupational Hygiene Society (BOHS) based on a study of the long-term effects of inhalation of white asbestos at measured levels, but the BOHS makes it clear that the standard is relevant only to asbestosis and is not necessarily protective against lung cancer or mesothelioma. At the time of writing this chapter, an official Committee of Enquiry chaired by the Chairman of the Health and Safety Commission is at an advanced stage of its deliberations and a new and possibly more stringent standard may possibly be one of its recommendations.

The severer standard already in use for crocidolite, which has been largely instrumental in leading to the abandonment of crocidolite for new use, is arbitrary but recognizes the much greater risk of mesothelioma when using that particular type of asbestos. In the U.S.A., a standard of 5 fibres/cm^3 is currently in force but is likely to be reduced to 1 fibre/cm^3. Whatever one may think of the present U.K. standard, it is salutary to remember that as recently as 10 years ago it was regarded as impossible to attain and, even when the Regulations were made, there were doubts held by some whether such a standard was enforceable. Such is the progress since then that in most asbestos factories, dust levels well below 2 fibres/cm^3 are now regularly achieved.

In general, the Regulations require all asbestos processes liable to produce asbestos dust as defined to be fitted with local exhaust ventilation or a not less effective means of dust control. Where for any reason, this is impracticable, e.g. in the stripping of old asbestos insulation work, approved respiratory protection and protective clothing are required for the workers. Other methods for the control of asbestos dust are also required, with special arrangements for the laundering of the protective clothing which may not be taken home by the workers. This measure is aimed at eliminating the problem of home contact with asbestos dust. Notice must be given to the Factory Inspector not less than 28 days before commencing any process involving the use of or exposure to crocidolite. There is an obligation on employees to comply with such Regulations as are relevant to their employment. The Regulations do not require medical examinations of the workers, but this aspect is dealt with in the description of the asbestos survey in the next section.

(7) THE ASBESTOS SURVEY

This survey (Employment Medical Advisory Service, 1974) is a long-term one of the prospective type i.e. it begins with people at work and follows them onwards in time until death to determine their fate and in what ways they differ from a control population. It contrasts with a retrospective study such as that on mesothelioma already described, which begins with the disease and traces it backwards to find out the past exposures of its victims. The latter method is seldom as satisfactory as a prospective study, mainly because records are rarely complete, but it is quicker. The asbestos survey, which is being carried out by the Employment Medical Advisory Service and the Factory Inspectorate with the cooperation of the medical officers in the asbestos industry, is an ambitious one whose main long-term objectives are:

1. To find out the varied responses to different measured levels of asbestos dust exposure and of known type. This involves periodic medical tests and chest X-rays (normally at intervals of 2 years) of all asbestos workers and relating these to known levels of dust exposure. Because there are so many places where asbestos is used, many being small, a 2-year cycle is necessary on the grounds of practicability. All people so examined are entered into a computerized asbestos register.

2. To find out the mortality experience (i.e. age at death and cause of death) of this population and to compare this with a matched control group to find out differences, if any.

A survey of this type requires many years of operation before final conclusions can be drawn, but trends will be apparent at an earlier stage. When carried to a successful conclusion it should provide answers to many present questions such as whether asbestos is a carcinogen or whether the lung cancer so common in asbestosis is no more than a complication of that disease. It should also make possible a better appreciation of the importance of mesothelioma as a risk to the asbestos worker and, of no less importance, to members of the public, particularly those with occasional and fortuitous exposure to dust of asbestos products. Finally, it should provide an answer to the importance to be attached to the frequent occurrence of asbestos bodies in people's lungs.

(8) RETROSPECT AND PROSPECT

It may be useful to summarize briefly some of the points in this chapter and then to take a look into the future.

Asbestos potentially is one of the most hazardous substances in widespread use today, but with common sense and conscientious observance of the provisions of the Regulations, the risks from its use can be contained. We have consistently, in the past, partly through ignorance and partly through complacency, underestimated its dangers and if possibly we are over-reacting today, this is a desirable reaction while uncertainty persists. As we have noted, many other occupations also carry some associated risk and this may be expected by Society. It is easy to call for the banning of asbestos but, as in all such cases, there is a benefit factor to set off against the cost, and the undoubted benefits to Society of asbestos are such that it would be difficult to envizage a society deprived of it. Asbestos fire control applicances alone may well save more lives than asbestos claims as its victims, whereas, to take an everyday example, the motor car (whose own annual toll of 7000 lives is accepted by Society) could scarcely operate as we know it without asbestos.

One important safeguard the asbestos worker holds in his own hands. Everybody today is aware of the dangers to health, including the risk of lung cancer, arising from smoking, particularly cigarette smoking. There is now well established evidence that the asbestos worker who is also a cigarette smoker increases his already not inconsiderable risk of developing such a cancer. A leaflet *Asbestos and You* (Health and Safety Executive, 1975), available free from any Employment Medical Advisory Service office, sets out this fact and others relevant to the health of the asbestos worker very clearly. The message is clear—if you are an asbestos worker, do not smoke cigarettes.

One other point deserves mention here. From time to time, accounts appear in the press of children exposed to asbestos, perhaps escaping from asbestos material used in buildings and often damaged by acts of vandalism. The children are usually taken to hospital for 'medical tests' and much alarm and worry is undoubtedly caused to their parents. A better understanding of the nature of the effects of asbestos would result in simple reassurance and not tests which are meaningless. There may be, in the remoter future, perhaps 30 or 40 years later, a slight risk of mesothelioma—we cannot tell at this stage, but many other things may also happen in that period of time.

Active research is in progress to find safer substitutes for asbestos. Glass fibre, rock wool, and slag wool have all been used for such a purpose. However, we must bear in mind that these too are fibrous in nature, even though artificially produced. Recently announced findings that, under experimental and admittedly artificial conditions, glass wool can induce pleural mesotheliomas in rats just like asbestos must necessarily strike a note of caution that the asbestos problem can easily be solved.

(9) REFERENCES

Employment Medical Advisory Service (1974) *A study of Asbestos Workers*, Occasional Paper No. 3.

Greenberg, M., and Lloyd Davies, T. A. (1974) Mesothelioma Register *Brit. J. Ind. Med.*, **31,** 91.

Health and Safety Executive (1975) *Asbestos and You, MS(A)*3.

Health and Safety Executive (1977) *Asbestos—Hygiene Standards and Measurements of Air-borne Dust concentrations*, Guidance Note EH 10, HMSO, London.

Merewether, E. R. A. (1947) *Annual Report of H.M. Inspector of Factories*, HMSO, London, 79–81.

Merewether, E. R. A., and Price, C. W. (1930) *Report on the Effects of Asbestos Dust on the Lungs and Dust Supression in the Asbestos Industry*, HMSO, London.

Wagner, J. C., Sleggs, C. A., and Marchand, P. (1960) Diffuse pleural mesothelioma and asbestos exposure in the North Western Cape Province, *Brit. J. Ind. Med.*, **17,** 260.

(i) Further Reading

Asbestos: Health Precautions in Industry, Health and Safety at Work Booklet No. 44, HMSO, London, 1974.

Biological effects of asbestos, *Ann. N.Y. Acad. Sci.*, 1965, **32**.

Problems Arising from the Use of Asbestos, Memorandum of the Senior Medical Inspector's Advisory Panel, HMSO, London, 1968.

CHAPTER 13

Asbestos-related diseases Part 2: Pathology and experimental pathology of asbestos dust exposure

J. S. P. Jones

Nottingham City Hospital

(1) INTRODUCTION

During the first half of the 20th Century, the only pathological entity that was generally associated with asbestos was Asbestosis. The first case of this disease was diagnosed at post-mortem examination at the Charing Cross Hospital, London, in 1900. Pulmonary fibrosis was noted in an asbestos textile worker and the case was reported by Murray (1907) to a Governmental Committee on Compensation for Industrial Diseases. Further evidence of the role of asbestos in the causation of pulmonary fibrosis was reported by Cooke (1924). He described the presence of 'curious bodies' in the lung tissue. Simson and Strachan (1931) described similar bodies in the lungs of African asbestos miners. It was Gloyne (1932) who suggested that they should be called 'asbestos bodies', as the central core of these structures appeared to be made up of an asbestos fibre.

As it became apparent that there was a significant hazard to man's exposure to asbestos dust, an official investigation was undertaken by the Factory Department of the Home Office by Merewether and Price in 1928 (Merewether and Price, 1930).

Asbestos, like other forms of pneumoconiosis, was also believed to increase the possible development of Pulmonary tuberculosis (Wood and Gloyne, 1934). While this was true during the period when tuberculosis was relatively common—up to the 1940s—it is no longer so, and the occurrence of tuberculosis in asbestotic patients is currently not more than that encountered in the general population (Wyers, 1949; Enterline, 1965; Smither, 1965).

In 1935, the first cases of asbestosis and carcinoma of the bronchus were described by Lynch and Smith (1935) and by Gloyne (1935), but it was not until 1955 that Doll (1955) showed the true significance of this association. Further evidence of the increased risk of bronchial carcinoma in asbestos workers was provided by Newhouse (1969, 1973). In addition to the increased risk of bronchial carcinoma in asbestos workers, a detailed study of insulation workers in the U.S.A. revealed an increased risk to the development of alimentary tract carcinomas (Hammond *et al.*, 1965; Selikoff *et al.*, 1973).

A considerable widening of interest into the possible carcinogenic properties of asbestos was given further impetus when Wagner *et al.* (1960) published details of a series of patients who had developed mesotheliomas of the pleura and peritoneum, following occupational or environmental exposures to asbestos dust in the North-West Cape Province of South Africa. Up until this time the mesothelioma was regarded as being an extremely rare tumour (Willis, 1960) but, following the South African observations, numerous reports followed of mesotheliomas developing in relation to asbestos dust exposure. (Hourihane, 1964; Enticknap and Smither, 1964; Hammond *et al.*, 1965; Elmes and Wade, 1965; Anspach *et al.*, 1965; Leiben and Pistawka, 1967). By the mid-1960s, it had become firmly established that there was an association between the development of mesothelioma and a previous exposure—often 20–40 years before—to asbestos dust, and that the exposure need not have been sufficiently heavy to have caused associated fibrosis of the lung (asbestosis).

A further pathological entity associated with asbestos dust exposure was described in Finland by the radiological observations of Kiviluoto (1960) and the pathological observations of Meurman (1966). They noted that pleural plaques were a frequent finding in adults who lived in the vicinity of an anthophyllite asbestos mining area, and there was a strong correlation between asbestos dust exposure and the development of these zones of thickening of the parietal pleura.

The pathological changes associated with the inhalation of asbestos dust may be summarized as:

1. asbestosis;
2. carcinoma;
3. mesothelioma;
4. pleural plaques.

(2) SOURCE OF ASBESTOS FIBRES

Asbestos causes a potential hazard to health only when fibres are inhaled into the respiratory system or are ingested into the alimentary system. In order for this to be accomplished, free dust particles must be present. Asbestos does not cause a hazard if the fibres are firmly bonded together and incorporated into structures such as asbestos–cement sheets, or into various composite materials. However, if these products are sawn, drilled, broken, or manipulated in such a way that the asbestos fibres are released, then a potential dust hazard does exist. Particular areas of risk also occur during mining, processing, manufacturing, installing, transporting, demolishing, removing, and dumping of any waste products which contain asbestos. High dust levels can be achieved during the manipulation of asbestos products and, in order to prevent asbestos inhalation, very high standards of dust suppression and occupational hygiene must be practised. When considering the pathological effects of asbestos it is important to know the concentration and duration of exposure, and also the type of asbestos to which individuals have been exposed.

Zielhuis (1977) has subdivided the types of dust exposure into the following types:

Ia. *Direct occupational exposure*—asbestos dust created by primary asbestos workers.

Ib. *Indirect occupational exposure*—for workers in the vicinity of asbestos workers who are generating dust.

Ic. *Occupational exposure in agriculture*—mainly in Eastern Europe (Burilkov and Babadjov, 1970).

IIa. *Para-occupational domestic exposure*—spouses of asbestos workers inhaling dust from work clothes.

IIb. *Para-occupational exposure associated with leisure-time activities*—handyman home repairs and construction.

III. *Neighbourhood exposure*—living in the vicinity of asbestos mines, factories, dumps, etc.

IV. *True general environmental exposure*—not due to specific sources, as listed above. This category represents the true public health exposure risk.

It should be noted that the release of the relatively indestructible fibres of asbestos into the environment implies that not only is the primary asbestos worker at risk, but also other workers in the vicinity—who may be unprotected and unaware of the potential hazards (Skidmore and Jones, 1975). Casual passers-by may also inhale the dust, and current investigations are under way to assess the risk to ambient exposure to asbestos. The Zielhuis Report (Zielhuis, 1977) states that there is no convincing evidence in Western Europe that an increased risk exists following casual exposure to asbestos in the urban air, water, drugs, beverages, or food, but the possibility cannot be denied. The IARC Monograph *Evaluation of Carcinogenic Risk of Chemicals to Man: Asbestos* (International Agency for Research on Cancer, (1977), which considers both experimental and epidemiological evidence, concludes more pessimistically that 'it is not possible to assess whether there is a level of exposure in humans below which an increased risk of cancer would not occur' (*Lancet*, 1977).

(3) MODE OF INHALATION

On the inhalation of dust-laden air into the respiratory system, most large asbestos fibres (10 μm or larger in diameter) are trapped in the mucus and hair of the nasal passages. Fibres of smaller size may enter the bronchial tree and most particles are cleared by the mucociliary process, particularly if they are larger than 5 μm in diameter. Smaller fibres (particularly if less than 3 μm in diameter) which are of respirable size may enter the terminal bronchioles and alveoli (Timbrell, 1970). It is these smaller fibres, particularly those with a diameter of less than 1 μm, which are considered to be the most significant in inducing biological effects in man (Zielhuis, 1977).

However, it is not possible to measure the effect of asbestos dust exposure on the lung with a high degree of certainty because of the differing aerodynamic properties of different types of asbestos fibre. A fibre which is straight and free from adherent particles (predominantly a fibre of the amphibole type) will tend to align itself with its major axis lying in the direction of air flow. A curved fibre, or one with adherent fragments or 'cabling' (predominantly a fibre of the chrysotile type), shows no such preferred axial orientation. There is therefore a greater possibility of chrysotile fibres coming into contact with the wall of the respiratory tract and of them being removed by the mucociliary escalator than would be the case with amphibole fibres (Skidmore, 1973). Amphibole fibres therefore are more likely to reach the periphery of the lung (Timbrell *et al.*, 1970).

It is not suggested, however, that chrysotile fibres cannot reach the pleura, and indeed although they are more readily eliminated from the lung parenchyma, those fibres of chrysotile which do reach the pleura remain there. Le Bouffant, *et al.* (1976) have noted that amphibole fibres are present in the pleura in lower concentrations than in the lung. This may be due to the fact that amphiboles are more penetrant and are also less liable to disintegration, and, having left the lung, they become more widely dispersed and less liable to remain in the pleura (Peto, 1978).

Following the inhalation of asbestos dust, the following possibilities may therefore occur:

1. The majority of the dust is removed by:

(a) *exhalation*;

(b) *the mucociliary escalator*, causing the particles to be either expectorated or swallowed;

(c) if *swallowed* the majority of the fibres pass through the alimentary tract to be excreted in the faeces. Holmes and Morgan (1968), using radioactive techniques, showed that after intratracheal injection of asbes- dust into rats, 20–50% was eliminated in the faeces during the first week, and thereafter there was continuing clearance from the lung, but at a much lower rate.

2. Asbestos retained in the lung enters the bronchioles and alveoli. Some of the needle-sharp fibres may penetrate bronchiolar and alveolar walls to remain in the interstitial tissue, or they may migrate to the pleural surface or enter lymphatic vessels.

3. A high proportion of inhaled asbestos fibres that are retained in the lung become coated with a ferro-protein complex, to become an asbestos body (Blount *et al.*, 1966). The stages of development of these structures have been studied in guinea pigs and have been described by Botham and Holt (1967) as follows:

> 'Asbestos fibres in the alveoli cause haemorrhage. Erythrocytes which leave the capillary are haemolysed and the haemoglobin which is released is converted into haemosiderin

granules. Haemosiderin is taken up by macrophages in which it dissolves, producing a ferro-protein in the cytoplasm. A macrophage takes up an asbestos fibre as well as haemosiderin and the ferro-protein is deposited on one of the fibres. The macrophage shrinks on to this structure and the ferro-protein takes up a beaded form, and the membrane of the cell disintegrates. The ferro-protein coat on the asbestos fibre may be identified experimentally in as little as 16 days after exposure'.

(4) ASBESTOS BODIES

Asbestos bodies may be formed in man following the inhalation of all types of asbestos fibre (chrysotile and amphiboles). They are elongated, golden brown structures 20–100 μm long and 3–5 μm wide which usually have a beaded appearance with many segments and a rounded or spear-shaped end (Figure 13.1). The central core of an asbestos body is an asbestos fibre, and

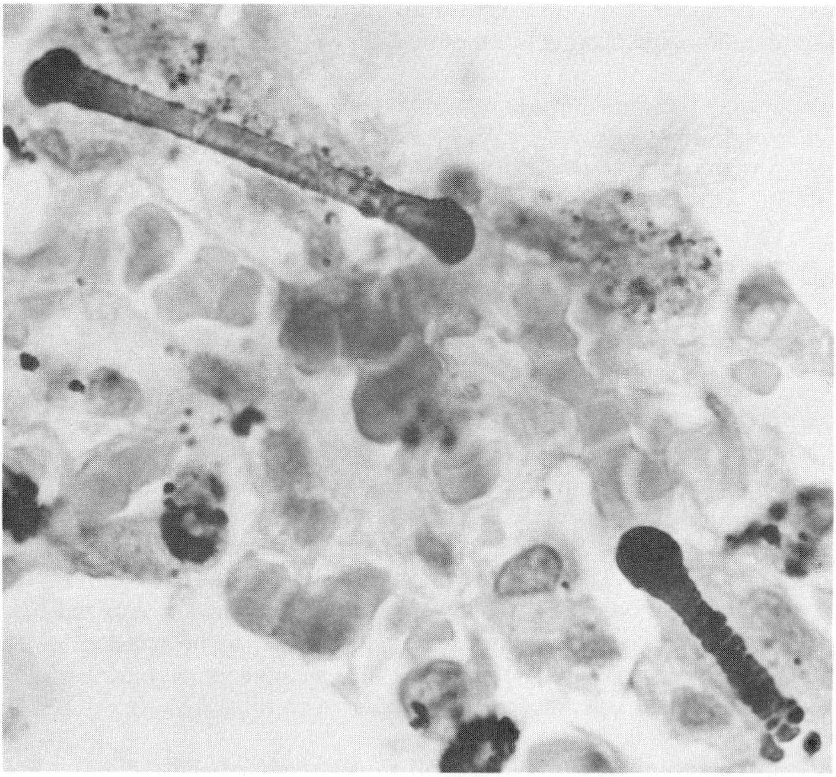

Figure 13.1 Asbestos bodies (×1512, unstained)

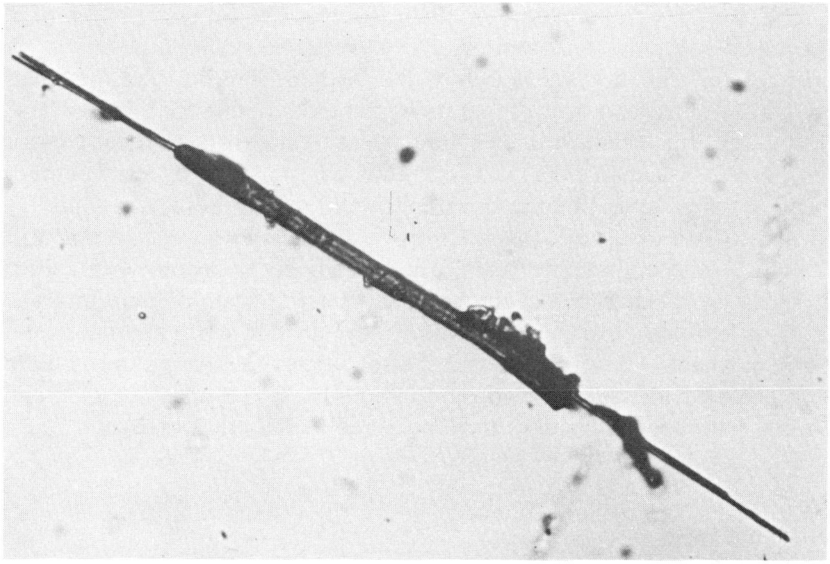

Figure 13.2 An asbestos fibre partly coated with ferro-protein complex (×1800, unstained). (Illustration by courtesy of Mr. D. Ranson)

this may frequently be seen running along the longitudinal axis of the structure. Sometimes the fibre is only partly covered by the ferro-protein complex (Figure 13.2).

Botham and Holt (1967) have demonstrated that sub-micron diameter glass fibres can produce early changes very similar to asbestos body formation. Cralley *et al.* (1968) have shown that respirable fibres other than asbestos may be present within the lung. Gross *et al.* (1967) have demonstrated iron-coated bodies in the lung which are not necessarily due to asbestos fibres and have suggested that these be designated 'ferruginous bodies'. However, Pooley *et al.* (1970) were able to confirm that in general the classical asbestos bodies seen by light microscopy do in fact contain an asbestos fibre core.

(5) PATHOLOGICAL INVESTIGATIONS WHICH INDICATE EVIDENCE OF ASBESTOS DUST EXPOSURE

5(i) Sputum

Asbestos fibres and/or asbestos bodies may be found in the sputum of patients who have experienced previous asbestos dust exposure. These are usually readily visible under the light microscope using a smear of the centrifuged concentrate of sputum. The asbestos bodies stain positively with Perl's reagent but they arc rcadily visible without staining, and the sputum may be examined as a wet preparation. Asbestos bodies may be present in the sputum after as short a time interval as 4 months after dust exposure (Simson and Strachan, 1931). The author has personal experience of the detection of asbestos bodies in the sputum of workers who had been exposed to asbestos dust 30 years previously and who have had no further exposure in the intervening years. The most likely occasion when asbestos particles may be detected in the sputum after this scale of time interval is during an episode of acute respiratory tract infection when sputum production is maximal. The presence of asbestos bodies in the sputum is merely confirmatory evidence of previous dust exposure, and it does not itself imply that any pathological changes have occurred in the lungs or pleura.

5(ii) Lung tissue

(a) After cutting into fresh or formalin-fixed lung tissue, the parenchymal surface may be scraped with a knife and the scrapings examined as a wet preparation by light microscopy. No staining is needed as the fibres and golden brown asbestos bodies can easily be detected, but permanent preparations may be stained with Perl's reagent.

(b) The cut lung tissue may be squeezed, the exuded fluid centrifuged and the deposit examined for the presence of asbestos fibres and bodies (Whitwell and Rawcliffe, 1971).

(c) Portions of lung tissue may be dissolved in heated 40% potassium hydroxide solution by the method of Gold (1967). This satisfactorily removes most of the lung tissue, leaving any asbestos fibres or bodies in their original state and available for microscopic examination and counting.

(d) Histological sections of lung tissue, 30 μm thick and unstained, allow asbestos bodies to be demonstrated. Thinner sections are not so successful as they often only allow portions of an asbestos body to remain in the section, and these are more difficult to detect. Also, some of the fibres may be torn out during the cutting of thinner sections. In scanning sections where the concentration of asbestos fibres and bodies is relatively low, they can usually be most readily found at sites of carbon dust deposition.

It should be remembered that uncoated small fibres will not be visible under a light microscope and attempts to quantify the number of fibres or bodies can only be a relative and not an absolute estimation. Pooley (1976) has found that in one post-mortem series of lung tissue from people who had been exposed to crocidolite asbestos 95% of the fibres identified by electron microscopy were of sub-light-microscope size.

(6) PATHOLOGICAL CHANGES ASSOCIATED WITH THE INHALATION OF ASBESTOS DUST

6(I) Asbestosis

Definition: Asbestosis means a fibrosis of the lungs caused by asbestos dusts which may or may not be associated with fibrosis of parietal or visceral pleura (Parkes, 1974).

It is important *not* to use the term asbestosis in a loose sense. It is *not* the generic term for all diseases associated with asbestos dust exposure; it should not be used to describe pleural thickening which is unaccompanied by lung parenchymal fibrosis; it should not be used if only pleural plaques are present or if asbestos bodies are detected in the sputum, or are found in the lung, without the presence of parenchymal fibrosis. Much confusion has arisen in the literature and in clinical practice because of the indiscriminate use of the term 'asbestosis', and it should be reserved for the specific pathological entity of parenchymal fibrosis of the lung associated with exposure to asbestos dust.

Asbestosis is a disease of the professional asbestos worker in that its onset is generally dependent on prolonged and relatively heavy exposures to asbestos dust, rather than more intermittent casual episodes of light exposure. Though shorter exposures may be reported, most cases of asbestosis have had industrial exposures of 10 years or more, and cases are documented in which there is an association with all commercially used types of asbestos (chrysotile, crocidolite, amosite, and anthophyllite). The efficacy of dust control is of paramount importance if the incidence of asbestosis is to be reduced (Smither and Lewinsohn, 1973). The disease is progressive, even after contact with asbestos dust has ceased, but if exposure ceases at an early stage the progression may be slowed down in some cases (Newhouse, 1967). The number of new cases of asbestosis in the U.K. is still increasing, despite the considerable awareness of potential dust hazards, and the incidence rate is approximately 5 per 1000 of those occupationally exposed (McVittie, 1965). Formerly tuberculosis was regarded as being a common complication of asbestosis (Gloyne, 1951; Bonser *et al.*, 1955), but it is now regarded as a rare accompanying condition (Buchanan, 1965).

The classical pathological features of asbestosis were described in a series of papers by Gloyne and co-workers (Wood and Gloyne, 1930, 1934; Gloyne, 1932–33, 1938).

6(ii) Macroscopic appearances

In all cases of suspected asbestosis at post-mortem examination it is recommended that one lung be inflated with formalin and allowed to fix before cutting. Whole lung sections (Gough and Wentworth, 1960) are useful to demonstrate the extent of fibrosis, but this technique is time consuming and is not available in all pathology laboratories. Barium sulphate impregnation of a slice of lung by the technique of Heard (1969) is a more simple way of demonstrating lung fibrosis. Black and white photographs may form a permanent record.

The classical uncomplicated asbestotic lung (Figure 13.3) is small and firm with a dry cut surface (Hourihane and McCaughey, 1966). Asbestosis is most commonly found in bilateral distribution at the bases of the lungs, but the mid and upper zones may also be affected. When the severity of the disease is slight, the lungs will be of normal size, but the greater the amount of fibrosis, the greater will be the degree of lung shrinkage. The prevalence of fibrosis of the lower lobes in asbestosis is helpful in differentiating the fibrosis associated with tuberculosis, allergic alveolitis and sarcoidosis which is predominantly in the upper zones, and in fibrosing alveolitis where it is more generalized. In mixed cases of asbestosis and silicosis the fibrosis may occur in the upper zones (Constantinidis, 1977) and, very rarely, uncomplicated asbestosis may predominate in the upper, rather than the lower zones (Gough, 1965).

There is almost invariably some fibrotic thickening of the visceral pleura (Figure 13.4), predominantly in the basal region. This may vary from a slight cloudiness of the pleural surface to dense fibrotic thickening with hyalinized zones which have the consistency of cartilage. Small pleural effusions may be present, especially during the early stages of asbestosis. Fusion of the visceral and parietal layers of pleura may be present, especially in the basal region. In addition, in more severe cases, the combination of the thickened visceral pleura with the fibrotic contraction of the underlying lung parenchyma may cause a 'rounding' of the lung in the costophrenic region instead of the normal, rather sharp-angled contour of the lung.

Because of the fibrous thickening of the visceral pleura covering the surface of lung and the interlobar septa, on cutting the lung surface the pleural margins stand out prominently from the contracted, fibrotic parenchymal tissue, which tends to be concentrated in the basal segments of the lower lobes. The greater the degree of asbestosis, the higher up in the lung is the involvement, so that the middle and upper zones become increasingly involved with increasing severity of the condition. Gough (1965) described two anatomical forms—diffuse and solid fibrosis.

The *diffuse form* shows 'honeycomb' (cystic) areas of varying sizes, depending on whether the smaller air passages or large bronchi are dilated. The distribution is not extensive and the changes are not uniform, thus making it different from the predominantly peripheral changes of idiopathic interstitial fibrosis.

In the *solid form*, solid fibrotic masses, which mimic other

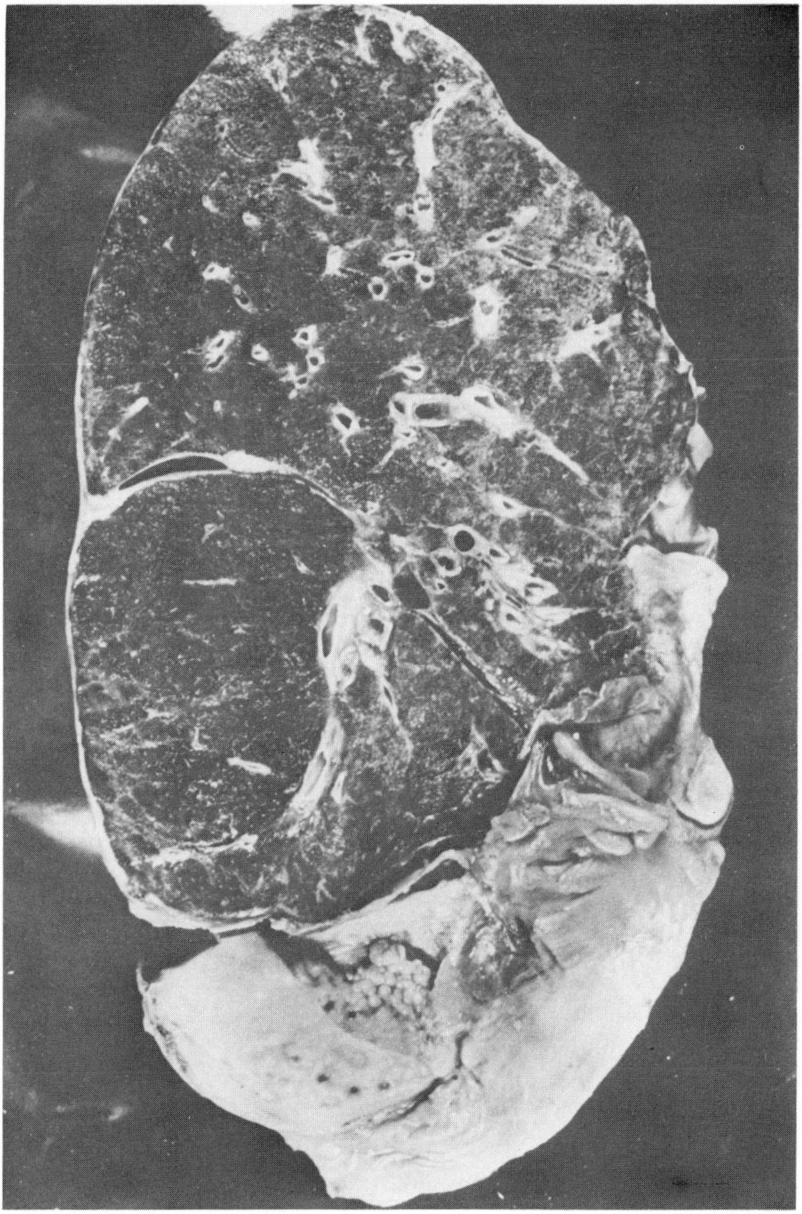

Figure 13.3 Asbestosis. The saggital section shows a contracted lung with rounding of the base in the costo-phrenic region. The adherent diaphragm shows a partly calcified pleural plaque on the central tendon

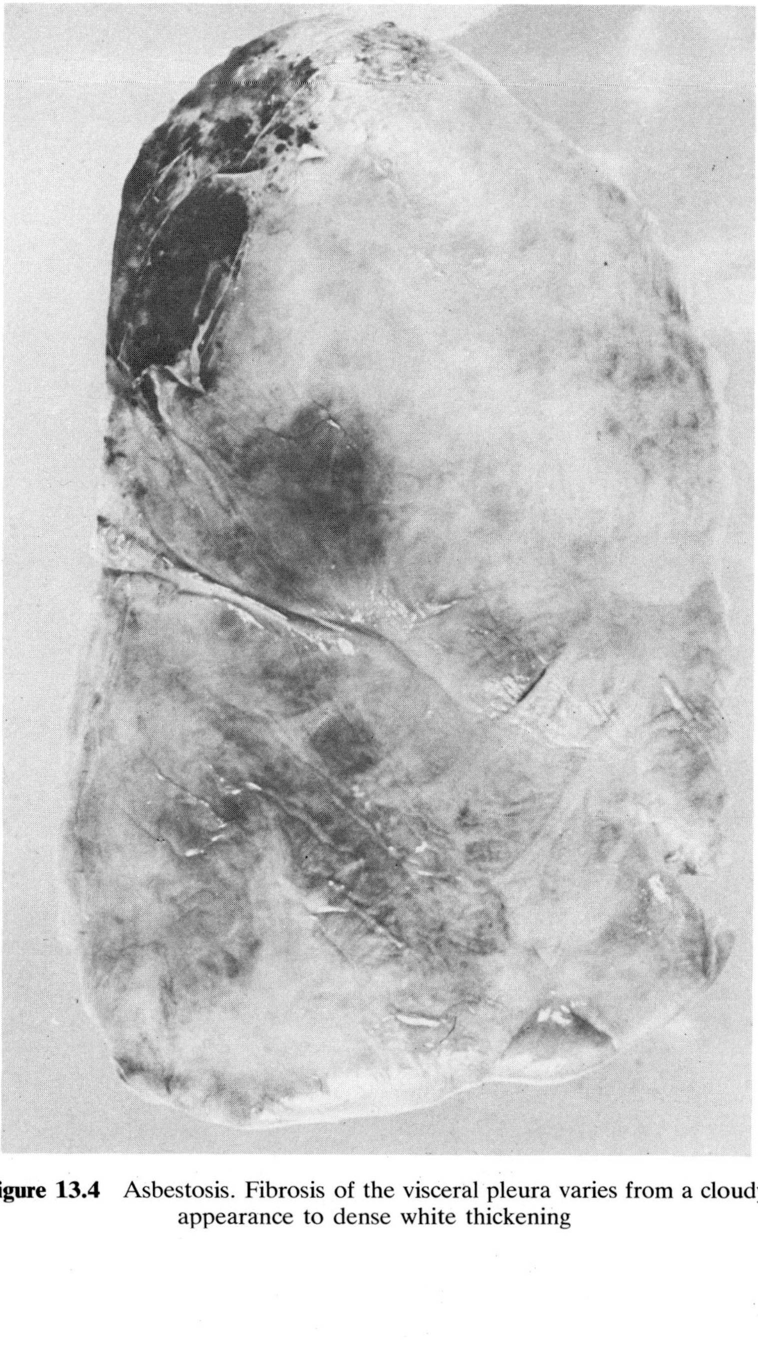

Figure 13.4 Asbestosis. Fibrosis of the visceral pleura varies from a cloudy appearance to dense white thickening

pneumoconioses or other conditions such as carcinoma, may reach several centimetres in diameter and may occur in any part of the lung.

Bronchiectasis is occasionally present in areas of severe fibrosis, but it is not a dominant feature. The presence of emphysema with asbestosis is uncommon (Parkes, 1974).

In established cases of asbestosis, it is usual to see right ventricular hypertrophy at autopsy (Hourihane and McCaughey, 1966).

6(iii) Microscopic appearances

As a result of deposition of asbestos fibres and accumulation of macrophages in the respiratory bronchioles and alveoli, a peribronchial fibrosis develops, which obliterates surrounding alveoli as it spreads outwards from the bronchiole. The fibrosis of the alveoli occurs from within the lumen of the respiratory bronchioles and thus the reaction is not an interstitial fibrosis (Wagner, 1965). Elastic stains show an intact alveolar wall and demonstrate that much of the fibrosis is intra-alveolar (Webster, 1970) (Figure 13.5).

Alveolar cell hyperplasia may be prominent in zones of severe fibrosis, and the blood vessels in such areas are frequently sclerotic (Hourihane and McCaughey, 1966). Where there is established fibrosis and replacement of

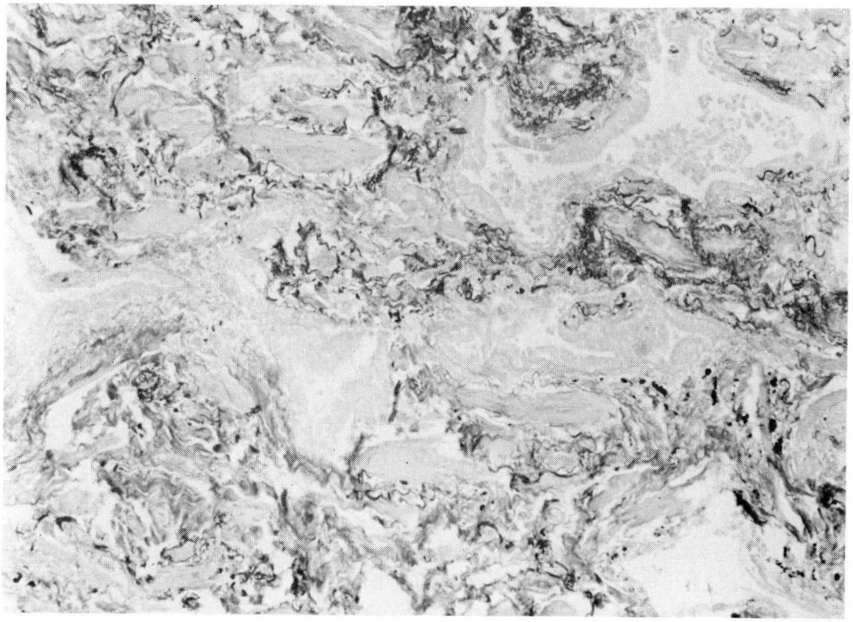

Figure 13.5 Asbestosis. There is retention of the elastic structure of the alveolus and obliteration of the lumen by fibrosis (×285, Van Gieson/elastic stain)

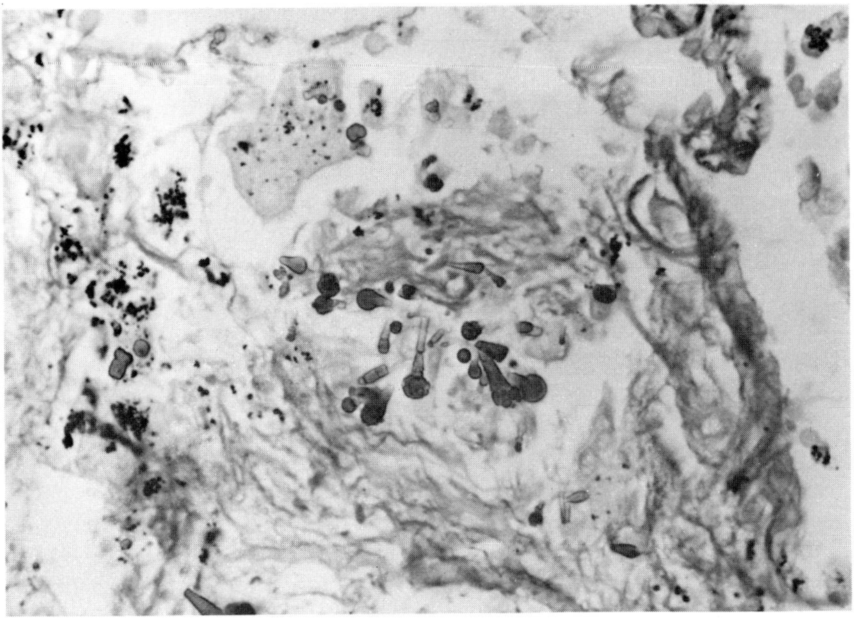

Figure 13.6 Asbestosis. Asbestos bodies and fibrous tissue within the alveolus and in the surrounding tissue (×375, unstained)

alveolar architecture by dense collagen, the tissues stain strongly with haematoxylin, and there is a positive dark granular appearance around lumens of small blood vessels. This material does not stain as DNA or calcium, but it reacts weakly for iron and gives intense reactions for neutral and acid mucoploysaccarides (Hourihane and McCaughey, 1966).

Asbestos bodies will usually be present in large numbers, either singly or in clumps, in asbestotic lungs. They may be partially obscured by dense fibrosis or by aggregates of other dusts, particularly carbon particles. They may be found in the walls of respiratory bronchioles (Heard and Williams, 1961) but may also be lying free within alveoli or be partly penetrating alveolar walls (Figure 13.6). In cases where there is considerable difficulty in identifying asbestos particles due to carbon deposits, incineration will drive off the obscuring material, leaving the less destructible asbestos visible on the slide (Hourihane, 1965).

It is important that microscopic examination be carried out to confirm the distribution of fibrosis in lung tissue as this may not always be obvious on macroscopic examination. In addition, early pathological changes precede clinical, physiological, and radiological evidence of disease (Parkes, 1974). The International Union against Cancer (1965) has recommended that tissue blocks should be taken from at least six separate sites from the lung:

1. apex of right upper lobe;
2. right middle lobe (lateral pleural surface);
3. right lower lobe (middle of basal surface);
4. left upper lobe (central section);
5. lingula (central section);
6. left lower lobe (central basal section).

The Working Group on the Biological Effects of Asbestos (Hinson *et al.*, 1973) has proposed that the severity of fibrosis and degree of asbestosis be graded as follows:

1. *Extent of lung involvement*
   (a) none;
   (b) less than 25%;
   (c) 25–50%;
   (d) over 50%.
2. *Severity of fibrosis*
   (0) none;
   (1) slight focal fibrosis around respiratory bronchioles, associated with the presence of asbestos bodies;
   (2) lesions confined to respiratory bronchioles of scattered acini; fibrosis extends to alveolar ducts and atria;
   (3) there is a further increase and condensation of the peribronchiolar fibrosis with early widespread interstitial fibrosis;
   (4) few alveoli are recognizable in the widespread diffuse fibrosis; bronchioli are distorted.

(7) CARCINOMA ASSOCIATED WITH ASBESTOS DUST EXPOSURE

7(i) Bronchial carcinoma

The first report of a relationship between asbestos dust exposure and bronchial carcinoma (Figure 13.7) was made by Lynch and Smith (1935) and by Gloyne (1935). In 1949 Merewether quoted that 14.7% of male workers with asbestosis died of malignant disease of the respiratory system. In 1965, Buchanan reported that this figure had risen to over 50%. A high mortality from bronchogenic carcinoma was found in insulation workers in the U.S.A. when it accounted for 20% of all deaths and 45% of fatal neoplasms (Selikoff *et al.*, 1973). Histological studies of tumours from these patients have shown the distribution of cell types to be the same as that found in the general population (Kannerstein and Churg, 1972). There is no difference in the clinical behaviour of carcinomas in asbestos workers as compared with non-asbestos workers. Of great interest and of considerable importance

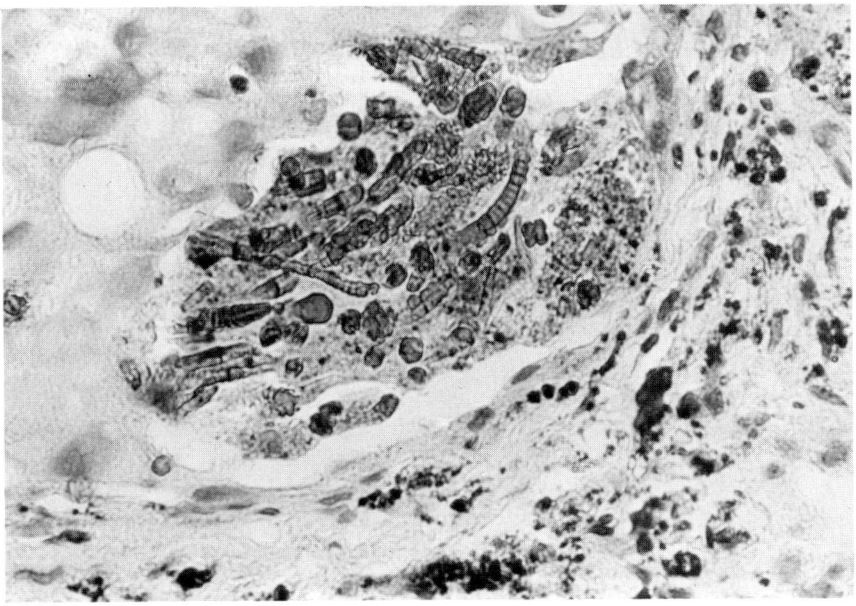

Figure 13.7 Carcinoma of the bronchus with asbestosis. The nest of asbestos bodies is situated between undifferentiated carcinoma cells and a zone of fibrosis (×525, H. & E. stain). Illustration by courtesy of Dr. A. MacFarlane)

from the aspect of preventative medicine is the finding that the combined carcinogenic properties of asbestos and cigarette tobacco produces a multiplicative effect on cancer incidence (Doll, 1971; Berry *et al.*, 1972). Selikoff *et al.* (1968) have shown that asbestos insulation workers with a history of regular cigarette smoking had an 8-fold greater risk of bronchial carcinoma deaths than cigarette smokers who did not work with asbestos, and approximately a 90-fold greater risk than men who neither smoked cigarettes nor worked with asbestos. This multiplicative effect of the two carcinogens has been confirmed by studies in a London asbestos factory (Newhouse, 1973).

In a series from Tyneside, asbestos exposure was found to increase the risk of carcinoma whatever the level of smoking (Martischnig *et al.*, 1977).

All commercially used types of asbestos have been shown to cause an increase in the occurrence of bronchogenic carcinoma (Wagner, 1971). This includes anthophyllite—a type of asbestos mined mainly in Finland (Meurman *et al.*, 1973).

Most bronchial carcinomas arise in the epithelium of the major bronchi. More rarely, peripheral carcinomas—mainly of adenocarcinomatous type—arise in more peripherally situated scar tissue in the lung.

7(ii) Alimentary tract carcinoma

More than the expected number of cases of alimentary tract carcinoma (in the oesophagus, stomach, colon, and rectum) have been reported in asbestos workers in the U.S.A. (Enterline, 1965; Mancuso, 1965; Selikoff *et al.*, 1970, 1973). More than a 2-fold increase in gastroinstestinal tumours over the expected numbers have been recorded in asbestos insulation workers there, and a similar excess has been recorded in a group of Belfast shipyard workers (Elmes and Simpson, 1971). Newhouse (1973) has not been able to reach a conclusion on this issue in the London asbestos factory series, mainly owing to a lack of histological material in the cases under surveillance.

There are no differences in the pathological features of the alimentary tract carcinomas of asbestos workers compared with others.

7(iii) Other neoplasms

There are no firm data available to link asbestos expsoure with other neoplasms. Suggestions that carcinoma of the ovary may be asbestos-related (Keal, 1960; Graham and Graham, 1967) have not been proved, but accurate and substantial epidemiological information is still required in order to prove or disprove the various possibilities of connected conditions.

(8) MESOTHELIOMA

It is only in recent years that it has been generally accepted that primary serosal tumours of the pleura and peritoneum exist as a distinct pathological entity. The early description by Klemperer and Rabin (1931) divided the tumours into *localized* (predominantly fibrous) and *diffuse* (predominantly epithelial) types. More definite criteria were added by Godwin (1957) for pleural tumours, and by Winslow and Taylor (1960) for peritoneal tumours. However, it was not clear in some cases whether these tumours were indeed of primary serosal origin or whether they could be diffuse metastatic deposits from primary carcinomata of undetermined source (Willis, 1967 Until the mid-1960s it was also thought that true primary malignant mesotheliomata of the pleura or peritoneum did not metastasize to distant organs and therefore, if such metastases were present, this tended to negate the diagnosis. However, diagnostic criteria later became more firmly established (McCaughey, 1958, 1965; Webster, 1965; Hourihane, 1965; Churg *et al.*, 1965) as this tumour began to excite considerable interest following the important observations made by Wagner *et al.* (1960) of an association between asbestos dust exposure and the subsequent development of mesothelioma, usually many years later. As more confidence was gained in diagnosis, so reports appeared in the medical literature of cases of mesothelioma arising in areas where asbestos was mined (McDonald *et al.*, 1970; McNulty, 1970; Webster, 1970), where it was used industrially (Jacob and Anspach, 1965; Hammond *et al.*, 1965), and particularly in shipbuilding cities (Elmes and Wade, 1965; Glyn Owen, 1965; Ashcroft and Heppleston, 1970). With increasing experience it has been possible to refine pathological criteria further. It should be remembered, however, that while many examples of mesothelioma show very characteristic features, there are still many serosal tumours in which universal agreement among pathologists cannot be obtained (McCaughey and Oldham, 1973).

8(i) Macroscopic features

8(i).1 *Pleural mesothelioma*

Characteristically, the dense white or grey–yellow tumour is usually found initially in one pleural cavity, and both the visceral and parietal layers of the pleura are involved (Figure 13.8). The tumour appears in sheets and spreads in a selective way around the pleural cavity, confining itself initially to the serosal surfaces so that eventually the entire lung may be encased in tumour. Sometimes the cut surface of the tumour has a slimy, gelatinous consistency. There is also selective spread along the interlobar fissures involving the visceral pleura (Figure 13.9). Spread around the pericardial sac is common. Selective spread to the pleura of the contralateral side may occur in more advanced cases (Figure 13.10). At a relatively early stage of tumour formation there is a pleural effusion. Indeed, this is a common presenting feature of the disease, the patient complaining of persistent localized chest pain in association with the effusion. Tumour cells from the pleural surface may be shed into the fluid of the effusion, and an early diagnosis by cytological means may be possible in some cases (Naylor, 1968; Butler and Berry, 1973). As the pleural cavity becomes progressively obliterated by tumour spread, so the effusion becomes loculated and eventually dries up. The visceral and parietal pleural layers become fused into one homogeneous tumour mass. Extension of the tumour into the lung parenchyma, if it occurs, usually has a nodular appearance, and compresses rather than invades the lung tissue. On the chest wall the tumour becomes very firmly fixed to the intercostal muscles, and while there is usually no deep invasion of the chest wall, the diffuse anchoring of neoplasm into the superficial zone of the musculature makes a clean enucleation of tumour an impossibility, in that there is no plane of cleavage for the surgeon or pathologist to strip out the neoplasm. Within the lung, spread into the peribronchial tissue may occur, usually via lymphatic vessels.

Should an open biopsy or a previous thoracotomy have been carried out,

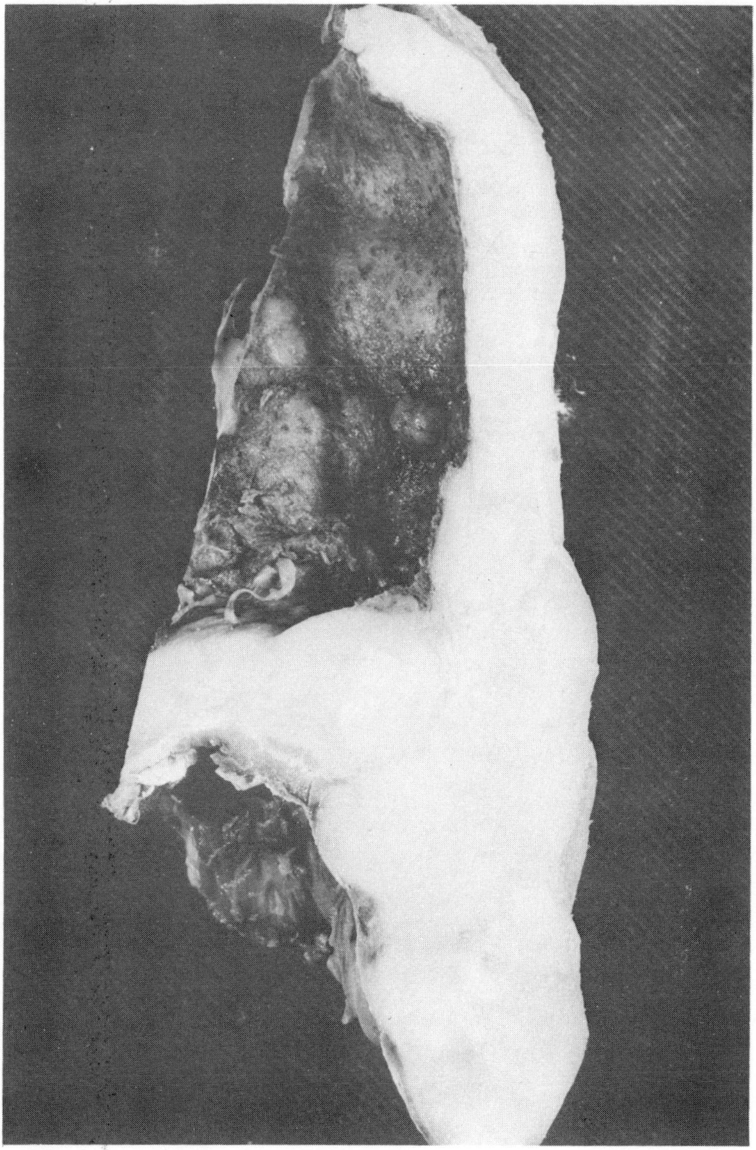

Figure 13.8 Surgically resected specimen of a pleural mesothelioma involving the left costo-phrenic region. Note how the pleural tumour spreads along the serosal surface

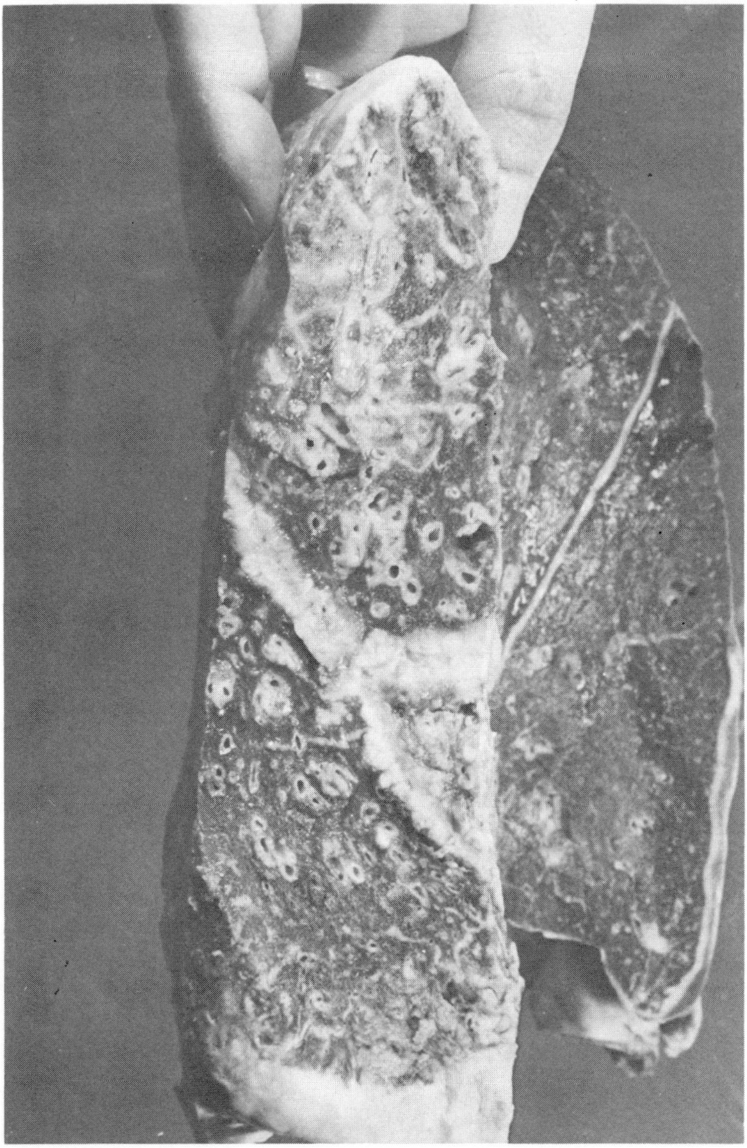

Figure 13.9 Mesothelioma of the pleura, spreading along interlobar pleural surfaces. There is also lymphatic spread, particularly in peribronchial distribution

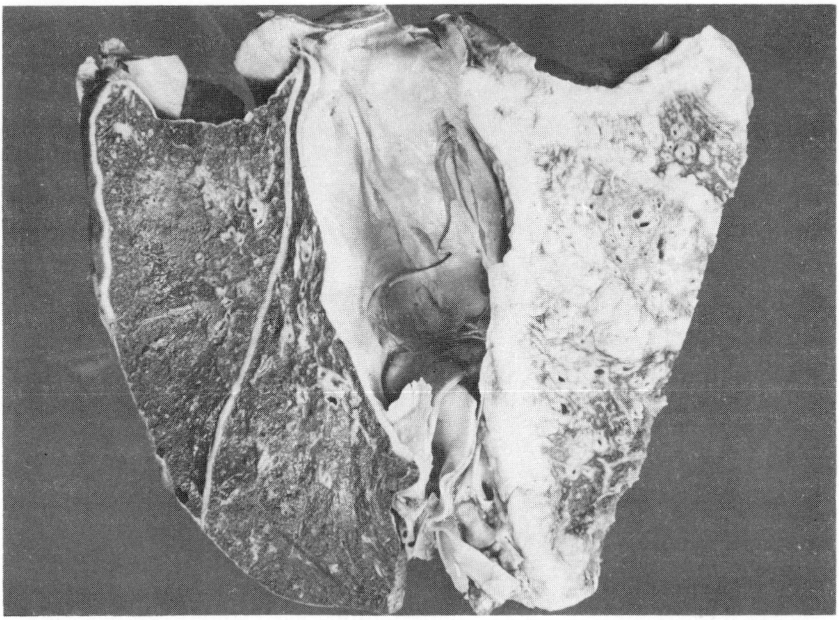

Figure 13.10 Mesothelioma of the pleura, commencing in the right pleural cavity and spreading along serosal surfaces to involve the pericardium and pleura of the left lung

it is not uncommon for the mesothelioma to track through the chest wall at the site of the old incision, thus causing a tumour mass to form in the subcutaneous tissue, and even to ulcerate through the skin. Because of this known hazard, which only adds to a patient's already considerable distress, surgical exploration of patients with mesothelioma is generally discouraged, and at most a needle biopsy is recommended for diagnostic purposes. Even after this simple procedure, cases have been recorded of tumour invading down the needle track (Elmes, 1973). It may therefore be necessary to limit the pathological diagnostic confirmation during life to 'a malignant tumour of pleura, strongly suggestive of/or consistent with a mesothelioma' because of the limited amount of scope in obtaining biopsy material. In many cases a confident diagnosis of mesothelioma has to await an autopsy, when it is possible not only to exclude other potential sources of primary neoplasm, but also to examine adequate samples of this histologically variable tumour. In addition to excluding possible primary neoplasms, a diligent search for metastatic tumour deposits should be made and it should be remembered that the finding of distant metastases by no means negates the diagnosis of mesothelioma, as was previously thought. In addition to the characteristic spread of tumour along serosal planes, there may be direct spread through

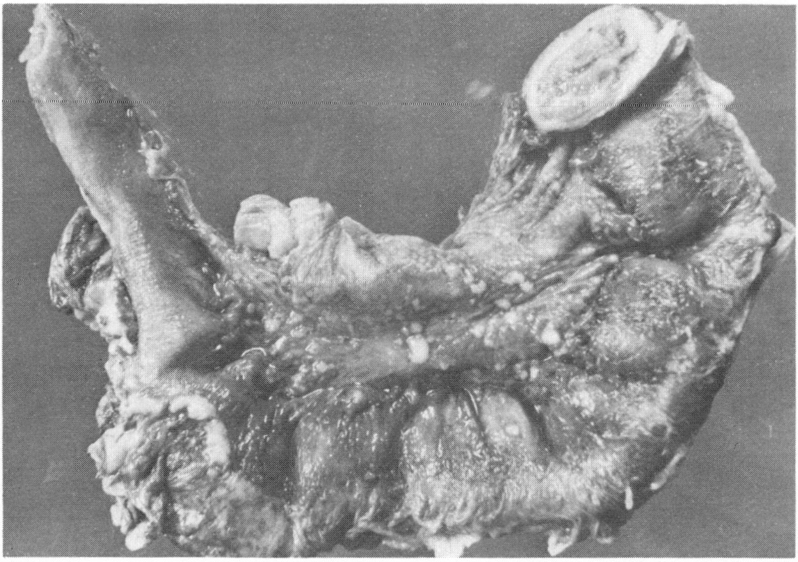

Figure 13.11 Metastatic spread from a primary pleural mesothelioma on to the peritoneal surface of the small bowel

the diaphragm with invasion of the adjacent liver, or multiple seedling deposits may be found on the surface of the peritoneum—on any of the abdominal viscera (Figure 13.11) as well as the parietal peritoneal wall. Distant metastases tend to occur late in the natural course of events, and in many cases the patients die of intercurrent lung infection when the tumour is still localized to the chest cavity. Spread by lymphatics to the mediastinal lymph nodes and abdominal aortic nodes is not uncommon. Metastatic deposits in the liver, kidney, adrenal, brain, bone, and even myocardium have been recorded.

The most common neoplasm which mimics the macroscopic appearance of a pleural mesothelioma is a peripheral carcinoma of the lung. Histological examination will usually readily distinguish these tumours, but not in every case. Large tumour masses which tend to be localized to the chest wall and which have not spread along the serosal planes are unlikely to be mesotheliomas, but are more likely to be fibrosarcomas, leiomyosarcomas, or rhabdomyosarcomas. Histological differentiation can usually distinguish between these different neoplasms. Strict criteria are essential for the diagnosis of mesothelioma and in the first place a carefully conducted post-mortem examination should have failed to have found a primary carcinoma in any of the viscera (Hourihane, 1964). Suspicious lesions, such as zones of bronchial thickening, or stenosis, gastric ulcers, and nodules in the prostate and thyroid glands and pancreas must be examined histologically before these potential sites of primary carcinoma can be excluded.

8(i).2 *Peritoneal mesothelioma*

Like the pleural mesothelioma, the peritoneal counterpart is a primary tumour which spreads along the serosal membranes within the abdominal cavity, and in the early and intermediate stages of development it may produce large amounts of ascitic fluid. Clumps of tumour cells may be recovered from this fluid, and made available for cytological examination. During the later stages, the fluid formation may diminish as matting together of visceral and parietal layers proceeds, and eventually a 'frozen abdomen' may ensue, with the entire visceral contents encased in dense yellow–white tumour (Figure 13.12). Alternatively, the tumour may manifest itself as multiple isolated plaques of neoplasm which mimic macroscopically the appearances of secondary carcinoma deposits. A positive diagnostic feature of the peritoneal mesothelioma however is the way in which the tumour, like

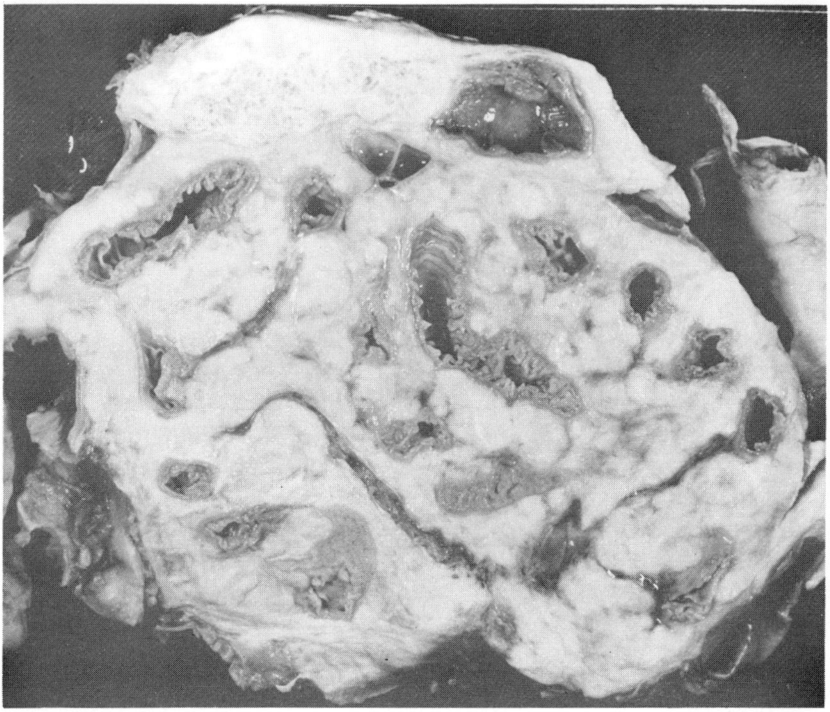

Figure 13.12 Mesothelioma of the peritoneum. A transverse section through the abdominal viscera to show the selective spread of tumour along the serosal planes. Note the normal mucosal folds within the alimentary tract

the pleural mesothelioma, spreads selectively along serosal planes, and does not invade the walls of the bowel. The mucosal folds within the lumen of the alimentary tract are unaffected by the extensive tumour which encases the peritoneal aspects of the viscera.

Metastatic spread to mesenteric and aortic lymph nodes is not uncommon (Hourihane, 1964; Whitwell and Rawcliffe, 1971) and distant metastases may occur, especially in the more long-standing cases.

The major difficulty in the diagnosis of peritoneal mesothelioma is the similarity in macroscopic appearance to secondary carcinoma deposits, and in the post-mortem examination a positive exclusion of all possible primary carcinoma sites must be made. Histological examination may resolve the diagnostic difficulties, but some tumours, particularly ovarian carcinomas, may be very difficult to distinguish from mesotheliomas.

8(ii) Histological features of mesothelioma of pleura and peritoneum

The most striking characteristic of the mesothelioma is the variability of structure that occurs within the same tumour. Variations from a mesenchymal appearance to an epithelial structure are seen, the tumour resembling a sarcoma on the one hand and an adenocarcinoma on the other (McCaughey, 1965). This dimorphic pattern with its variety of tumour appearances is a helpful diagnostic feature, but when tissue samples are very limited it may not be easy to demonstrate. Typically, the mesothelioma consists of two components, an 'epithelial' element and a 'sarcomatous' connective tissue element.

8(ii).1 *'Epithelial' element*

The tumour has a tubulo-papillary structure, strongly mimicking an adenocarcinoma. Usually the structures consist of a single layer of cuboidal or flattened epithelium of relatively uniform cell size and appearance. In some areas these epithelial-like cells line clefts, the supporting cellular matrix consisting of fibrocellular material of banal appearance (Figure 13.13). In other areas the connective tissue has a more cellular, sarcomatous appearance. The papillary structures seen in the epithelial components may cause the shedding of superficial cells into the tubule or, in the case of surface lesions, these desquamated clumps, or individual cells, may be shed into the pleural or peritoneal fluid.

The stromal cells sometimes have acidophilic cytoplasm and they frequently have nuclei which are indistinguishable from those of the epithelial cells. Reticulin stains often show a more intimate supporting framework than is otherwise present in areas of a carcinoma (McCaughey, 1965).

The malignant mesothelial cell of the epithelial type is large and polygonal

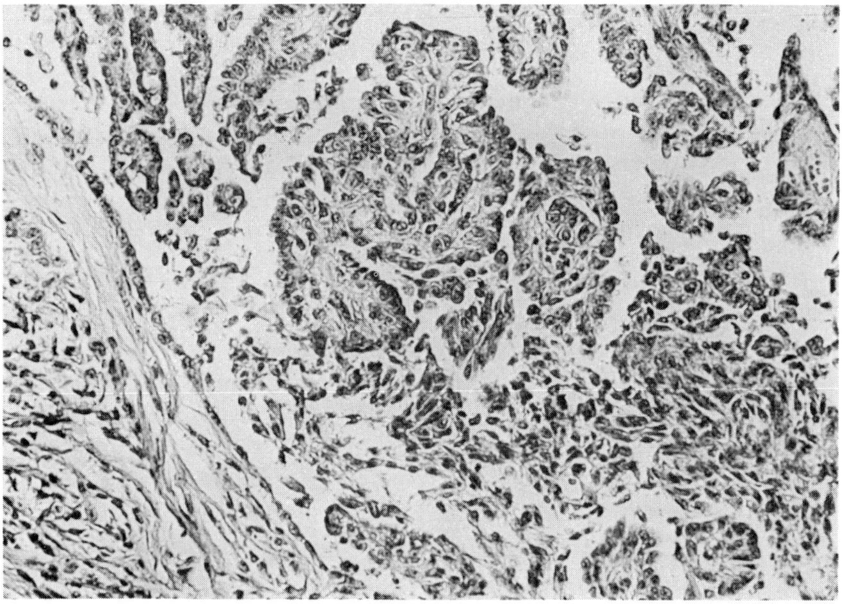

Figure 13.13 Mesothelioma, showing papillary areas and a cleft lined by a single layer of cuboidal or flattened cells (×285, H. & E. stain)

with amphophilic cytoplasm—occasionally eosinophilic—with clear cytoplasmic membranes. The round nucleus occupies about one third of the cell and is often eccentrically placed in the cell. The chromatin has a loose vesicular arrangement. The nuclear membrane may be folded or crenated, and a round eosinophilic nucleus is a common feature. Occasional vacuoles are seen in the cytoplasm and these do not stain as epithelial mucin with diastase–periodic acid Schiff reagent, or mucicarmine, but they may react metachromatically with various stains (Hourihane, 1964). Massive necrosis is uncommon, even in large tumours, but smaller foci of eosinophilic necrosis are frequently seen. Mitoses are few, but occasional multinucleated giant cells are formed. Generally the tumour cells are of uniform shape and size, and even the larger types of cell conform to the description of the smaller ones.

Stromal calcification or ossification within the tumour is not seen. Psammoma bodies have generally been regarded as negating a diagnosis of mesothelioma, and in the female a secondary carcinoma of the ovary is the commonest cause for this finding, but very rarely these bodies are present in otherwise well established mesotheliomas. Lymphatic permeation is not uncommon (Hourihane, 1964).

Many mesotheliomas contain large amounts of dense hyaline collagen as well as loose fibrous tissue. These fibrous elements may be indistinguishable

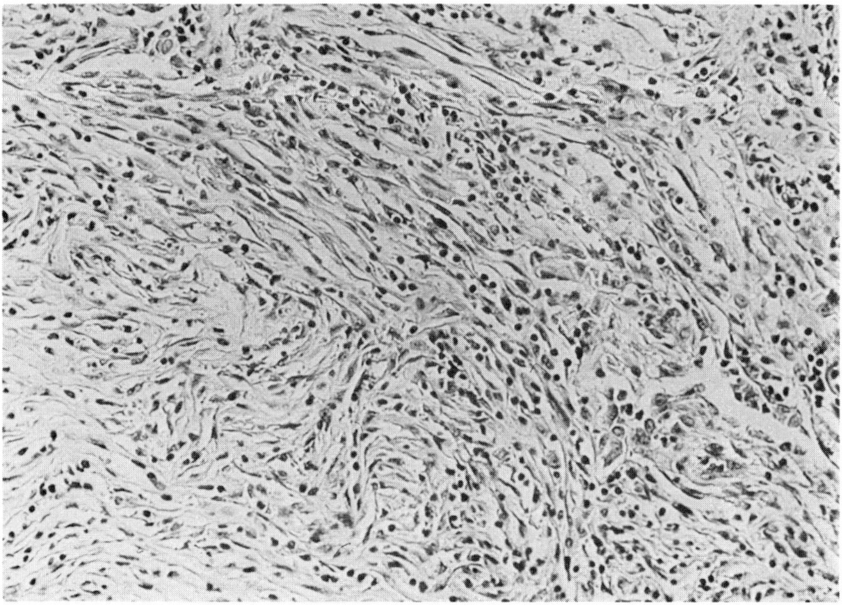

Figure 13.14 Mesothelioma. Sarcomatous area of tumour (×285, H. & E. stain)

from those associated with inflammatory processes or with pleural plaques. When hyaline collagen is formed it may assume patterns which are particularly distinctive of mesothelioma, such as irregularly orientated spindle-shaped clefts, some of which may be lined by, or occupied by, tumour cells. Loosely arranged fine strands of hyaline collagen may also form complex meshworks which are rarely seen in other neoplasms or fibrous reactions (Figure 13.14).

8(ii).2 *'Sarcomatous' connective tissue element*

The other component to the mesothelioma is the neoplastic change seen in part or all of the stromal tissue (Figure 13.15). The spindle cells of sarcomatous appearance frequently have nuclei identical with those cells of the epithelial component. When cells occur in solid nests, it is sometimes difficult to tell whether one is observing epithelium-like cells or whether the appearance is due to a bundle of spindle cells cut perpendicular to their long axis (Churg *et al.*, 1965). Rarely, the mesothelioma consists entirely of this type of cell, but most tumours have both components present, and indeed this dimorphic pattern is one of the cardinal microscopic diagnostic criteria (Figure 13.16). The importance of a cyto-architecture which is epithelial in

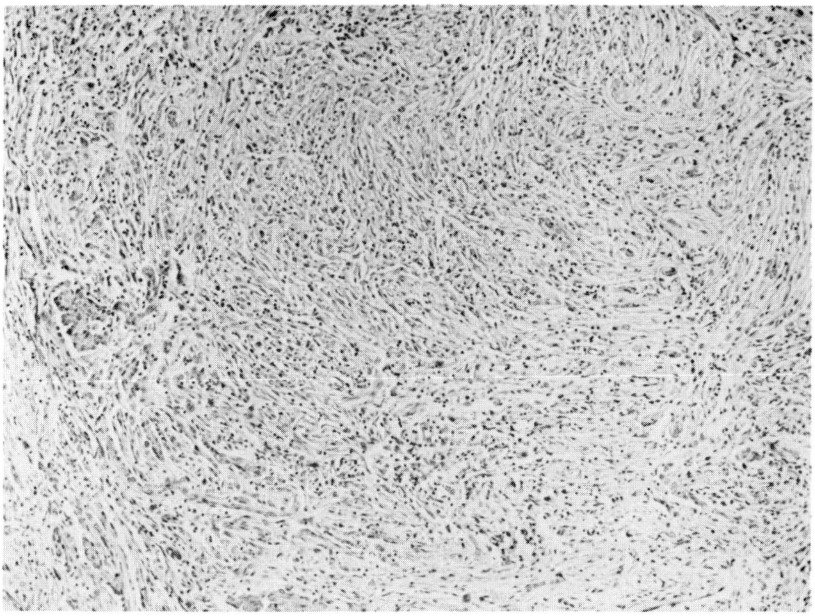

Figure 13.15 Variegated appearance of a mesothelioma (×56, H. & E. stain)

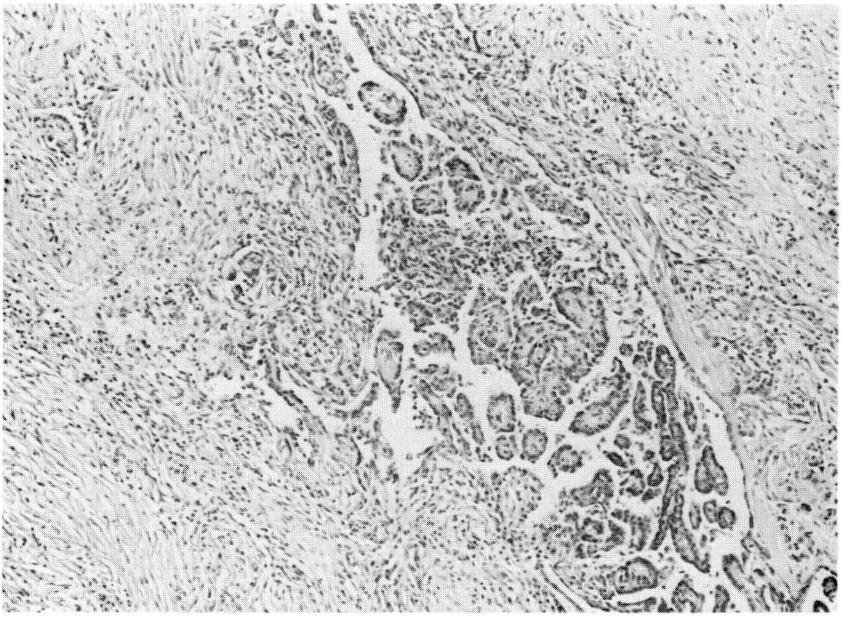

Figure 13.16 Mesothelioma, showing mixed pattern of stromal and epithelial elements (×83, H. & E. stain)

some areas and mesenchymal in others is stressed (McCaughey, 1958; Hourihane, 1964). Metastatic tumours which have spread to serosal membranes do not show this dimorphic pattern. The widely diverging elements are not necessarily seen close together in the neoplasm, but may be found in different areas of the same tumour. This characteristic emphasizes.

1. the need for sampling *multiple* blocks from different parts of the tumour;
2. the difficulty of making a confident histologic diagnosis from a single, small sample, e.g. a needle biopsy.

8(iii) Special stains

Having studied the tumour architecture by H & E stained sections, the differential diagnosis often lies between an adenocarcinoma and a mesothelioma. In order to resolve this dilemma, the most useful stain to employ is the diastase–periodic acid Schiff method, according to McManus (1948). Many mesotheliomas have a high glycogen content and it is therefore essential to carry out preliminary diastase digestion. The absence of mucosubstance increases the certainty of diagnosis of mesothelioma, as most of these tumours how no reactivity to diastase–periodic Schiff reagent (Kannerstein *et al.*, 1973). Occasionally a fine granularity is seen in the cytoplasm, and the brightly staining stroma between tumour cells can mimic positivity of this stain, but in general there is a clear differentiation between positive and negative reactions. If there is a strongly positive diastase–periodic acid Schiff reaction within the cell cytoplasm, and evidence of mucus secretion within the acinar lumen, then this is indicative of a diagnosis of adenocarcinoma, rather than mesothelioma.

Hyaluronic acid forms a large part of the mucin produced by these tumours. It is water soluble and, when tissue is left in formalin fixative, hyaluronic acid may be lost and cytoplasmic vacuoles may appear empty. If there is to be a delay between the time of sampling tissue and processing the blocks, it is recommended that a sample is placed in alcohol so that the hyaluronic acid is retained in the tissue. During dehydration in processing however, the water-soluble hyaluronic acid will become precipitated, thus leaving the mucin in blobs or spots on the section, an artefact which is difficult to avoid. This material stains strongly with alcian blue, metachromatically with toluidine blue and does not stain with Southgate's mucicarmine (Smith, 1973).

8(iv) Pleural and peritoneal fluid

In a proportion of cases, mesotheliomas produce hyaluronic acid (Meyer and Chaffee, 1940; Boersma *et al.*, 1973), and this may enter the pleural or ascitic fluid. A simple precipitation reaction (Wagner *et al.*, 1962) can identify the presence of hyaluronic acid in pleural or ascitic fluids. While this reaction is not positive in all cases of mesothelioma, and some false positives can be obtained, there is sufficient merit to consider it as a contributory screening test in the investigation of pleural and ascitic fluids. In the Nottingham Pathology Laboratories it is used as a routine screening test and several cases of mesothelioma have been detected before clinical suspicion was apparent.

8(v) Cytology

Cytological examination of pleural and ascitic fluid is well worth pursuing, but apart from certain centres (Klempman, 1962; Naylor, 1963; Berge and Grontöft, 1965) diagnostic experience in examining specimens has been somewhat limited. Butler and Berry (1973) noted that the diagnosis depends on the recognition of cells which show the cytoplasmic differentiation of mesothelial cells and have nuclei which show the usual criteria of malignancy (Figure 13.17). It is often of considerable value to make histological blocks of the centrifuged deposit of the pleural or ascitic fluid, and in some cases the characteristic architecture of the desquamated clumps of cells may assist the diagnosis.

8(vi) Classification of mesotheliomas

Whitwell and Rawcliffe (1971) have classified mesotheliomas into four main types:

1. *Tubopapillary*, in which the predominant pattern consists of acini—often of serpiginuous form—lined by low columnar or cuboidal cells of relatively uniform appearance, while other areas show branching papillary projections covering a fine reticulin core. Mitoses are few. (This is the type of mesothelioma which most closely mimics an adenocarcinoma.)
2. *Sarcomatous*, in which the predominant pattern varies from a cellular fibrosarcomatous appearance to an acellular collagenous appearance. Cell nuclei are regular and mitoses few. (This type of tumour must be differentiated from a fibrosarcoma at one end of the spectrum to a pleural plaque at the other.)
3. *Undifferentiated polygonal (epithelial) type*, in which the predominant cell is similar to that which lines the tubulo-papillary structures (in type 1), but in this group the cells may be arranged in sheets or clumps. These cells are usually surrounded by a more intricate framework of reticulin fibres than would be found in similar areas of a carcinoma (McCaughey, 1965).
4. *Mixed.* This group has a mixed appearance of all of the preceding types, though one type may predominate.

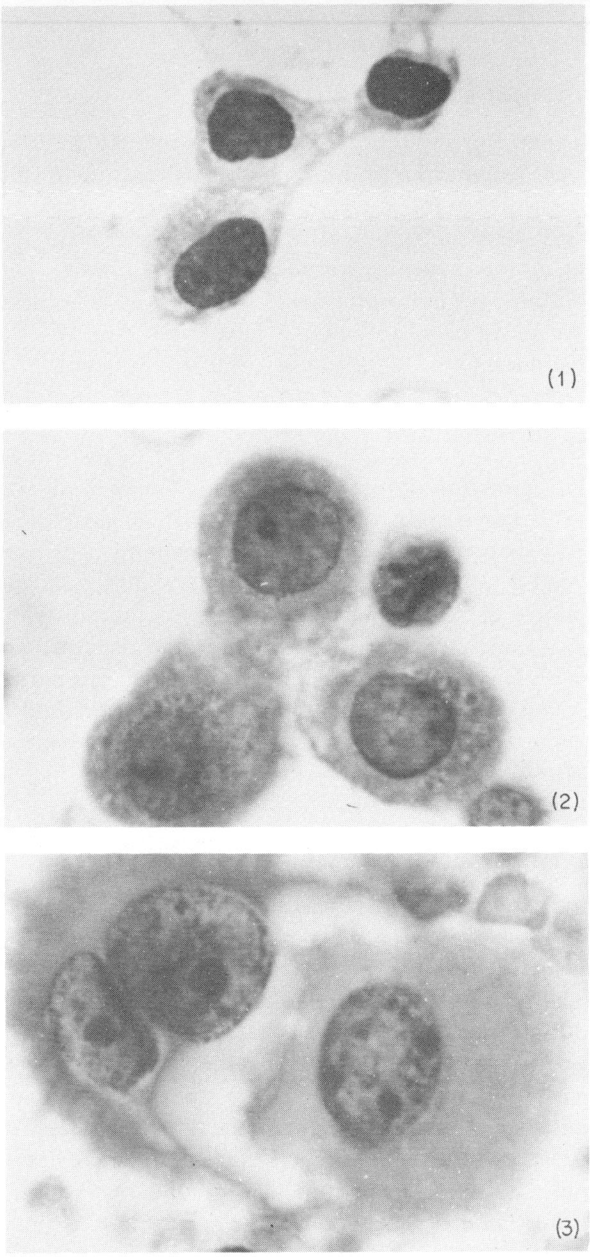

Figure 13.17 Cytology of mesothelial cells. (1) Normal mesothelial cells from the peritoneal surface of a woman undergoing laparoscopic sterilization (×1490, Papanicolau stain). (2) Benign active mesothelial cells from the pleural fluid of a man with pulmonary infarction (×1490, Papanicolau stain). (3) Malignant mesothelial cells from the pleural fluid of a man with mesothelioma (×1490, Papanicolau stain). (Illustrations by courtesy of Dr. E. B. Butler)

The typing of mesotheliomas has limitations, as an accurate classification will depend on a large number of samples being taken from different parts of the tumour. Because of the variability of appearance in different parts of the same tumour, absolute categorization is difficult.

8(vii) Relationship of asbestos exposure to mesothelioma formation

Most cases (85–90%) of mesothelioma, whether of the pleura or peritoneum, are associated with occupational exposure to asbestos, and there is considerable supporting evidence that crocidolite carries the greatest risk of subsequent mesothelioma formation (Wagner *et al.*, 1971). However, amosite and chrysotile also predispose to the tumour (*Lancet.*, 1977) but at lower risk rates. No cases of mesotheliomas have been recorded in association with pure anthophyllite exposure. One of the major difficulties in drawing clear conclusions on the relative risks of different types of asbestos and on dose–response relationships is the fact that most workers have been exposed to a variety of different types of asbestos during their lifetime (Gilson, 1973). Considerable progress to the resolution of this problem is currently being made following the development of techniques for the identification of asbestos fibres in post-mortem lung tissue by electron microscopy and by microprobe methods (Pooley, 1973; Langer and Pooley, 1973) (Figure 13.18).

Studies involving groups of workers who have been exposed to a single type of asbestos are few. In the U.K., wartime gas-mask workers who were exposed only to crocidolite have shown a high incidence of mesotheliomas in that 26 out of 1600 workers have developed this tumour 20–35 years after exposure (Figure 13.19). In 12 of these cases, examination of post-mortem lung tissue (Figure 13.20) has shown that only one type of fibre is present, crocidolite, and that its chemical structure is unchanged after its many years in the lung tissue. None of these workers who developed mesothelioma had a sufficiently high dust content in their lungs to develop asbestosis, and some had been only marginally exposed to the dust during their working life (Jones *et al.*, 1976). A high incidence of mesotheliomas has also been found in Canadian wartime gas-mask workers (McDonald and McDonald, 1977).

Selikoff (1976) has studied a group of 870 wartime workers in a factory where amosite was the exclusive type of asbestos fibre and by 1973 nine of the deaths (1.5%) were due to mesothelioma.

McDonald (1977) studied 11 000 Canadian chrysotile miners and millers and of the 4547 deaths which occurred between 1910 and 1975, 11 (0.24%) were due to mesothelioma.

On the vexed question of dose–response relationship with regard to mesothelioma formation, it is too early to draw firm conclusions. Whitwell *et al.* (1977) carried out light microscopy counts of asbestos bodies and fibres,

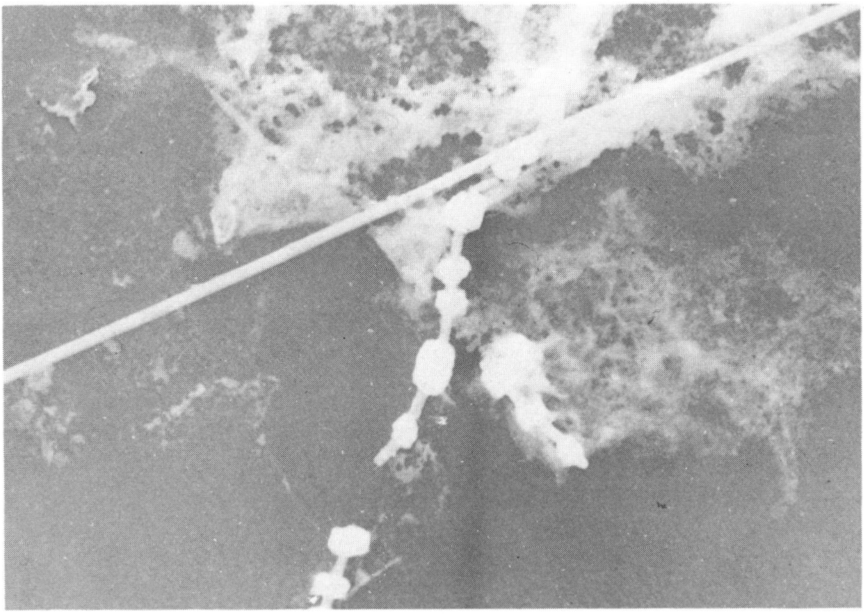

Figure 13.18 An asbestos fibre and an asbestos body identified in post mortem lung tissue by electron microscopy (×3285). (Illustration by courtesy of Dr. F. D. Pooley)

and concluded that a dose–response relationship does exist between asbestos exposure and mesothelioma formation. Newhouse and Berry (1976) also found a dose–response relationship for mesothelial tumours in workers who had been exposed to mixed types of asbestos dust. Further work is necessary, particularly in the field of detailed post-mortem lung mineral content

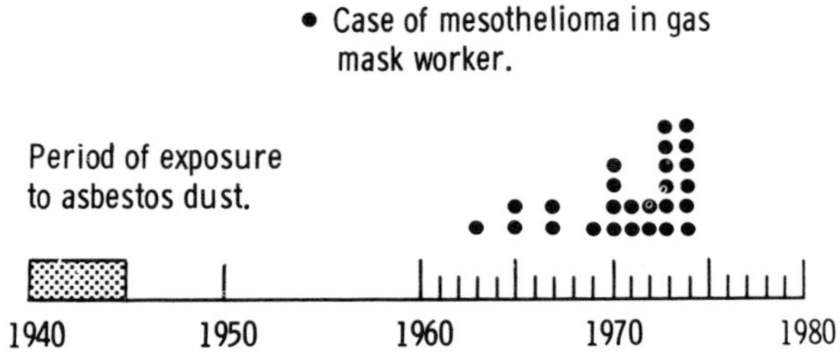

Figure 13.19 The chart illustrates the long interval of 20 to 30 years between exposure to crocidolite asbestos and the development of mesotheliomas (Jones *et al.*, 1976). (Reproduced by courtesy of INSERM and IARC scientific publications)

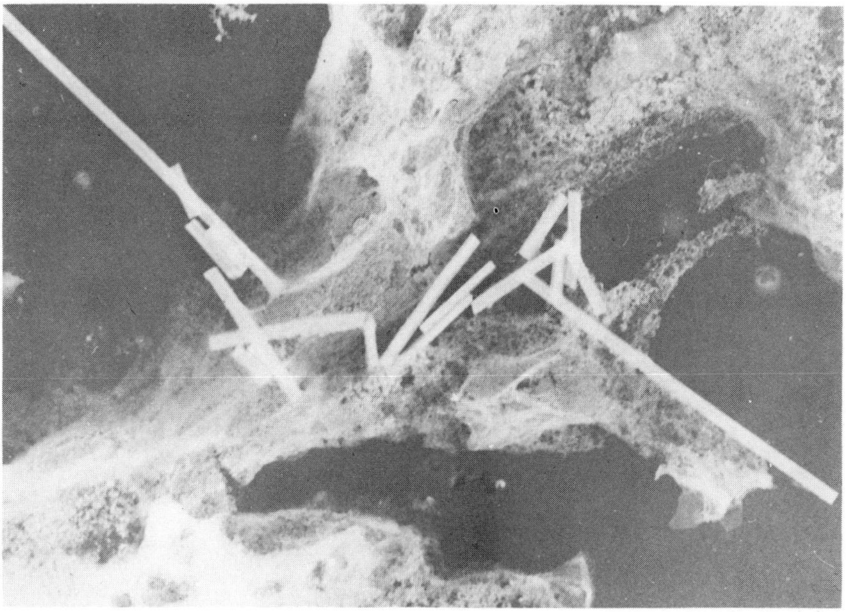

Figure 13.20 Asbestos fibres identified in post mortem lung tissue by electron microscopy (×3285). (Illustration by courtesy of Dr. F. D. Pooley)

analysis, and judgement should be withheld on this issue until further information is available.

8(viii) Immunological aspects of mesothelioma

Because of the probability of individual susceptibility with development of mesotheliomas in some people, but not others, even though their asbestos exposures were similar, interest has been shown in pursuing immunological aspects of tumour formation. So far, the results have been disappointing. Embleton *et al.* (1976) concluded that little or no tumour-directed, cell-mediated immunity is detectable against mesothelioma by microcytotoxicity methods. However, further work is in progress in many centres where an immunological approach to the problem is is being actively pursued.

(9) PLEURAL PLAQUES

9(i) Macroscopic appearance

Pleural plaques are yellow–white areas of patchy thickening, projecting slightly above the contour of the parietal pleura which lines the chest cavity.

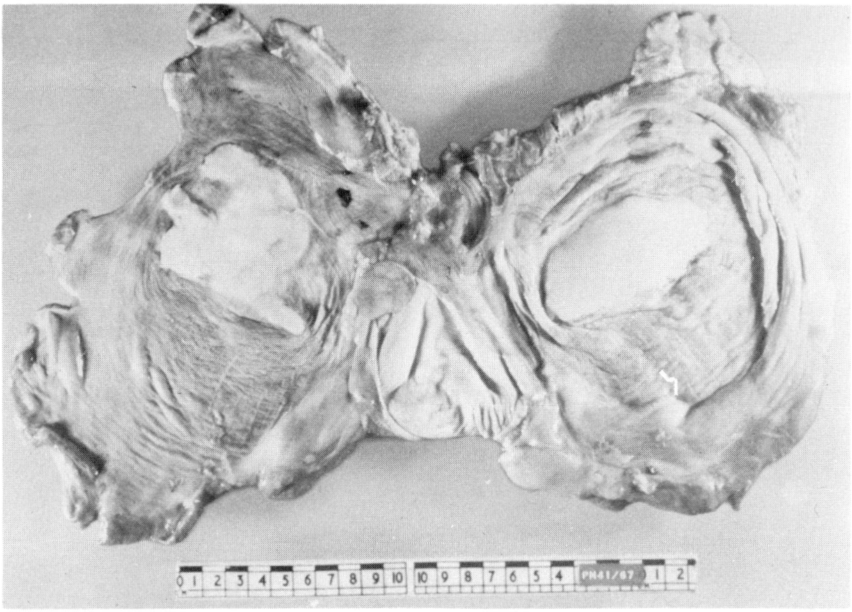

Figure 13.21 Pleural plaques. Note the selective bilateral distribution on the central tendons of the diaphragm. The plaques are smooth and shiny with a clear-cut edge which shows nodularity in places. (Illustration by courtesy of Dr. P. G. Smith)

They appear as shiny porcelain-like plateaux with either smooth or nodular edges. There is often a shiny, bead-like quality to the surface of the plaque. There is a great variation in the size of plaques, but they conform to a relatively consistent pattern of distribution in that they tend to be localized to the parietal pleural surface which overlies the central tendons of the diaphragm (Figure 13.21) and they also overlie the lines of the lower ribs, particularly from the seventh to the tenth (Meurman, 1966). Most plaques are bilateral and are fairly symmetrically arranged (Figure 13.22). Confluence may be seen, especially in the paravertebral regions. Their thickness varies from a few millimetres to 1 cm. Occasionally plaques are found at other sites, e.g. on the anterior mediastinal pleura (Selikoff, 1965) and on the left cardiac border (Fletcher and Edge, 1970). In general, pleural plaques are not associated with fibrous adhesions between the visceral and parietal layers.

9(ii) Microscopic appearance

Pleural plaques consist of dense bundles of collagen arranged so as to give a 'basket-weave' appearance. They are avascular and free from inflammatory infiltrate within the body of the plaque, although lymphocytes and

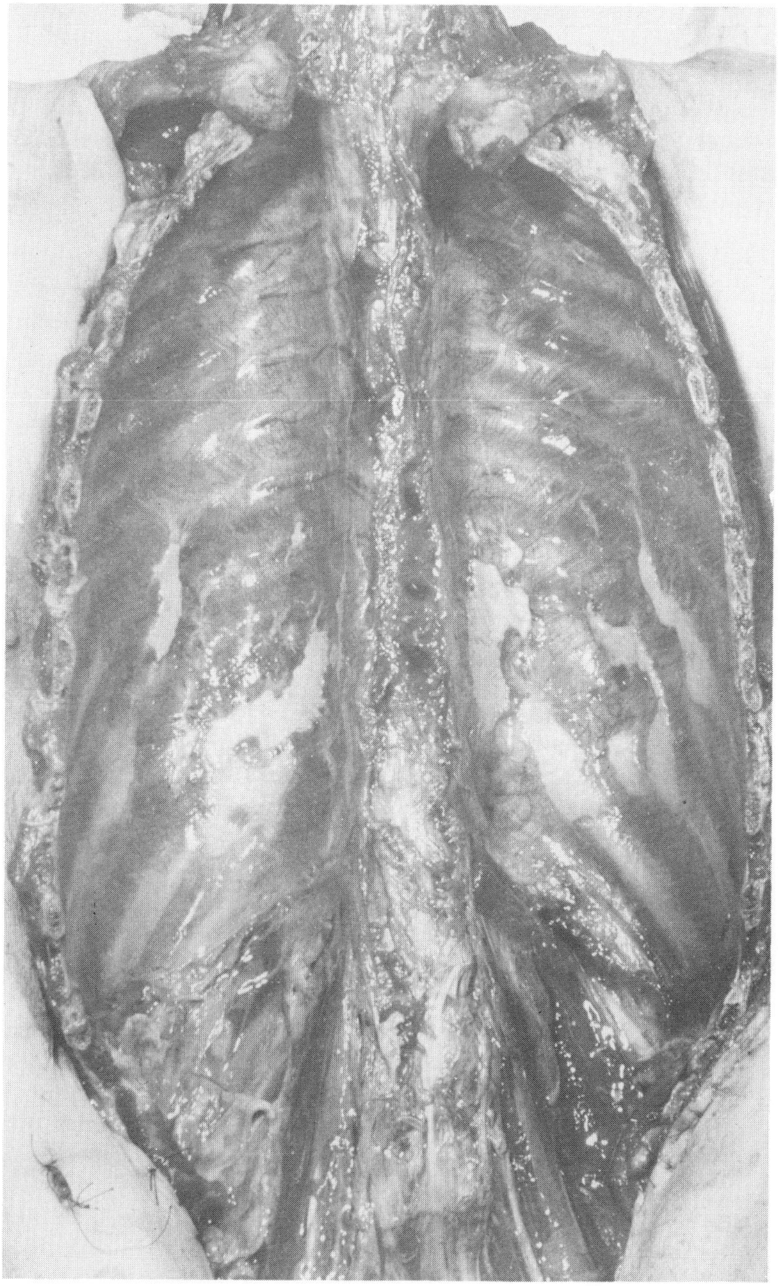

Figure 13.22 Pleural plaques. Note the bilateral distribution on the parietal pleura of the lower thoracic wall. Plaques tend to follow the lines of the ribs and some confluence is seen. (Illustration by courtesy of Dr. P. G. Smith)

fibroblasts are occasionally seen in the deeper aspects of a developing plaque (Thomson, 1969). The mesothelial cell lining of the pleural cavity extends over the plaque surface in continuity, making the plaque in effect an extra-pleural structure. Calcium is found in fine granules in irregular distribution in different layers of the plaque (Meurman, 1966). Well calcified plaques may therefore be seen radiologically during life, the appearances depending on the extent of calcification (Kiviluoto, 1960).

9(iii) Relationship of pleural plaques to asbestos exposure

Much of the fundamental investigation of plaques was carried out by Meurman (1966, 1968) when he studied people who were exposed to anthophyllite asbestos throughout their lives as a result of living in mining communes in Finland. Asbestos bodies were present in 86% of subjects with pleural plaques and, if asbestos was present in abundance, then plaques were regularly seen. This observation has been confirmed by other investigators (Hourihane *et al.*, 1966 Roberts, 1967; McPherson and Davidson, 1969; Mattson and Ringqvist, 1970; Le Bouffant *et al.*, 1973).

Meurman (1966) was unable to detect asbestos bodies within the plaques, but asbestos fibres have been identified by Hourihane *et al.* (1966) and by Thomson (1969). A period of about 20 years usually elapses between the occasion of the first exposure to asbestos and the development of pleural plaques (Selikoff, 1965). The youngest recorded case of plaque formation is in an 18-year-old Finnish subject (Meurman, 1966), but radiological evidence (by calcium deposition) rarely occurs under the age of 30 years in persons who have been environmentally exposed to asbestos since birth (Kiviluoto, 1969). Pleural plaque formation does not seem to depend on particularly heavy exposure to asbestos dust, and any of the commercially used types of asbestos can lead to this pleural change. In the absence of any other abnormalities, pleural plaques do not give rise to clinical symptoms or to abnormal physical signs, and they are not accompanied by defects of lung function (Leathart, 1968). Pleural plaques are regarded as a 'visiting card' of asbestos exposure (Meurman, 1966) and, while they are not harmful clinically, they provide a useful marker to draw the attention of the radiologist, clinician, and pathologist to the possibility of exposure to asbestos (Jones and Sheers, 1973). Asbestos is not the sole cause of pleural plaques, but it is certainly the most common. Because of the frequent finding of plaques following exposure to asbestos, they are usually present in cases of asbestosis, lung cancer associated with asbestos exposure (Fletcher, 1972) (Figure 13.23), and pleural and peritoneal mesothelioma. They are frequently found in workers who have an indirect occupational exposure to asbestos, e.g. carpenters, electricians, plumbers, and fitters in the building construction industry who, although not 'asbestos workers', are nonetheless exposed to the dust on building sites (Skidmore and Jones, 1975).

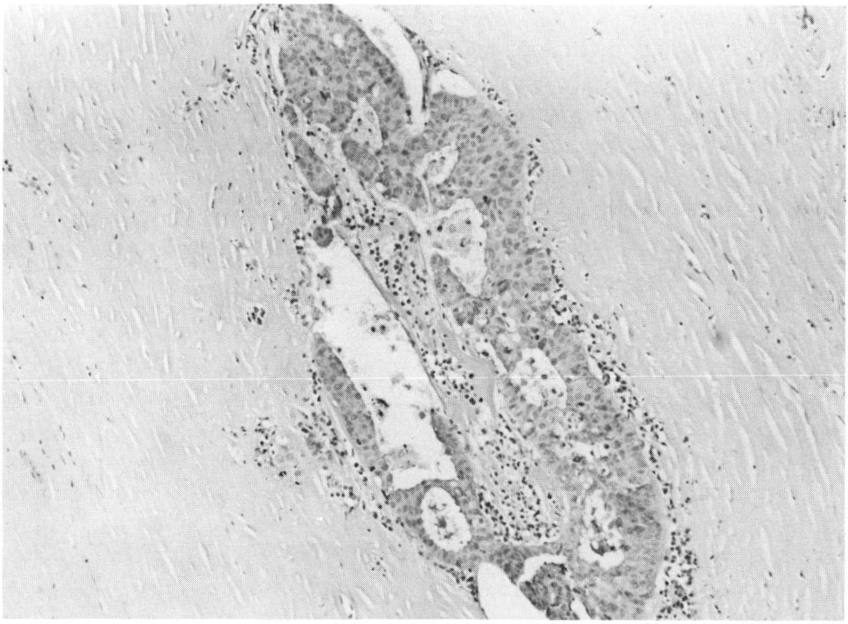

Figure 13.23 A pleural plaque, with 'basket weave' appearance, invaded by adenocarcinoma (×105, H. & E. stain)

9(iv) Pathogenesis of pleural plaques

Various possibilities have been suggested as to how the plaques are formed. Kiviluoto (1960) postulated that they have resulted from an inflammatory reaction in the parietal pleura following the mechanical irritation by fibres in the lung projecting through the visceral pleura. Thomson (1969) noted that mesothelioma cells play no part in plaque formation as any reaction appears to originate in its deepest aspect. He thought that the downward and outward gravitational movement of the sharp asbestos fibres was of relevance, and that the relatively extensive lesion arising from a very small number of fibres might be due to a sensitivity reaction. In support of an immunological basis for plaque formation, Pernis *et al.* (1965) reported an increased prevalence of rheumatoid factor in asbestos workers. Turner Warwick and Parkes (1970) confirmed an increased prevalence of rheumatoid factor and also antinuclear factor in asbestos workers. Serum gamma-globulin levels are higher in asbestos-exposed people who have pleural plaques than in those who do not, and they are even higher than the level in non-exposed controls (Navratil, 1970).

Enticknap and Smither (1964) and Hourihane *et al.* (1966) have postulated the intracellular carriage of asbestos particles within lymphatic vessels to the parietal pleura. The sites of plaque formation do coincide with the pathways of lymphatic drainage (Lemon and Higgins, 1932) and the specific sites are possibly also zones of maximum stasis. Jones and Sheers (1973) suggested that the massaging effect of the diaphragm and muscles of the chest wall might result in the squeezing of dust particles along lymphatic vessels until they are 'washed up on the shore' of the inert central tendon of the diaphragm and on to the pleura overlying the ribs.

(10) EXPERIMENTAL PATHOLOGY

Until 1964, investigations to determine the biological effects of asbestos in animals were carried out by a number of workers in various centres, using a variety of different animals and a variety of asbestos samples from different sources.

Among these were inhalation experiments in guinea pigs (Holt *et al.*, 1965) when four types of asbestos dust—chrysotile, crocidolite, amosite, and anthophyllite—were used. It was shown that animals killed early in the experiment revealed few asbestos fibres in their lungs on examination by light microscopy, yet animals killed later showed large numbers of asbestos bodies. Dust reached the terminal bronchioles after a few days and trace amounts entered the alveoli. A bronchiolitis with damage to the surface epithelium followed, and the lumen became filled with cellular debris and dust-laden macrophages. Collagen formation then sealed off many alveoli bordering the alveolar ducts or respiratory bronchioles. There was later extension of the inflammatory infiltrate from the bronchioles into the lung, with thickening of alveolar walls and spreading fibrosis. The tracheal lymph nodes contained dust-laden macrophages in which asbestos fibres and bodies were identified.

Davis (1965) used the guinea pig lung material from the above experiments for electron microscopy studies. He noted that the bulk of the dust which entered the lung was of very small particle size, usually below 1 μm in length. Asbestosis was seen to be basically an intracellular process, and the pathogenesis of asbestosis is concerned with the reaction of the dust on the lung macrophage.

Wagner (1963) had noted that in animal dusting experiments using rats, there were differences in the fibrogenic properties of different types of asbestos. In a further series of dusting experiments (Wager and Skidmore, 1965), an equal amount of dust of a particular type was inhaled by a series of rats. The dusts tended to accumulate in the alveoli arising directly from the respiratory bronchioles. The elimination rate of Rhodesian chrysotile was three times greater than that of amosite or crocidolite, and this finding suggests an explanation for the reduced fibrogenicity of this type of fibre. Similar results were obtained by Timbrell (1970).

Using hamsters, Millen *et al.* (1965) tested the theory that the failure to find malignant tumours in the respiratory tract after inhalation of asbestos fibres in these animals, as opposed to the experience in man, was possibly due to the fact that the carcinogenic effect was not so much due to the asbestos itself, but rather that the asbestos had a co-carcinogenic effect on some other factor present in human lungs, but not in animals. The hypothesis was explored using benzo(*a*)pyrene in conjunction with two types of asbestos—chrysotile and amosite. The results showed that in the hamsters, papillomas and carcinomas occurred in the respiratory tract after intratracheal injection of benzo(*a*)pyrene. Addition of chrysotile asbestos gave results which supported the hypothesis that this type of fibre promoted the benzo(*a*)pyrene carcinogenic effect. Addition of amosite, however, did not increase the yield of tumour. Experiments using rats have been undertaken in the U.S.S.R. on the significance of benzo(*a*)pyrene in asbestos carcinogenesis (Pylev and Shabad, 1973).

Harington and Roe (1965) investigated the possibility of the carcinogenicity of asbestos fibres being due to their natural oils.

Intrapleural injection of asbestos in rats (Wagner, 1965; Wagner *et al.*, 1970), hamsters (Smith *et al.*, 1965), and rats and mice (Harington and Roe, 1965) all resulted in mesothelioma formation. Injection of asbestos into the axillary air sacs of white leghorn fowls (Peacock and Peacock, 1965) resulted in localized tumour formation in a proportion of birds.

Subcutaneous injection of asbestos into mice (Roe *et al.*, 1967) showed initially that a considerable amount of the inoculated fibre tended to track through the tissues to mesothelial surfaces. More sophisticated subsequent studies (Kanazawa *et al.*, 1970; Morgan and Holmes, 1970; Morgan *et al.*, 1971) have not confirmed this finding. The tagging of asbestos fibres with fluorescent dyes has been carried out by Berkley *et al.* (1965) so that they may be detected subsequent to their insertion into experimental animals. Morgan and Holmes (1970), using neutron-activation techniques, have studied the passage of irradiated asbestos fibres through the alimentary tracts of rats.

10(i) Standardization of animals and asbestos samples

Following the UICC working party in 1964, it was generally agreed that in order to make results from various laboratories more comparable, it would be of benefit to limit the range of animals to be used, and that the UICC standard reference samples of asbestos prepared by Timbrell (1973) be used. The SPF rat was generally regarded as the most widely used experimental animal. In South Africa, Harington (1973) used baboons with good comparable results with humans.

The length of the various experiments has been found to be of crucial value in the study of asbestos carcinogenicity (Wagner and Berry, 1973). Animals should live their full life span. In the MRC Pneumoconiosis Unit at Penarth, following a series of intrapleural inoculations with asbestos, only 20% of tumours appeared within 18 months of inoculation, whereas two thirds of the total number of tumours were seen within 2 years, and 95% of the tumours had occurred 30 months after inoculation. Premature termination of the experiments would have deprived the observers of the bulk of the information that was to be derived from this investigation.

Experiments have been carried out in which SPF Wistar rats were exposed by inhalation to dust clouds of the UICC standard reference samples for periods varying from 1 day to 2 years (Wagner *et al.*, 1974), with the following results:

1. All of the samples of asbestos produced asbestosis. This continued to progress after removal of the animal from dust exposure. Only slight fibrosis was observed in control rats.
2. Lung tumours (ranging from adenomas to squamous carcinomas) were produced by all samples. In the control group there were only a few adenomata and no malignant tumours.
3. Of the 20 tumours which metastasized, 16 occurred after exposure to chrysotile asbestos.
4. In addition to the above, 11 mesotheliomas occurred, 4 following crocidolite exposure, and 4 following Canadian chrysotile exposure. Two of the mesotheliomas occurred after only 1 day's exposure to asbestos.
5. There was a positive association between asbestosis and lung tumours.

Using rats, an experiment was carried out by Wagner *et al.* (1977) in which the effects of Italian talc were compared with those of chrysotile asbestos. The results showed that with equal dosage, talc can produce a similar amount of fibrosis to asbestos. However, the rats which had been exposed to chrysotile asbestos also developed lung adenomas and an adenocarcinoma, whereas the only lung tumour seen in the rats exposed to talc was a small adenoma.

10(ii) Cell and organ culture

Asbestos has two types of cytotoxic effect on cells in culture (Allison, 1973):

(i) an early cytotoxic effect due to interaction with the cell membrane—this is markedly inhibited by the presence of serum or some other biological macromolecule;

(ii) a late cytotoxic effect due to an interaction of ingested asbestos particles with the membranes around secondary lysosomes. Chrysotile and anthophyllite are more active than crocidolite and amosite with respect to both of these reactions.

The effect of asbestos on human embryonic lungs and on adult pleura maintained in organ culture has been investigated by Rajan and Evans (1973). Exposure to crocidolite has shown a proliferation of fibroblasts and mesothelial cells.

(11) ACKNOWLEDGEMENTS

I acknowledge the considerable help that I have derived from the many authors whose work I have studied and quoted in compiling this chapter, and also to those who have kindly allowed me to use their illustrations. Most of the other photographs were taken by Mr. G. B. Gilbert of the Department of Medical Photography, City Hospital, Nottingham, to whom I give my thanks. I owe a special debt of gratitude to my secretary, Mrs. Valerie Bolton, for her considerable help in preparing the manuscript.

(12) REFERENCES

Allison, A. C. (1973) Experimental methods—cell and tissue culture: effects of asbestos particles on macrophages, mesothelial cells and fibroblasts, in *Biological Effects of Asbestos*, IARC Scientific Publications No. 8, International Agency for Research on Cancer, Lyon, 89–93.

Anspach, M., Roitzsch, E., and Clausnitzer, W. (1965) Ein Beitrag zur Atiologie des Diffusion malignen Pleura-Mesothelioms, *Arch. Gewerbepath. Gewerbehyg.*, **21,** 392–407.

Ashcroft, T., and Heppleston, A. G. (1970) Mesothelioma and asbestos on Tynside—a pathological and social study, in *Pneumoconiosis: Proceedings of the International Conference, Johannesburg*, 1969, (Ed. H. A. Shapiro), Oxford University Press, Cape Town, 177–179.

Berge, T., and Gröntoft, O. (1965) Cytologic diagnosis of malignant pleural mesothelioma, *Acta Cytol.* **9,** 207–212.

Berkley, C., Churg, J., Selikoff, I. J., and Smith, W. E. (1965) The detection and localization of mineral fibres in tissue, *Ann. N.Y. Acad. Sci.*, **132,** 48–63.

Berry, G., Newhouse, M. L., and Turok, M. (1972) Combined effect of asbestos exposure and smoking on mortality from lung cancer in factory workers, *Lancet*, **ii,** 476–479.

Blount, M., Holt, P. F., and Leach, A. A. (1966) The protein coating of asbestos bodies, *Biochem. J.*, **101,** 204–207.

Boersma, A., Degand, P., and Havez, R. (1973) Diffuse mesothelioma: biochemical stages in the diagnosis, detection and measurement of hyaluronic acid in the pleural fluid, in *Biological Effects of Asbestos*, IARC Scientific Publications No. 8, International Agency for Research on Cancer, Lyon, 65–67.

Bonser, G. M., Foulds, J. S., and Stewart, M. J. (1955) Occupational cancer of the urinary tract in dye-stuffs operatives, and of the lung in asbestos textile workers and iron-ore miners, *Am. J. Clin. Pathol.*, **25,** 126.

Botham, S. K., and Holt, P. F. (1967) *Kongressbericht V. Internationale Staublungentagung, Munster.*

Buchanan, W. D. (1965) Asbestosis and primary intrathoracic neoplasm, *Ann. N.Y. Acad. Sci.*, **132,** 507–518.

Burilkov, T., and Babadjov, L. (1970) Endemic occurrence of bilateral pleural calcification, *Prax. Pneumonol.*, **24,** p. 7.

Butler, E. B. B., and Berry, A. V. (1973) Diffuse mesotheliomas: diagnostic criteria using exfoliative cytology, in *Biological Effects of Asbestos*, IARC Scientific Publications No. 8, International Agency for Research on Cancer, Lyon, 68–73.

Churg, J., Rosen, S. H., and Moolten, S. (1965) Histological characteristics of mesothelioma associated with asbestos, *Ann. N.Y. Acad. Sci.*, **132,** 614–622.

Constantinidis, K. (1977) Asbestos exposure—its related disorders, *Brit. J. Clin. Pract.*, **31,** 89–101.

Cooke, W. E. (1924) Fibrosis of the lungs due to inhalation of asbestos dust. *Brit. Med. J.*, **2,** 147.

Cralley, L. J., Keenan, R. G., Lynch, J. R., and Lainhart, W. S. (1968) Source and identification of respirable fibers, *Am. Ind. Hyg. Ass. J.*, **29,** 129–135.

Davis, J. M. G. (1965) Electron microscope studies of asbestosis in man and animals, *Ann. N.Y. Acad. Sci.*, **132,** 98–111.

Doll, R. (1955) Mortality from lung cancer in asbestos workers, *Brit. J. Ind. Med.*, **12,** 81–86.

Doll, R. (1971) The age distribution of cancer. Implications for models of carcinogenesis, *Jl. R. Statist. Soc.*, **134,** 133–155.

Elmes, P. C. (1973) Therapeutic openings in the treatment of mesothelioma, in *Biological Effects of Asbestos*, IARC Scientific Publications No. 8, International Agency for Research on Cancer, Lyon, 277–280.

Elmes, P. C., and Simpson, M. J. C. (1971) Insulation workers in Belfast Mortality 1940–1966, *Brit. J. Ind. Med.*, **28,** 226–236.

Elmes, P. C., and Wade, O. L. (1965) Relationship between exposure to asbestos and pleural malignancy in Belfast, *Ann. N.Y. Acad. Sci.*, **132,** 549–557.

Embleton, M. J., Wagner, J. C., Wagner, M. M. F., Jones, J. S. P., Sheers, G., Oldham, P. D., and Baldwin, R. W. (1976) Assessment of cell-mediated immunity to malignant mesothelioma by microcytotoxicity tests, *Int. J. Cancer*, **17,** 597–601.

Enterline, P. E. (1965) Mortality among asbestos products workers in the United States, *Ann. N.Y. Acad. Sci.* **132,** 156–165.

Enticknap, J. B., and Smither, W. J. (1964) Peritoneal tumours in asbestosis, *Brit. J. Ind. Med.*, **21,** 20–31.

Fletcher, D. E. (1972) A mortality study of shipyard workers with pleural plaques, *Brit. J. Ind. Med.*, **29,** 142–145.

Fletcher, D. E., and Edge, J. R. (1970) The early radiological changes in pulmonary and pleural asbestos, *Clin. Radiol.*, **21,** 355–365.

Gilson, J. C. (1973) Progress in epidemiology, in *Biological Effects of Asbestos*, IARC Scientific Publications No. 8, International Agency for Research on Cancer, Lyon, 5-10.

Gloyne, S. R. (1932) The asbestos body, *Lancet*, **i,** 1351–1355.

Gloyne, S. R. (1932–33) The morbid anatomy and histology of asbestos, *Tubercle*, **14,** 445–451, 493–497, 550–558.

Gloyne, S. R. (1935) Two cases of squamous carcinoma of the lung occurring in asbestosis, *Tubercle*, **17,** 5.

Gloyne, S. R. (1938) *Silicosis and Asbestosis*, (Ed. A. J. Lanza), Oxford University Press, London.

Gloyne, S. R. (1951) Pneumoconiosis. A histological survey of necropsy material in 1205 cases, *Lancet*, **i,** 810.

Glyn Owen, W. (1965) Mesothelial tumours and exposure to asbestos dust, *Ann. N.Y. Acad. Sci.*, **132,** 674–679.

Godwin, M. C. (1957) Diffuse mesotheliomas, *Cancer*, **10,** 28.

Gold, C. (1967) A simple method of detecting asbestos in tissues, *J. Clin. Pathol.*, **20,** 674.

Gough, J. (1965) Differential diagnosis in the pathology of asbestosis, *Ann. N.Y. Acad. Sci.*, **132,** 368–372.

Gough, J. and Wentworth, J. E. (1960) in *Recent Advances in Pathology*, 7th Ed., J. & A. Churchill, London, 80.

Graham, J., and Graham, R. (1967) Ovarian cancer and asbestos, *Environ. Res.*, **1,** 115–128.

Gross, P., Cralley, L. J., and De Treville, R. T. P. (1967) Asbestos bodies: their non-specificity, *Am. Ind. Hyg. Ass. J.*, **28,** 541–544.

Hammond, E. C., Selikoff, I. J., and Churg, J. (1965) Neoplasia among insulation workers in the United States with special reference to intraabdominal neoplasia, *Ann. N.Y. Acad. Sci.*, **132,** 519–525.

Harington, J. S. (1973) Discussion summary, in *Biological Effects of Asbestos*, IARC Scientific Publications No. 8, International Agency for Research on Cancer, Lyon, 106.

Harington, J. S., and Roe, F. J. C. (1965) Studies of carcinogenesis of asbestos fibres and their natural oils, *Ann. N.Y. Acad. Sci.*, **132,** 439–450.

Heard, B. E. (1969) *Pathology of Chronic Bronchitis and Emphysema*, J. & A. Churchill, London, 9.

Heard, B. E., and Willimas, R. (1961) The pathology of asbestosis with reference to lung function, *Thorax*, **16,** 264–281.

Hinson, K. F. W., Otto, H., Webster, I., and Rossiter, C. E. (1973) Criteria for the diagnosis and grading of asbestos, in *Biological Effects of Asbestos*, IARC Scientific Publications No. 8, International Agency of Research on Cancer, Lyon, 54–57.

Holmes, A., and Morgan, A. (1968) The use of radioactive techniques in studies of the composition of asbestos and its translocation in rats following intrapleural, subcutaneous and intratracheal injection, *Proceedings of International Conference, Dresden, Biologische Wirkungen des Asbestos*, 56–64.

Holt, P. F., Mills, J., and Young, D. K. (1965) Experimental asbestosis with four types of fibres: importance of small particles, *Ann. N.Y. Acad. Sci.*, **132,** 87–97.

Hourihane, D. O'B. (1964) The pathology of mesotheliomata and an analysis of their association with asbestos exposure, *Thorax*, **19,** 268–278.

Hourihane, D. O'B. (1965) A biopsy series of mesotheliomata and attempts to identify asbestos within some of the tumours, *Ann. N.Y. Acad. Sci.*, **132,** 647–673.

Hourihane, D. O'B., Lessof, L., and Richardson, P. C. (1966) Hyaline and calcified pleural plaques as an index of exposure to asbestos, *Brit. Med. J.*, **i,** 1069–1074.

Hourihane, D. O'B., and McCaughey, W. T. E. (1966) Pathological aspects of asbestosis, *Postgrad. Med. J.*, **42,** 613–622.

International Agency for Research on Cancer (1977) Evaluation of the Carcinogenic Risk of Chemicals to Man, Vol. 14, Asbestos, International Agency for Research on Cancer, Lyon.

International Union against Cancer (1965) Working group on asbestos and cancer, *Arch. Envir. Hlth.*, **11,** 221–229.

Jacob, G., and Anspach, M. (1965) Pulmonary neoplasia among Dresden asbestos workers, *Ann. N.Y. Acad. Sci.*, **132,** 536–548.

Jones, J. S. P., Pooley, F. D., and Smith, P. G. (1976) Factory populations exposed to crocidolite asbestos—a continuing survey in *Environmental Pollution and Carcinogenic Risks*, in INSERM Symposia Series Vol. 52, IARC Scientific Publications No. 13, International Agency for Research on Cancer, Lyon, 117–120.

Jones, J. S. P., and Sheers, G. (1973) Pleural plaques, in *Biological Effects of Asbestos*, IARC Scientific Publications No. 8, International Agency for Research on Cancer, Lyon, 243–248.

Kanazawa, K., Birbeck, M. S. C., Carter, R. L., and Roe, F. J. C. (1970) Migration of asbestos fibres from subcutaneous injection sites in mice, *Brit. J. Cancer*, **24,** 96–106.

Kannerstein, M., and Churg, J. (1972) Pathology of carcinoma of the lung associated with asbestos exposure, *Cancer*, **30,** 14–21.

Kannerstein, M., Churg, J., and Magner, D. (1973) Histochemical studies in the diagnosis of mesothelioma, in *Biological Effects of Asbestos*, IARC Scientific Publications No. 8, International Agency for Research on Cancer, Lyon, 62–64.

Keal, E. E. (1960) Asbestosis and abdominal neoplasms, *Lancet*, **ii,** 1211–1216.

Kiviluoto, R. (1960) Pleural calcification as a roentgenolic sign of non-occupational anthophyllite asbestosis, *Acta Radiol.*, **194,** suppl.

Kiviluoto, R. (1969) Asbestosis: aspects of its radiological features, in *Pneumoconiosis: Proceedings of the International Conference*, Oxford University Press, Cape Town, Johannesburg, 1969, (Ed. H. A. Shapiro), 253–255.

Klemperer, P., and Rabin, C. B. (1931) Primary neoplasms of the pleura, *Arch. Pathol.*, **11,** 385.

Klempman, S. (1962) The exfoliative cytology of diffuse pleural mesothelioma, *Cancer*, **15,** 691–704.

Koss, L. (1968) *Diagnostic Cytology in its Histopathologic Bases*, Lippincott, Philadelphia, Toronto, 505–510, 537–540.

Lancet (1977) Asbestos, *Lancet*, **ii,** 1211–1212.

Langer, A. M., and Pooley, F. D. (1973) Identification of single asbestos fibres in human tissues, in *Biological Effects of Asbestos*, IARC Scientific Publications No. 8, International Agency for Research on Cancer, Lyon, 119–125.

Leathart, G. L. (1968) Pulmonary function tests in asbestos workers, *Trans. Soc. Occup. Med.*, **18,** 49–55.

Le Bouffant, L., Bruyère, S., Martin, J. C., Tichoux, G., and Normand, C. (1976) *Rev. Fr. Malad. Resp.*, **4,** suppl. 2, 121.

Le Bouffant, L., Martin, J. C., Durif, S., and Daniel, H. (1973) Structure and function of pleural plaques, in *Biological Effects of Asbestos*, IARC Scientific Publications No. 8, International Agency for Research on Cancer, Lyon, 249–257.

Lemon, W. S., and Higgins, G. M. (1932) Absorption from the pleural cavity of dogs, *Am. J. Med. Sci.*, **184,** 846–858.

Leiben, J., and Pistawka, H. (1967) Mesothelioma and asbestos exposure, *Arch. Envir. Hlth.*, **14,** 559–563.

Lynch, K. M., and Smith, W. A. (1935) Pulmonary asbestosis; carcinoma of the lung in asbestos silicosis, *Am. J. Cancer*, **24,** 56–64.

McCaughey, W. T. E. (1958) Primary tumours of the pleura, *J. Path. Bact.*, **76,** 517.

McCaughey, W. T. E. (1965) Criteria for diagnosis of diffuse mesothelial tumors, *Ann. N.Y. Acad. Sci.* **132,** 603–613.

McCaughey, W. T. E., and Oldham, P. D. (1973) Diffuse mesotheliomas; observer variation in histological diagnosis, in *Biological Effects of Asbestos*, IARC Scientific Publications No. 8, International Agency for Research on Cancer, Lyon, 58–61.

McDonald, A. D., Harper, A., El Attar, O. A., and McDonald, J. C. (1970)

Epidemiology of primary malignant mesothelial tumours in Canada, in *Pneumoconiosis: Proceedings of the International Conference, Johannesburg*, 1969, (Ed. H. A. Shapiro), Oxford University Press, Cape Town, 197–200.

McDonald, J. C. (1977) Asbestos Symposium, Johannesburg.

McDonald, J. C., and McDonald, A. D. (1977) Personal communication.

McManus, J. F. A. (1948) Histological and histochemical uses of periodic acid, *Stain Technol.*, **23,** 99–108.

McNulty, J. C. (1970) Asbestos exposure in Australia, in *Pneumoconiosis: Proceedings of the International Conference, Johannesburg*, 1969, (Ed. H. A. Shapiro), Oxford University Press, Cape Town, 201–203.

McPherson, P., and Davidson, J. K. (1969) Correlation between lung asbestos count at necroscopy and radiological appearances, *Brit. Med. J.*, **i,** 355–357.

McVittie, J. C. (1965) Asbestosis in Great Britain, *Ann. N.Y. Acad. Sci.*, **132,** 128–138.

Mancuso, T. F. (1965) Discussion, *Ann. N.Y. Acad. Sci.*, **132,** 589–594.

Martischnig, K. M., Newell, D. J., Barnsley, W. C., Cowan, W. K., Feinmann, E. L., and Oliver, E. (1977) Unsuspected exposure to asbestos and bronchogenic carcinoma, *Brit. Med. J.*, **i,** 746–749.

Mattson, S., and Ringqvist, T. (1970) Pleural plaques and exposure to asbestos, *Scand. J. Resp. Dis.*, suppl., 75.

Merewether, E. R. A. (1930) The occurrence of pulmonary fibrosis and other pulmonary affections in asbestos workers, *J. Ind. Hyg. Toxicol.*, **12,** 198–222, 239–257.

Merewether, E. R. A. (1949) *Annual Report, Chief Inspector of Factories*, 1947, HMSO, London.

Merewether, E. R. A., and Price, C. W. (1930) *Report on the Effects of Asbestos dust on the Lungs and Dust suppression in the Asbestos Industry*, HMSO, London.

Meurman, L. (1966) Asbestos bodies and pleural plaques in a Finnish series of autopsy cases, *Acta Path. Microbiol. Scand.*, **181,** suppl.

Meurman, L. (1968) Pleural fibrocalcific plaques and asbestos exposure, *Envir. Res.*, 30–46.

Meurman, L. O., Kiviluoto, R., and Hakama, M. (1973) Mortality and morbidity of employees of an anthophyllite asbestos mine in Finland, in *Biological Effects of Asbestos*, IARC Scientific Publications No. 8, International Agency for Research on Cancer, Lyon, 199–202.

Meyer, K., and Chaffee, F. (1940) Hyaluronic acid in the pleural fluid associated with a malignant tumour involving the pleura and peritoneum, *J. Biol. Chem.*, **133,** 83–91.

Miller, L., Smith, W. E., and Berliner, S. W. (1965) Tests for effect of asbestos on benzo(*a*)pyrene carcinogenesis in the respiratory tract, *Ann. N.Y. Acad. Sci.*, **132,** 489–500.

Morgan, A., and Holmes, A. (1970) Neutron activation techniques in investigations of the composition and biological effects of asbestos, in *Pneumoconiosis: Proceedings of the International Conference, Johannesburg*, 1969, (Ed. H. A. Shapiro), Oxford University Press, Cape Town, 52–56.

Morgan, A., Holmes, A., and Gold, C. (1971) Studies of the solubility of constituents of chrysotile asbestos *in vivo* using radioactive tracer techniques, *Envir. Res.*, **4,** 558–570.

Murray, M. (1907) *Departmental Commission on Compensation for Industrial Diseases*, Cmd. 3495, p. 14; Cmd. 3496, pp. 127–128, HMSO, London.

Navratil, M. (1970) Pleural calcification due to asbestos exposure compared with relevant findings in the non-exposed population, in *Inhaled Particles III, Proceedings of the British Occupational Hygiene Society Symposium* (ed. W. H. Walton). London, Old Woking, Unwin, 695–701.

Naylor, B. (1963) The exfoliative cytology of diffuse malignant mesothelioma. *J. Path. Bact.*, **86,** 293–298.

Naylor B. (1968) The role of the cytology laboratory in the diagnosis of diffuse malignant mesothelioma. *Proceedings of International Conference, Dresden.* Biologische Wirkungen des Asbestos, 288–295.

Newhouse, M. L. (1967) The medical Risks of Exposure to Asbestos. *The Practitioner,* **199,** 285.

Newhouse, M. L. (1969) A study of the mortality of workers in an asbestos factory. *Brit. J. Ind. Med.*, **26,** 294–301.

Newhouse, M. L. (1973) Cancer among workers in the asbestos textile industry, in *Biological Effects of Asbestos.* I.A.R.C. scientific publications No. 8. Lyon, 203–208.

Newhouse, M. L., and Berry, G. (1976) *Brit. J. Ind. Med.*, **33,** 147.

Parkes, W. R. (1974) Occupational Lung Disorders, Butterworths, London, 270–357.

Peacock, P. R., and Peacock, A. (1965) Asbestos-induced tumors in white leghorn fowls, *Ann. N.Y. Acad. Sci.*, **132,** 501–503.

Pernis, B., Vigliani, E. C., and Selikoff, I. J. (1965) Rheumatoid factor in serum of individuals exposed to asbestos, *Ann. N.Y. Acad. Sci.*, **132,** 112–120.

Peto, J. (1978) The hygiene standard for chrysotile asbestos, *Lancet*, **i,** 484–489.

Pooley, F. D. (1973) Methods for assessing asbestos fibres and asbestos bodies in tissue by electron microscopy, in *Biological Effects of Asbestos*, IARC Scientific Publications No. 8, International Agency for Research on Cancer, Lyon, 50–53.

Pooley, F. D. (1975) Personal communication.

Pooley, F. D. (1976) See Jones, J. S. P. J., Pooley, F. D. and Smith, P. G. (1976).

Pooley, F. D. Oldham, P. D., Chang-Hyun Um, and Wagner, J. C. (1970) The detection of asbestos in tissues, in *Pneumoconiosis: Proceedings of the International Conference, Johannesburg,* 1969, (Ed. H. A. Shapiro), Oxford University Press, Cape Town, 108–116.

Pylev, L. N., and Shabad, L. M. (1973) Some results of experimental studies in asbestos carcinogenesis, in *Biological Effects of Asbestos*, IARC Scientific Publications No. 8, International Agency for Research on Cancer, Lyon, 99–105.

Rajan, K. T., and Evans, P. H. (1973) Experimental methods—organ culture, in *Biological Effects of Asbestos*, IARC Scientific Publications No. 8, International Agency for Research on Cancer, Lyon, 94–98.

Roberts, H. (1967) Asbestos bodies in lungs at necropsy, *J. Clin. Pathol.*, **20,** 570–573.

Roe, F. J. C., Carter, R. L., Walters, M. A., and Harington, J. S. (1967) The pathological effects of subcutaneous injections of asbestos fibres in mice: migration of fibres to submesothelial tissues and induction of mesotheliomata, *Int. J. Cancer,* **2,** 628–638.

Selikoff, I. J. (1965) The occurrence of pleural calcification among asbestos insulation workers, *Ann. N.Y. Acad. Sci.*, **132,** 351–367.

Selikoff, I. J. (1976) *Rev. Fr. Malad. Resp.*, **4,** suppl. 1, 7.

Selikoff, I. J., Hammond, E. C., and Churg, J. (1968) Asbestos exposure, smoking and neoplasia, *J. Am. Med. Ass.*, **204,** 106–112.

Selikoff, I. J., Hammond, E. C., and Churg, J. (1970) Mortality experiences of asbestos insulation workers, in *Pneumoconiosis: Proceedings of the International*

Conference, Johannesburg, 1969, (Ed. H. A. Shapiro), Oxford University Press, Cape Town, 180–186.

Selikoff, I. J., Hammond, E. C., and Seidman, H. (1973) Cancer risk of insulation workers in the United States, in *Biological Effects of Asbestos,* IARC Scientific Publications No. 8, International Agency for Research on Cancer, Lyon, 209–216.

Simson, F. W., and Strachan, A. S. (1931) Asbestosis bodies in the sputum; a study of specimens from fifty workers in an asbestos mill, *J. Path. Bact.,* **34,** 1.

Skidmore, J. W. (1973) In Symposium on Diseases Associated with Asbestos Dust Exposure, *Int. Pathol.,* pp. 9–17.

Skidmore, J. W., and Jones, J. S. P. (1975) Monitoring an asbestos spray process, *Ann. Occup. Hyg.,* **18,** 151–156.

Smith, P. G. (1973) In Symposium on Diseases Associated with Asbestos Dust Exposure, *Int. Pathol.,* 9–17.

Smith, W. E., Miller, L., Elsasser, R. E., and Hubert, D. D. (1965) Tests for carcinogenicity of asbestos, *Ann. N.Y. Acad. Sci.,* **132,** 456–488.

Smither, W. J. (1965) Secular changes in asbestosis in an asbestos factory, *Ann. N.Y. Acad. Sci.,* **132,** 166–181.

Smither, W. J., and Lewinsohn, H. C. (1973) Asbestosis in textile manufacturing, in *Biological Effects of Asbestos,* IARC Scientific Publications No. 8, International Agency for Research on Cancer, Lyon, 169–174.

Thomson, J. G. (1969) The pathogenesis of pleural plaques, in *Pneumoconiosis: Proceedings of the International Conference, Johannesburg,* 1969, (Ed. H. A. Shapiro), Oxford University Press, Cape Town, 138–141.

Timbrell, V. (1970) The inhalation of fibres, in *Pneumoconiosis: Proceedings of the International Conference, Johannesburg,* 1969, (Ed. H. A. Shapiro), Oxford University Press, Cape Town, 3–9.

Timbrell, V. (1973) Physical factors in aetiological mechanisms, in *Biological Effects of Asbestos,* IARC Scientific Publications No. 8, International Agency for Research on Cancer, Lyon, 295–303.

Timbrell, V., Pooley, F. D., and Wagner, J. C. (1970) Characterics of respirable asbestos fibres, in *Pneumoconiosis: Proceedings of the International Conference, Johannesburg,* 1969, (Ed. H. A. Shapiro), Oxford University Press, Cape Town, 120–125.

Turner Warwick, M., and Parkes, W. R. (1970) Circulating rheumatoid and antinuclear factors in asbestos workers, *Brit. Med. J.,* **iii,** 492–495.

Wagner, J. C. (1965) Discussion, *Ann. N.Y. Acad. Sci.,* **132,** 505.

Wagner, J. C. (1963) Asbestosis in experimental animals, *Brit. J. Ind. Med.,* **20,** 1–12.

Wagner, J. C. (1965) The sequelae of exposure to asbestos dust, *Ann. N.Y. Acad. Sci.,* **132,** 691–695.

Wagner, J. C. (1971) Asbestos cancers, *J. Natn. Cancer Inst.,* **46,** V–IX, 5.

Wagner, J. C., and Berry, G. (1973) Investigations using animals, in *Biological Effects of Asbestos,* IARC Scientific Publications No. 8, International Agency for Research on Cancer, Lyon, 85–88.

Wagner, J. C., Berry, G., Cooke, T. J., Hill, R. J., Pooley, F. D., and Skidmore, J. W. (1977) Animal experiments with talc, *Inhaled Particles,* **IV,** Pt. 2, Pergamon Press, Oxford, 647–658.

Wagner, J. C., Berry, G., Skidmore, J. W., and Timbrell, V. (1974) The effects of the inhalation of asbestos in rats, *Brit. J. Cancer,* **29,** 252–269.

Wagner, J. C., Berry, G., and Timbrell, V. (1970) Mesotheliomas in rats, following intrapleural inoculation of asbestos, in *Pneumoconiosis: Proceedings of the International Conference, Johannesburg*, 1969, (Ed. H. A. Shapiro), Oxford University Press, Cape Town, 216–219.

Wagner, J. C., Gilson, J. C., Berry, G., and Timbrell, V. (1971) Epidemiology of asbestos cancers, *Brit. Med. Bull.*, **27,** 71.

Wagner, J. C., Munday, D. E., and Harington, J. S. (1962) Histochemical demonstration of hyaluronic acid in pleural mesotheliomas, *J. Path. Bact.*, **84,** 73–78.

Wagner, J. C., and Skidmore, J. W. (1965) Asbestos dust deposition and retention in rats, *Ann. N.Y. Acad. Sci.*, **132,** 77–86.

Wagner, J. C., Sleggs, C. A., and Marchand, P. (1960) Diffuse pleural mesothelioma, *Brit. J. Ind. Med.*, **17,** 260–271.

Webster, I. (1965) Mesotheliomatous tumours in South Africa: pathology and experimental pathology, *Ann. N.Y. Acad. Sci.*, **132,** 623–646.

Webster, I. (1970) Asbestos exposure in South Africa, in *Pneumoconiosis: Proceedings of the International Conference, Johannesburg*, 1969, (Ed. H. A. Shapiro), Oxford University Press, Cape Town, 209–212.

Webster, I. (1970) The Pathogenesis of asbestosis, in *Pneumoconiosis: Proceedings of the International Conference, Johannesburg*, 1969, (Ed. H. A. Shapiro), Oxford University Press, Cape Town, 117–119.

Whitwell, F., and Rawcliffe, R. M. (1971) Diffuse malignant pleural mesothelioma and asbestos exposure, *Thorax*, **26,** 6-22.

Whitwell, F., Scott, J., and Grimshaw, M. (1977), *Thorax*, **32,** 377.

Willis, R. A. (1960) *Pathology of Tumours*, 3rd Ed., Butterworths, London, 184.

Winslow, D. J., and Taylor, H. B. (1960) Malignant peritoneal mesotheliomas: a clinico-pathological analysis of 12 cases, *Cancer*, **13,** 127.

Wood, W. B., and Gloyne, S. R. (1930) Pulmonary asbestosis, *Lancet*, **ii,** 445–448.

Wood, W. B., and Gloyne, S. R. (1934) pulmonary asbestosis—a review of one hundred cases, *Lancet*, **ii,** 1383–1385.

Wyers, H. (1949) Asbestosis, *Postgrad. Med. J.*, **25,** 631–638.

Zielhuis, R. L. (1977) *Public Health Risks of Exposure to Asbestos*, Commission for the European Communities, Pergamon Press, Oxford.

(13) BIBLIOGRAPHY

Bogovski, P., Gilson, J. C., Timbrell, V., and Wagner, J. C. (Eds.), *Biological Effects of Asbestos*, IARC Scientific Publication No. 8, International Agency for Research on Cancer, Lyon, 1973.

Parkes, W. R., *Occupational Lung Disorder*, Butterworths, London, 1974.

Shapiro, H. A. (Ed.), *Pneumoconiosis: Proceedings of the international Conference, Johannesburg*, 1969, Oxford University Press, Cape Town, 1970.

Whipple, H. E. (Ed.), *Biological Effects of Asbestos, Ann. N.Y. Acad. Sci.*, 1965, **132,** 1–766.

CHAPTER 14

Asbestos-related diseases Part 3: Asbestos associated thoracic disorder: Clinical features

Leslie H. Capel

London Chest Hospital, E2 9JX

There are no clinical features specific for thoracic disorder associated with inhalation of asbestos dust: the diagnosis of asbestos dust-associated lung and pleural disease is by inference. The inference is based on the clinical and industrial history, the findings on skilled physical examination of the chest and other parts, especially the fingers, the appearance of the chest radiograph, tests of lung function, examination by microscope of sputum and sometimes pieces of the lung and its coverings surgically removed via an incision in the chest or by aspirating needle, and in the future by examination of blood serum.

According to the pattern of findings, so the probability that any disorder found is wholly or partly due to asbestos dust inhalation or not due to such inhalation can be assessed. It will be convenient first to describe the pieces which go to make the pattern, and then attempt to put them together in the shape accepted as pulmonary asbestosis: pulmonary fibrosis caused by inhalation of asbestos dust.

(1) THE HISTORY

Asbestos-associated lung and pleural disease may be considered as a diagnosis because of abnormality discovered on a routine radiological examination of symptom-free persons, of those with symptoms, and of those known to have been at risk. The enquiry into the medical history of the subject should include an account of work and environment. Above all, life-long smoking habits must be recorded.

Symptoms are the patient's account of his or her experience of disability. He or she may be symptom-free, however, despite radiological change and abnormality on physical examination. If present, symptoms may not necessarily indicate that disorder results from known asbestos exposure; for example, there may be cigarette bronchitis in an asbestos dust exposed person. If symptoms are present and are believed to be due to asbestos dust inhalation, then the severity of the symptoms is not necessarily proportionately related to assessments of the magnitude of radiological and functional change.

The symptoms of asbestos-associated lung disorder, when present, include breathlessness on effort, cough with little expectoration and pain in the chest. Any chest pain may be due to coughing; rarely is it due to the development of lung cancer or the rare and lethal mesothelioma of the pleura. If pain is due to mesothelioma, then there will be radiological evidence of tumour. Occasionally pain is due to pneumothorax, escape of air from the lung into the pleural space, an occasional complication of asbestos-lung disease.

Assessment of the relative contribution to the severity of symptoms of any asbestos-induced lung disorder on the one hand and that of any other lung disorder such as bronchitis, emphysema, and lung fibrosis on the other which might not be asbestos-induced can be very difficult. The cigarette smoker who may on that account suffer cigarette smoker's bronchitis or emphysema presents a special problem. In passing it must be mentioned that some who have studied the subject claim that lung fibrosis and lung cancer are much more likely to develop in exposed cigarette smokers than in exposed non-smokers.

(2) THE PHYSICAL SIGNS

The relevant physical signs are those discovered from inspection of the patient, from percussion and auscultation of the chest, and from examination of the hands.

2(i) Inspection

In advanced thoracic disorder of any cause the breathing may be seen to be laboured at rest, but serious thoracic disorder can be present without this.

2(ii) Clubbing

In certain forms of serious lung and heart disorder clubbing of the fingers develops. This is a thickening mainly in the region of the nail beds. These serious disorders include lung cancer, congenital heart disease, and two conditions with very similar structrual lung changes: fibrosing alveolitis, also called diffuse pulmonary fibrosis of unknown cause, and pulmonary asbestosis. In fibrosis of the lung the alveoli are in part destroyed and replaced by scar tissue. Pulmonary asbestosis is advanced if clubbing of the fingers is a feature.

2(iii) Percussion

Percussion of the thorax with the fingers tests whether its resonance is normal or impaired. If impaired, then thickening of the lining of the lungs (the pluera) or of the lung substance (as in pneumonia), or effusion of liquid between the chest wall and the lungs ('wet pleurisy') is inferred to be present. Thickening of the pleura in asbestos-associated thoracic disorder can be advanced enough to be obvious by percussion. One fortunately rare form of this thickening is the pleural mesothelioma.

2(iv) Auscultation

Auscultation of the thorax is the most important part of the physical examination when asbestosis of the lungs is in question. It can help in two ways. It can detect impairment of the natural free movement of the air in the airways of the lungs, and it can detect impairment of transmission of breath sounds when pleural thickening interrupts their transmission to the stethoscope from the large airways where they are generated. Impairment of free movement of air in the smaller airways in the lower parts of the lung is associated with the generation of soft crackling sounds. This can occur in a number of quite different disorders. These include heart failure and pulmonary fibrosis from all causes, including asbestosis. All are conditions with this in common: they can cause partial deflation of lung in dependent regions, so that at the end of expiration smaller airways in those dependent parts of the lungs are closed. Towards the end of inspiration sufficient retractive forces developed around these closed airways to pull them open: air rushes in and a charactaristic soft crackling noise is generated. These crackles may be heard as showers of sound towards the end of inspiration. They persist after coughing and repeat from breath to breath. In the development of those lung disorders with which they are associated crackles are heard first in the lower third of the chest posterior (along the posterior axillary line). If carefully listened for at the end of a long slow inspiration after a full expiration it is possible that their presence is the only evidence of involvment of the lungs in asbestosis dust-induced lung disorder.

(3) THE CHEST RADIOGRAPH

Examination of the chest radiograph can give information on gross changes in the airways and blood vessels serving the alveoli of the lungs, on changes in the alveoli, and on changes in the pleura covering the lungs. Deciding whether early changes are present, and whether any such changes are likely to have arisen from asbestos dust inhalation is not straightforward, and even the most experienced observers would agree that their conclusions from their observations must often be tentative.

All the radiological changes to be described may be found in those with no known exposure to asbestos dust.

Inhalation of asbestos dust can cause structural and eventually radiological change. The structural changes are due mainly to fibrosis, a response to damage by the formation of scar tissue replacing the functioning tissue and by its shrinking and pulling on surrounding lung tissue, so stiffening the lungs. The pleural coverings of the lungs also are characteristically involved, and there can be an effusion of liquid between the pleural coverings lining the lungs and the inner surface of the chest wall.

Calcification of the pleural lesion also is characteristic. Distinguishing radiological shadows due to lung fibrosis from those due to pleural thickening, and detecting and assessing minor (but none the less important) changes due to lung fibrosis when these may be masked or partly masked by pleural changes can be difficult.

3(i) Lung fibrosis

Radiological changes due to lung fibrosis are usually symmetrical and nearly always in the lower zones of the lung fields. The earliest abnormalities include an increase in the number and an extension further into the periphery of those linear shadows cast by blood vessels. Further development is to a net-like appearance including more and more of the lung volume with shrinking of the lung. Shrinking in the lower zones of the radiograph may be associated with rarefaction of the lung substance and consequent increase in transradiency in the upper zones. Occasionally the outline of the heart shadow is shaggy. The nature and severity of all these changes can be classified and recorded according to an international standard.

3(ii) Pleural fibrosis and plaques

Pleural fibrosis causes pleural thickening, and this throws radiological shadows. This thickening may be localized in plaques or more extensive.

The shadowing of pleural fibrosis may be just detectable from careful inspection and from angled radiological views and it may be extensive obscuring the lung fields. The characteristic calcification, when present, is usually easily seen.

3(iii) Lung cancer

The characteristic localized dense lung shadows of lung cancer, usually in the lower lobes, are sadly not uncommon where there has been prolonged exposure both to asbestos dust and cigarette smoke.

3(iv) Assessment of radiological changes

Decisions on the presence of any radiological change and, if present, assessment of the nature and magnitude of the change and, in the light of clinical and functional information, judgement on whether lung changes and pleural change are due to asbestos dust inhalation is a matter for disinterested expert opinion. Since some radiological appearances are shared both by asbestosis and other disorders, for example fibrosing alveolitis (diffuse pulmonary fibrosis), such expert opinion may finally be uncertain.

(4) LUNG FUNCTION

Assessment of lung function is concerned with the mechanical efficiency with which the lungs can breath air in and out, and with the efficiency with which they can transfer oxygen to the blood. The patterns of functional abnormality on testing consistent with lung damage from asbestos dust inhalation are found also in other conditions.

The simplest index of the mechanical efficiency of the lungs is the Vital Capacity. This is the largest breath which can be expired after a full inspiration. It is perhaps the most useful index of lung damage. The volume of the Vital Capacity is related to age, sex, and stature. If the Vital Capacity is smaller than expected or is falling from year to year more than would be expected in health, then lung damage is likely to be present. Any such damage to lung function can in whole or part be due to cigarette smoking and to causes other than asbestos dust inhalation. Simple and convenient as it is, the usefulness of measurement of the Vital Capacity is limited by uncertainty on what the Vital Capacity of a particular individual should be in health. The figures available apply to groups. It is generally accepted that the normal range is 20% above and below the predicted mean value for an individual of a given age, sex, stature, and race. These measurements may be obtained from tables and graphs: from these graphs the expected rate of decline in the Vital Capacity with the passing of the years may be obtained.

A fit young man of average stature will have a Vital Capacity of 5000 ml (a woman about 500 ml less). He will lose about 50 ml per year as he ages: more if he smokes cigarettes. If an individual's Vital Capacity is less than say, 20% of its predicted value or falling faster than, say, 20% more than the predicted amount of annual fall, then some disorder of lung function may be expected. The larger the deficit, the more likely is it that it indicates lung abnormality. It will at once be appreciated that measurements in the 'normal' range may nevertheless go with lung abnormality: an individual with values initially 10 or 20% above normal will fall far before entering the 'abnormally' low range.

Pulmonary asbestosis develops without that major increase in lung airway resistance to airflow which is characteristic of bronchitis, asthma, and emphysema. With such major increase in airflow resistance the speed at which the lungs can be emptied of air is reduced. In health about 75 to 80% of the Vital Capacity can be expelled in one second. Airway narrowing reduces this ratio. Though some airway narrowing can occur with pulmonary asbestosis the ratio is not characteristically reduced in pulmonary asbestosis: the ratio can be in the low normal range and even abnormally high where lung stiffening supports the airways abnormally well during expiration. Where pulmonary asbestosis and chronic bronchitis and emphysema (characteristically in cigarette smokers) are found together, then the ratios are reduced because of the bronchitis and emphysema.

The Vital Capacity is customarily abbreviated as VC, the Forced Expiratory Volume in one second as the FEV, and the ratio as FEV%VC.

Complementing assessment of the efficiency of the ventilatory function of the lungs from the recording of the VC, FEV, and FEV%VC is assessment of the efficiency of transfer of gasses from the lungs to the blood. The Transfer Factor for carbon monoxide is an index of this. The Transfer Factor is derived from an estimate of the volume of the gas exhanging region of the lungs (called the alveolar volume) and an estimate of the rate at which carbon monoxide (at unit partial pressure) is taken up by each unit of this alveolar volume (called the Transfer Coefficient). The alveolar volume multiplied by the transfer coefficient is the Transfer Factor.

Carbon monoxide in small dilution is used as the index gas because it is so avidly taken up by the red cells of the blood that there is no back pressure from the blood in the alveolar capillaries. Hence the alveolar carbon monoxide partial pressure (which can be estimated) is the effective pressure driving the gas from alveoli to capillaries.

In general changes in the Transfer Factor tend to parallel changes in the Vital Capacity. The Transfer Factor tends also very approximately to reflect the clinical and radiological condition of the patient. Its predicted values for age, sex, and stature have the same range of uncertainty as those for the Vital Capacity. It is a convenient objective index of the functional state of the lungs, and can clearly indicate loss of lung function proportionately as its value may be found to be reduced: when this value is clearly reduced, say to 60% of predicted or less, it is clear that lung function is impaired. Loss of functional capacity may, of course, have occurred even if values are in the predicted range for a particular individual.

Whether loss of lung function arises from lung fibrosis (due to asbestosis or otherwise) or to some other structural lesion, such as emphysema, cannot be decided from functional studies alone. An inference can be made only after taking all clinical, radiological, and experimental information into account.

Other methods of experimental investigation include estimation of lung compliance (for stiffening of the lungs) and estimation of total lung capacity. Emphysema makes the lung less stiff. Most other lung abnormalities reduce total lung capacity and stiffen the lungs. Estimation of lung volume may underline findings from spirometry and gas transfer. Estimation of stiffness is likely to be academic. Further, its values tend to reflect the Vital Capacity (except in emphysema).

Measurement of blood gas tensions for oxygen and carbon monoxide during rest, and during exercise if the magnitude of the exercise load is taken into account can (provided heart and blood function are normal, and the subject is breathing pure air near sea level) indicate lung and ventilatory efficiency in satisfying the metabolic needs of the body. If pulmonary asbestosis is advanced and no other disorder is present, then blood carbon dioxide partial pressure will tend to be low because the ventilation of the stiffer lungs of asbestosis tends to be marginally increased, and the load of carbon dioxide in the blood and tissues of the body is hence marginally reduced. This increase in lung ventilation eventually fails to maintain the oxygen partial pressure of the blood: this then drops below the normal range. If arterial blood oxygen is reduced, then the capacity of the body for exercise and stress is reduced. A relatively small reduction in arterial blood oxygen saturation means a relatively large loss in oxygen carrying reserves of the blood. If arterial blood oxygen is reduced at rest this is clinically very important: it indicates serious impairment. If detected only on excercise, then the importance depends on the magnitude of the exercise load generating the reduction.

4(i) Assessment of the individual

Suspicion that an individual has inhaled asbestos dust in the distant past, that such inhalation has caused structural damage in the lungs, and that such damage is sufficient to impair lung function so that the enjoyment and expectation of life is curtailed tends towards certainty only as some or all of the described features are present:

1. Exposure to asbestos dust, prolonged and more than (say) 10 years distant.

2. Certain radiological changes, commonly those indicating fibrosis in the lower lung fields, usually with some changes in the pleura.

3. Breathlessness on more severe effort.

4. Crackling sounds heard with the stethoscope at the lung bases towards the end of inspiration.

5. Reduction in the Vital Capacity, Total Lung Capacity, and the transfer Factor for carbon monoxide. Slight reduction in both arterial blood oxygen and arterial blood carbon dioxide.

6. Asbestos bodies in the sputum. (When present these are suggestive but *not* conclusive: they indicate only that exposure has occurred.)

7. Pathological changes of fibrosis with asbestos bodies seen on special microscopic examination of lung removed by trephine (needle) biopsy or through surgical incisions. (See page 000)

Such features as severe breathlessness, cyanosis (blue hue to the skin) indicating very low arterial blood oxygen and clubbing of the fingers are fortunately rare late manifestations of severe disease. Commonly a diagnosis will be made without these features, and without some of the seven features listed. Indeed, a diagnosis may be made when only the first three are clearly present.

The diagnosis of asbestosis is required to help manage the individual, to secure compensation either by civil action or, in the United Kingdom, through the Pneumoconeosis Panel or both, and as a monitor of the long term success of the management of the work environment. The diagnosis of pulmonary asbestosis and the assessment and apportioning of disability is a clinical opinion. In the long run the interests of the exposed individual and his or her workmates might be best served if this opinion is unbiased by any natural resentment that the individual has suffered avoidably in doing the world's work. Where prevention fails, early detection of the disorder and rapid retraining and return to paid work may perhaps lead to a better quality to life than absence from paid work and the worries and preoccupations of prolonged legal dispute.

CHAPTER 15

Asbestos-related diseases Part 4: Epidemiology of asbestos-related disease

Muriel Newhouse

London School of Tropical Medicine and Hygiene

(1) INTRODUCTION

Epidemiology has been defined as the study of the personal and environmental factors that determine the incidence and distribution of disease, that is, the factors which influence the probability of a given person contracting a particular condition. Originally the techniques were used to study the causes of epidemics of infectious diseases, but in recent years they have been expanded and applied to many chronic diseaseas such as coronary heart disease, hypertension, and chronic bronchitis, and have frequently been used to define particular industrial risks.

In principle, in an industrial setting if a health risk is suspected, the experience of the particular group is compared with that of a different group where the risk or hazard is not present. However, if the condition is very rare in the general population, such as the mesothelial tumour, a number of cases occurring in one particular group is in itself significant, but if it is a tumour commonly found in the general population, such as cancer of the lung, special techniques are required to define an excess risk in the particular group under examination.

The objectives of epidemiological investigations are to detect any abnormal degree of disease, to define the type of disease, to measure how severely the group under investigation is affected, and to determine what factors, such as age, sex, and degree of exposure, influence the amount and severity of the condition under investigation, and secondly by long-term or repeated studies to determine the effectiveness of any preventive or regulatory measures.

The investigation of asbestos-related diseases has particular problems as the results of exposure are not immediately manifest. In this respect, both the pneumoconiosis asbestosis, which is an interstial fibrosis of the lungs, leading to chronic respiratory disability, and the asbestos-related tumours, mesothelioma, cancer of the bronchus, and possibly gastro-intestinal and other tumours behave similarly, there is a long interval known as the latent period between first exposure and development of the disease. Workers may thus become affected after many years in employment or indeed years after leaving the job where exposure occurred. Therefore, although actual surveys of workers in industry have been made, the most commonly used technique is to examine the causes of death of groups of asbestos workers and determine whether the causes and number of deaths differ from what might be expected in the general population.

Further complexities are that there are four different types of asbestos (chrysotile, crocidolite, amosite, and anthophyllite), all of which are in commercial use, and as inhalation of asbestos dust may occur at any stage between the mining of asbestos and the handling of the finished asbestos products, the methods of exposure are very numerous.

(2) TYPES OF ASBESTOS EXPOSURE

Chrysotile asbestos is chiefly mined by open-cast mining in the Quebec Province of Canada and in the U.S.S.R. There are also chrysotile mines in Southern Africa, Northern Italy, and Cyprus. Crocidolite is mined only in South Africa, with deep mines in the Cape Province and a further group in the Transvaal. There was an important mine in Perth, Australia, but this has now ceased production. Amosite is mined only in South Africa. The chief source of anthophyllite asbestos was Finland but the mine, which had a small production, is now closed.

After the ore is extracted from the mine, it is crushed and milled and the fibre is extracted and bagged for export. Milling of the ore may be a dusty process and bagging of the fibre before transport is a difficult process to control. Fibre from the mines used to be exported in hessian bags which emitted clouds of dust; polythene bags are now most commonly used but transport operatives such as railway workers and dockers could suffer considerable exposure on the journey from the mine to the factory.

Exposure in factories is an obvious industrial hazard. The first factories, dating from the 1880s, were for the manufacture of asbestos textiles. The process was essentially similar to others in textile manufacturing, the dustiest jobs being in opening and disintegrating the fibre and in carding, but the original factories were small and the atmosphere of the workshops was laden with asbestos fibre; contamination of the vicinity of the factory from deliveries and emissions from any exhaust ventilation installed was inevitable. Other asbestos products were soon developed. Asbestos insulation products and asbestos–cement became important products with their own factories, and asbestos boarding and asbestos tiles of various types were soon developed. Asbestos is also an essential component of brake linings and clutches and there are large factories manufacturing friction materials in all countries with automobile industries.

Outside the factory there are important groups of workers who handle asbestos materials. Probably the largest group are the laggers and workers in the building industry, and men employed in shipbuilding and repairing. Laggers or insulators are particularly at risk, often working in confined spaces where old lagging must be torn off before new is applied. Asbestos spraying was also a common method of applying insulation, where not only the sprayer but also other workers in the vicinity were exposed. Exposure of workers such as electricians and carpenters, not recognized as asbestos workers, frequently occurs in shipbuilding where various craftsmen are working together in the hold of a ship (Harries, 1968). Leisure-time activities must also be considered: 'do-it-yourself' jobs involving sawing or cutting of asbestos boarding or roofing is common; recently an amateur potter revealed that he insulated his kiln with an asbestos material, renewing it every 4 or 5 weeks.

In the past few years, there has been increasing concern about environmental contamination arising from the widespread manufacture and use of asbestos products and their inevitable deterioration. Considerable efforts have been made to determine whether contamination of the atmosphere or of water supplies affects the community at large.

It can be seen that with such a variety of methods of contact with asbestos fibres, epidemiological investigations to provide definite information about the severity of exposure and the type of asbestos used are difficult to design, and this information is usually necessary in order to produce data on which control measures can be based.

(3) STATUTORY RECOGNITION OF ASBESTOS

Within a few years of asbestos factories opening in the U.K., it was recognized that respiratory complaints had become unduly common among the workers. During the first 30 years of the 20th Century, tuberculosis was extremely common among all social classes, and the relationship between the specific disease, asbestos, and tuberculosis was not entirely clear.

In the 1920s, the Factory Department of the Home Office, which was responsible for health and safety in factories, determined to review the situation in the asbestos industry. The work was undertaken by Dr. E. R. A. Merewether, a Medical Inspector of Factories, and Mr. C. W. Price, a Factory Inspector. There were then about 2000 men and women employed in the asbestos industry, of whom 363 were selected for examination; the results of this survey are summarized in Table 15.1.

The diagnosis of pulmonary fibrosis was made by interpretation of X-rays of the chest. The results clearly show that approximately a third of the workers were affected and the proportion of those affected increased with length of service in the factory. This early piece of epidemiology was published as a Command paper in 1930 (Merewether and Price, 1930) and not only covered the health aspects of asbestos exposure as far as was realized at the time but also gave a careful account of the various jobs and processes then in use. Analysis of the prevalence of fibrosis according to the job of the worker was also made, and it was noted that pulmonary fibrosis was relatively uncommon among spinners. This report formed the basis of the Asbestos Regulations of 1931. These Regulations were in two parts: firstly to control the manufacturing processes in order to lessen the production of dust and secondly to provide for medical surveillance of all workers in the so-called scheduled jobs and for compensation for workers who had developed asbestosis. It is important to realize that the Regulations were designed for the industry as it was in the 1920s and, although it expanded rapidly and its products diversified and the usage of asbestos or asbestos products increased in many trades and occupations, the Regulations did not apply to any workers who were not itemized in these regulations. Although

Table 15.1 Incidence of fibrosis relative to length of employment (Merewether and Price, 1930)

Years employed	Number examined	Cases of fibrosis	
		Number	Group incidence (%)
0–4	89	0	—
5–9	141	36	25.5
10–14	84	27	32.1
15–19	28	15	53.6
20 and over	21	17	80.9
TOTALS	363	95	26.2

the Reports of the Medical Inspector of Factories listed the number of new cases of asbestosis occurring each year and there was a growing awareness, both in the U.S.A. and the U.K. that cancer of the lung was unduly common among workers suffering from asbestosis, there was little epidemiological investigation of any group of workers until the 1950s. Renewed interest into the asbestos-related diseases was aroused by the investigations of Doll (1955), who measured the great excess of cancer of the lung in asbestos textile workers, and by the survey of Wagner *et al.* (1960) in South Africa, where the occurrence of mesothelial tumours of the pleura and peritoneum was found to be associated with exposure to crocidolite asbestos.

In 1964, the first International Conference on the Biological Effects of Asbestos was held in New York. A comprehensive account of recent and current research into the subject was presented (Whipple, 1965).

(4) ASBESTOSIS

Any worker, whether he is employed in a mine, mill, or factory, or uses asbestos products, is at risk of contracting asbestosis if he is exposed to fibres fine enough to penetrate to the alveoli of the lung. When deposited in the lung parenchyma, asbestos fibres are retained almost indefinitely. Men and women are equally susceptible and age at first exposure appears to have little influence. There is no evidence to suggest that any one of the different types of asbestos is more fibrogenic than the other. The determining factor in development of the disease is the amount of respirable asbestos fibre inhaled. The onset of detectable asbestosis in life is therefore related both to the dustiness of the atmosphere and the duration of exposure.

Since the implementation of the Asbestos Regulations, all workers with certified asbestosis who have received compensation are medically examined each year and, after death, a coroner's inquest and autopsy are required. Records are kept of these examinations. Also, since the early recognition of the health hazards of the industry, the larger companies have maintained in-plant medical services, in an effort to protect the workers and detect early signs of disease. A large amount of information is therefore available.

Smither (1965) reviewed the medical history of the large textile factory where he was the physician in charge. The natural history of asbestosis had changed. Up to 1950 many workers suffering from asbestosis died prematurely from pulmonary tuberculosis with an average age of 40.8 years. In the subsequent 15 years, the average age at death improved to 56.5 years. The number of cases of pulmonary tuberculosis, as in the general population, decreased, but cancer of lung became an increasingly common cause of death. It was also notable that in the later period the average length of employment in the factory before diagnosis of asbestosis increased from about 7 years in the 1930s to over 17 years in cases diagnosed between

1960 and 1964, suggesting considerable improvement in the factory environment.

Before turning to the epidemiology of the asbestos-related cancers, the epidemiology based on the distribution of asbestos bodies and asbestos fibres in the lungs in the population generally and of the detection of asbestos plaques of pleuras will be described.

(5) ASBESTOS BODIES AND ASBESTOS FIBRES IN THE LUNGS

In their 1930 report, Merewether and Price described asbestos bodies as 'peculiar bodies, yellowish brown in colour and elongated or beaded in form often with bulbous ends'. They are visible under the light microscope. These bodies had been noted as early as 1900 both in lung tissues at autopsy and in the sputum of asbestos workers. They aroused considerable interest and were first known as 'asbestosis' bodies, but it was gradually realized that these bodies were not invariable associated with asbestosis but were present in nearly all workers exposed to asbestos, and they became regarded as markers of exposure not disease. In the early 1960s Thompson (Thompson *et al.*, 1963), a pathologist in South Africa, investigated the incidence of asbestos exposure in the population of Cape Town while a parallel investigation was undertaken in Miami in the U.S.A. Smears from the lungs of all patients coming to autopsy at the principal hospitals in these towns were examined. Asbestos bodies were found in 26% of the specimens, with similar results in both cities. This type of investigation was repeated in the U.K., the U.S.A., and Europe. The techniques used for preparing the specimens and counting the bodies varies, and the results of the different investigations are not strictly comparable, but Becklake (1976) has listed the results of all of the reported surveys. The prevalence of asbestos bodies in the lungs varies between about 20 and 60%. Generally, higher counts are found in men than in women, and in urban rather than rural populations. The highest counts are found in asbestos workers. In New York City, asbestos bodies were found in 60% of the lungs examined. Investigations in London showed asbestos bodies in none of the lungs examined in 1936 but in 20% of the lungs in 1966.

In 1975, a detailed investigation was reported from the London Hospital (Doniach, *et al.*, 1975) where the prevalence of these bodies was considered in relation to job, residence, and disease (patients with asbestosis, or known as asbestos factory workers, were excluded). The highest counts (up to 61%) were found in heavy manual workers in the shipping, electrical, and engineering industries. Those living in the industrial and dockland areas of East London had higher counts than those living in less industrial areas to the North East of the hospital. Compared with patients with asbestosis, the number of asbestos bodies per cubic millimetre was low, but a higher than expected number of cases of carcinoma of the stomach and of the breast had asbestos bodies in the lung tissue, but there was no excess of positive cases in patients dying of carcinoma of the lung.

More modern methods of microscopy will reveal uncoated asbestos fibres and fibrils in lung tissue, and fibre type can now also be identified with certainty. Investigations in New York have identified asbestos fibrils by means of electron microscopy in nearly all lung tissue examined.

The investigations considered above are positive evidence of environmental contamination. The sources of the asbestos are probably building operations, demolition work, and heavy users of asbestos such as asbestos factories and shipyards, and also possibly the dust from wear of brake and clutch linings.

(6) PLEURAL PLAQUES

The term pleural plaque is reserved for a localized thickening of the parietal pleura. The plaques are often multiple and bilateral, and calcification is common. These plaques occur in healthy people and are easily visible on x-ray. They have been noted when occupational groups are surveyed, but also in non-occupational groups. Kiviluoto (1960) drew attention to their common occurrence among people resident near two anthophyllite mines in Finland, and they were also found in a rural population in Bulgaria who were growers of tobacco, the soil in the region being found to contain anthophyllite asbestos (Zolov *et al.*, 1967). Although it has been shown that lung function may be slightly diminished and there is evidence to suggest that carcinoma of the lung will develop more commonly among asbestos workers with plaques than among workers with no x-ray changes, they may, like the asbestos body, be regarded as a marker of exposure rather than as evidence of disease, and their common occurrence in non-occupational groups is also evidence of environmental exposure.

Asbestos fibres have been found in water either from contamination by discharge from mining operations or from natural asbestos in ore in the area. Asbestos fibres have also been demonstrated in beverages such as beer, presumably from the asbestos filters used in clarification of the drink, but there is no convincing evidence of increased risk to populations due to ingestion of fibres.

(7) ASBESTOS-RELATED TUMOURS

The tumours which have been shown to be related to exposure to asbestos dust are mesotheliomas of the pleura and peritoneum, which is a very rare form of tumour except among asbestos workers, cancer of the lung, and possibly gastro-intestinal tumours and cancer of the larynx, all of which occur commonly in the general population. Special epidemiological methods are required to show that these later groups of tumours are associated with or casually related to asbestos exposure. The method most commonly employed is the study of the causes of death or mortality in groups of workers. The mortality in the group under investigation is compared with that in an unexposed group. Often comparison is made with the mortality of the general population. In the U.K. the national statistics for England and Wales prepared by the Registrar General are usually used and in the U.S.A. the data provided by the U.S. Branch for Vital Statistics. Death rates vary with age and sex and the period of death. For example, the death rate for cancer of lung has been rising steeply for the past 45 years, but the death rate for cancer of stomach has been declining. Therefore, for purposes of comparison, the figures for deaths for selected causes in the standard population are adjusted by means of a computer program to relate to the same age, sex, and period of death as the subjects in the study. The derived figure is referred to as the 'expected' number of deaths.

A convenient method of estimating comparative risks in different populations is to calculate the standard mortality ratio (SMR):

$$\text{SMR} = \frac{\text{Observed number of deaths} \times 100}{\text{Expected number of deaths}}$$

Alternatively, the relative risk may be calculated, which is the ratio of observed to expected deaths. An SMR of 100 or a relative risk of 1 indicates no excess risk in the group being studied. Appropriate statistical tests are also used to test whether an excess number of deaths might have been caused by chance. A probability of more than 20:1 is accepted by convention as excluding a chance finding.

(8) CANCER OF THE LUNG

The first systematic study of the mortality in groups of factory workers was undertaken by Doll in the early 1950s (Doll, 1955). The mortality of workers in a North Country textile factory was examined in relation to their jobs and the period when they were employed (Table 15.2). The group particularly affected were heavily exposed workers who had had 20 years or more employment before the implementation of the Asbestos Regulations 1931. Among these workers, the risk of lung cancer was 10 times greater than in the general population; asbestosis was recorded as an underlying cause of death in all cases of lung cancer and was the main cause of death in a further 14 workers. The study continued and a second report was presented in 1968 (Knox *et al.*, 1968). This was cautiously optimistic in tone as it was clear that mortality from lung cancer was much reduced in men who had worked only in the post-Regulation period, but a further report in

Table 15.2 Causes of death among male asbestos workers compared with mortality experience of all men in England and Wales (Doll, 1955)

Cause of death	No. of deaths: No. observed	No. of deaths: Expected on England and Wales rates	Test of significance of difference between observed and expected (value of *P*)
Lung cancer:			
with mention of asbestosis	11	—	<0.000 001
without mention of asbestosis	0	0.8	
Other respiratory diseases (including pulmonary tuberculosis) and cardiovascular diseases:			
with mention of asbestosis	14	—	<0.001
without mention of asbestosis	6	7.6	
Neoplasms other than lung cancer	4	2.3	>0.1
All other diseases	4	4.7	
ALL CAUSES	39	15.4	<0.000 001

1977 (Peto *et al.*, 1977) showed that the risk of lung cancer was still 1.5 times greater than in the general population.

Mortality studies have been carried out in many different groups of workers. Selikoff *et al.* (1973) in New York studied insulators and then enlarged his study to include all insulators in the United States (1973). McDonald *et al.* (1974) studied the chrysotile miners and millers of Quebec Province and Newhouse (1969) the workers of another textile factory in East London. The relative risk of lung cancer in these and other studies varied between 1.5 for chrysotile miners to 5.6 or higher in groups exposed to amphibole asbestos. Such studies have increased value if information on the degree of exposure is available. Generally reliable data concerning the actual level of asbestos in air are not found before the late 1960s, but in Canada sufficient information was obtained to attach a 'dust index' to each individual. Newhouse classified the jobs in the factory she studied according to the severity of exposure to dust. Others have relied on length of exposure. In most studies there is evidence of a dose–response relationship, lung cancers being markedly more common in heavily exposed groups.

However, in 1968 important new information on the inter-relationship between asbestos exposure and cigarette smoking was published by Selikoff *et al.* (1968) (Table 15.3). They showed that non-smoking asbestos workers in the group studied had not died from lung cancer, but that the risk for asbestos workers who smoked was 90 times greater than for non-smoking workers not exposed to asbestos. The relationship between smoking and

Table 15.3 Mortality of asbestos workers by smoking habit (Selikoff *et al.*, 1968)

Smoking habits	No. of men	Observed deaths	Expected deaths
Never smoked regularly	40	0	0.05
History of pipe or cigar smoking only	39	0	0.13
History of regular cigarette smoking	283	24	3.16

asbestos exposure was further studied in London (Berry *et al.*, 1972). The smoking habits of over 1300 male and 480 female workers were obtained by postal questionnaires and visiting, and for the deceased from hospital and general practioner records. The mortality of the group was studied over a 10-year period.

In Table 15.4 the observed number of deaths in smoking and non-smoking men and women are contrasted with the numbers to be expected if the two carcinogenic factors acted firstly in an additive and secondly in a multiplicative or synergistic fashion. The figures, particularly for the females, favour the multiplicative hypothesis. Further evidence from both animal experiments and other mortality studies has now confirmed the synergistic effect of asbestos and cigarette smoke (Saracci, 1977). This finding has important implications in planning the control of this particular hazard. Health education and anti-smoking campaigns are of particular importance in the asbestos industry.

(9) MESOTHELIAL TUMOURS

Although in groups exposed to asbestos deaths from lung cancer are three or four times more common than deaths from mesothelial tumours, these tumours pose a serious risk to the health of the asbestos worker, and since the early 1960s have been the subject of active research. Although it is possible to recognize case reports of mesothelial tumours earlier, it was not until the mid-1950s that it was recognized by all pathologists as a definite pathological entity. The tumour spreads over the pleura or peritoneum or may affect both membrane. Microscopically it may resemble a secondary spread of cancer, the primary tumour being in other organs. Histologically it is pleomorphic, and again without special experience difficult to diagnose, and the diagnosis should not be accepted unless an autopsy has confirmed that no other tumour is present. Attention was first drawn to the association of this tumour with asbestos exposure by Wagner *et al.* (1960) in South Africa. They obtained the occupational and residential histories of 33

Table 15.4 Distribution of lung cancers in asbestos workers by smoking habit

Smoking habits	Observed deaths	Expected non-occupational deaths	Expected occupational and non-occupational deaths: Additive model	Multiplicative model
Males:				
Smokers	32	12.0	31.4	33.1
Non-smokers	0	0.0	1.2	0.1
Females:				
Smokers	18	1.9	14.0	17.8
Non-smokers	2	0.2	5.4	2.2

affected patients, and found that all but one had either worked in the crocidolite asbestos mines in the Cape Province or had lived near the mines or handled asbestos in transport or other ways. Both men and women were affected. All of these patients died of pleural tumours, although later peritoneal tumours were detected among workers in these mine fields.

In 1964, a case control study was made of patients dying of these tumours at the London Hospital (Newhouse and Thompson, 1965). At this hospital a diagnostic index of all autopsies had been maintained since 1916. All cases with the diagnosis of mesothelioma were reviewed and the diagnosis was confirmed in 83, consisting of 41 men and 42 women; 56 of the tumours were pleural and 27 peritoneal in origin. The occupational and residential histories of these patients were then obtained by visiting a surviving spouse or other relative, or from ward or other medical records. Forty of these patients (52.6%) had a positive history of asbestos exposure (Table 15.5); 25 of them had worked in asbestos factories, 19 in an East End factory within easy distance of the London Hospital. Nine had had relatives who were asbestos factory workers or dockers handling asbestos, and, of extreme interest, 11 (Table 15.6) had neither worked with asbestos nor had a relative

Table 15.5 Type of asbestos exposure in 76 patients with mesothelial tumours and 76 controls

Exposure	Mesothelioma series	Control series
Asbestos factory	24 (32%)	2 (3%)
Insulators and laggers	7 (9%)	4 (5%)
Relative worked with asbestos	9 (12%)	1 (1%)
Docker handling asbestos cargoes	0	2 (3%)
No such history	36 (47%)	67 (88%)

$\chi^2 = 27.11$, $P < 0.001$.

Table 15.6 Neighbourhood cases

Category	No occupational or home contact	Lived within $\frac{1}{2}$ mile of asbestos factory
Mesothelioma series	36	11 (30.6%)
Control series	67	5 (7.6%)

$\chi^2 = 7.85$, $P < 0.01$.

working with asbestos, but lived in the immediate vicinity of the factory. This investigation showed that in addition to occupational exposures, domestic contacts from dust brought home on clothes or hair, and exposure outside the factory, possibly by dust from deliveries or waste dumps, were of importance. This factory was a heavy user of crocidolite asbestos.

This particular factory had kept a file of all past employees from the time it opened in 1913. Sufficient information was present on the individual cards of the workers to classify them by category of exposure, whether lightly, moderate, or heavily exposed to dust, and by length of exposure. A study of the mortality of all males who worked there before 1965 has been made and still continues (Newhouse, 1969, 1973). It was also possible to identify and trace a cohort, or group, of women, mainly textile workers who started work between 1936 and 1942 (Newhouse *et al.*, 1972). The study was made with the cooperation of the Office of Population Censuses and Surveys (OPCS), who, through their registers, were able to identify the individuals in the study and supply copies of death certificates. Further enquiries were made to obtain autopsy reports and histological specimens. It was a sub-group of these cohorts which was included in the smoking study already referred to. In addition to data on all causes of death, the studies have revealed important information about the mesothelial tumours. Obtaining additional pathological information about as many deaths as possible has shown that some patients certified on the death certificate as dying from gastro-intestinal tumours, cancer of the pancreas, or carcinomatosis from an unknown primary tumour, may have been wrongly certified and have died from peritoneal mesothelial tumours (Table 15.7). Pleural mesothelioma

Table 15.7 Peritoneal mesothelioma reclassified after histological diagnosis

Certified cause of death	No.
Carcinoma of pancreas	3
Carcinoma of gall bladder	1
Carcinoma of colon	1
Carcinoma of rectum	1
Fibrosarcoma of diaphragm	1
Carcinomatosis	3

Table 15.8 Pleural tumours reclassified after histological diagnosis

Certified cause of death	No.	Histological diagnosis
Carcinoma of lung	7	Pleural mesothelioma
Asbestosis	1	Pleural mesothelioma
Pleural mesothelioma	2	Adenocarcinoma of lung

may also be under-recognized; in seven of these the certified cause of death was cancer of the lung (Table 15.8). Mis-certification has been less common in recent years, as pathologists have become more familiar with the appearance of mesothelioma.

Although in the absence of reliable data on dust levels in the factory it was possible only to divide exposure into four categories, a dose–response relationship was demonstrated. By 1970, 32 mesothelioma had occurred among approximately 4500 men followed since 1933, and 11 among 700 women. The numbers of tumours occurring is related to the number of years of follow-up in the cohort; it can be seen (Table 15.9) that the rate increases both with increasing severity of exposure and with increasing length of exposure. The implications of a dose–response relationship are important as one of the more disturbing features that occur not infrequently in the history of patients is the story of short and apparently low levels of exposure, as for instance in the neighbourhood cases in the London Hospital series, but it must be remembered that exposure was 30 or 40 years before development of the tumour, and histories may be unreliable and exposure may have been heavier than was realized. Examination of the asbestos fibre content of patients dying of mesothelioma also suggests that more than neglible exposure has occurred in these patients (Whitwell *et al.*, 1977).

(10) FACTORS INFLUENCING THE INCIDENCE OF MESOTHELIAL TUMOURS

The incidence of mesothelial tumours varies widely in different groups of workers studied, even comparing groups who have been followed for 30

Table 15.9 Mesothelioma rate per 100 000 subject years at risk

Exposure	<2 years in job	>2 years in job
Low to moderate	25	102
Severe	71	154
Laggers	157	261

years or longer. McDonald and McDonald (1977) have studied the proportion of all deaths due to mesothelial tumours in 15 different mortality series. It varies between zero among anthophyllite workers in Finland to nearly 9% among American and Canadian shipyard workers. Insulators or laggers usually have the highest rates (between 6 and 9% of all deaths), but the rate may vary widely among factory workers. In the East End of London factory nearly 7% of the deaths were due to these tumours, but in another English textile factory it was 3.2%. The factors which can influence the incidence, as has been shown, are firstly, the concentrations of dust to which workers were exposed, and secondly, the type of asbestos fibre predominantly used. The insulators in the U.S.A. certainly used amphibole asbestos from South Africa, but there is no clear evidence that crocidolite asbestos was actually used. Shipyard workers used to work in very high concentrations of dust in unventilated holds of ships, predominantly fireproofing and insulating the ship, and probably used the same materials. In the manufacturing industries, it is known that the East End of London factory, although using all types of asbestos, was a heavy user of crocidolite asbestos, but the other English factory was predominantly a user of chrysotile. Studies of the chrysotile mining industry where there was exposure to pure chrysotile show a proportional mortality rate of less than 1%. There is one study of a factory where only amosite asbestos was used, and this factory has an intermediate rate for mesothelial tumours.

There is other supporting evidence which suggests that crocidolite asbestos is particularly implicated in the aetiology of these tumours. In South Africa, 75 out of 78 patients suffering from mesothelial tumours had worked in the Cape Province crocidolite mines, and the other 3 in the Penge amosite mines. Exposure to pure crocidolite asbestos was experienced by women making army gas-mask filters in Nottingham during the Second World War (Jones *et al.*, 1976). About 1600 women were employed on this job and so far 26 are known to have died of the tumour. Among a small group of workers employed in similar work in Canada during the Second World War, nine died of mesotheliomas with a proportional mortality rate of 16%.

The weight of the epidemiological evidence, supported by experimental evidence, suggests that there is a gradient from crocidolite through amosite to chrysotile asbestos in potency of mesothelial induction.

It is clear from a number of studies that the incidence of mesothelial tumours is rising. This is undoubtedly due to the rapid expansion of the industry and the increased use of asbestos products during the Second World War, particularly in shipbuilding. Asbestos usage continues to expand, in spite of increasing emphasis on substitution with other materials. Owing to the long latent period before development of the tumour, those first exposed during the 1940s often do not develop the tumour until the 1970s. Newhouse and Berry (1976) have predicted that the number of deaths from

Table 15.10 Asbestos exposure and mesothelial tumours: summary of 22 papers from 10 countries (Bohlig, 1974, personal communication)

Type of exposure	No.	%
Occupational	401	49.4
Environmental	153	18.8
Unknown	258	31.8
TOTAL	812	100.0

the tumour will continue to rise among the asbestos factory population until the 1980s.

A history of occupational exposure is not obtained in all cases of mesothelial tumours. About 18–20% give a history of para-occupational exposure, that is, exposure in the home, through contact with an asbestos worker, or through living in the vicinity of an asbestos mine, processing factory, shipyard, or other large commercial user of asbestos. In about 30% no contact can be traced (Table 15.10). Asbestos bodies or fibres are not always found in the lung tissues of patients dying of mesothelial tumours, even in those with known occupational exposure. Where no possible contact can be demonstrated, environmental pollution might be postulated, but at present it must be accepted that mesothelial tumours are not invariably related to asbestos exposure.

(11) OTHER ASBESTOS-RELATED TUMOURS

Among the insulation workers that Selikoff (1976) has studied, cancers of the stomach, colon, and rectum appear to be unduly common, the ratio of observed to expected deaths from these causes being about 3:1. Similar excesses have been observed in other studies.

It is possible that in some of the investigations part of this excess is due to under-diagnosis of peritoneal mesothelioma. Autopsy examinations are by no means invariable and in not more than 40–60% of deaths can the death certificate be validated, but the excess of deaths from these causes in several studies cannot be dismissed.

(12) CANCER OF THE LARNYX

It has been shown both in Liverpool and in Canada by case control studies (Stell and McGill, 1973; Shettigara and Morgan, 1975) that asbestos exposure is more common among patients with cancer of the larynx than among controls. Mortality studies also showed a slight excess risk. Smoking is a strong aetiological factor in this cancer, as in cancer of lung, and there may be a synergestic factor. Leukaemia and multiple nyeloma have also been implicated, although it remains to be proved whether asbestos itself is the carcinogen responsible or a co-carcinogen or potentiator of cigarette smoke (Becklake, 1976).

(13) CONTROL OF THE HAZARD

The functions of occupational epidemiology are to detect risks or hazards, to measure the size of the risk, to provide evidence on which preventive measures can be based, and to monitor the exposed populations in order to assess whether preventive measures are effective. Although the risks of asbestos exposure were slowly identified, the last two functions were sadly neglected after the Asbestos Regulations 1931 were formulated and implemented. However, after the first International Congress in New York in 1964, the subject received new impetus. In the U.K. it was decided to review all available evidence from the industry in order to have a sound basis to provide a standard for asbestos in air in factories and other places where asbestos is used. This standard should be low enough to prevent the development of asbestos-related diseases. Only one factory could provide the necessary data, namely the North Country factory studied by Doll (1955). Here there were the necessary records giving measurements of the levels of asbestos in air, and medical records of routine medical examinations of workers. It was possible to link these two sets of data and find the lowest levels of asbestos in air at which the earliest signs of asbestosis could be detected. The sign chosen was the development of adventitious sound (rales) in the chest. The work was undertaken by a Committee of the British Occupational Hygiene Society, with Dr. S. Roach as Secretary, which published its conclusions in 1968 (Roach, 1968). The Committee considered that the appropriate standard was one of 2 fibres/cm^3. The standard was proposed for chrysotile and amosite asbestos. It is pointed out that as long as there is any airborne chrysotile in the air, there may be some small risk to health, but with the suggested standard the risk should be reduced to 1% for an accumulated exposure of 100 fibre-years/cm^3, which is equivalent to 2 fibres/cm^3 for 50 years (50 years being the longest working life in modern times).

Crocidolite, in view of the evidence of special association with mesothelial tumours, was excluded and a standard 10 times as rigid (i.e. 0.2 fibres/cm^3) was adopted. The 2 fibres/cm^3 standard was designed for the prevention of asbestosis, not for lung cancer; however, with the strong association of lung cancer and asbestosis it was believed that prevention of asbestosis would to a large extent diminish the incidence of asbestos-related lung cancers, although it is admitted that there is no certain evidence to support this hypothesis. Dose–response curves obtained by investigations such as those

Table 15.11 Lung cancer (excluding mesothelioma) in asbestos workers in London

Exposure		Observed	Expected	Ratio × 100 (SMR)
Light or moderate	<2 years	7	7.9	88
	>2 years	12	6.3	189
Severe	<2 years	20	8.9	225
	>2 years	40	7.8	510

of McDonald suggest that lung cancers occurring below certain levels of exposure are probably those related only to cigarette smoking and are a general risk in any group with the usual smoking habits (McDonald *et al.*, 1974). There is other supporting evidence, from the East End of London factory studied by Newhouse, that there is no excess mortality for cancer of lung in those with short and light exposure (Table 15.11), but a five-fold excess in the most severely exposed groups.

New Regulations controlling the use of asbestos were passed through Parliament in 1969 and were enforced 1 year later. They embodied the new standard and now apply to all asbestos workers, not only to those specifically mentioned in the 1931 Regulations. Under the Health and Safety at Work etc. Act 1974, the health and safety of workers is now the responsibility of the Health and Safety Commission, whose medical branch is the Employment Medical Advisory Service (EMAS), which replaced the Medical Inspectorate of Factories. One of the first objectives of this new Service has been the control of the asbestos hazards. A nationwide survey of all asbestos workers has been in progress since 1971 and, starting with the larger factories, it is gradually being extended to all known asbestos workers. Each worker is examined clinically and has a chest X-ray every 2 years. Job details and the levels of dust in air to which he is exposed are added to the clinical records. The scheme is voluntary for the workers, but nearly all participate. Ultimately about 30 000 workers will be included in the scheme. EMAS also has a mortality study of these workers in progress. Reports of the Factory Inspectorate indicate that the current standards for asbestos are being maintained in nearly all factories. Ultimately firm data will be available for the scientific basis for standards both for asbestosis and asbestos-related lung cancer, but this cannot be expected from this survey for another 15–20 years; however, in the meantime it ensures that the health of the individual worker is carefully monitored.

The important epidemiological tasks currently are to continue to investigate the relative carcinogenicity of the different types of asbestos and to determine from records already available in industry the degree of risk of low level exposures.

(14) REFERENCES

Becklake, M. R. (1976) Asbestos related disease of the lung and other organs, *Am. Rev. Resp. Dis.*, **114,** 187–227.

Berry, G., Newhouse, M. L., and Turok, M. (1972) Combined effect of asbestos exposure and smoking on mortality from lung cancer, *Lancet*, **ii,** 476–479.

Doll, R. (1955) Mortality from lung cancer in asbestos workers, *Brit. J. Ind. Med.*, **12,** 81–86.

Doniach, I., Swettenham, K. V., and Hawthorn, M. K. (1975) Prevalence of asbestos bodies on a necropsy series in East London, *Brit. J. Ind. Med.*, **32,** 16–34.

Harries, P. G. (1968) Asbestos hazards in Naval Dockyards, *Ann. Occup. Hyg.*, **11,** 135–145.

Jones, J. S. P., Pooley, F. D., and Smith, P. G. (1976) Factory populations exposed to crocidolite asbestos, in *Environmental Pollution and Carcinogenic Risks*, (Ed. C. Rosenfeld and W. Davis) IARC Scientific Publications No. 13, International Agency for Research on Cancer, Lyon, 117–120.

Kiviluoto, R. (1960) Pleural calcification as a sign of non-occupational endemic anthophyllite asbestosis, *Acta Radiol., Suppl.*, **194,** 1–67.

Knox, J. F., Holmes, S., Doll, R., and Hall, F. D. (1968) Mortality from lung cancer and other causes among workers in an asbestos textile factory, *Brit. J. Ind. Med.*, **25,** 293–303.

McDonald, J. C., Becklake, M. R., Gibbs, G. W., McDonald, A. D., and Rossiter, C. F. (1974) The health of chrysotile mine and mill workers, *Arch. Envir. Hlth.*, **28,** 61–68.

McDonald, J. C., and McDonald, A. D. (1977) Epidemiology of mesothelioma from estimated incidence, *Prev. Med.*, **6,** 426–446.

Merewether, E. R. A., and Price, C. W. (1930) Report on the Effects of Asbestos Dust on the Lungs and Dust Suppression in the Asbestos Industry, HMSO, London.

Newhouse, M. L. (1969) A study of the mortality of workers in an asbestos factory, *Brit. J. Ind. Med.*, **26,** 294–301.

Newhouse, M. L. (1973) Asbestos in the work place and the community, *Ann. Occup. Hyg.*, **16,** 97–107.

Newhouse, M. L., and Berry, G. (1976) Predictions of mortality from mesothelial tumours in asbestos factory workers, *Brit. J. Ind. Med.*, **33,** 147–151.

Newhouse, M. L., and Thompson, H. (1965) Mesothelioma of the pleura and peritoneum following exposure to asbestos in the London area, *Brit. J. Ind. Med.*, **22,** 261–269.

Newhouse, M. L., Berry, G., Wagner, J. C., and Turok, M. (1972) A study of the mortality of female asbestos workers, *Brit. J. Ind. Med.*, **29,** 134–141.

Peto, J., Doll, R, Howard, S. V., Kinlen, L. J., and Lewisohn, H. C. (1977) A mortality study among workers in an English asbestos factory, *Brit. J. Ind. Med.*, **34,** 169–173.

Roach, S. A. (1968) Hygiene standards for chrysotile asbestos dust, *Ann. Occup. Hyg.*, **11,** 47–60.

Saracci, R. (1977) Asbestos and lung cancer; an analysis of the epidemiological evidence on the asbestos/smoking interaction, *Int. J. Cancer*, **20,** 323–331.

Selikoff, I. J. (1976) Asbestos disease in the United States, *Rev. Fr. Mal. Resp.*, **4,** Suppl., 7–24.

Selikoff, I. J., Hammond, E. C., and Churg, J. (1968) Asbestos exposure, smoking and neoplasm, *J. Am. Med. Ass.*, **204,** 106–112.

Selikoff, I. J., Hammond, E. C., and Seidman, H. (1973) Cancer risk of insulation workers in United States, in *Biological Effects of Asbestos*, IARC Scientific Publications No. 8, International Agency for Research on Cancer, Lyon, 209–216.

Shettigara, P. T., and Morgan, R. W. (1975) Asbestos, smoking and laryngeal carcinoma, *Arch. Envir. Hlth.*, **30,** 517–518.

Smither, W. J. (1965) Secular changes in asbestosis in an asbestos factory, *Ann. N.Y. Acad. Sci.*, **1,** 166–182.

Stell, P. M., and McGill, T. (1973) Asbestos and laryngeal cancer, *Lancet*, **ii,** 416–417.

Thompson, J. G., Kashula, R. O. C., and MacDonald, R. R. (1963) Asbestos as a modern urban hazard, *S. Afr. Med. J.*, **37,** 77–81.

Wagner, J. C., Sleggs, C. A., and Marchand, P. (1960) Diffuse pleural mesothelioma, *Brit. J. Ind. Med.*, **17,** 260–271.

Whipple, H. E. (Ed.), *Biological Effects of Asbestos, Ann. N.Y. Acad. Sci.*, 1965, **132,** 1–766.

Whitwell, F., Scott, J., and Grimshaw, M. (1977) Relationship between occupations and asbestos fibre content in patients with pleural mesothelioma by cancer and other diseases, *Thorax*, **32,** 377–386.

Zolov, C., Burhlikov, T., and Michailova, L. (1967) Pleural asbestosis in agricultural workers, *Envir. Res.*, **1,** 287.

CHAPTER 16

Asbestos-related diseases Part 5: The prevention of asbestos-related diseases*

Richard J. Levine†, M. David Gidley, Michael Feuerstein, Margaret Chesney and Paul M. Giever

Center for Resource and Environmental Systems Studies, SRI International, Menlo Park, California

Jean Spencer Felton

Long Beach Naval Shipyard, Long Beach, California

SYNOPSIS

Four programme approaches for controlling the adverse health effects of asbestos fibres in man's environment are presented in this chapter:

(1) Physical control of human exposure to asbestos—i.e. reducing the contact between man and asbestos fibres.

(2) Medical management—measures taken by physicians to protect persons at greater risk of asbestos-related diseases.

(3) Smoking cessation programmes—given the link between smoking and increased risk of cancer in asbestos workers.

(4) Education—in order to maximize the effectiveness of the other control approaches.

The various sections are entitled:

* Based on material included in *Asbestos: An Information Resource*, Ed. by Richard J. Levine, U.S. Dept. Health, Education, and Welfare DHEW Publication No. (NIH) 78-1681, May 1978, and supported under contract number NO1-CN-55176 from the U.S. National Cancer Institute.

† Present address: Chemical Industry Institute of Toxicology, Research Triangle Park, North Carolina 27709.

(1) PHYSICAL CONTROL

Reducing the extent of contact between asbestos fibres and man can be accomplished by actions which include: (1) protective engineering of materials, processes, and industrial facilities; (2) substitution of alternative materials for asbestos; (3) instituting safe work practices; (4) using administrative means to regulate persons who might be exposed; and (5) properly treating and disposing of emissions and wastes.

1(i) Engineering controls

Controlling airborne asbestos fibres by engineering techniques is not greatly different from controlling other solid particulate matter of similar aerodynamic size. A variety of methods are reviewed in the sections that follow.

1(i).1 *Enclosure*

All dust-producing machine activities should be provided with the maximum enclosure consistent with operational requirements. Enclosures must often be designed with an opening to permit normal operation of a machine. Hence, there must be a continual inflow of air into the enclosed unit at a sufficient velocity to prevent the escape of asbestos dust.

Owing to their aerodynamic characteristics and small mass, unless borne by wind, asbestos fibres travel only a short distance in air (a few inches at most) even when impelled at extremely high velocities. Asbestos dust generated by machines can, therefore, be effectively controlled by enclosure with exhaust ventilation. In many cases, a minimum intake velocity of 200 ft/min at the face of an enclosure is adequate.

1(i).2 *Exhaust ventilation*

In order to remove airborne fibres, adequate exhaust ventilation must be provided to enclosed areas. Exhaust ducts for asbestos operations are designed to carry air at velocities ranging from 3000 to 5500 ft/min. The lower velocities are used to convey well opened fibres, as in the dust control systems at asbestos textile plants. Higher velocities are needed for larger particles, as in exhaust systems for machines that cut and shape brake shoes or asbestos–cement pipe. Most systems function well at duct velocities of 4000–4500 ft/min. A number of variables affect the amount of air to be exhausted, and only experience will indicate the amount of exhaust needed at a given dust-producing operation or machine.

All parts of asbestos control systems should be maintained under negative pressure. This is necessary to prevent dust from leaking into the plant at loose joints and open seams as well as to ensure adequate dust collection at enclosure faces.

A major consideration in planning a ventilation system is to provide for make-up air. The amount of air introduced into a building should slightly exceed the amount of air exhausted. A mechanical air supply system is preferred to 'natural' room ventilation and, in most cases, a mechanical system is necessary to achieve the desired ventilation performance.

Many specifications for duct construction design may be found in the industrial ventilation manual issued by the American Conference of Governmental Industrial Hygienists (1978).

Even with a ventilation system built according to good engineering principles, periodic measurements must be made in order to ensure that the system is adequately balanced and performing according to design. Measurements of static pressure, air flow, supply, capture, and conveying velocities, and fan performance are essential, since plugging or wear can cause variations in the balance of the system and therefore in its efficiency.

1(i).3 *Isolation*

Isolating and restricting access to 'dirty' operations is employed extensively in the asbestos industry. Present engineering technology has not been able to reduce airborne asbestos concentrations to acceptable levels in certain situations. In these instances, enclosure or isolation is mandatory. Negative pressure should be maintained within isolated areas to prevent escape of asbestos fibres.

In addition to reducing fibre levels in other parts of the plant, there are additional advantages to isolating asbestos fibre-emitting operations. An employee working at a dusty job is not as apt to relax adherence to restrictive work practices if he is separated from employees who are working freely at operations that present no exposure hazard. Isolation can also reduce costs for local exhaust ventilation and facilitate good housekeeping.

The bagging operation in asbestos mills, which may be a dusty activity, can be isolated and enclosed. Further control may be gained by limiting the number of operators and restricting access to the area.

Facilities for unloading and storing asbestos fibre should be kept isolated because of dust that may leak from broken bags. This problem can be minimized by pressure packing of asbestos at the mill. Inevitably, some bags will have been perforated by in-transit shifting or careless loading. These should be repaired (or the asbestos re-bagged) and vacuum-cleaned before being transferred to the warehouse. Needless to say, careful handling is necessary to prevent further damage.

Bags of asbestos fibre are opened and emptied at a bag-opening station (usually a hopper), which should be enclosed, ventilated, and isolated from other operations in the mixing or compounding area. An in-draft ventilation of about 250 ft/min must be maintained at the face of the bag-opening enclosure. Emptied bags should be rolled up within the hood and placed into a shredder, a large bag-conveying pipe, or a clean bag and sealed. In some industries, bags can be mixed with other materials and submersed directly into the manufacturing process. Bags that have been conveyed to an isolated, enclosed, central collection point should be placed into sealed containers for disposal.

1(i).4 *Plant design*

Industry is not often afforded the opportunity of designing a plant to specifications that place hazard control measures at the forefront of priorities. Management is aware, however, that factors which relate to efficiency, product quality, and cost savings do not necessarily conflict with appropriate design for health and safety.

An ideal method for isolating a dusty work area is one in which entry can be made only through locker rooms. Two locker rooms, separated by showers, should be provided—one room for street clothes (clean conditions locker room), the other for work clothes and protective equipment (working conditions locker room). Interposing showers between the locker rooms enhances the likelihood that workers will shower at the end of a shift (see Figure 16.1).

The working conditions locker room should be maintained under negative pressure. An air flow between the locker rooms should be directed towards

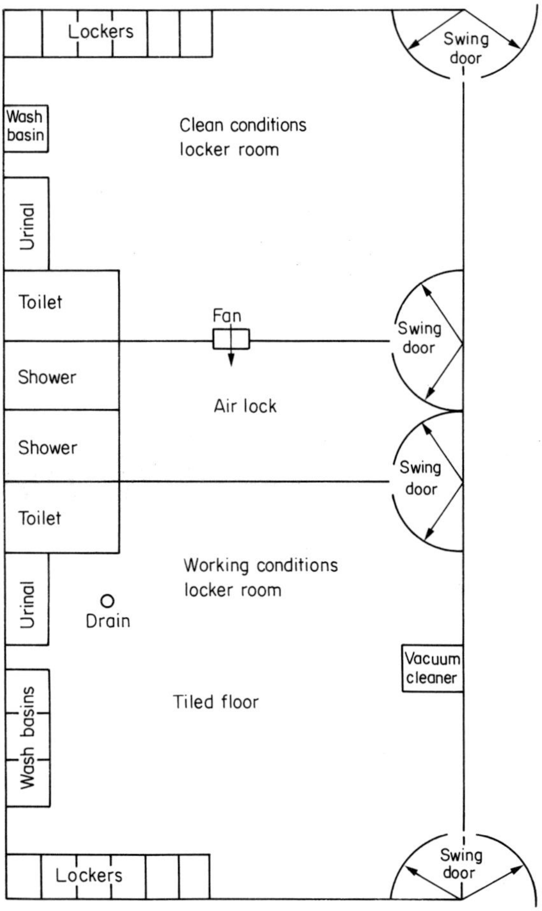

Figure 16.1 Plan of locker rooms and showers for isolated plant areas. (Source: Engineering Equipment Users Association, 1969)

the working conditions room, with exhaust air vented to a suitable collecting system. If connecting doors between change-rooms and showers are self-closing and well sealed, it may be possible to use the shower area as an air lock.

Some other important considerations in designing for asbestos fibre control are:

(a) engineering a dusty operation so that it can be handled by as few employees as possible;

(b) including a protected observation site next to an isolated work area so that entry of supervisory personnel and visitors can be kept to a minimum;

(c) planning the layout so that airflow into hoods, enclosures, and other exhaust equipment is not disturbed by draughts from fans, windows, and doors. Off-set doors with indirect, right-angle entries help to deflect and diffuse incoming air currents.

(d) constructing interiors without exposed beams, pipes, and ledges onto which airborne asbestos might settle.

1(i).5 *Treatment of asbestos*

There are various methods of treating loose asbestos so as to reduce fibre emissions. One of the most effective is *wetting*. Although not applicable to operations that will not tolerate moisture, such as in the manufacture of friction products, wetting can be used judiciously in mining and milling and in the construction industry.

At present in the U.S.A., there is one asbestos mill that utilizes *wet processing*, separating asbestos from heavier rock through a series of flotation devices. The purified asbestos is collected between pillow filters and extruded and dried in pellet form or fluff-dried and packaged as loose fibrous material. Dust control is needed only for pelletizing and for bagging the finished product.

The use of *pelletized asbestos*, which can be pumped pneumatically into enclosed railway waggons and unloaded through gravity release, has been tried in the manufacture of friction products and textiles, but with little success. These industries require long fibres, but pelletizing breaks asbestos fibres into shorter lengths. The use of pelletized asbestos, however, should present less of a problem to other asbestos industries.

A recent method of treatment to reduce airborne fibres in the asbestos textile industry is by *application of a polymer* to asbestos yarn (U.S. Environmental Protection Agency, 1976). Such a coating cannot be used where the inherent surface characteristics of asbestos fibres are required, particularly when the fibres are to be bound in a matrix with other materials.

Treatment of asbestos with *anti-dusting agents* may be helpful. These agents are viscous liquids that are applied to dry asbestos by spraying or mixing. The fibres are then dried at room temperature. The procedure retains the performance criteria of untreated asbestos (Smith and White, 1975).

1(ii) Substitution of alternative materials

Although asbestos is well established in many important commercial applications, it is likely that substitution of alternative materials will have a future role in reducing asbestos health hazards. A number of materials have been investigated as possible alternatives, including carbon fibres, cellulose, ceramics, fibrous glass, potassium titanate, silica, sintered metals, kaolin wool, rock wood, slag wool, steel wool, and exfoliated vermiculite. Few alternative materials have proved as satisfactory as asbestos owing to lack of strength, heat resistance, flexibility, or durability, or because of cost. Moreover, since attention has been drawn to the possibility that inhaled fibres other than asbestos may be carcinogenic (Stanton, 1974), it is essential to evaluate the toxicity of proposed asbestos substitutes.

In certain industrial processes where asbestos is used as a binder, less toxic materials have been substituted with little effect on the quality of the product. This has been the case in the manufacture of rubber, plastics, and various adhesives and cements. For similar reasons, less asbestos might be used in paints, coatings, caulks, sealants, and joint fillers.

Satisfactory asbestos substitutes have been developed for a variety of reinforced plastics, resins, and insulating materials. For critical insulating applications, however, the strength and heat resistance of asbestos cannot be duplicated economically.

Not many replacements have been found for asbestos in paper products in which the heat resistance, chemical inertness, and electrical and insulating properties of asbestos are highly valued—products such as mill board, roofing felts, pipe coverings, fine-quality electrical and insulating papers, and asbestos–latex flooring felts. Glass cloth, felt, and thread have found limited application in roofing and flooring underlayments, but in most of these paper products asbestos fibres are still used.

A satisfactory substitute for asbestos in friction materials has not been developed.

Two of the more successful asbestos substitutes, soda-lime–silica and high-silica glass fibres, are reviewed briefly below.

1(ii).1 *Soda-lime–silica glass fibres*

Soda-lime–silica glass fibres, made by highly refined processes involving the use of platinum dies, are of high quality and uniform size. Some are less than 0.5 μm in diameter and are adaptable to highly specialized uses, such as weaving into fabrics.

Glass fibres will not burn, but they will soften and coalesce at temperatures which vary according to the composition of the glass. Generally, the fibres will withstand temperatures up to 650 °C. Heat and moisture resistance is limited by the organic film applied to the fibres during manufacture to improve processing and to reduce breakage during subsequent plying and weaving operations. Without this film, the fibres tend to be brittle and self-abrasive.

Exposure of very fine glass fibres to water vapour results in relatively rapid deterioration, making them less resistant than asbestos to the effects of steam and moisture. Attempts to use glass fibres in place of asbestos in asbestos–cement products have been unsuccessful because of a chemical reaction between the glass and cement that decomposes the fibres.

Glass fibres are efficient thermal insulators and may be used in equipment such as stoves and refrigerators, where conditions are not corrosive. Their high tensile strength, electrical resistance, and greater thermal stability compared with organic fibres make them suitable for electrical insulation.

Glass or combined glass–asbestos fabrics have some advantages over asbestos textile products because they are lighter and stronger, but they are generally less resistant to flexure, abrasion, and chemical action. Glass fibre is used in conjunction with asbestos, or as an optional alternative material, in Navy shipboard cable (U.S. Bureau of Mines, 1959). A glass–asbestos cloth woven with a plied yarn using alternating strands of glass and asbestos is used to cover thermal insulation applied to piping on naval vessels. Fabrics made of interwoven glass and asbestos yarns are being made in many weights and colours for use as theatre curtains and fireproof draperies.

As a substitute for asbestos in friction equipment, glass fibre has been unsatisfactory, chiefly because of its abrasive qualities.

1(ii).2 *High-silica glass fibres*

Glass fibres approximating vitreous silica in composition are superior to soda-lime–silica glass fibres in resistance to the action of water vapour and high temperatures. However, because fused silica is extremely viscous at its melting point (940 °C), these fibres are difficult to manufacture.

1(iii) Work practices (including housekeeping and use of personal protective equipment)

Changes in work practices often may be the most cost-effective means of reducing occupational exposure to asbestos. Some of the ways in which work practices may be modified include:

(a) mixing asbestos mortar in closed polythylene bags rather than in mortar boxes or buckets;

(b) maintaining central fabrication shops from which pre-cut insulating materials can be sent to the field for installation, thereby minimizing the need for on-site cutting or sawing.

(c) prohibiting power tools, except in central shops;

(d) using single-point cutting and chipping tools, rather than saws or abrasive cutting equipment;

(e) jettisoning bags into the product mix whenever possible;

(f) substituting vacuuming for blowing off machines and equipment with compressed air,

(g) good housekeeping (as discussed below);
(h) use of personal protective equipment (as discussed below);

1(iii).1 *Housekeeping*

Good housekeeping is essential to reducing levels of airborne asbestos. Waste material such as rejects, scrap, shavings, or other debris should be picked up and placed into plastic bags. The bags should be taped shut at the end of a shift, labelled as to the hazard contained, and disposed of.

Asbestos dust on floors, ledges, equipment, overheads, and other plant surfaces can become airborne when disturbed by draughts or work activity and should be removed. Dry sweeping is not an effective means of removing dust, as it tends to waft fibres into the air, from which they may become deposited on remote ledges, pipes, and other inaccessible surfaces. Nor is wet mopping a satisfactory way to clean, as it tends only to spread the wetted dust around. The recommended method is vacuum cleaning, using a central vacuum system consisting of a suction source with filtration and settling units. From the central system, suction pipes run to parts of the plant where cleaning is necessary. Hoses of vacuum-cleaning implements can be connected to the pipes through access openings.

1(iii).2 *Personal protective equipment*

Personal protective equipment must always be available, but never to serve as a replacement for appropriate engineering control measures. If, however, an unexpected event creates a potential for exposure greater than the maximum permitted, or if engineering control measures are insufficient, personal protective equipment will be necessary. Such equipment may always be needed for some operations, such as the cleaning and repair of exhaust ductwork, the manual shakedown of collection bags in baghouses, and the removal of thermal insulation.

(a) *Respirators.* Respirators require proper fitting, maintenance, and cleaning to be effective. Beards may prevent proper fitting and in such cases must be shaved off.

The elements of an acceptable respirator programme are set out by the American National Standards Institute (Washington, D.C.) in ANSI Standard No. 288.2–1969, entitled *American National Standard Practice for Respiratory Protection.* The type of respirator needed—dust, mist, or fume, air-line, or abrasive blasting—will be determined by the concentration of airborne asbestos fibre (U.S. Federal Register, 1975a). The concentration should always be re-checked whenever there are significant changes in process, control, work site, or climate.

(b) *Protective clothing.* Special work clothing, not to be worn outside the plant, has the dual purpose of protecting both asbestos workers and their families from fibres that might adhere to the body or clothing.

A satisfactory form of basic protection is overalls, preferably made of cotton–polyester material. Cotton alone cannot be used, because static build-up causes fibres to adhere to the cloth tenaciously. Disposable paper overalls, although comparatively inexpensive and a means of avoiding the potential for exposure of laundry workers, are easily perforated by body movement, chemical action, or sparks. The overall should be one-piece, without pockets, cuffs, or rolled edges, and with adequate closures for necessary openings. Overalls should not be worn over street clothes, and they must be provided clean each day.

A head covering is required to prevent fibres from lodging in the hair. Paper surgical-type caps are satisfactory for this purpose, but hard hats alone are not. Where hard hats are necessary, a paper cap should be worn underneath.

Also needed are underwear, socks, and foot coverings, in the form of canvas boots, rubber galoshes, or safety shoes.

Street clothes and personal effects should be kept in a clean conditions locker room, and work clothes and protective equipment in a working conditions locker room, as noted above. Whenever persons leave a restricted work area, they should enter the working conditions room, remove adherent asbestos fibres using a vacuum equipped with a filtered exhaust, and take off protective equipment (respirators last). Showering should be mandatory before putting on street clothes in the clean conditions room.

If work clothing is laundered by an outside service, rather than at the plant, the laundry should be advized in writing of the asbestos hazard. Clothing to be laundered should be vacuum-cleaned, dampened, and packed and sealed in plastic bags clearly marked 'Asbestos-Contaminated Clothing—Wet Before Handling'.

1(iv) Administrative controls

Administrative means for reducing exposure to asbestos fibres can complement the vigorous application of engineering control measures and assist in protecting persons in situations where engineering methods have failed to reduce airborne concentrations to acceptable levels. Suggested administrative controls include limiting the number of employees exposed, limiting the duration of exposure for any given person, and restricting smoking, eating, drinking, and the like in the workplace*.

* Another administrative control measure, which has been used in the British dyestuffs industry in connection with cancer control (Engineering Equipment Users Association, 1969), is to give preference to job applicants of an advanced age not previously exposed to asbestos in the course of work. The reasoning is that older employees may die of natural causes before completing the latency period for asbestos-related cancer.

1(iv).1 *Limiting the number of employees exposed*

The number of employees exposed to excessive airborne concentrations of asbestos may be limited by:

(a) restricting access to contaminated areas [see also Sections1(i).3 and 1(i).4];

(b) reducing to a minimum the number of persons handling asbestos;

(c) conducting particularly dusty operations during shifts where the number of persons in the plant is at a minimum.

Smaller numbers of continuously exposed employees are more easily trained, controlled, and protected than a larger group that is only occasionally exposed.

1(iv).2 *Limiting the duration of exposure for any employee*

The present U.S. occupational standard for asbestos is an 8-h time-weighted-average not to exceed 2 optical-microscope-visible fibres per cubic centimetre of air. Concentrations must never exceed 10 fibres/cm^3 (U.S. Federal Register, 1972). This means that during a single shift, employees may be exposed to airborne asbestos levels above 2 fibres/cm^3 provided that such excursions are compensated for by equivalent reductions in exposure, except that in no instance can the exposure exceed 10 fibres/cm^3.

For example, employees on a shift can be exposed for 4 h to an airborne asbestos level of 4 fibres/cm^3, which is equivalent to an exposure of 2 fibres/cm^3 for 8 h, provided that they are subject to zero exposure for the remainder of the day. The remaining hours of the shift would presumably be covered by employees who had received zero exposure during the first 4 h.

Control of exposure using averaging is deficient in a number of respects, since it assumes that: (1) assigned exposure levels will remain constant; (2) the worker receives no exposures other than those assigned; and (3) the submicroscopic fibres present but not counted are unimportant from a toxicity standpoint*. Further, there must be a sufficient work force in order to permit the required alternation of personnel; the economic considerations of possibly needing to hire additional workers must be relatively unimportant.

1(iv).3 *Restrictions on smoking, eating, drinking, etc.*

A strong corporate stand should be established against the practice of smoking, eating, drinking, chewing gum or tobacco, or taking medicine while working. These activities must be restricted to designated clean

* It has been noted recently that optical-microscope-visible asbestos fibres longer than 5 μm with a length-to-width ratio greater than 3 account for only approximately 2% of all asbestos fibres present in industrial settings (Bruckman and Rubino, 1975).

locations visited only after established decontamination procedures have been carried out. Smoking cessation should be encouraged and, if possible, preference when hiring should be given to persons who pledge not to smoke.

1(v) Air pollution control

The most useful means of controlling asbestos emissions to community air is fabric filtration, and design parameters for successful systems have been published (U.S. Environmental Protection Agency, 1973; Harwood *et al.*, 1974, 1975; Seibert *et al.*, 1976). The efficiency of fabric filter units used to collect asbestos fibres has ranged from 95 to 99.9% on the basis of weight. Although the fibres are collected dry, many may be fractured, thereby limiting opportunities for recycling. Used fabric mats and collected fibres should be placed into suitable sealed containers and disposed of.

Wet collectors—wet dynamic scrubbers and Venturi-type collectors—range in efficiency from 50 to 90%. The fibres collected are in a slurry and may pose a water pollution problem. Usually, the slurry is filtered and the wet fibres disposed of in a suitable container.

Mechanical collectors (cyclones) generally operate with the same range of efficiency as wet scrubbers, the actual efficiency depending on the size, design, and amount of energy expended. In general, mechanical collectors are economical to operate and maintain since they require no power other than to move air, and apart from the collector shell, there are no parts that can wear out (U.S. Environmental Protection Agency, 1976).

Electrostatic precipitation is a less effective means of asbestos air pollution control, yielding 70% efficiency at best.

Since operations that generate asbestos fibres are usually conducted under negative pressure, cleaning of the gas in the ventilating system will generally provide adequate control of air pollution.

While the procedures described above may be satisfactory for many asbestos-related activities, they are usually not practical for demolition or rip-out. Control of air pollution in such activities is limited to the use of water sprays.

1(vi) Water pollution control

Until recently, little attention had been directed towards asbestos-contaminated industrial waste waters, and there is virtually no published information. The number of plants involved is not large, and the volumes of waste have been relatively small, the contamination of Lake Superior water by tailings from a taconite mining operation being an exception (see Chapter 5). A significant amount of process water in manufacturing operations is recirculated. Moreover, most plants have some form of waste treatment, and many are situated where they can discharge process waste waters into municipal sewers. Increased concern over asbestos fibres in air, however, has resulted in the conversion of some dry processes into wet ones and has increased the use of water sprays to control dust, thereby enhancing the potential for water pollution.

1(vi).1 *Treatment processes*

The standard processes for removing suspended solids from water have been found to be satisfactory for removing asbestos fibres, viz.:

(a) pre-treatment—removal of oil, grease, and the larger aggregates of solid matter;

(b) primary treatment—sedimentation and chlorination;

(c) secondary treatment—biological degradation;

(d) tertiary treatment—required if the effluent from secondary treatment is unsatisfactory (Krenkel, 1974).

The several means of removing suspended solids in tertiary treatment include microstraining, diatomaceous earth filtration, chemical clarification, and deep-bed granular media filtration. All of these methods, except chemical clarification, involve physically straining out finely divided solids, as well as adsorbing them on to the surface of the filter media in diatomaceous earth and deep-bed granular media filtration. Concentrated asbestos fibres and used filter media must be disposed of in a landfill.

In *microstraining*, the waste water is strained through a woven mesh screen on the surface of a rotary drum. As the drum rotates, the solids are strained out of the water to a position from which they are removed, from the drum.

Diatomaceous earth filtration is a form of mechanical separation that utilizes a diatomaceous earth filter aid on a supporting medium to trap solid material.

In the *chemical clarification* process, chemicals such as aluminium, iron, or calcium oxides are added to the water to coagulate fine solids. Coagulation is followed by a flocculation phase, in which particulates are aggregated into larger flocs. During sedimentation, the flocs that have been formed are allowed to settle to the bottom of a settling tank. While some of the solid material will have been removed by sedimentation, the remainder must be removed by *filtration*, usually in beds of a porous medium such as sand or coal, or a combination of media such as sand and coal, or sand, coal, and garnet.

1(vi).2 *Control of asbestos fibres in potable water supplies*

It has not been established to what extent, if any, ingested asbestos may be harmful. In the meantime, it is the opinion of some researchers that the oral intake of asbestos should be reduced as much as possible (Levy *et al.*, 1976).

One method of reducing asbestos concentrations in potable water involves minor modification of standard coagulation/filtration techniques as practised in most water treatment plants. Preliminary results indicate that the number of asbestos fibres can be consistently reduced to below detectable limits (<20 000 fibres/l.). Even simple filtration systems have been shown to be partially effective and could prove useful as a low-cost interim measure in areas of high fiber concentration (Lawrence *et al.*, 1974). According to one study, both alum and polyelectrolyte coagulation optimize fibre removal and can be used with sand filters (Lawrence *et al.*, 1975). This is a considerable advantage, since sand filters are used in most filtration plants in North America and Europe.

Diatomite-filter pilot plants operating at 10 gall/min removed over 80% of amphibole asbestos fibre from a drinking water source. The turbidity of the finished water was 0.05–0.06 FTU (formazin turbidity units), with over 95% of the fibre removed. Amphibole fibre seemed easier to remove than chrysotile. The following were concluded (applicable to the filtration of Lake Superior water at Duluth):

(a) Several filter operating conditions can result in removal of 95–98% of asbestiform fibres. Conditions providing the best filtered water would involve the use of either alum-coated Hyflow Super Cel (or equivalent grade) as body feed and pre-coat, or alum-coated C-512 filter aid (or equivalent grade) as pre-coat plus a continuous coagulant feed of Cat-Floc B polymer to the filter influent water.

(b) Using vacuum diatomite filters is significantly more expensive and would be difficult under conditions of high turbidity.

(c) A plant designed for a 20-year life and to yield water of turbidity 1.9 FTU equalled or exceeded only 5% of the time could produce 30 000 000 gallons of water per day at a cost of U.S. $0.0556 (*ca.* £0·03) per 1000 gallons. Designing a filtration plant to produce water of turbidity 2.5–3.5 FTU would be the best protection against increases in the price of filter aid (Baumann, 1975).

1(vii) Solid waste disposal

The greatest hazard associated with asbestos solid waste is the potential for air emissions arising from improper handling and from improper final disposal. At each step in the handling of the solid waste material—whether the waste is to be concentrated, isolated, disposed of, re-used, or otherwise treated—hazards to the waste handlers may arise. Asbestos-containing wastes must be treated with the same respect accorded asbestos and asbestos products.

The most desirable waste management options, in order of importance, are (U.S. Federal Register, 1975b):

(a) waste reduction—by reducing the amount of asbestos used, substituting less hazardous materials, and increasing process efficiency;

(b) waste separation and concentration—segregating hazardous from non-hazardous wastes;

(c) waste recovery—re-using the material;

(d) secure ultimate disposal—disposal to landfill in a way that precludes future re-entrainment.

The most important aspects of disposing of asbestos solid waste are identification, separation, secure transport, and secure ultimate disposal. These are discussed in the following paragraphs.

1(vii).1 *Identification*

No control measures can be directed at asbestos solid waste issuing from a source if asbestos has not been identified as part of the solid waste stream. Identifying asbestos-containing wastes and tracking their sources is a necessary first step.

1(vii).2 *Separation*

Having identified asbestos-containing wastes and their sources, those wastes should be separated from non-hazardous wastes, taking precautions to prevent exposures to workers in the process. The most generally satisfactory containers for small amounts of dry asbestos wastes are heavy-gauge, impervious plastic bags. Asbestos may also be disposed of as a slurry, provided that the slurry does not dry between collection and disposal.

1(vii).3 *Secure transport*

Waste must be removed to the ultimate disposal site without producing emissions. Persons who dispose of asbestos wastes must be aware of the hazards and informed about proper handling procedures. Care should be taken not to tear plastic bags, and waste containers such as cans or bins must have tight-fitting lids that will not come off during transit. Closed conveyance to the ultimate disposal site is necessary.

These requirements are more likely to be met by hazardous-waste disposal companies than by ordinary public or private disposal services. If waste is disposed of on the premises of an asbestos plant, employees of that facility will probably be familiar with the necessary handling precautions.

1(vii).4 *Secure ultimate disposal*

The only method of disposing of asbestos that can be considered as secure ultimate disposal is disposal in a landfill site with an adequate covering layer of non-asbestos-containing waste or earth. If an acceptable vegetative cover is established and maintained, the covering layer must be at least 15 cm deep, or 60 cm deep without such a vegetative cover. Emissions from waste may also be controlled by maintaining a resinous or petroleum-based dust-suppression cover at the landfill site, or by wetting the wastes with water and sealing it in an impermeable container before disposal (U.S. Federal Register, 1975). Covering waste with soil and vegetation does not require as much subsequent care as is needed to maintain a disposal site with dust-suppression agents; however, obtaining sufficient soil to cover a very large site might in itself create an environmental problem. An excellent discussion of establishing stabilized waste piles and of controlling emissions from landfill sites has been published (Harwood *et al.*, 1976).

(2) MEDICAL MANAGEMENT†

Efforts to prevent asbestos-related diseases assume special significance since opportunities for cure are very limited. The best medical means of preventing such diseases is to recognize particularly susceptible persons during a pre-employment examination and to recommend that they not be hired for jobs involving asbestos exposure. Workers necessarily exposed to asbestos should be enrolled in a medical screening programme. In this way, hopefully, asbestos-related diseases (and acquired disorders predisposing to these diseases) can be detected early enough so that removal from exposure or medical intervention may successfully limit their course. Nevertheless, asbestosis may progress even after removal from asbestos exposure, and screening programmes for the early detection of lung cancer have not yet proved to be more than marginally beneficial.

The industrial doctor must be well acquainted with the nature of particular jobs and should meet with plant industrial hygienists on a regular basis to exchange information. He should be responsible for maintaining uniform medical and industrial hygiene data files for individual employees to facilitate epidemiological surveillance*. An important continuing function will be to provide accurate information to plant personnel and their families on the possible health effects of asbestos exposure.

* Name, social security number, date of birth, date of hire, date of termination, race (if available), and job titles with dates held must be included as a minimum to facilitate epidemiological studies of mortality.

† Portions reprinted from Levine (1978) by courtesy of the editors.

2(i) Prevention of asbestos-related diseases

The extent to which preventive medical practices can be put into effect in any particular situation will be limited, inevitably, by social and political considerations; and each industrial doctor must develop his own programme for preventing asbestos-related diseases. However, a number of suggestions can be made based upon the hypothesis that, as with smoking, all other risk factors act synergistically with asbestos. Although evidence for these recommendations is incomplete, the heavy toll of asbestos-related diseases requires the physician to follow the most reasonable course of action available to him now.

2(i).1 *Lung cancer*

Lung cancer is the most frequent of cancers related to asbestos exposure and today exacts the heaviest mortality of any asbestos-related disease. Several risk factors for lung cancer have been identified, as follow; while the inter-relation of these factors with asbestos exposure has been documented only for cigarette smoking, it would be wise to assume that they increase the risk of lung cancer in asbestos workers:

(a) *Cigarette smoking* greatly increases the mortality from lung cancer, and, based on present knowledge, it is the most important single risk factor.

(b) *A history of lung cancer in a first-degree relative*—parents, siblings, offspring—has been shown to confer almost the same magnitude of risk of lung cancer for the general population as cigarette smoking, and together both factors are synergistic (Tokuhata, 1976).

(c) *Prior occupational exposure to carcinogens affecting the lung* also may reasonably be expected to enhance the risk of lung cancer in an asbestos worker. Such carcinogens include arsenic, chloromethyl ethers, chromates, coal tar, petroleum, and their by-products, creosote, mustard gas, nickel, radium, and uranium (Cole and Goldman, 1975).

(d) *Non-malignant respiratory disease* may increase the risk of lung cancer. Pulmonary fibrosis has been noted in association with lung cancer in persons with scleroderma (Montgomery *et al.*, 1964; Godeau *et al.*, 1974) and certain rare hereditary diseases such as fibrocystic pulmonary dysplasia (Koch, 1965; Swaye *et al.*, 1969) and congenital cystic disease of the lung (McKusick and Fisher, 1958). Several reports have associated an excess risk of lung cancer with chronic bronchitis, after taking into account differences in smoking habits (Boucot *et al.*, 1966; Campbell and Lee, 1963; Dean, 1966; Van Der Wal *et al.*, 1966). However, such studies have often employed broad smoking categories and have neglected such important variables as duration of smoking and extent of inhalation (Wynder and Fairchild, 1966). Perhaps most convincing was the finding of a substantial excess mortality attributed to non-malignant respiratory diseases among non-smoking blood relatives of lung cancer cases that was not present among case spouses (Tokuhata and Lilienfeld, 1966). This suggests a propensity for non-malignant respiratory disease independent of smoking behaviour among lung cancer cases themselves. Significantly higher rates of impaired forced expiration have recently been found among never-smoking relatives of patients with lung cancer or chronic obstructive pulmonary disease when compared with neighbourhood controls (Cohen *et al.*, 1977). Experiments with animals indicate that individuals with asbestosis may be at increased risk of lung cancer, regardless of the level and duration of asbestos exposure (Wagner *et al.*, 1974).

(e) *Immunodeficiency* may play a role in the pathogenesis of lung cancer. Immunosuppressed renal transplant recipients have been reported to have an excess risk, and two cases of lung cancer were observed among members of a family with a striking array of lymphoproliferative disorders attributed to anomalous genetic control of IgM-bearing lymphocytes (Blattner, 1977).

(f) *Vitamin A intake* at all levels of cigarette smoking may be negatively related to incidence of lung cancer, according to a Norwegian study (Bjelke, 1975). High doses of vitamin A and its synthetic analogs (together called retinoids) have been shown to be useful for preventing lung cancer in laboratory animals (Sporn, 1976).

Possible laboratory measures of risk of lung cancer include inducible levels of lymphocyte aryl hydrocarbon hydroxylase (AHH) and the condition of exfoliated cells in sputum. AHH is an inducible enzyme thought to be responsible for converting hydrocarbon carcinogens such as are found in tobacco smoke into an active form. One team of investigators has reported that levels of inducible AHH activity in human tissues appear to be genetically regulated and that patients with bronchogenic carcinoma have higher levels of inducible activity than controls (Kellermann *et al.*, 1973a,b). Inducible enzyme levels, therefore, might indicate among cigarette smokers those individuals with the greatest risk of developing lung cancer. At present, however, there are numerous difficulties with the assay technique even in the most experienced laboratories, and the method is not ready for general use (Shaw, 1975). Further, the association between levels of inducible AHH activity and cancer has not yet been firmly established (Paigen *et al.*, 1977).

Examination of exfoliated cells found in sputum has been used to assess the state of the human tracheobronchial tree. Cancer has been observed to develop after a series of gradual cytological changes occurring over several years. These changes have been categorized into stages of regular squamous cell metaplasia, various degrees of atypical squamous cell metaplasia, carcinoma *in situ*, and invasive carcinoma (Saccomanno *et al.*, 1974). The more severe the cellular changes, the greater is the likelihood of developing lung cancer (Lilienfeld *et al.*, 1966). Cytological examination of sputum can be used to complement the use of risk factors listed above since, at any one time, the condition of exfoliated cells must reflect the interrelated effects of all operating risk factors, age, and latency. A great limitation to the widespread use of this technique at present is the paucity of laboratories capable of accurate cytologic diagnosis of sputum samples (Saccomanno, 1976).

Miners and millers of chrysotile asbestos may present a special situation for measurement of risk since their cumulative residual pulmonary asbestos may be able to be estimated magnetically by means of ferromagnetite contaminants present in the ore (Cohen, 1973). This technique is still under development.

Persons with any of the risk factors for lung cancer or whose sputum examination reveals cytopathology of severity equal to or greater than moderate atypical squamous cell metaplasia should preferably not be hired for jobs involving asbestos exposure.

Although no evidence exists to show that the risk of lung cancer may be reduced among smoking asbestos workers who stop smoking, it is reasonable to assume that this indeed would be the case, as it is for the general population. Asbestos workers who smoke, therefore, should be encouraged to stop smoking and to enroll in smoking cessation programmes.

Individuals who develop stigmata of asbestos-related diseases (e.g. pleural thickening) should be removed from asbestos exposure, and a thorough epidemiological investigation should ensue. Deficient working conditions or work practices, when found, must be remedied. Consideration should also be given to removing from exposure any person who develops moderate or severe atypical squamous cell metaplasia of exfoliated cells in sputum (although it is not known if removal from exposure at this point will influence the progression of cellular atypias to carcinoma).

Persons working with asbestos should be advised to maintain an adequate daily intake of vitamin A. Physicians are urged to keep abreast of developments in the synthetic chemistry of retinoids, which may make structures of great potency and little toxicity available for purposes of cancer prevention in high-risk groups.

2(i).2 *Laryngeal cancer*

Except for family history, the risk factors identified for lung cancer generally apply to cancer of the larynx. Risk of laryngeal cancer has in addition been correlated with alcohol consumption (Hammond, 1975; Rothman, 1975; Lynch, 1976a). This cancer is rare, and most highly susceptible persons will have been screened out in efforts to prevent lung cancer.

2(i).3 *Mesothelioma*

Little is known about susceptibility to mesothelioma, an important cause of mortality in some groups of asbestos workers. Persons who have had an asbestos-related pleural effusion may be at greater risk (Gaensler, 1976) and should be removed from asbestos exposure.

2(i).4 *Cancers of the alimentary tract*

These cancers occur about twice as frequently among heavily exposed asbestos workers as among the general population. Cigarette, pipe, and cigar smoking, the consumption of alcoholic beverages, and the chewing of tobacco and betel have been linked to excessive mortality from cancers of the oropharynx and oesophagus (Hammond, 1966, 1975; Rothman, 1975). In addition, the consumption of beer has been correlated with mortality from rectal cancer (Rothman, 1975). Persons with the hereditary skin condition tylosis palmaris et plantaris have extraordinarily high rates of cancer of the oesophagus (Lynch, 1976b) and high rates of cancer of the entire gastrointestinal tract, especially of the colon and rectum, are associated with the familial polyposis coli syndromes or a family history of juvenile polyposis (Lynch *et al.*, 1976; Sherlock and Winawer, 1977). There is an increased risk of gastric or colon cancer in general among persons with a history of a close relative having had the same cancer, and persons with previous colon polyps or colon cancer or with multiple sebaceous tumours are at greater risk of subsequent colon cancer (Lynch *et al.*, 1976; Sherlock and Winawer, 1977; Gorlin, 1977). Achlorhydria, pernicious anaemia, and a family history of ataxia telangiectasia are known risk factors for gastric cancer, while colon cancer occurs frequently among persons with ulcerative colitis and possibly other chronic inflammatory bowel diseases (Lynch, 1976a; Lynch *et al.*, 1976; Zamcheck *et al.*, 1955; Swift, 1977).

Since lung cancer has been found to be a far more important cause of excessive mortality among all groups of asbestos workers studied to date, *individual* risk factors of importance only to cancers of the alimentary tract, except in instances where the risk is very great, are probably insufficient grounds for denying employment to job applicants. It must be stressed, however, that the epidemiology on which this judgment is based comes exclusively from studies conducted in Europe and North America. Excess mortality from alimentary tract cancers among asbestos workers may be far more significant in regions of the world such as Japan (Miyagi Prefecture) where the incidence of these cancers in the general population is high and contrasts with a low rate of lung cancer.

Persons highly susceptible to gastrointestinal cancer, such as those with tylosis, familial polyposis, or ulcerative colitis, and persons with *multiple* risk factors relating only to alimentary tract cancer (such as pipe or cigar smoking, heavy alcohol consumption, and a history of colon cancer in a close relative) should preferably not be hired for jobs involving asbestos exposure.

2(i).5 *Asbestosis*

Despite suggested effects of cigarette smoking (Weiss, 1971), immunological responses (Turner-Warwick, 1973; Kang *et al.*, 1974) and specific HL-A antigens (Merchant *et al.*, 1975), risk factors for asbestosis have not been established (Becklake, 1976). The previously mentioned electromagnetic technique being developed to estimate residual pulmonary asbestos in miners and millers of chrysotile asbestos may prove useful for measuring the risk of asbestosis in addition to lung cancer.

Because of the potential adverse effects of further respiratory disability in individuals with existing respiratory disease, as well as their possible increased risk of lung cancer, it would be prudent not to hire such persons for jobs involving asbestos exposure.

2(ii) Early detection and treatment of asbestos-related diseases

Treatment of asbestos-related diseases, even when detected early, is far from satisfactory. Cure is rarely possible, and often death supervenes despite the best of efforts. Furthermore, programmes for early detection and treatment run the risk of instilling a false sense of complacency, which may detract from more important efforts to prevent disease. This must be kept in mind during the subsequent discussion.

2(ii).1 *Lung cancer*

Screening programmes which rely on roentgenograms and symptom questionnaires at intervals of 6 months have been notably unsuccessful in improving the chances of survival from lung cancer (Brett, 1968; Boucot and Weiss, 1973). The Philadelphia Pulmonary Neoplasm Research Project reported a 5-year survival rate of only 12% in individuals whose tumours were detected within 6 months of a negative roentgenogram, compared with 4% in those whose tumours were detected more than 6 months afterwards (Boucot and Weiss, 1973).

A semi-annual screening programme conducted among residents of Veterans Administration domiciliaries and consisting of stereoroentgen films, questionnaires, and sputum cytology slides reported a 3-year post-operative survival of only 12% (Lilienfeld *et al.*, 1966). This study documented a considerable amount of inter- and intra-observer variability in the interpretation of sputum smears and chest X-ray films, but noted that the addition of positive and suspect sputum cytopathology increased the sensitivity of the screening method by about 50% without substantially compromising its specificity (Lilienfeld *et al.*, 1966; Lilienfeld and Kordan, 1966; Archer *et al.*, 1966).

Currently, large detection and follow-up programmes are underway at the Mayo Foundation, Johns Hopkins University, and Memorial Sloan-Kettering Cancer Center to evaluate the efficacy of sputum cytological examinations, chest X-rays, and questionnaires administered every 4 months. At the Mayo Lung Project, each chest X-ray is reviewed by three doctors individually; chest roentgenography for patients having follow-up examinations at the Mayo Clinic itself consists of posteroanterior (PA) stereoroentgenograms and 350-kV PA views, as opposed to conventional PA and lateral films (Fontana *et al.*, 1975; Fontana, 1977).

In the presence of normal chest X-rays, if a single sputum specimen contains frankly cancerous cells or if repeated specimens from the same individual contain markedly atypical cells, procedures to localize a tumour are begun. Detailed radiological and radioisotope studies are undertaken, and a thorough otolaryngological examination is made to rule out cancer of the upper respiratory tract. If localization is not achieved, a meticulous endoscopic investigation follows, utilizing fibre-optic bronchoscopy, followed if necessary by bronchographic studies (Fontana *et al.*, 1975).

Initial results from the Mayo Lung Project suggest that not more than a third of detected cases of lung cancer may be expected to survive 5 years or more. However, more observation is needed in order to determine actual rates of survival. Roentgenographically occult tumours, which are likely to be centrally placed, were generally smaller and had a better post-operative prognosis. Most newly diagnosed lung cancers (64%) were detectable by chest X-ray alone, and only 13% of cancers detected after an initially negative screen were first noted as the result of clinical symptoms (Fontana *et al.*, 1975; Fontana, 1977).

Whatever the outcome of the current detection and follow-up programmes, certain limitations will apply:

(a) Some individuals should be ineligible for screening because of inability to tolerate pulmonary resection or unwillingness to undergo operation if it becomes necessary (Weiss *et al.*, 1975; Baker *et al.*, 1974; Grzybowski and Coy, 1970). To screen such persons not only would be wasteful, but also might ultimately contribute to lowering the morale of other participants in the screening programme—that is, whether or not the death of a screening participant was due to inability or unwillingness to be operated on, it might be regarded by others as a failure of the programme.

(b) A considerable number of persons may discontinue participation in the screening programme because of retirement, termination of employment, or other reasons. Since the incidence of lung cancer increases with age (Lilienfeld *et al.*, 1966) and has been found in one study to be higher among screening dropouts (Weiss *et al.*, 1975), there is reason to believe that persons who discontinue participation in a screening programme may in fact be at greater risk.

(c) A certain proportion of screening examinations will be found to be incomplete or technically unsatisfactory (Lilienfeld *et al.*, 1966).

(d) Once a cancer has been detected, there will inevitably be a delay until localization and operative resection. The greater this delay, the less the potential benefit that can accrue from the screening programme.

(e) It will be difficult to extrapolate from results obtained at premier centres of medical care to results likely to obtain at an average industry programme. (The difficulty in finding laboratories capable of accurate sputum cytological diagnosis has already been mentioned.)

(f) The considerable financial expense of a screening programme and the drain on available time of medical care personnel should not be overlooked.

(g) Despite screening, some cancers will not be detected early and, despite early detection, certain cancers will be inoperable or have a poor prognosis.

(h) A not inconsiderable percentage of persons who will have been successfully operated on to remove a lung cancer will develop a second tumour (Fontana *et al.*, 1975).

Despite these limitations, medical screening remains the only possible means of assisting the unfortunate individuals destined to develop asbestos-related diseases and should not be abandoned. The use of sputum cytology to distinguish persons of heightened susceptibility to lung cancer for removal from asbestos exposure, motivation to give up smoking, or treatment with high doses of retinoids deserves to be tested. Screening examinations remind the worker of the health hazards of his job and may be used to enlist his cooperation in improving work practices as well as in changing detrimental personal habits.

It is suggested that programmes for early detection of lung cancer in asbestos workers should consist of periodic sputum cytological examinations, chest X-rays, and symptom questionnaires administered according to a schedule which varies with age, risk of lung cancer, and time elapsed since first exposure to asbestos. Available medical facilities that are competent in localizing and resecting lung cancer should be identified in advance of the programme to minimize the time taken from detection of cancer to operation. Emphasis must be placed on proper chest roentgenography, since chest X-rays alone should detect the majority of lung tumours (Lilienfeld *et al.*, 1966; Fontana, 1977)—sputum cytology is more effective in detecting centrally placed tumours (Fontana *et al.*, 1975), whereas asbestos workers may more likely develop peripheral lung cancers (adenocarcinomas) Whitwell *et al.*, 1974). Each chest film must be read independently by more than one doctor especially trained to detect early lung cancer and qualified in the ILO-U/C classification of radiographs of pneumoconioses (International Labour Office, 1971).

A protocol of medical examinations for asbestos-exposed workers is presented in Table 16.1. It must be emphasized that the real benefits of such a protocol, if any, are unknown at present.

2(ii).2 *Laryngeal cancer*

Asbestos workers with clinical symptoms of hoarseness or pain or soreness of the throat should be referred to an ear, nose, and throat specialist for a detailed otolaryngological examination of the upper respiratory tract.

2(ii).3 *Mesothelioma*

At present, mesotheliomas are uniformly fatal. Neither radical surgery, radiation, nor chemotherapy prolongs survival; in fact, these modes of treatment may be harmful. Since no useful therapy is available, screening for early detection beyond what may be done to detect lung cancer is of no clinical value. Invasive diagnostic procedures should be kept to a minimum, and management restricted to the relief of pain and breathlessness (Elmes, 1973; Elmes and Simpson, 1976).

2(ii).4 *Cancers of the alimentary tract*

Faecal occult-blood testing has been used as an annual screening device to detect colorectal cancer in asymptomatic men and women 40 years or more old (Winawer *et al.*, 1977), and has been recommended at more frequent intervals for the surveillance of gastric cancer in patients with pernicious anaemia (Zamcheck *et al.*, 1955). Positive tests have been obtained from persons with cancer of the stomach, small intestine, colon, and rectum, and, to a lesser extent, from persons with benign gastrointestinal lesions (Stephens and Lawrenson, 1970). False positives can be reduced considerably by using guiac-impregnated filter-paper slides (Hemoccult, Smith, Kline and French Laboratories, Philadelphia) in conjunction with a diet high in residue (to stimulate bleeding from existing lesions) and free of red meat and high peroxidase foods (e.g. horseradish and beets). Vitamins and aspirin-containing medications should also be avoided (Winawer *et al.*, 1977; Greegor, 1971; Ostrow *et al.*, 1973).

Persons with positive results should be referred to a gastroenterologist for

Table 16.1 Medical examinations for asbestos-exposed workers: an experimental protocol*

Pre-employment

1. Questionnaire: medical history, family history, history of smoking† and consumption of alcoholic beverages, occupational history.
2. Physical examination: concentrating on the oral cavity, chest, and abdomen, and including a digital examination of the rectum.
3. Spirometry: including measurements of vital capacity, forced vital capacity, and forced expiratory volume at 1 second.
4. Chest X-ray: 14 × 17 inch posteroanterior and lateral films.
5. Sputum cytology.

Follow-up (*a*) *Non-smokers, ex-smokers, and smokers who do not inhale*

(i) No more than mild atypical sputum cytopathology:
a yearly questionnaire, spirometry, chest X-rays, and sputum cytology.

(ii) More than mild atypical sputum cytopathology:
a yearly questionnaire and spirometry; chest X-rays and sputum cytology every 4 months.

(iii) 40 years old and older, at least 20 years from onset of asbestos exposure:
add faecal occult-blood testing and an examination of the oral cavity every 6 months.

(*b*) *Smokers who inhale*

(i) Less than 15 years from onset of asbestos exposure:
No more than mild atypical sputum cytopathology:
a yearly questionnaire, spirometry, chest X-rays, and sputum cytology.
More than mild atypical sputum cytopathology:
a yearly questionnaire and spirometry; chest X-rays and sputum cytology every 4 months.

(ii) 15–20 years from onset of asbestos exposure:
No more than mild atypical sputum cytopathology:
a yearly questionnaire and spirometry: chest X-rays and sputum cytology every 6 months.
More than mild atypical sputum cytopathology:
a yearly questionnaire and spirometry; chest X-rays and sputum cytology every 4 months.

(iii) More than 20 years from onset of asbestos exposure:
Less than 40 years old:
a yearly questionnaire and spirometry; chest X-rays and sputum cytology every 4 months.
40 years old and older:
add faecal occult-blood testing and an examination of the oral cavity every 6 months.

* Source: protocol modified from the Mt. Sinai School of Medicine Environmental Sciences Laboratory Pulmonary Surveillance Programme for Asbestos-Exposed Workers.

further studies, which may include endoscopy, cytology, biopsy, and radiology. Early detection and excision of colorectal and gastric cancers, it has been noted, may result in 5-year survival rates as high as 90%, compared with overall national averages of 40% and 10%, respectively (Sherlock and Winawer, 1974; Prolla *et al.*, 1969).

Biennial faecal occult-blood testing and an examination of the oral cavity may be considered in the case of asbestos workers 40 years or more old, especially those with 20 or more years since their first exposure to asbestos (see Table 16.1).

2(ii).5 *Asbestosis*

Periodic comparative chest X-rays and pulmonary function tests (see Table 16.1) will improve the chances of detecting early asbestosis. Many abnormalities, however, are non-specific, and it will be difficult to determine if these reflect early asbestosis or are merely related to smoking or ageing (Weiss, 1967). Pleural thickening or plaques in an asbestos worker must always be suspected as evidence of a biological effect related to inhaled asbestos (Selikoff, 1965). Persons with early asbestosis or with pleural thickening should be removed from asbestos exposure and referred to a chest specialist for careful follow-up.

(3) SMOKING CESSATION PROGRAMMES

As noted previously, smoking is the most important single predictor of enhanced carcinogenic risk in asbestos workers. Hence, it is reasonable to assume the risk of cancer would be reduced among asbestos workers who stop smoking and that, therefore, smoking cessation programmes might assist in achieving this goal.

Unfortunately, although a number of different programme strategies may reduce smoking rates, significant long-term cessation is rare. Most reviews of the effects of smoking-cessation techniques indicate that the greatest rate of recidivism occurs between 1 and 5 weeks following treatment, when only 30% of those who had been abstinent at the end of treatment report continued abstinence. Follow-up at 3–18 months indicates approximate cessation rates of 20–30%. One report noted 18% cessation at 5 years after treatment (West *et al.*, 1977).

Most smoking cessation programmes have enrolled volunteers, who may have been motivated to quit on their own. Since 16% of persons who

---

† Since smoking is such an important risk factor, breath should be sniffed for tobacco odour; and in situations where the reliability of smoking histories is in doubt, levels of expired air carbon monoxide or serum thiocyanate may be used to distinguish cigarette smokers from non-smokers (Vogt *et al.*, 1977).

quit by themselves have been reported to remain abstinent 1 year (Guilford, 1966), the 20–30% rates of abstinence achieved 3–18 months after smoking cessation programmes suggest that formal cessation techniques may be only of modest benefit. Such programmes, however, may be substantially beneficial for persons who might not have been able to give up smoking on their own.

The various approaches to smoking cessation are discussed in this section with a view to providing a broad perspective of the available options*. While it is difficult to recommend particular methods, it is hoped that the discussion will assist the health care worker to select an appropriate programme. All programmes must emphasize maintaining abstinence in the long term.

3(i) Approaches to smoking cessation

3(i).1 *Medical advice*

When they advise patients to give up or reduce smoking, many health professionals, particularly doctors, serve as agents of behaviour change. One report described the results of a counselling programme in which 100 surgical patients were seen for a brief interview during which the health hazards and financial burdens of smoking were discussed. After 1 year, 39% of males and 11% of females had stopped smoking. Following another such programme, which included medical lectures, physical examinations, group discussions, and films, the cessation rate was 58%, and 29% were still abstinent 1 year later (Handel, 1973; Delarue, 1973).

A number of health organizations have developed resource material to assist doctors in counselling patients who smoke (see Appendix).

3(i).2 *Self-help programmes*

A recent Gallup poll (1974) indicated that only 34% of smokers expressing a desire to quit were interested in attending cessation clinics. The majority of smokers who became abstinent gave up without the use of formal smoking-cessation interventions (National Cancer institute, in press).

There appears, however, to be a great deal of interest in self-help manuals and kits as judged by the amount of such materials requested during anti-smoking campaigns (Dubren, 1977). A typical self-help manual includes a behavioural interpretation of smoking, a general explanation of the

* There are several published works that provide a more comprehensive review of the subject (Bernstein, 1969; Bernstein and McAlister, 1976; Hunt and Matarazzo, 1973; Schwartz, 1969; Schwartz and Rider, 1977).

principles of behaviour change, and specific instructions for implementing self-control procedures directed at smoking cessation (Lichtenstein and Danaher, 1976). For example, the smoker might be instructed to:

(a) list positive reasons for quitting;
(b) record the time each cigarette is smoked;
(c) note feelings and behaviour prior to and during each smoke;
(d) reduce gradually the number of cigarettes smoked;
(e) impose circumstantial barriers to smoking, such as not carrying matches;
(f) change to a brand lower in tar and nicotine, twice weekly;
(g) increase physical activity;
(h) avoid situations most closely associated with smoking;
(i) find substitute behaviours for smoking.

The effectiveness of self-help programmes has not been established.

3(i).3 *Group therapy and five-day plan*

Various health organizations have sponsored community group smoking cessation clinics, which provide health information, encouragement, and group therapy. Groups typically involve 8–18 persons and a group leader, meeting once or twice weekly for a month. Participants in the group are informed of smoking risks, asked to describe why they smoke and to detail their smoking habits, and are then encouraged to follow one of several procedures for giving up. Estimates of the effectiveness of these programmes after 1 year range from 18 to 25% (National Clearinghouse for Smoking and Health, 1976; Schwartz and Dubitzky, 1968). Higher success rates result from programmes stressing formal long-term maintenance support.

One variant of group therapy is the Five-Day Plan, which consists of five daily meetings of $1\frac{1}{2}$–2 h each. Up to several hundred volunteer participants may be treated at these sessions. Intervention strategies range from lectures and inspirational messages to fear-arousing stimuli and behaviour-modification procedures. One such programme—a live-in clinic including lectures, exercise, and individual and group therapy—reported cessation rates of 21–40% at a 3-month follow-up (Schwartz and Rider, 1977).

Isolated use of the non-specific treatment factors characteristic of the smoking clinic approach (e.g. suggestion, high expectation of success in quitting) result in post-treatment cessation rates similar to those of clinics that use specific planned intervention strategies (Bernstein, 1969; Lichtenstein *et al.*, 1973). A follow-up of a group of 559 volunteers who had attended such a smoking clinic (pharmacological agents, health education, and brief suggestions to use certain techniques for quitting) indicated that only 18% remained abstinent at 5 years (West *et al.*, 1977).

3(i).4 *Behaviour therapies*

The behaviour therapy approach assumes that problematic behaviour is a function of a person's learned pattern of interacting with the environment. The goal is to teach more adaptive means of responding. The initial step is a 'behavioural assessment', an evaluation that includes identification of antecedent events that trigger smoking, determination of thoughts and feelings that influence smoking behaviour, analysis of personal smoking behaviour (frequency, situations), and identification of the psychological consequences of smoking. A number of techniques may then be used to teach the individual new patterns of interacting with environmental smoking stimuli (Bernstein and McAlister, 1976; Lichtenstein and Danaher, 1976). These techniques can be grouped into five major categories: systematic desensitization, punishment and aversive conditioning, stimulus control, reinforcement of non-smoking, and multicomponent interventions.

Assuming that it is anxiety that elicits the urge to smoke, attempts have been made using relaxation techniques to *desensitize smokers systematically* to anxiety-evoking stimuli (Koenig and Masters, 1965; Pyke *et al.*, 1966; Wagner and Bragg, 1970). However, no substantial effect on smoking behaviour has been reported.

A number of *aversive conditioning* techniques have been employed to modify smoking behaviour. Loud noises or electric shocks have been coupled to smoking or to the urge to smoke, generally with little effect (Lichtenstein and Danaher, 1976).

Two techniques incorporate cigarette smoke as the aversive stimulus:

(a) *rapid smoking*, which requires the individual to smoke rapidly and continuously, sometimes in conjunction with drafts of warm smoky air;

(b) *satiation*, which requires increasing cigarette consumption over a certain period of time (e.g. smoking double or triple the usual amount for a week).

In the context of a persuasive personal relationship between patient and therapist, the rapid smoking technique appears to result in about 50% abstinence 3–6 months following termination (Lichtenstein and Danaher, 1976). While one report of a satiation programme indicated a 62% cessation rate 4 months following treatment (Resnick, 1968), other programmes have not been as successful (Lichtenstein and Danaher, 1976). Before implementing the rapid smoking method, medical screening of participants should be required owing to the possible deleterious effects of increased carbon monoxide levels (Miller *et al.*, 1977).

Another behavioural technique is *stimulus control*. This may involve forbidding smoking in situations where smoking would habitually occur (e.g. no smoking while drinking coffee, watching television, or following a meal), as well as gradually restricting the number of situations where smoking is permitted. While there appears to be no clear cessation effect from using this method and a high attrition has been frequently reported, reduction in the rate of cigarette consumption does result (Bernstein and McAlister, 1976).

The *reinforcement of abstinence* through the use of social or monetary incentives has been relatively successful. An example of a technique incorporating a monetary incentive is the use of a deposit made prior to initiation of the programme. The money is returned in portions made contingent upon progressively longer periods of abstinence (Tiche and Elliott, 1967; Winett, 1973). A cessation rate of 50% has been reported at 6 months compared with 24% in a group not employing the incentive.

Multicomponent interventions have been designed by combining these techniques. A number of reports indicate that this approach may yield high abstinence rates (65–100% immediately and 55–65% after 1 year (Harris and Rothberg, 1972; Chapman *et al.*, 1971; Morrow *et al.*, 1973; Pomerleau and Ciccone, 1974; Tooley and Pratt, 1967).

3(i).5 *Filters for gradual smoking cessation*

These are a series of cigarette filters designed to assist smoking cessation by gradually reducing levels of inhaled tar and nicotine. The effectiveness of this method has not been evaluated (National Cancer Institute, in press).

3(i).6 *Use of pharmacological agents*

A diverse array of medications, including stimulants as well as tranquillizers, nicotine substitutes as well as antagonists, and anticholinergics, have been prescribed to assist smokers in overcoming withdrawal symptoms. With regard to actual cessation, however, none of these pharmacological agents has shown any more promise than placebos (Gritz and Jarvik, 1977). In fact, several studies have suggested that placebo groups may have higher cessation rates (Schwartz, 1969).

3(i).7 *Acupuncture*

Very little research is available on the effects of acupuncture as a smoking cessation technique. In one study, 50% of subjects were reported to be abstinent 6 weeks following auricular acupuncture (Globglas, 1975). This cessation rate is similar to those obtained using other techniques.

3(i).8 *Hypnosis*

Hypnotic techniques have included attempts to reveal personality conflicts assumed to be major underlying causes of the smoking habit as well as direct suggestions to give up smoking (Bryan, 1964; Johnston and Donoghue, 1971). While certain investigators have reported positive initial results (Kroger, 1963; von Dedenroth, 1968), there have been few controlled studies. Furthermore, initial successes have not been maintained at follow-up. One report of the results of a single-session self-hypnotic treatment noted a cessation rate at 1 year of 20% (Spiegel, 1970). Using this approach, another investigator has observed significantly higher rates of recidivism than with group counselling (Shewchuk, 1976a).

3(ii) Issues in smoking cessation

3(ii).1 *Monitoring smoking behaviour*

As with any intervention, monitoring of treatment effectiveness is critical. Although self-report has been the major measure used to date, it lacks precision for a variety of reasons, which include the desire to please, denial of shortcomings, or inconsistent motivation to maintain accurate records over reporting intervals. The use of such objective measures as serum thiocyanate or expired-air carbon monoxide along with self-report may assist in this regard (Vogt *et al.*, 1977).

3(ii).2 *The 'successful quitter'*

The 'successful quitter' is likely to be a man, to be concerned about his health, to be older, to report fewer neurotic or psychosomatic symptoms, to smoke less and to have begun smoking at a later age, to have a supportive social milieu, and to have tried to give up smoking on several previous occasions (Pederson and Lefcoe, 1976; West *et al.*, 1977). The behaviour of spouses appears to be a significant factor. Smokers with non-smoking spouses, spouses who also give up smoking, or spouses who 'made it easier to give up' were more likely to remain abstinent at 5-year follow-up (West *et al.*, 1977).

3(ii).3 *Weight gain*

A frequent concern expressed by those planning to stop smoking is whether or not they will gain weight. Indeed, three quarters of the persons in one study gained weight after giving up cigarettes (Hammond and Percy, 1958). Dietary or behavioural weight management may possibly be used to good advantage in cessation programmes.

3(ii).4 *Long-term maintenance*

While immediate cessation of smoking can be achieved with a variety of techniques, long-term abstinence is seldom observed. Earlier techniques for promoting long-term abstinence through the use of non-structured supportive groups and a telephone 'hot line' were found to be used infrequently by ex-smokers. Later efforts, which directed ex-smokers to telephone a specific number daily (more frequently, if necessary) to receive a 3-min supportive message, seem to have been more helpful. Additional techniques being investigated include the use of congratulatory messages (mailed or telephoned), structured group sessions at specified intervals, and monetary and social incentives (Shewchuk, 1976b). An example of mobilizing social support at the workplace is the use of a 'buddy system'. Under this plan, two workers giving up smoking together support each other's efforts to remain abstinent.

3(iii) Smoking cessation programmes in industry

Industry provides an ideal setting for smoking cessation programmes. Industrial programmes often have the advantages of:

(a) an existing system of occupational health care, which affords a means of careful health surveillance and follow-up;

(b) an established network of communications, which permits the rapid dissemination of health information;

(c) peer and management interaction, which provide for social incentives;

(d) readily perceived benefits in terms of diminished illness and absenteeism, which can be assessed against programme costs.

While the plant doctor may wish to refer workers to smoking cessation programmes outside the workplace, on-site programmes reduce lost time and inconvenience. The local Chapter of the American Cancer Society would be an excellent resource for information on available outside programmes, and lists of additional resources are given in the Appendix.

A number of imaginative industry-based smoking cessation programmes have been established, but as yet there have been no published reports of their effectiveness. Several programmes have been described in *Smoking Digest* (National Cancer Institute, in press):

(a) Intermatic Incorporated (Detroit) has a no-smoking 'parimutuel' window where employees can bet up to \$100 on their ability to stop smoking for 1 year. The company has contributed \$1000 to be divided among successful participants. Persons not remaining abstinent donate their bet to the American Cancer Society.

(b) The Aluminaire Standard Glass Company (Phoenix) has established a programme in which a dollar amount equivalent to what abstinent smokers would have spent for cigarettes is deducted from workers' salaries. At the end of 1 year the company matches the total deductions and pays the entire sum to the worker, provided that he has maintained abstinence.

(c) Sears Roebuck and Company (New York) encourages employees to take outside smoking cessation courses by rebating a portion of course fees to those remaining abstinent for 6 months or more.

(d) An interesting programme package including education, social, and monetary incentives was implemented at the Dow Chemical Company (Freeport, Texas) in collaboration with the American Cancer Society. Abstinence was rewarded by $1 each week, and abstinent workers were enrolled in monthly and quarterly lotteries for prizes that included a boat and engine as well as cash awards. Ex-smokers were used to recruit programme participants. For each recruited participant who remained abstinent for 1 month, the recruiter was awarded a chance in a lottery. [Recruiter incentive is thought to provide a useful source of social mobilization (Janis and Hoffman, 1970).] Of 395 participants, only 15 (less than 4%) continued to smoke at termination of the programme; however, the lack of adequate follow-up precludes an assessment of long-term effects.

While these programmes are useful in providing rationale and motivation for smoking abstinence, some persons do not possess the skills needed to quit. It would be useful to assist such persons through the use of behaviour therapies.

(4) EDUCATION

Asbestos was the subject of the first occupational safety and health standard issued by the U.S. Department of Labor following passage of the Occupational Safety and Health Act of 1970. Although the standard has been amended since then, no specific directive has yet been issued regarding the education of persons who work with asbestos. In standards for a number of other carcinogens, however, requirements for employee training and indoctrination have been specified. These have included the nature of the carcinogenic hazard, the nature of operations involving the carcinogen, the recognition of conditions which may release the carcinogen, the nature and purpose of decontamination practices, emergency practices and procedures and the specific role of the employee in an emergency, and the nature and purpose of the medical surveillance programme (U.S. Code of Federal Regulations, 1976). It would appear that a comparable mandate should exist for work with asbestos.

4(i) Goals

Persons who work with asbestos should be given sufficient information about the hazards to enable them (1) to decide whether they want to work or to continue to work in the face of these hazards, (2) to take steps to minimize personal exposure, and (3) to assist in monitoring and improving the work environment (Committee on Public Information in the Prevention of Occupational Cancer, 1977). Implicit in these fundamental goals are several secondary educational objectives:

(a) knowledge of the physical characteristics of asbestos;

(b) knowledge of the work processes involving asbestos and of the potential for fibre emissions;

(c) knowledge of the nature and purposes of engineering and work practice control methods, with emphasis on how the worker can assist in reducing exposure;

(d) knowledge of the reasons for, and methods of, environmental monitoring;

(e) knowledge of the diseases that may result from exposure to asbestos fibres and of the concepts of latency and risk (both relative risk and absolute risk);

(f) knowledge of the biologically significant routes of exposure to asbestos fibres;

(g) knowledge of the role of related factors such as smoking which may increase the risk of disease, and of the importance of smoking cessation;

(h) knowledge of the elements and purposes of medical surveillance.

4(ii) Modes of education

Of the two principal methods of delivering health messages, the written word and the spoken word, the written word is far less effective. It does not convey personal feeling, nor does it permit opportunity for discussion. Moreover, worker reading skills may be limited.

A good opportunity for oral communication is when a doctor and an employee review together the results of a periodic medical examination. Attention is then focused on a single employee, whose apprehension may be heightened by misunderstanding, inadequate information, folk beliefs, or rumour.

The orientation of new employees provides an excellent forum for group health education. The occupational health staff should use the occasion to discuss the purposes of the pre-employment medical examinations recently completed and to describe employee health services. Pursuit of a health problem in the future will be facilitated by having had this introduction. Industrial hygienists should also participate in the orientation to explain environmental controls, proper work practices, and the use of personal protective equipment.

4(iii) The educators

Persons from a variety of backgrounds may be involved in the education effort: doctors, nurses, industrial hygienists, health educators, safety specialists, and others. The more the educators understand employee values, beliefs, and behaviour, and the less their opinions are considered to reflect management bias, the more effective they will be. Industrial educators should work closely with union health and safety specialists, for they can ensure that important health messages will be delivered.

4(iv) The persons to be educated

There are many target groups within industry for whom instruction in the prevention of asbestos-related diseases is needed. Apart from actively employed workers, these include retired or terminated asbestos workers, families of asbestos workers, industrial managers, and trade union officials.

4(iv).1 *Actively employed workers*

While workers in the asbestos trades are obvious targets of education efforts, others whose work may involve exposure to asbestos—sometimes through mere juxtaposition to asbestos workers—need to be taught of the hazard and how to minimize it. These workers include construction and demolition crew for ships and buildings, welders, painters, electricians, sheet metal workers, carpenters, marine machinists, shipfitters, and automotive mechanics who install and repair clutch and brake linings.

4(iv).2 *Retired or terminated asbestos workers*

The asbestos worker cannot be permitted to depart with the misapprehension that if he no longer has contact with the material, he is no longer at risk of developing asbestos-related diseases. Exit interviews provide an opportunity to emphasize the need for lifetime contact and to stress the importance of smoking cessation and medical surveillance. It is good practice to provide a list of local chest specialists. Whenever possible, former asbestos workers should be included in plant programmes of medical surveillance, smoking cessation, and health education.

4(iv).3 *Families of asbestos workers*

Families of asbestos workers must be told about asbestos-related diseases, of the importance of smoking cessation, environmental controls, and good work practices, and of the potential hazards of asbestos fibres brought home on body, clothing, equipment, lunchboxes, and automobiles, and as souvenir samples of ore and fibre. Communication is good with mixed audiences, and programmes of prevention which require worker cooperation will be strongly reinforced by informed families.

4(v) Assessment of results

The value of programmes of health education must be assessed. Before and after educational programmes, target groups may be examined for appropriate knowledge of environmental hazards, occupational diseases, and control methods, for compliance in the use of safe work practices, personal protective equipment, and in adherence to administrative restrictions, and for participation in programmes of medical surveillance.

(5) APPENDIX

5(i) Educational aids for smoking cessation*

Pamphlets	
Cigarette Smoking	ALA†
Smoker's Self-Testing Kit	NCSH
If You Must Smoke (Tar and Nicotine Content)	NCSH
Smoker's Aid to Nonsmoking; a Scorecard	NCSH
Facts: Smoking and Health	NCSH
Me Quit Smoking: Why?	ALA
Me Quit Smoking: How?	ALA
Be Kind to Nonsmokers	ALA
Pipe and Cigar Smoking	ALA
Second Hand Smoke	ALA
If You Want to Give Up Cigarettes	ACS
Remember When:	ALA
Q & A of Smoking and Health	ALA
What's Your Cigarette Smoking I.Q.?	ALA
Slide Presentations	AHA, ACS
How to Stop Smoking	AHA
Cigarette Quiz	AHA
What Everyone Should Know about Smoking and Heart Disease	AHA
Reprints	
This Doctor is Firm	ALA
Films	
Everything You Always Wanted to Know About How to Stop Smoking but were Afraid to Ask	ALA
Is it Worth Your Life	ALA
Point of View (Teens and Adults)	ALA

* Source: adapted from National Cancer Institute (in press).
† ALA = American Lung Association; ACS = American Cancer Society; NCSH = National Clearinghouse for Smoking and Health; AHA = American Heart Association.

Trigger Films for Smoking Clinics—'Health Aspects of Quitting', 'Insights into the Quitting Process', and 'Helping and Being Helped'	ACS
The Benefits of Quitting Smoking	ACS
Posters	
Thank You for Not Smoking	ALA, AHA
Lungs at Work, No Smoking (also in Tentcards and Buttons)	ALA
We All Share the Same Air	ALA
Thank You for Not Smoking	ALA

5(ii) Smoking cessation: potential sources for additional information (North America)*

Action on Smoking and Health (ASH),
2000 H Street, N.W.,
Washington, D.C. 20006.
(202) 659–4310†.

American Cancer Society,
777 Third Avenue,
New York, N.Y. 10017.
(212) 371–2900.

American Health Foundation,
Department of Behavioral Sciences,
1370 Avenue of the Americas,
New York, N.Y. 10019.
(212) 489–8700.

American Heart Association,
7320 Greenville Avenue,
Dallas, Texas 75231,
(214) 750–5300.

American Lung Association,
1740 Broadway,
New York, N.Y. 10019.
(212) 245–8000.

National Association on Smoking and Health,
4155 East Jewel Avenue,
Denver, Colo. 80237.
(303) 753–0777.

National Clearinghouse for Smoking and Health,
Public Health Service,
U.S. Department of Health, Education and Welfare,
Center for Disease Control,
Atlanta, Ga. 30333.
(404) 633–3311.

National Interagency Council on Smoking and Health,
419 Park Avenue South,
New York, N.Y. 10016
(212) 532–6035.

Occupational Safety and Health Administration,
U.S. Department of Labor,
Constitution Avenue and Third Street, N.W.,
Washington, D.C. 20210.
(202) 523–7081.

* Adapted from National Cancer Institute (in press).
† Telephone number.

Canadian Council on Smoking and Health,
343 O'Connor,
Ottawa, Ontario K-2P-1V9.
(613) 236–6035.

Cancer Information Clearinghouse,
7910 Woodmont Avenue, Suite 1320,
Bethesda, Md. 20014.
(310) 565–5955.

General Headquarters,
5 Day Plan to Stop Smoking,
Seventh Day Adventist Church,
Narcotics Education Division,
6840 Eastern Avenue, N.W.,
Washington, D.C. 20012.
(202) 723–0800.

Kaiser Foundation Research Institute,
1956 Webster St., Room 310B,
Oakland, Calif. 94612.
(415) 645–5000.

Schick Center,
4101 Frawley Drive,
Fort Worth, Texas 76118.
(817) 268–1157.

Smoking and Health Information Program (Tyler Asbestos Workers Program),
P.O. Box 2003,
Tyler, Texas 75710.
(214) 877–3011.

SmokEnders,
Memorial Parkway,
Phillipsburg, N.J. 08865.
(201) 454-HELP.

The Tobacco Institute,
1776 K Street, N.W.,
Washington, D.C. 20006.
(202) 457–4800.

5(iii) Asbestos educational materials

Sources of Published Educational Materials

1. *Asbestos Information Association/North America*
   Materials: Brochures on work practices, fact books, reprints of scientific articles, conference proceedings, and films; reference library.
   Address: 1745 Jefferson Davis Highway,
   Arlington, VA 22202.
   (703) 9791150
2. *Asbestos Information Committee*
   Materials: Brochures on work, practices, health effects, and control procedures.
   Address: 10 Wardour Street,
   London WIV 3HG,
   England.

Sources of Published Educational Materials (*Continued*)

3. *Asbestosis Research Council*
   Materials: Brochures on work practices, ventilation, control procedures, protective devices, and disposal methods.
   Address: P.O. Box 18, Cleckheaton,
   West Yorkshire BD19 3UJ, England.
4. *Congress of the United States*
   Materials: Pertinent Public Laws.
   Address: Document Room, Congress of the United States,
   Washington, D.C.
5. *Johns-Manville Corporation*
   Materials: Brochures on work practices, health effects, films, slides with sound, slides with script, video tapes, newspapers, bulletins, letters, reports, and reprints of scientific articles.
   Address: Health, Safety and Environment Department,
   Ken-Caryl Ranch,
   Denver, Colo. 80217
   Telephone: (303) 770-1000, extension 2969.
6. *National Institute of Occupational Safety and Health, U.S. Department of Health, Education and Welfare*
   Materials: 'Criteria Document' on asbestos; reports of occupational disease research and epidemiological investigations.
   Address: Robert A. Taft Laboratories,
   4676 Columbia Parkway,
   Cincinnati, Ohio 45226.
7. *National Safety Council*
   Materials: Reprints of articles, safety data sheets; reference library.
   Address: 425 North Michigan Avenue,
   Chicago, Ill. 60611.
   Telephone: (312) 527-4800.
8. *Occupational Safety and Health Administration, U.S. Department of Labor*
   Materials: Occupational Safety and Health Standard:
   Subpart Z, Sec. 1910.1001, Asbestos
   Brochures on OSHA 1970.
   Address: Occupational Safety and Health Administration,
   U.S. Department of Labor,
   Washington, D.C. 20210.
9. *Oil, Chemical and Atomic Workers International Union, AFL-CIO*
   Materials: Poster, slide-tape cassette presentation.
   Address: Citizenship-Legislative Department,
   1126 16th Street, N.W.,
   Washington, D.C. 20036.

Sources of Published Educational Materials (Continued)

10. *Quebec Asbestos Mining Association*
    Materials: Brochures on work practices, health effects, and control procedures.
    Address: 5 Place Ville Marie,
    Montreal 113, Quebec, Canada.

Possible Sources of Published Educational Materials

1. *American Association of Poison Control Centers*
   c/o, Academy of Medicine of Cleveland,
   Poison Information Center,
   10525 Carnegie Avenue,
   Cleveland, Ohio 44106.
   Telephone: (216) 231-3500.
2. *American Medical Association*
   535 North Dearborn Street,
   Chicago, Ill. 60610.
   Telephone: (312) 751-6000.
3. *American Public Health Association*
   1015 18th Street, N.W.,
   Washington, D.C. 20036.
   Telephone: (202) 467-500.
4. *California State Department of Health*
   Occupational Cancer Control Unit,
   2151 Berkeley Way,
   Berkeley, Calif. 94704.
   Telephone: (415) 843-7900, extension 306.
5. *Center for Science in the Public Interest*
   1757 S Street, N.W.,
   Washington, D.C. 20009.
   Telephone: (202) 332-4250.
6. *Companies mining asbestos ore.*
7. *Consumer Federation of America*
   1012 14th Street, N.W., Suite 901,
   Washington D.C. 20005.
   Telephone: (202) 737-3732.
8. *Consumers Union*
   256 Washington Street,
   Mount Vernon, N.Y. 10550.
   Telephone: (914) 664-6400.
9. *Health Research Group*
   2000 P Street, N.W., Suite 708,
   Washington, D.C. 20036.
   Telephone: (202) 872-0320.

Possible Sources of Published Educational Materials (*Continued*)

10. *Insulation Industry Hygiene Research Program*
    Environmental Sciences Laboratory,
    Mount Sinai School of Medicine of the City University of New York,
    Fifth Avenue and 100th Street,
    New York, N.Y. 10029.
11. *Local affiliates of American Cancer Society, American Lung Association, and National Safety Council.*
12. *Manufacturers or fabricators of asbestos products.*
13. *Mining Enforcement and Safety Administration*
    U.S. Department of Labor,
    Washington, D.C. 20210.
14. *Scientist's Institute for Public Information*
    49 East 53rd Street,
    New York, N.Y. 10022.
    Telephone: (212) 688-4050.
15. *SOURCE, Inc.*
    P.O. Box 21066,
    Washington, D.C. 20009.
    Telephone: (202) 387-1145.
16. *State Bureau of Mines.*
17. *State Workers' Compensation Authorities.*
18. *Vitalograph Ltd.*
    8347 Quivira Road,
    Lenexa, Kansas 66215.
    Telephone: (913) 888-4221.
19. *Workers' Compensation insurance underwriters* (*casualty insurance companies*).

Publications Available

1. *Monographs*
   (a) *IARC Monographs on the Evaluation of Carcinogenic Risk of Chemicals to Man, Vol.* 14, *Asbestos,* International Agency for Research on Cancer, Lyon, 1977.
   (b) *Public Health Risks of Exposure to Asbestos,* Report of a Working Group of Experts prepared for the Commission of the European Communities, Directorate-General for Social Affairs, Health and Safety Directorate, Commission of the European Communities, Luxembourg, 1977.
   (c) *Criteria for a Recommended Standard—Occupational Exposure to Asbestos,* National Institute of Occupational Safety and Health, Washington, D.C., 1972, reproduced 1974.
   (d) *Recommended Work Practices—Shop and Field Fabrication of Asbestos Sheet Products,* Asbestos Information Association, Washington, D.C.

(e) Whipple, H. E. (Ed.), *Biological Effects of Asbestos, Ann. N.Y. Acad. Sci.*, 1965, **132,** 1–766.
(f) *Informing Workers and Employers about Occupational Cancer*, Committee on Public Information in the Prevention of Occupational Cancer, Division of Medical Sciences, Assembly of Life Sciences, National Research Council, National Academy of Sciences, Washington, D.C., 1977.
(g) Jacobson, G., and Lainhart, W. S. (Eds.), ILO U/C 1971 International Classification of Radiographs of the Pneumoconioses, Med. Radiogr. Photogr., 1972, **48** (No. 3), 65–110.
(h) *Early Detection of Health Impairment in Occupational Exposure to Health Hazards*, Report of a WHO Study Group, Technical Report Series 571, World Health Organization, Geneva, 1975.

2. *Proceedings of Pertinent Conferences*
(a) *Public Information in the Prevention of Occupational Cancer*, Proceedings of a Symposium, December 2–3, 1976, National Academy of Sciences, Washington, D.C., 1977.
(b) *Proceedings of the Cancer in the Workplace Conference*, November 16, 1976, College of Medicine and Dentistry of New Jersey, Piscataway, N.J., 1977.
(c) Bogovski, P., Gilson, J. C., Timbrell, V., and Wagner, J. C. (Eds.), *Biological Effects of Asbestos*, Proceedings of a Working Conference held at the International Agency for Research on Cancer, Lyon, France, October 2–6, 1972, IARC Scientific Publication No. 8, International Agency for Research on Cancer, Lyon, 1973.
(d) Asbestos Information Association/North America, Third Annual Industry—Government Conference, September 8–9, 1976, Arlington, Virginia.

(6) REFERENCES

American Conference of Governmental Industrial Hygienists (1978) *Industrial Ventilation—A Manual of Recommended Practice 15th Edition*, Lansing, Michigan.

Archer, P. G., Koprowska, I., McDonald, J. R., Naylor, B., Papanicolaou, G. N., and Umiker, W. O. (1966) A study of variability in the interpretation of sputum cytology slides, *Cancer Res.*, **26,** 2122–2144.

Baker, R. B., Marsh, B. R., Frost, J. K., Stitik, F. P., Carter, D., and Lee, J. M. (1974) The detection and treatment of early lung cancer, *Ann. Surg.*, **179,** 813–818.

Baumann, E. R. (1975) Diatomite filters for asbestiform fibre removal from water, in *Proceedings of the American Water Works Association, 95th Annual Conference*, American Water Works Association, Minneapolis.

Becklake, M. R. (1976) Asbestos-related diseases of the lung and other organs: their epidemiology and implications for clinical practice, *Am. Rev. Resp. Dis.*, **114,** 187–227.

Bernstein, D. A. (1969) Modification of smoking behavior: an evaluative review, *Psychol. Bull.*, **71,** 418–440.

Bernstein, D. A., and McAlister, A. (1976) The modification of smoking behavior: progress and problems, *Addict. Behav.*, **1,** 89–102.

Bjelke, E. (1975) Dietary vitamin A and human lung cancer, *Int. J. Cancer*, **15,** 561–565.

Blattner, W. A. (1977) Family studies: the interdisciplinary approach, in *Genetics of Human Cancer*, (Ed. J. J. Mulvihill, R. W. Miller, and J. F. Fraumeni, Jr.), Raven Press, New York, 269–279.

Boucot, K. R., Cooper, D. A., Weiss, W., and Carnahan, W. J. (1966) Cigarettes, cough, and cancer of the lung, *J. Am. Med. Ass.*, **196,** 985–990.

Boucot, K. R., and Weiss, W. (1973) Is curable lung cancer detected by semiannual screening?, *J. Am. Med. Ass.*, **224,** 1361–1365.

Brett, G. Z. (1968) The value of lung cancer detection by six-monthly chest radiographs, *Thorax*, **23,** 414–420.

Bruckman, L., and Rubino, R. A. (1975) Asbestos: rationale behind a proposed air quality standard, *J. Air. Pollut. Control Ass.*, **25,** 1207–1215.

Bryan, W. J. (1964) Hypnosis and smoking, *J. Am. Inst. Hypnosis*, **5,** 17–37.

Campbell, A. H., and Lee, E. J. (1963) The relationship between lung cancer and chronic bronchitis, *Brit. J. Dis. Chest*, **57,** 113–119.

Chapman, R. F., Smith, J. W., and Layden, T. A. (1971) Elimination of cigarette smoking by punishment and self-management training, *Behav. Res. Ther.*, **9,** 255–264.

Cohen, D. (1973) Ferromagnetic contamination in the lungs and other organs of the human body, *Science, N.Y.*, **180,** 745–748.

Cohen, B. H., Diamond, E. L., Graves, C. G., Kreiss, P., Levy, D. A., Menkes, H. A., Permutt, S., Quaskey, S., and Tockman, M. S. (1977) A common familial component in lung cancer and chronic obstructive pulmonary disease, *Lancet*, **ii,** 523–526.

Cole, P., and Goldman, M. B. (1975) Occupation, in *Persons at High Risk of Cancer*, (Ed. J. F. Fraumeni, Jr.), Academic Press, New York, 167–183.

Committee on Public Information in the Prevention of Occupational Cancer, Division of Medical Sciences, Assembly of Life Sciences, U.S. National Research Council (1977) *Informing Workers and Employers about Occupational Cancer*, U.S. National Academy of Sciences, Washington, D.C.

Dean, G. (1966) Lung cancer and bronchitis in Northern Ireland, 1960–2, *Brit. Med. J.*, **1,** 1506–1514.

Delarue, N. C. (1973) A study in smoking withdrawal: the Toronto smoking withdrawal study centre: description of activities, *Can. J. Publ. Hlth.*, **64,** 5-19.

Dubren, R. (1977) Evaluation of a televised stop-smoking clinic, *Publ. Hlth. Rep.*, **92,** 81–84.

Elmes, P. C. (1973) The natural history of diffuse mesothelioma, in *Biological Effects of Asbestos*, IARC Scientific Publications No. 8, International Agency for Research on Cancer, Lyon, 267–272.

Elmes, P. G., and Simpson, M. J. C. (1976) The clinical aspects of mesothelioma, *Q. J. Med., New Ser.*, **45**(179), 427–449.

Engineering Equipment Users Association (1969) *Recommendations for Handling Asbestos*, Handbook No. 33, Engineering Equipment Users Association, London.

Federal Register (1972) **37**(110), 11318–11322.
Federal Register (1975a) **40**(197), 47662.
Federal Register (1975b) **40**(6), 1873–1878.
Federal Register (1975c) **40**(199), 48292–48302.
Fontana, R. S. (1977) Early diagnosis of lung cancer, *Am. Rev. Resp. Dis.*, **116,** 399–402.
Fontana, R. S., Sanderson, D. R., Woolner, L. B., Miller, W. E., Bernatz, P. E., Payne, W. S., and Taylor, W. F. (1975) The Mayo Lung Project for early detection and localization of bronchogenic carcinoma, a status report, *Chest*, **67,** 511–522.
Gaensler, E. A. (1976) Personal communication.
Globglas, A. (1975) Auricular acupuncture and the smoking habit, *Nouv. Presse Med.*, **4,** 980.
Godeau, P., deSaint-Maur, P., Herreman, G., Rault, P., Cenac, A., and Rosenthal, P. (1974) Carcinome bronchioloalvéolaire et sclérodermie, *Sem. Hop. Paris*, **50,** 1161–1168.
Gorlin, R. J. (1977) Monogenic disorders associated with neoplasia, in *Genetics of Human Cancer*, (Ed. J. J. Mulvihill, R. W. Miller, and J. F. Fraumeni, Jr.), Raven Press, New York, 169–177.
Greegor, D. H. (1971) Occult blood testing for detection of asymptomatic colon cancer, *Cancer*, **28,** 131–134.
Gritz, E. R., and Jarvik, M. E. (1977) Pharmacological aids for the cessation of smoking, in *Proceedings of the Third World Conference on Smoking and Health*, (Ed. J. Steinfeld, W. Griffiths, K. Ball, and R. M. Taylor), DHEW Publication No. NIH 77-1413, U.S. Government Printing Office, Washington, D.C.
Grzybowski, S., and Coy, P. (1970) Early diagnosis of carcinoma of the lung. Simultaneous screening with chest X-ray and sputum cytology, *Cancer*, **23,** 113–120.
Guilford, J. (1966) *Factors Related to Successful Abstinence from Smoking: Final Report*, American Institute for Research, Los Angeles.
Hammond, E. C. (1966). Smoking in relation to the death rates of one million men and women, in *Study of Cancer and Other Chronic Diseases*, National Cancer Institute Monograph No. 19, National Cancer Institute, Washington, D.C., 127–204.
Hammond, E. C. (1975) Tobacco, in *Persons at High Risk of Cancer*, (Ed. J. F. Fraumeni, Jr.), Academic Press, New York, 131–137.
Hammond, E. C., and Percy, C. (1958) Ex-smokers, *N.Y. State J. Med.*, **58,** 2956–2959.
Handel, S. (1973) Change in smoking habits in general practice, *Postgrad. Med. J.*, **49,** 479–681.
Harris, M. B., and Rothberg, C. (1972) A self-control approach to reduced smoking, *Psychol. Rep*, **31,** 165–166.
Harwood, C. F., Siebert, P., and Blaszak, T. P. (1974) *Assessment of Particle Control Technology for Enclosed Asbestos Sources*, EPA Publication EPA-650/2-74-088, U.S. Environmental Protection Agency, Washington, D.C.
Harwood, C. F., Oestreich, D. R., Siebert, P., and Stockham, J. D. (1975) Asbestos emissions from baghouse-controlled sources, *Am. Ind. Hyg. Ass. J.*, **36,** 595–603.
Harwood, C. F., Ase, P., and Stinson, M. (1976) Study of the effect of asbestos waste piles on ambient air, in *Proceedings of the Symposium on Fugitive Emission Measurement and Control*, EPA Publication EPA-600/2-76-246, U.S. Environmental Protection Agency, Washington, D.C., 183–202.
Hunt, W. A., and Matarazzo, J. D. (1973) Three years later: recent developments in

the experimental modifications of smoking behaviour, *J. Abnorm. Psychol.*, **81,** 107–114.

International Labour Office (1971) ILO-U/C Classification of radiographs of pneumoconioses, International Labour Office, Geneva.

Janis, I. L., and Hoffman, D. (1970) Facilitating effects on daily contact between partners who make a decision to cut down on smoking, *J. Pers. Soc. Psychol.*, **17,** 25–35.

Johnston, E., and Donoghue, J. R. (1971) Hypnosis and smoking: a review of the literature, *Am. J. Clin. Hypnosis*, **13,** 265–272.

Kang, K.-Y., Sera, Y., Okuchi, T., and Yamamura, Y. (1974) T-lymphocytes in asbestosis, *N. Engl. J. Med.*, **291,** 735–736.

Kellermann, G., Luyten-Kellermann, M., and Shaw, C. R. (1973a) Genetic variation of aryl hydrocarbon hydroxylase in human lymphocytes, *Am. J. Hum. Genet.*, **25,** 327–331.

Kellermann, G., Shaw, C. R., and Luyten-Kellerman, M. (1973b) Aryl hydrocarbon hydroxylase inducibility and bronchogenic carcinoma, *N. Engl. J. Med.*, **289,** 934–937.

Koch, B. (1965) Familial fibrocystic pulmonary dysplasia: observations in one family, *Can. Med. Ass. J.*, **92,** 801–808.

Koenig, K. P., and Masters, J. (1965) Experimental treatment of habitual smoking, *Behav. Res. Ther.*, **3,** 235–243.

Krenkel, P. A. (1974) Waste treatment methodology, in *Industrial Pollution*, (Ed. N. I. Sax), Van Nostrand Reinhold, New York, 220–243.

Kroger, W. S. (1963) *Clinical and Experimental Hypnosis*, Charles C. Thomas, Springfield, Ill.

Lawrence, J., *et al.* (1974) Asbestos: its removal from potable water, *Can. Res. Dev.*, Nov/Dec, 29–30.

Lawrence, J., Tosine, H. M., Zimmermann, H. W., and Pang, T. W. S. (1975) Removal of asbestos fibres from potable water by coagulation and filtration, *Wat. Res.*, **9,** 397–400.

Levine, R. J. (1978) How the industrial physician can reduce mortality from asbestos-related diseases, *J. Occup. Med.*, **20,** 464–468.

Levy, B. S., Sigurdson, E., Mandel, J., Laudon, E., and Pearson, J. (1976) Investigating possible effects of asbestos in city water: surveillance of gastrointestinal cancer incidence in Duluth, Minnesota, *Am. J. Epidemiol.*, **103,** 362–368.

Lichtenstein, E., Harris, D. E., Birchler, G. R., Wahl, H. M., and Schmall, D. P. (1973) Comparison of rapid smoking, warm, smoky air, and attention placebo in the modification of smoking behaviour, *J. Consult. Clin. Psychol.*, **40,** 92–98.

Lichtenstein, E., and Danaher, B. G. (1976) Modification of smoking behavior: a critical analysis of theory, research and practice, in *Progress in Behavior Modification*, Vol. 3, Academic Press, New York.

Lilienfeld, A. (Chairman), *et al.* (1966) An evaluation of radiological and cytological screening for early detection of lung cancer. A cooperative pilot study of the American Cancer Society and the Veterans Administration, *Cancer Res.*, **26,** 2083–2147.

Lilienfeld, A. M., and Kordan, B. (1966) A study of variability in the interpretation of chest X-rays in the detection of lung cancer, *Cancer Res.*, **26,** 2145–2147.

Lynch, H. T. (1976a) Miscellaneous problems, cancer, and genetics, in *Cancer Genetics*, (Ed. H. T. Lynch), Charles C. Thomas, Springfield, Ill., 491–554.

Lynch, H. T. (1976b) Skin, heredity, and cancer, in *Cancer Genetics*, (Ed. H. T. Lynch), Charles C. Thomas, Springfield, Ill., 424–470.

Lynch, H. T., Lynch, J., and Guirgis, H. (1976) Heredity and colon cancer, in *Cancer Genetics*, (Ed. H. T. Lynch), Charles C. Thomas, Springfield, Ill. 326–354.

McKusick, V. A., and Fisher, A. M. (1958) Congenital cystic disease of the lung with progressive pulmonary fibrosis and carcinomatosis, *Ann. Int. Med.*, **48,** 774–790.

Merchant, J. A., Klouda, P. T., Soutar, C. A., Parkes, W. R., Lawler, S. D., and Turner-Warwick, M. (1975) The HL-A system in asbestos workers, *Brit. Med. J.*, **1,** 189–191.

Miller, L. C., Schilling, A. F., Logan, D. L., and Johnson, R. L. (1977) Potential hazards of rapid smoking as a technique for the modification of smoking behavior, *N. Engl. J. Med.*, **297,** 590–592.

Montgomery, R. D., Stirling, G. A., and Hamer, N. A. (1964) Bronchiolar carcinoma in progressive systemic sclerosis, *Lancet*, **i,** 486–487.

Morrow, J., Gmeinder, S., Sachs, L. B., and Burgess, H. (1973) Elimination of cigarette smoking behavior by stimulus satiation, self-control techniques, and group therapy, Paper presented to the meeting of the Western Psychological Association, Los Angeles, April.

National Cancer Institute (in press) Progress report on a nation kicking the habit, *Smoking Digest.*

National Clearinghouse for Smoking and Health (1976) Adult use of tobacco: 1975, DHEW, Public Health Service.

Ostrow, J. D., Mulvaney, C. A., Hansell, J. R., and Rhodes, R. S. (1973) Sensitivity and reproducibility of chemical tests for fecal occult blood with an emphasis on false-positive reactions, *Am. J. Digest. Dis.*, **18,** 930–940.

Paigen, B., Gurtoo, H. L., Minowada, J., Houten, L., Vincent, R., Paigen, K., Parker, N. B., Ward, E., and Hayner, N. T. (1977) Questionable relation of aryl hydrocarbon hydroxylase to lung cancer risk, *N. Engl. J. Med.*, **297,** 346–350.

Pederson, L. L., and Lefcoe, N. M. (1976) A psychological and behavioral comparison of ex-smokers and smokers, *J. Chron. Dis.*, **29,** 431–434.

Pomerleau, O. F., and Ciccone, P. (1974) Preliminary results of a treatment program for smoking cessation using multiple behavior modification techniques, Paper presented at the meeting of the Association for Advancement of Behavior Therapy, Chicago, Ill., November.

Prolla, J. C., Kobayashi, S., and Kirsner, J. B., (1969) Gastric cancer, *Arch. Int. Med.*, **124,** 238–246.

Pyke, S., Agnew, N. McK., and Kopperud, J. (1966) Modification of an overlearned maladaptive response through a learning program: a pilot study on smoking, *Behav. Res. Ther.*, **4,** 197–203.

Resnick, J. H. (1968) Effects of stimulus satiation on the overlearned maladaptive response of cigarette smoking, *J. Consult. Clin. Psychol.*, **32,** 501–505.

Rothman, K. J. (1975) Alcohol, in *Persons at High Risk of Cancer*, (Ed. J. F. Fraumeni, Jr.), Academic Press, New York, 139–148.

Saccomanno, G., Archer, V. E., Auerbach, O., Saunders, R. P., and Brennan, L. M. (1974) Development of carcinoma of the lung as reflected in exfoliated cells, *Cancer*, **33,** 256–270.

Saccomanno, G. (1976) personal communication.

Schwartz, J. L., and Dubitzky, M. (1968) One-year follow-up results of a smoking cessation program, *Can. J. Publ. Hlth.*, **59,** 161–165.

Schwartz, J. L. (1969) A critical review and evaluation of smoking control methods, *Publ. Hlth. Rep.*, **84,** 489–506.

Schwartz, J. L., and Rider, G. (1977) Smoking cessation methods in the United States and Canada: 1969–1974, in *Proceedings of the Third World Conference on*

Smoking and Health, (Ed. J. Steinfeld, W. Griffiths, K. Ball, and R. M. Taylor), DHEW Publication No. NIH 77–1413.

Seibert, P. C., Ripley, T. C., and Harwood, C. F. (1976) Assessment of particle control technology for enclosed asbestos sources—Phase II, EPA Publication EPA 600/2-76-065, U.S. Environmental Protection Agency, Research Triangle Park, North Carolina.

Selikoff, I. J. (1965) The occurrence of pleural calcification among asbestos insulation workers, *Ann. N.Y. Acad. Sci.*, **132,** 351–367.

Shaw, C. R. (1975) The microsomal mixed function oxidases and chemical carcinogens, in *Isozymes. III. Developmental Biology*, Academic Press, San Francisco, 809–819.

Sherlock, P., and Winawer, S. J. (1974) Modern approaches to early identification of large-bowel cancer, *Am. J. Digest. Dis., New Ser.*, **19,** 959–964.

Sherlock, P., and Winawer, S. J. (1977) The role of early diagnosis in controlling large-bowel cancer, *Cancer*, **40,** 2609–2615.

Shewchuk, L. A. (1976a) A comparison of smoking cessation techniques: initial success and eventual recidivism, L. A. Shewchuk, 21 Fairview Avenue, Tuckahoe, N.Y. 10707.

Shewchuk, L. A. (1976b) Smoking cessation programs of the American Health Foundation, *Prevent. Med.*, **5,** 454–476.

Smith, M. K., and White, R. W. (1975) Composition and method for inhibiting asbestos fiber dust, *U.S. Pat.*, 3 928 060, December 23.

Spiegel, H. (1970) A single-treatment method to stop smoking using ancillary self-hypnosis, *Int. J. Clin. Exp. Hypnosis*, **18,** 235–249.

Sporn, M. B. (1976) Prevention of chemical carcinogenesis by vitamin A and its synthetic analogs (retinoids), *Fed. Proc.*, **35,** 1332–1338.

Stanton, M. F. (1974) Fiber carcinogenesis: is asbestos the only hazard? *J. Natn. Cancer Inst.*, **52,** 633–634.

Stephens, F. O., and Lawrenson, K. B. (1970) The pathologic significance of occult blood in feces, *Dis. Col. Rect.*, **13,** 425–428.

Swaye, P., Van Ordstrand, H. S., McCormack, L. J., and Wolpaw, S. E. (1969) Familial Hamman-Rich syndrome: report of eight cases, *Dis. Chest*, **55,** 7–12.

Swift, M. (1977) Malignant neoplasms in heterozygous carriers of genes for certain autosomal recessive syndromes, in *Genetics of Human Cancer*, (Ed. J. J. Mulvihill, R. W. Miller, and J. F. Fraumeni, Jr.), Raven Press, New York, 209–213.

Tiche. T. J., and Elliott, R. (1967) Breaking the cigarette habit: effects of the technique involving threatened loss of money, Paper presented at the annual meeting of the American Psychological Association.

Tokuhata, G. K., and Lilienfeld, A. M. (1966) Familial aggregation of lung cancer in humans, *J. Natn. Cancer Inst.*, **30,** 289–312.

Tokuhata, G. K. (1976) Cancer of the lung: host and environmental interaction, in *Cancer Genetics*, (Ed. H. T. Lynch), Charles C. Thomas, Springfield, Ill., 213–232.

Tooley, J. T., and Pratt, S. (1967) An experimental procedure for the extinction of smoking behavior, *Psychol. Rec.*, **17,** 209–218.

Turner-Warwick, M. (1973) Immunology and asbestosis, in *Biological Effects of Asbestos*, IARC Scientific Publications No. 8, International Agency for Research on Cancer, Lyon, 258–263.

U.S. Bureau of Mines (1959) *Asbestos—A Materials Survey*, Information Circular No. 7880, U.S. Bureau of Mines, Washington, D.C.

U.S. Code of Federal Regulations (1976) Title 29, Chapter XVII, Subpart Z, Sections 1910.1003–1910.1029.

U.S. Environmental Protection Agency (1973) *Control Techniques for Asbestos Air Pollutants*, EPA Publication No. AP-117, Environmental Protection Agency, Washington, D.C.

U.S. Environmental Protection Agency (1976) Asbestos: *A Review of Selected Literature Through* 1973 *Relating to Environmental Exposure and Health Effects*, EPA Publication No. 560/2-76-001, Environmental Protection Agency, Washington, D.C.

Van Der Wal, A. M., Huizinga, E., Orie, N. G. M., Sluiter, H. J., and de Vries, K. (1966) Cancer and chronic nonspecific lung disease, *Scand. J. Resp. Dis.*, **47,** 161–172.

Vogt, T. M., Selvin, S., Widdowson, G., and Halley, S. B. (1977) Expired air carbon monoxide and serum thiocyanate as objective measures of cigarette exposure, *Am. J. Publ. Hlth.*, **67,** 545–549.

von Dedenroth, T. E. A. (1968) The use of hypnosis in 1000 cases of 'tobaccomaniacs', *Am. J. Clin. Hypnosis*, **10,** 194–197.

Wagner, J. C., Berry, G., Skidmore, J. W., and Timbrell, V. (1974) The effects of inhalation of asbestos in rats, *Brit. J. Cancer*, **29,** 252–269.

Wagner, J. K., and Bragg, R. A. (1970) Comparing behavior modification approaches to habit decrement-smoking, *J. Consult. Clin. Psychol.*, **34,** 258–263.

Weiss, W. (1967) Cigarette smoking and pulmonary fibrosis: A preliminary report, *Arch. Envir. Hlth.*, **14,** 564–568.

Weiss, W. (1971) Cigarette smoking, asbestos, and pulmonary fibrosis, *Am. Rev. Resp. Dis.* **104,** 223–227.

Weiss, W., Seidman, H., and Boucot, K. R. (1975) The Philadelphia Pulmonary Neoplasm Research Project: thwarting factors in periodic screening for lung cancer, *Am. Rev. Resp. Dis.*, **111,** 289–297.

West, D. W., Graham, S., Swanson, M., and Wilkinson, G. (1977) Five year follow-up of a smoking withdrawal clinic population, *Am. J. Publ. Hlth.*, **67,** 536–544.

Whitwell, F., Newhouse, M. L., and Bennett, D. R. (1974) A study of the histological types of lung cancer in workers suffering from asbestosis in the United Kingdom, *Brit. J. Ind. Med.*, **31,** 298–303.

Winawer, S. J., Miller, D. G., Schottenfeld, D., Leidner, S. D., Sherlock, P., Befler, B., and Stearns, M. W., Jr. (1977) Feasibility of occult-blood testing for detection of colorectal neoplasia: debits and credits, *Cancer*, **40,** 2616–2619.

Winett, R. A. (1973) Parameters of deposit contracts in the modification of smoking, *Psychol. Rec.*, **23,** 49–60.

Wynder, E. L., and Fairchild, E. P., Jr. (1966) The role of a history of persistent cough in the epidemiology of lung cancer, *Am. Rev. Resp. Dis.*, **94,** 709–720.

Zamcheck, N., Grable, E., Ley, A., and Norman, L. (1955) Occurrence of gastric cancer among patients with pernicious anemia at the Boston City Hospital, *N. Engl. J. Med.*, **252,** 1103–1110.

Author Index

Subject Index